AF527835

Herbicides

HERBICIDES

Shobha Sondhia
Jay G. Varshney

2010

SATISH SERIAL PUBLISHING HOUSE
403, Express Tower, Commercial Complex, Azadpur, Delhi-110033 (India)
Phone : 011-27672852, Fax : 91-11-27672046
E-mail :info@satishserial.com, hkjain1975@yahoo.com
Website : www.satishserial.com

Published by :

SATISH SERIAL PUBLISHING HOUSE
403, Express Tower, Commercial Complex, Azadpur, Delhi-110033 (INDIA)
Phone : 011-27672852 Fax : 91-11-27672046
E-mail : info@satishserial.com, hkjain1975@yahoo.com

ISBN 81-89304-71-2

Typeset at: Laxmi Art Creations, Delhi # 98 11 48 23 28
Printed at: Salasar Imaging Systems, Delhi-35

डा. मंगला राय
सचिव एवं महानिदेशक
DR. MANGALA RAI
SECRETARY & DIRECTOR-GENERAL

भारत सरकार
कृषि अनुसंधान और शिक्षा विभाग एवं
भारतीय कृषि अनुसंधान परिषद
कृषि मंत्रालय, कृषि भवन, नई दिल्ली 110 114

GOVERNMENT OF INDIA
DEPARTMENT OF AGRICULTURAL RESEARCH & EDUCATION
AND
INDIAN COUNCIL OF AGRICULTURAL RESEARCH
MINISTRY OF AGRICULTURE, KRISHI BHAVAN, NEW DELHI 110 114
TEL.: 23382629; Fax: 91-11-23384773 E-MAIL: mrai@icar.delhi.nic.in

FOREWORD

Weeds constitute a serious limitation to crop production in all agricultural systems. Losses due to weed competition have been known since mankind first started agriculture. Traditionally it has been a practice in India to remove weeds by hand as there was an abundance of labour on the land to remove these weeds. There are many reasons that weeds need to be controlled. The direct competition that leads to yield loss is the most obvious. The application of herbicides in crop and non-cropped situations is an effective approach for management of weeds. Herbicide use is increasing day by day and large numbers of herbicides are being introduced every year to cater the needs of farmers. Excessive and uncontrolled use of herbicids may pose problem of phytotoxicity to crop plants, residual effect on susceptible suceeding crops and health hazards due to accumulation of residues in crop produce and ground water. Therefore the detailed knowledge of herbicides regarding their chemistry, classification, use, mode of action and persistence is of great relevance in making meaningful recommendations of herbicide application.

In this book, entitled "Herbicides" an attempt has been made by the authors to give a comprehensive account of herbicides regarding their chemistry, classification, mode of action, metabolism in plant soil, adsorption and translocation, application in major field crops, residues in soil, water, food chain and its management, effect on non-target organisms, soil health and succeeding crops, etc. I appreciate the efforts of authors, Dr. Shobha Sondhia and Dr. Jay G. Varshney for bringing out this book, which I hope will be of immense use to the students and researchers, scientists and policy planners in the field of agriculture.

(Mangala Rai)
Director General, ICAR
New Delhi

Date : 24th june, 2008
New Delhi

PREFACE

Weed control has always been a basic indispensable and integral part of agriculture from time immemorial. In India the concept of weed control by herbicides came in 1937. In 1946, 2, 4-D were tested for weed control in *Carthamus oxycantha*, since then a number of herbicides have been imported and tested for their effectiveness for control of many weed species. Now there has been a tremendous growth in herbicide industry over the last decade and herbicides are increasingly used for control of weeds in field and plantation crops than ever before. From 1950 onwards, industrial research began to produce many new herbicides. Now the sulfonylurea group herbicides are more selective and effective at very low dose of few grams per hectare. As new and better molecules being continuously synthesized and developed, it seems that improvements will continue for the foreseeable future. At global level presently nearly 400 herbicides are registered or are in the process of registration. Herbicides industry at present has developed from almost nothing to a multi billion-dollar business.

Large numbers of herbicides applied annually create a need to study the fate of these herbicides in soil, crop produce and ground water. The recognized persistence of several substituted ureas, uracil, s-triazines, the benzoic acid derivatives, picloram and other herbicides in soils has received considerable attention. It is being recognised that some persistence is necessary property of all herbicides; with no persistence soil applied herbicide would not control weeds. Thus data on residual toxicity of recommended herbicide become indispensable for approving a herbicide for large scale commercial use.

In this book, there are chapters dealing with basic information of herbicide chemistry, classification, mode of action and metabolism of herbicides, herbicide selectivity, persistence of residues and their management, dissipation, herbicide formulations, toxicology etc. There are chapter on effect of herbicides on non-target organisms, soil health and succeeding crops etc. in detail. Information is also provided on weed management in major field crop. At the end of this book lists of herbicides common name, chemical name along with toxicity class, maximum residue level of different herbicide in various food commodity are also provided. At the last glossary of different words used in herbicide and weed management practice is also given. We hope that this publication will serve as reference book for advanced course on weed control among academicians, researchers and students etc.

There are many persons to whom we would like to thank for their contributions over the last two years. We specially thank Mr. Ghanshyam Viswakarma, Dr. Ms. Benu Singhai, Ms. Vineeta Parmar and Mr. Sanjay Tiwari for their help. We are also thankful to Mr. M. K. Bhatt for some of the drawing made by him.

Shobha Sondhia
Jay G. Varshney

CONTENTS

Chapter 1

Introduction

Weeds have been a nemesis to agricultural endeavours since ancient times. Over the centuries, attempts to control or eradicate weeds have ranged from spiritual to pragmatic. Weeds have been a part of the agriculture since man first started cultivating crops. Weeds are competitive and adaptable to all adverse environments, harmful to crop and exert a marked effect on crop yields. Of the total annual loss of agricultural produce from various pests in India, weeds account for 37 percent, insect 30 percent, disease 20 percent and other pests 13 percent (Fig. 1).

Herbicides are the chemicals which are employed to kill or control vegetation. Over the years herbicides have emerged as an important tool in the management of weeds. Common salt, ash, smelter waste etc. have been used for centuries to control weeds, but selective control of weeds in agriculture was first conceived in 1896 in France, when Bordeaux mixture sprayed on grapevines for protecting it from downy mildew damaged certain broadleaf weeds. Soon it was found that copper sulfate present in the Bordeaux mixture was responsible for its weed killing effect. Herbicides are the fastest growing chemicals in recent years. Between 1889 and 1908 several other inorganic salts such as sodium chlorate, carbon bisulfide, sodium arsenite, kainite, calcium cynamide and sulfuric acid were developed for non-selective control of perennial weeds. Between 1930 to 1940 some boron compounds, thiocyanates, dinitrophenols, ammonium sulfate and certain mineral salts were developed for selective and non-selective weed control.

The discovery of the herbicidal activity of 2,4-D (2, 4-dichlorophenoxyacetic acid) first synthesized in 1941, triggered the development of modern herbicide technology. 2,4-D proved to be an outstanding herbicide. The commercial success of 2,4-D led to the development of other herbicides such as MCPA, Silvex, 2, 4, 5-T, phenylurea herbicides such as monuron and linuron.

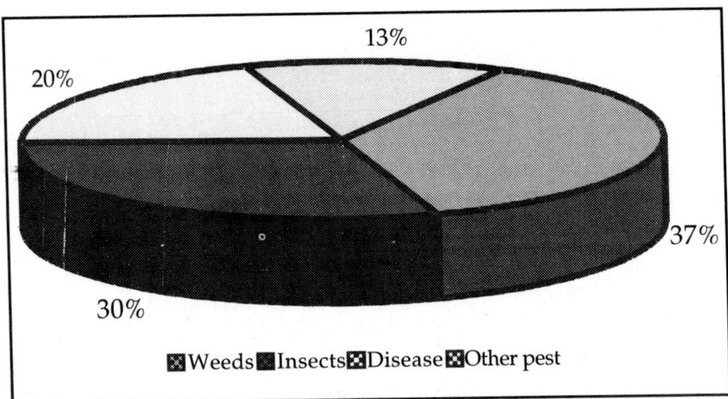

Fig. 1. Total annual loss of agriculture produce from various pests in India

It was during the 1950s and 1960s that the modern practices of using relatively low rates of synthetic herbicides for selective weed control in field crops was adopted in developed countries of the world. The introduction of glyphosate a non-selective herbicide in the late 1970's provided outstanding control of most perennial grasses and many perennial broadleaf weeds. In modern herbicide development, the discovery of organic compounds that exhibit phytotoxic properties often lead to the synthesis of related compounds, where chemist attempts to optimize herbicidal activity of the original compound by altering or modifying its chemical structure. Herbicide development in the 1980s was marked by the introduction of selective post-emergence treatments in major crops such as sulfonylureas, imidazolinones and aryloxy phenoxy propionate which not only provide excellent selectivity but are used at extremely low dosage.

Consequently there are often several herbicides developed within a chemical family that are structurally related and have essentially the same mode of action in plants. However, it cannot be stated exclusively that all members of a chemical family have the same mode of action, as there are a few notable exceptions (Zimdhal, 1993). Herbicides within a chemical family also often vary in selectivity, a result of physico-chemical differences that cause them to behave differently in the soil or plant system.

In India, the concept of weed control by herbicides was made in 1937 in Punjab for controlling wild safflower (*Carthamus oxycantha*). 2, 4-D was first tested in India in 1946 (Mani, 1977). Since then a number of herbicides have been imported and tested for their effectiveness for control of many weeds species. In 1978, Indian Council of Agricultural Research initiated a scheme on management of weeds in different crops with an objective to investigate the weed flora of major crops and the relative effectiveness of herbicides and economization of their usages by setting up of 6 centres of All India Coordinated Research Project on Weed Control which increased to 16 later. At present there are twenty-two centres located in different states throughout the country. The research on weed and weed management got further boost with the establishment of the National Research Centre for Weed Science (NRCWS) at Jabalpur in 1989. The NRCWS is a nodal centre for research and training in the field of weed science in the country and provides leadership mandated to act as repository of information in the field of weed sciences. NRCWS is also responsible for coordinating the research carried at under AICRP on Weed control since its inception.

Herbicides help farmers in increasing their crop yield, with efficient weed control. Specific recommendations have been developed by the scientists all over the India, on the dose, time and methods of application for each situation. The consumption of technical grade herbicides was 15 MT in 70's, which increased to 480 MT in 1991 and is currently about 7364 MT (2004-05). Punjab and Haryana lead in herbicide usage in wheat and rice.

Table 1. Trends in consumpion pattern of different groups of pesticides in India (MT).

Year	Insecticide	Herbicide	Fungicide	Total
1995-96	38788	6040	10563	61260
1996-97	34665	7060	9969	56114
1997-98	33376	7103	10054	52239
1998-99	30469	7292	10428	49157
1999-00	28926	7369	8435	46195
2000-01	26756	7299	8307	43584
2001-02	29839	6979	9222	47348
2002-03	28197	7857	10712	48146
2003-04	25627	5610	9087	40762
2004-05	25929	7364	9397	41350

Source: Directorate of PP and Q & S, Faridabad.

At present nearly 400 herbicides are available worldwide for weed control. In India, the chemical concept of weed control struck deeper roots, when tea growers began using herbicides on a commercial scale in late 1970's. With the discovery and development of 2, 4-D as a herbicide following II world War, there has been a phenomenal growth in development of new molecules as herbicides. Triazines (1956), carbamates (1954-58), ureas (1960), toludine (1965), amides (1966), diphenyl ethers (1970) and triazines (1974) were introduced. Trifluralin and fluchloralin were particularly useful in a wide variety of crops. Dual action herbicides like pendimethalin were introduced during 1980's. Now the sulfonylurea group herbicides are more selective and effective at very low dose of few grams per hectare. The world consumption of herbicides is 36 % followed by the insecticides (25 %), fungicides (10 %) and others (29 %) (Fig. 2).

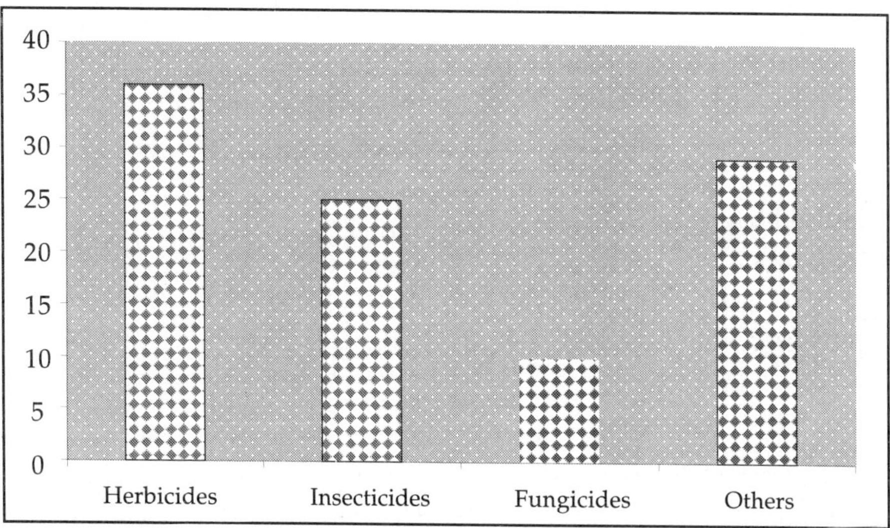

Fig. 2: Consumption (%) of pesticides active ingredient (a.i.) at user level in the World (1999-2000) *Source:* EPA estimates based on Crop Life America Annual surveys, USDA/NASS.

In India, the herbicides market has grown eight fold in the last 15 years. Herbicide consumption increases approximately 30 percent between 1995 to 2003 (Fig. 3). More new herbicides are being introduced to cater to the needs of farmers. Under this situation excessive and repeated use of herbicides may pose problems such as phytotoxicity to crop plants, residual effect on susceptible inter-crops or succeeding crops and health hazards due to accumulation of residues

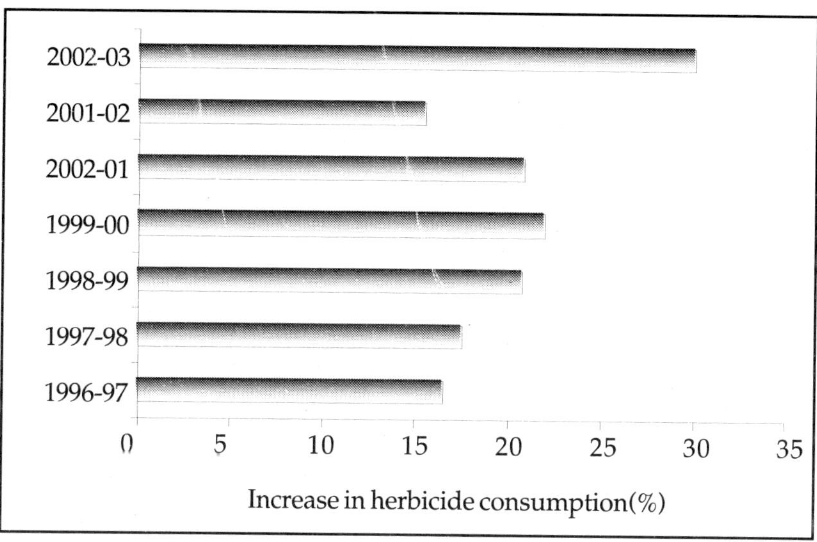

Fig. 3: Percent increase in herbicide consumption in India since 1996 to 2003.

in the soil, crop produce and ground water. With increased awareness on environmental pollution and health hazards, the data on residual toxicity of recommended herbicides become indispensable for approving a herbicide for large scale commercial use. The knowledge of herbicide and residues is of great concern to all research and field workers to make meaningful recommendations to the farmers on the application of herbicides. Hence an effort has been made in this publication to give a comprehension version of herbicide classification, chemistry, entry into the plant by absorption, translocation, mode of action, herbicide metabolism, formulation, adjuvant, toxicology, residues in soil, water, crop produce, herbicide persistence and residue management, effect of herbicides on soil health, non-target organisms and succeeding crop *etc.*

❑❑❑

Chapter 2

Chemistry and Classification

Every herbicide is named in three ways, i) chemical name to describe its chemical structure. The chemical constituent that makes up the herbicide active ingredient can be determined and similarities to other chemicals can be found in this way. ii) trade name to distinguish it from other products and assists in its sale, herbicides are manufactured by several companies and each give its product a different trade name iii) common name, herbicides are manufactured by several companies and each give its product a different trade name. To avoid confusion herbicides are also preferred by common name by the Weed Science Society of America (WSSA). The common name refers to all herbicide products that have the same active ingredients. The herbicides in the WSSA Herbicide Handbook are organized according to common name, although the chemical name is also provided.

For Example:

(i) Glyphosate (common name)
Roundup (trade name)
N-(phosphonomethyl) glycine (chemical name)

(ii) Metribuzin (common name)
Sencor (trade name)
4-amino-6-tert-butyl-3-methylthio-1,2,4-triazine-5-(4H)-one (chemical name)

Basic herbicide chemistry

Most herbicides are organic compounds, i.e. they contain carbon. The major elements contained in organic herbicides in order of decreasing occurrence are carbon, hydrogen, oxygen, nitrogen, chlorine, phosphorus, sulfur, and fluorine. The basic structural components of most organic herbicides are carbon chains (aliphatic groups) and rings (aromatic groups). There are several inorganic herbicides too such as: sodium chlorate, carbon bisulfide, sodium arsenite, kainite, calcium cynamide.

Chemistry and nomenclature of simple organic herbicides

Aliphatic compounds

The simplest aliphatic compound consists of one carbon and four hydrogens (CH_4) and is a gas called methane. Similarly based on number of carbon and hydrogen different aliphatic compound are formed as below:

Name	Number of carbon atoms	Number of hydrogen atoms	Chemical name	Radical
CH_4	1	4	Methane	Methyl
C_2H_6	2	6	Ethane	Ethyl
C_3H_8	3	8	Propane	Propyl
C_4H_{10}	4	10	Butane	Butyl
C_5H_{12}	5	12	Pentane	Pentyl
C_6H_{14}	6	14	Hexane	Hexyl
C_7H_{16}	7	16	Heptane	Heptyl
C_8H_{18}	8	18	Octane	Octyl

A two carbon aliphatic compound is called ethane (C_2H_6); three carbons, propane; four carbons, butane, etc (Fig. 1). When the radicals of these compounds are included as a portion of a large molecule they are similarly named but with *-yl* as the ending, e.g. methyl, ethyl, propyl, butyl, etc. An aliphatic group with a carboxyl group (- COOH) at the end of the chain is called a carboxylic acid, a carbon chain in which one or more of the hydrogen is replaced by a hydroxyl group (-OH) is called an alcohol. A carboxylic acid with two carbons in the chain is called ethanoic acid (or acetic acid), the two carbon alcohol chain is called ethanol (ethyl alcohol).

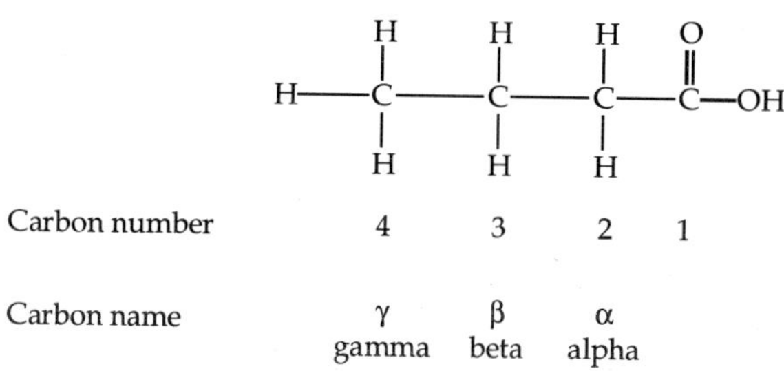

Fig. 1. Chemistry and nomenclature of simple aliphatic compounds.

Aromatic compounds

Ring compounds or aromatics are derivatives of benzene (C_6H_6). Removal of one hydrogen from a benzene ring is termed as a phenyl group and in this form can accept an appropriate substitution. Replacement of the hydrogen atom with a hydroxyl (-OH) group results in the formation of phenol; substitutions can include benzoic acid, toluene, and aniline. The positions on the benzene ring are numbered from 1 to 6 and also may be named. The positions 2 and 6 are the ortho positions; 3 and 5, the meta positions; and 4, the para position (Fig. 2). Derivatives of organic compounds are named according to the functional groups and substitutions present.

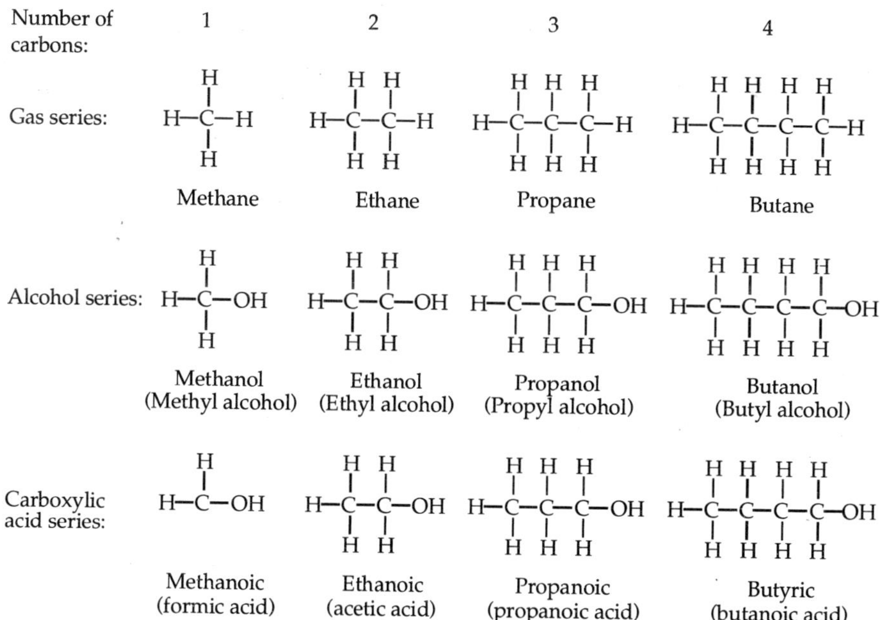

Fig. 2. Chemistry and nomenclature of simple aliphatic compounds.

benzene C_6H_6

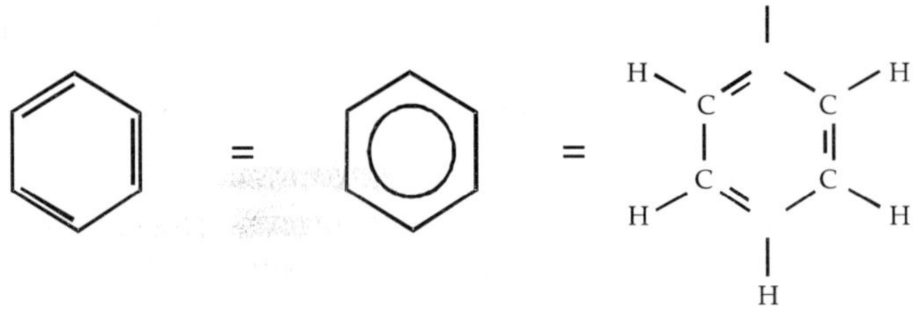

The chemistry of a compound determines how the herbicide will act in biological and physical systems such as plants, animals, soils, and water. Compounds made up of simple chains of carbon, hydrogen, nitrogen, sulfur, and oxygen tend to be degraded readily by microbes e.g. glyphosate, glufosinate and several thiocarbamate herbicides. Compounds with aromatic structures and halogen (chlorine, bromine, fluorine, iodine) substitutions tend to be longer lived than straight chains, e.g chloroacetamide, triazine, sulfonylurea, and dinitroaniline herbicides. The more chlorinated a compound the longer it will persist in soil. The substitutions and alterations that can be made to an organic acid illustrate the effects that chemical structure can have on important herbicidal and mixing properties. The acid form of 2,4-D is only slightly soluble in water and oil so use is limited in commercial formulations, the acid can be reacted with bases to form salts. Common salt formulations include sodium, potassium, ammonium, lithium, and several amine salts. These compounds ionize (dissociate) in water to form charged particles (see Fig. 5; major forms of 2,4-D). All salts listed above are soluble in water with the amine salt being the most soluble and the most commonly used salts are not soluble in oil and when used as foliar treatments must be applied with wetting agents to enhance penetration into the leaf. The acid can be reacted with alcohols to form esters: esters of 2,4-D are nonpolar molecules and do not ionize (Fig. 6). They are insoluble in water, but highly soluble in oil during formulation. They are usually diluted in oil-based solvents and mixed with an emulsifying agent (keeps the oil-ester droplets suspended in water). Oils may be used directly as carriers for application of esters particularly for treatment of woody species. Ester formulations more readily wet and penetrate plant cuticles and are more toxic to plants than the salt formulations. The salts of phenoxy herbicides are non-volatile, whereas the ester formulations vary in their degree of volatility.

Volatility of an ester is determined by molecular weight of the alcohol. Short-chain, low-molecular-weight alcohols of five or fewer carbons lead to the formation of highly volatile esters. Low volatile esters are heavier and consist of long-chained alcohols with an ether linkage (although called low volatile, they are still somewhat volatile. Hard water have high calcium or magnesium content. Calcium content of 125 ppm can affect herbicide stability.

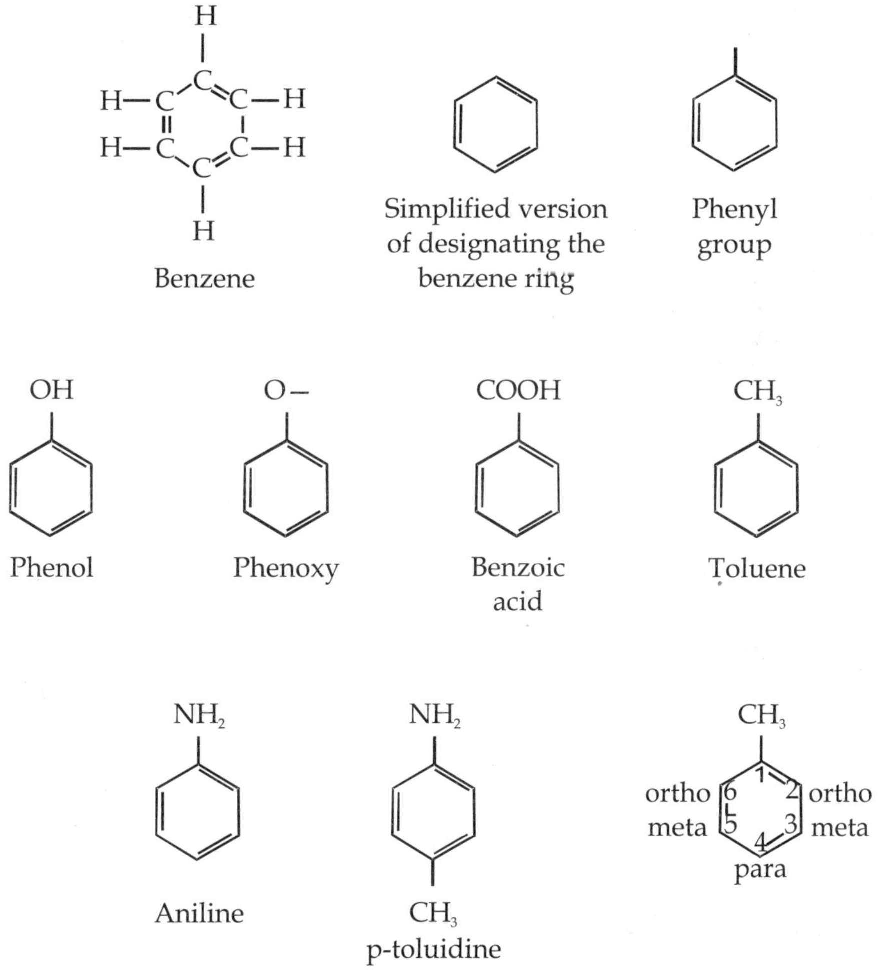

Benzene

Simplified version of designating the benzene ring

Phenyl group

OH — Phenol

O– — Phenoxy

COOH — Benzoic acid

CH_3 — Toluene

NH_2 — Aniline

NH_2, CH_3 — p-toluidine

CH_3 — 1, 2, 3, 4, 5, 6 — ortho, meta, para, meta, ortho

Fig. 3. Chemistry and nomenclature of simple aromatic compounds.

For example, 2, 4-D dissociates in water and the sodium, potassium, lithium, or amine groups (cations) are released during the process of dissociation of the 2,4-D. Calcium and magnesium bind to the 2,4-D anion causing formation of precipitates and of clog filters and nozzles. Amine salts are less susceptible than the alkali salts in forming precipitates; the esters have excellent stability in water such as Roundup (glyphosate). In hard water, the isopropyl amine salt of glyphosate reacts with $CaCO_3$ to form a calcium salt of glyphosate, which is not acceptable to the plant (does not readily pass through the cell membrane) resulting in decreased activity. To remedy the situation, ammonium sulfate (NH_4SO_4) is added to the spray solution. The reaction of NH_4SO_4 with $CaCO_3$ forms $CaSO_4$ and the ammonium salt of glyphosate. This removes Ca as a factor affecting glyphosate activity. The ammonium salt of glyphosate is acceptable to the plant (readily passes through the cell membrane) and weeds are controlled.

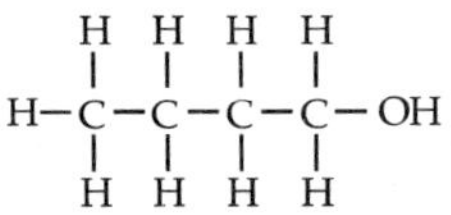

1- Butanol

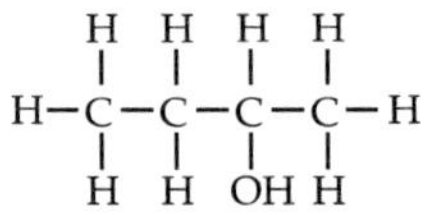

2- Butanol (sec-butyl alcohol)

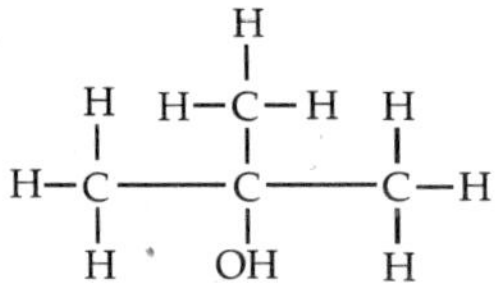

2- Methyl-2-propanol (tert-butyl alcohol)

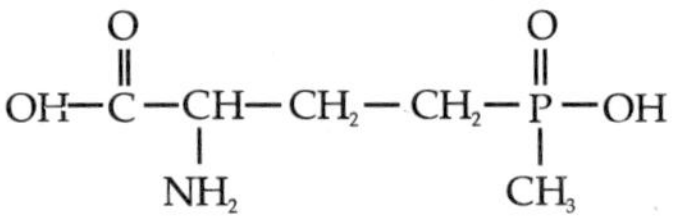

2- Amino-4-(hydroxymethyl-phosphinyl) bulanoic acid (glufosinate)

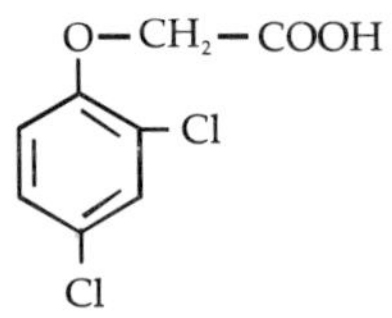

(2, 4-dichlorophenoxy) acetic acid (2, 4-D)

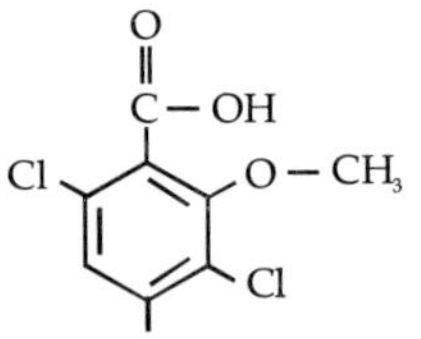

3, 6-Dichloro-2-methoxybenzoic acild (dicamba)

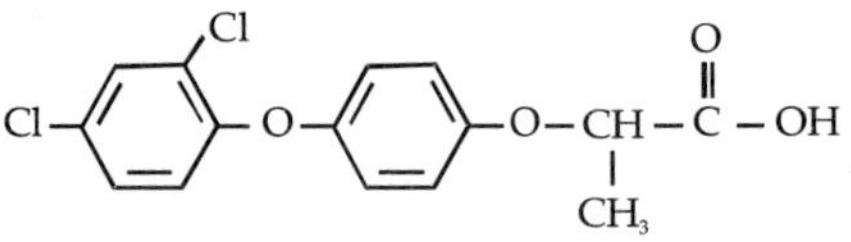

(±)-2-[4-(2, 4-dichlorophenoxy) phenoxy] propanoic acid (diclofop)

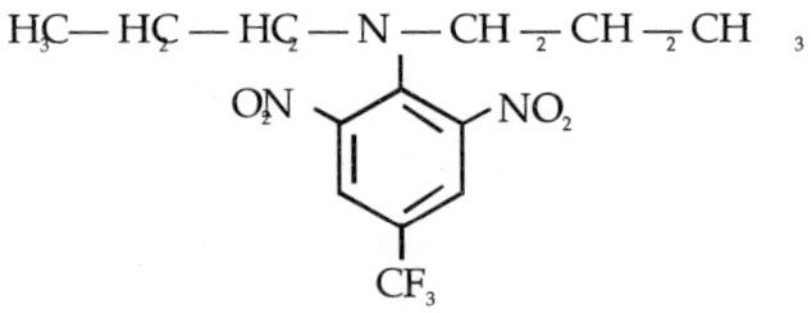

2, 6-Dinitro-N, N'-diprepyl-4-(trifluoromethyl) benzeneamine (trifluralin)

Fig. 4. Structural formulae of some aliphatic and aromatic compounds

High pH of water used as a herbicide carrier (pH of 8 or more) can result in the breakdown (degradation) of some herbicides through alkaline hydrolysis. This has been documented for Asulox/Asulam, which is used extensively for johnsongrass (*Sorghum halepense*) control in sugarcane. To alleviate this problem, a buffer is added to the spray solution, lowering the pH of the spray solution.

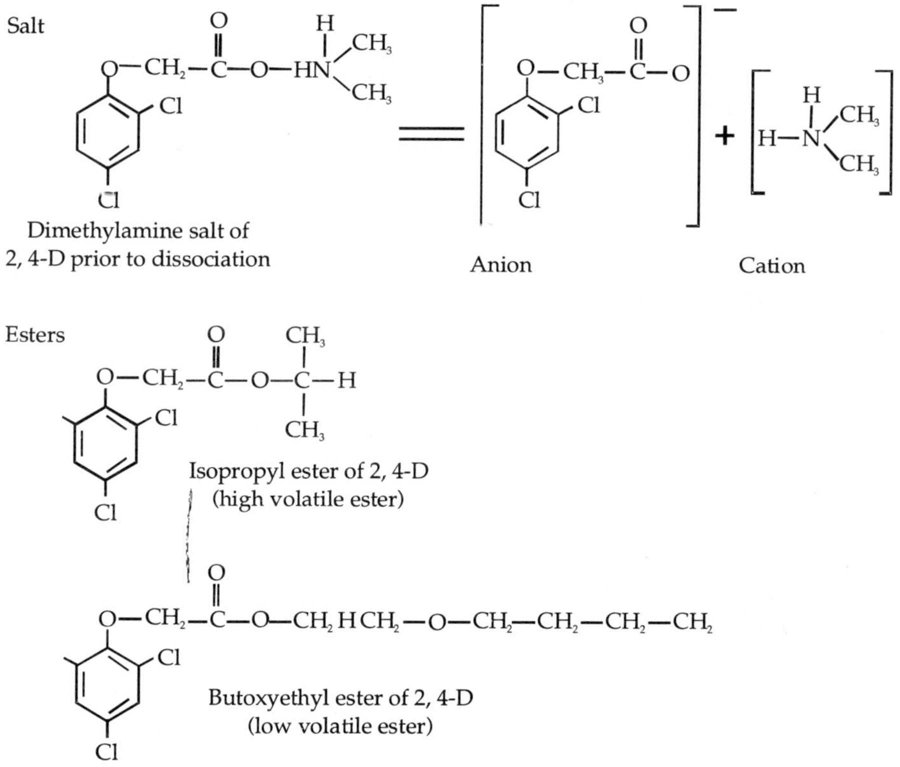

Salt

Dimethylamine salt of 2, 4-D prior to dissociation

Anion

Cation

Esters

Isopropyl ester of 2, 4-D (high volatile ester)

Butoxyethyl ester of 2, 4-D (low volatile ester)

Fig. 5. Major forms of 2, 4-D

Classification of herbicides

Herbicides most often are classified according to 1) chemical structure 2) use and 3) effect on plants. Herbicides are also classified according to toxicity or hazard level. Thorough discussion of herbicide classification is also provided by Zimdhal (1993).

(A) Based on chemical structure

Classification systems based on chemical structure catalog herbicides by chemical similarities. This classification system is used in the Weed Science Society of America (WSSA) Herbicide Handbook

(1994), which provides a brief description of the various herbicides that are used in the United States. The primary use, formulations, water solubility and acute oral toxicity for each herbicide is usually provided by this method of classification. Frequently herbicides of the same chemical group have common physiological characteristics that allow on to predict how a new herbicide of the group may be used. Minor difference in chemical structure often led to significant difference in selectivity. The classification of herbicides based on carbon atoms is listed below:

1. Inorganic herbicides

These herbicides were used between 1869 and 1930 before the introduction of organic herbicides. Presently, these herbicides have been discouraged for use as herbicides owing to their non selective nature, high mammalian toxicity and /or extended persistence in soil.

Ammonium sulphamate (AMS)	$(NH_4)OSO_2.NH_2$
Borax	$B_4H_{20}Na_2O_{17}$
Calcium cynamide	$CaCN_2$
Copper sulfate	$CuSO_4.\ 5H_2O$
Hexaflurate	AsF_6K
Potassium azide	KN_3
Sodium azide	NaN_3
Sodium chlorate	$NaClO_3$
Sulfuric acid	H_2SO_4

2. Organic herbicides

Aliphatics

Chlorinated acids

These are usually formulated as the sodium salt and are particularly effective on grasses. By substituting one chlorine by a methyl group to TCA (2,2,2 trichloro propionic acid), one can obtain 2,2-dichloropropionic acid named as dalapon. These are used for the control of annual and perennial weeds. These herbicides inhibit the growth of both shoots and roots and cause leaf chlorosis.

$$Cl-C(Cl)_2-C(=O)-OH \xrightarrow[+CH_3]{-Cl} CH_3-C(Cl)_2-C(=O)-OH$$

TCA → Dalapon

Organic arsenicals

MSMA and DSMA are contact herbicides with some systemic properties used for control of grass weeds in cotton and on non-crop area as post-emergence. These herbicides cause foliar necrosis followed by total necrosis.

$$\begin{array}{c} CH_3 \\ | \\ CH_3-As-OH \\ \| \\ O \end{array}$$

Cacodylic acid

$$CH_3-As(=O)(OH)(ONa)$$

MSMA

$$\begin{array}{c} O \\ \| \\ CH_3-As-OH \\ | \\ OH \end{array}$$

MAA

$$\begin{array}{c} O \\ \| \\ CH_3-As-ONa \\ | \\ ONa \end{array}$$

DSMA

$$\begin{array}{c} O \\ \| \\ CH_3-As-OH \\ | \\ O\text{-}NH_4 \end{array}$$

MAMA

Others

Acrolein	$CH_2{=}CH\text{-}CHO$
Allyl alcohol	$CH_2{=}CH\text{-}CH_2OH$
Methyl bromide	CH_3Br
Glyphosate	$HO\text{-}C(=O)\text{-}CH_2\text{-}N(H)\text{-}H_2C\text{-}P(=O)(OH)(OH)$

Amides

Chloroacetamides

They have a monochlorinated methyl ($Cl\text{-}CH_2\text{-}$) group in the R_1 position which includes alachlor, butachlor, CDAA, metolachlor, propachlor, terbuchlor. The chemical structure of remaining amide herbicides have little in common except for the central amide structure although some have a hydrogen (H-) substitution on the R_3 position. They are used as selective herbicides in variety of crops as pre-emergence or preplant-soil incorporated herbicides. Benzadox, cypromid and propanil are applied to the foliage of the weeds to be controlled.

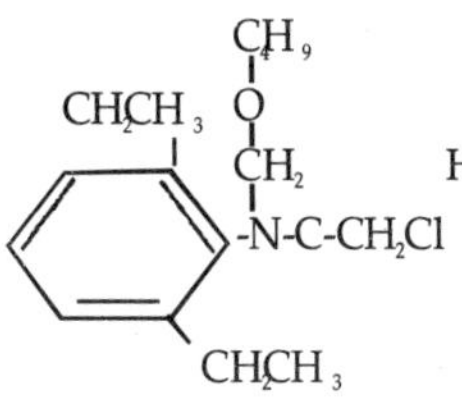

Butachlor

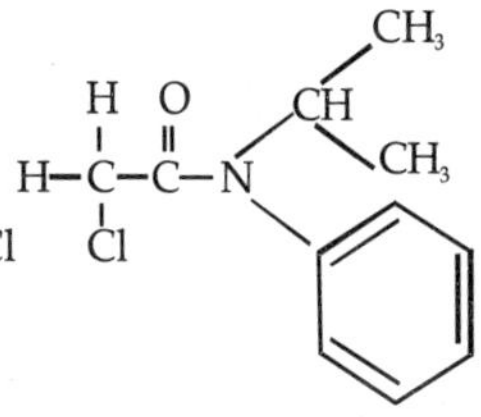

Propachlor

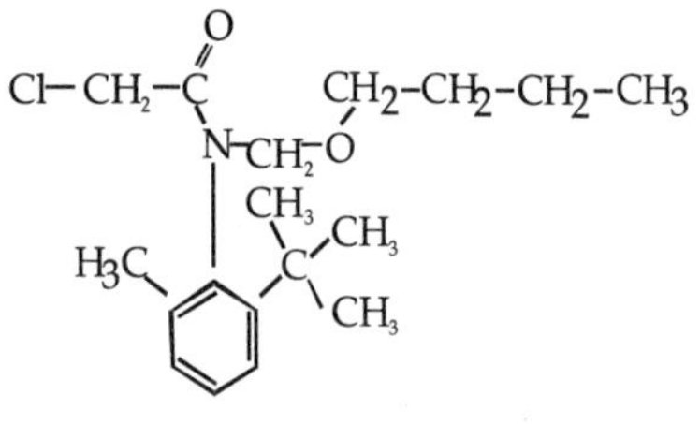

Terbuchlor

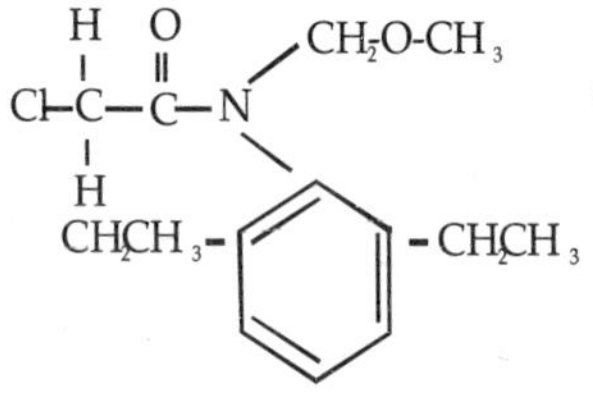

Alachlor

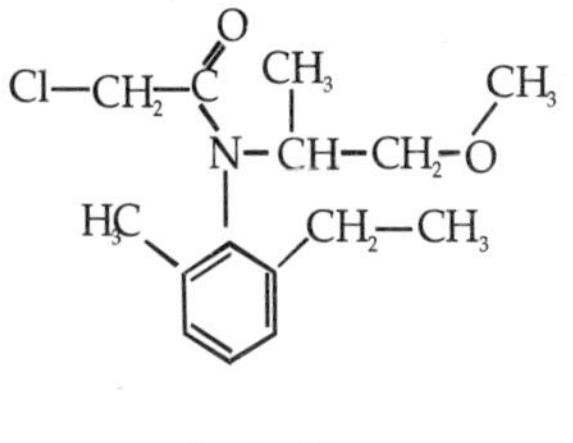

Metolachlor

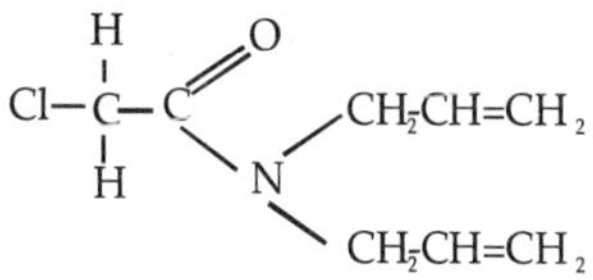

CDAA

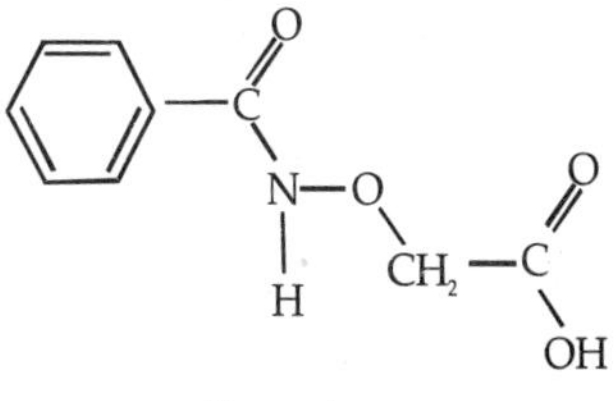

Benzadox

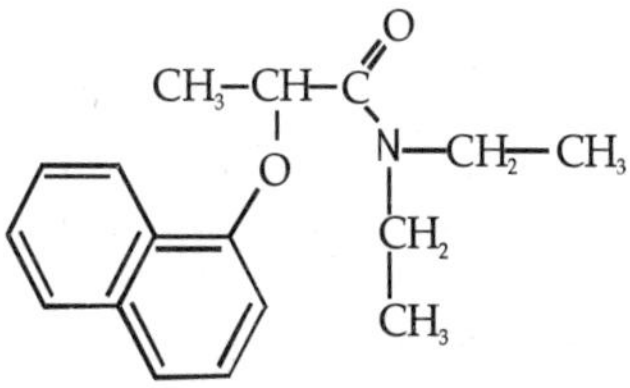

Napropamide

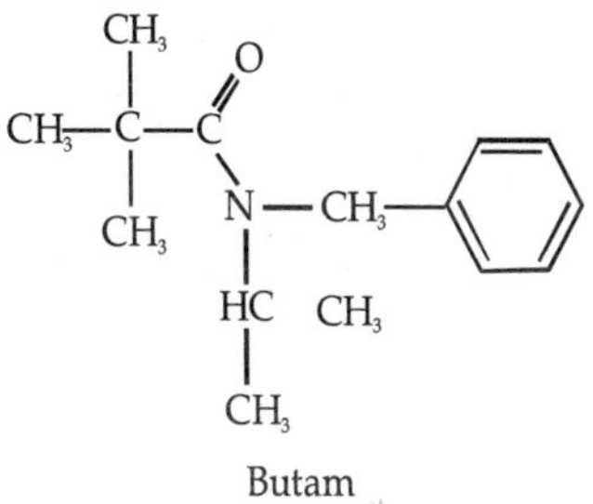

Butam

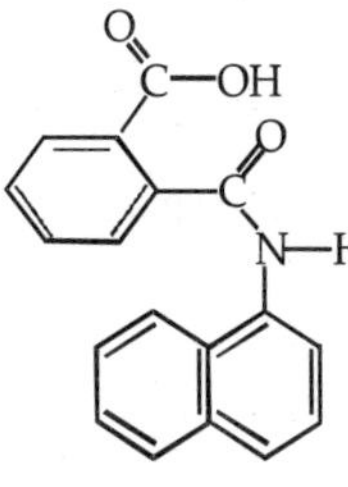

Naplatam

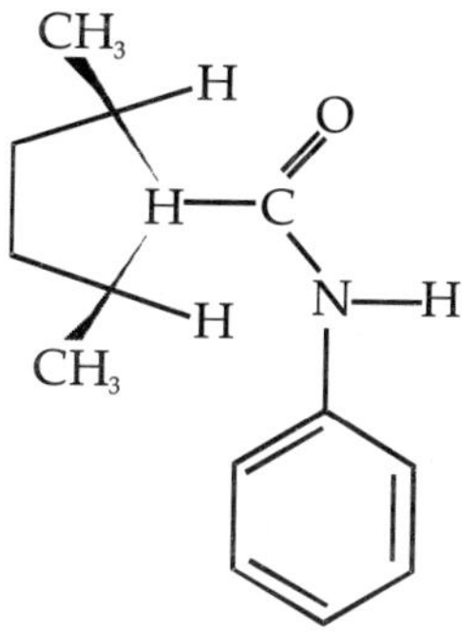

Cisanilide

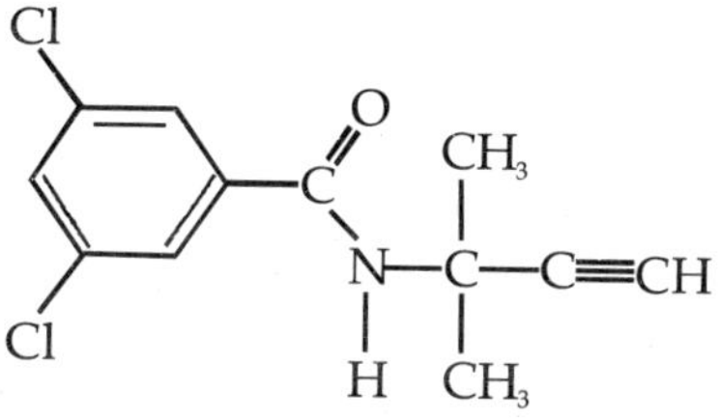

Propanamide

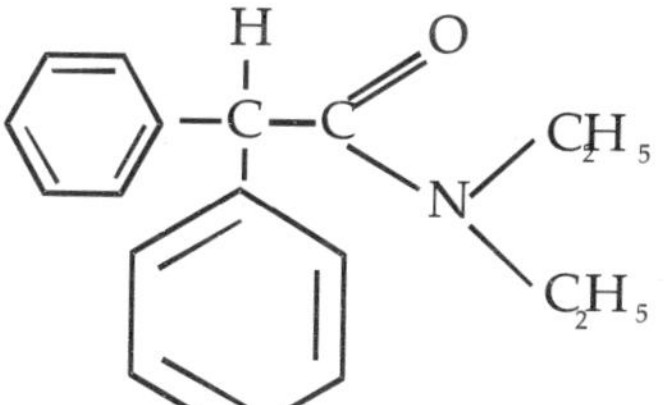

Diphenamid

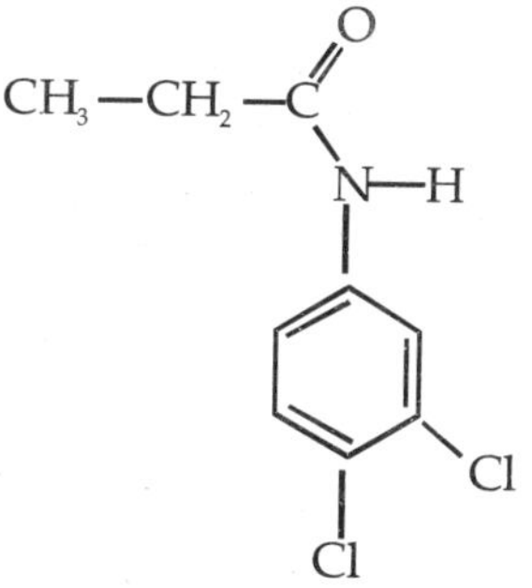

Propanil

Benzoics

Activity of benzoic herbicides is similar to phenoxy type herbicides but is more persistent in soil and are especially useful for the control of deep rooted perennial weeds and other weeds that are not controlled by phenoxy acids. Acceptable selectivity to several crop species has been obtained with certain chlorinated benzoic acids by substituting a methoxy (-OCH_3) or an amino (-NH_2) group for one of the chlorine atoms. They may be formulated as salts and esters similar to 2, 4-D. Dicamba is used in barley, maize, grasses, oats, sorghum and wheat. Chloramben is used in asparagus, beans, maize, peanuts, peppers, pumpkins, soybeans, squash, sunflower, sweet potatoes and tomatoes.

Dicamba and 2,3,6-TBA are readily absorbed by leaves and roots and translocated *via* both the symplastic and apoplastic systems with accumulation in areas of high metabolic activity. Chloramben although absorbed by roots and leaves is translocated very little from either.

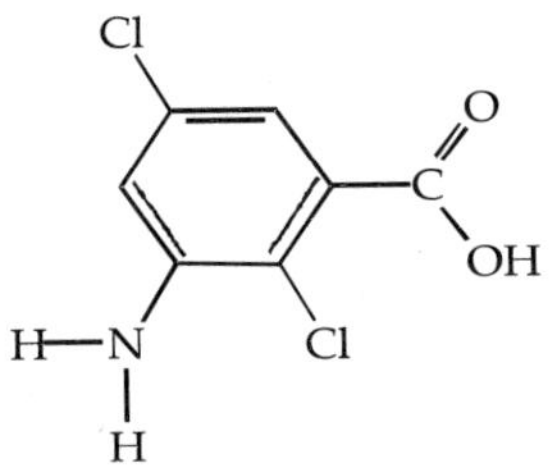

Chloramben

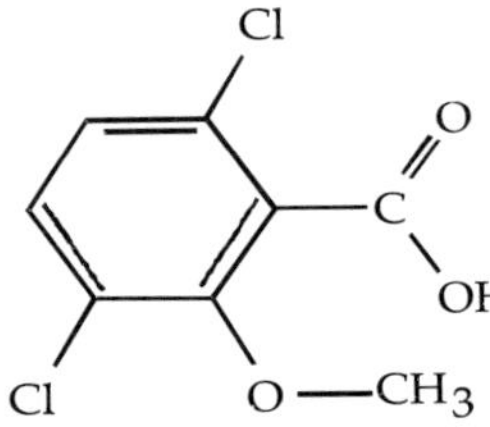

Dicamba

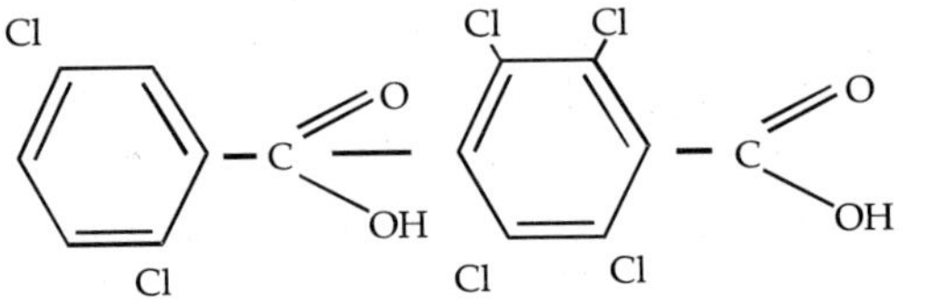

PBA

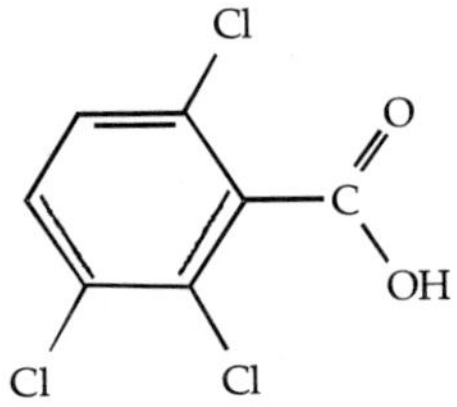

2, 3, 6-TBA

Bipyridiums

Diquat and paraquat are the most widely used member of bipyridium quaternary ammonium class. Diquat and paraquat are usually used as general contact herbicides and are applied to foliage. Selectivity may be obtained with certain crops by avoiding contact with the desired species.

These are highly water soluble and strongly absorbed by means of base exchange in most of the soil that they are considered as biologically inactive in soil. Paraquat can be used as preplant, pre-emergence or directed post-emergence in many crops, whereas diquat is used for preharvest desiccation or aquatic weed control. This is followed by necrosis and ultimate death of the entire leaf. Diquat and paraquat are also used as non-selective herbicides. The contact action of these herbicides is very rapid, often visible within a few hours.

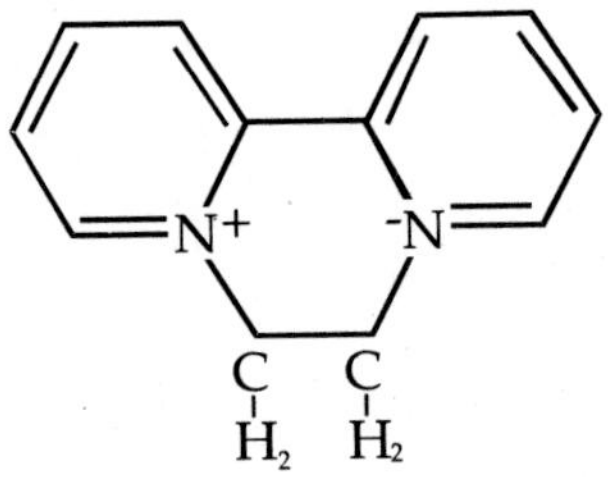

Diquat

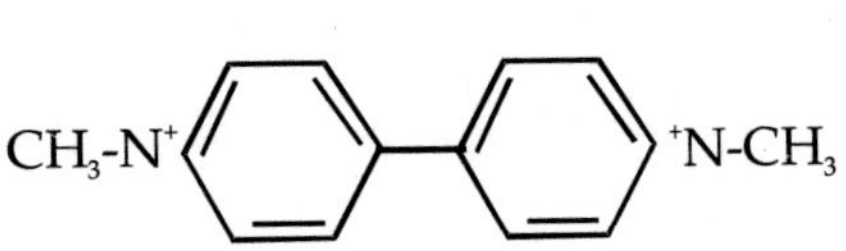

Paraquat

Carbamates

Carbamate herbicides derive their basic structure from carbamic acid (NH_2COOH). Herbicidal properties of propham were described by Templeman and Sexton in 1945. It has been widely used to control grasses in sugarbeet, soybean, onion, garlic, peas, flax, sunflower, rape and mustard. Propham is toxic to barley, oat, wheat, red rice, maize and timothy. Chlorpropham has been used to control grassy weed and more persistent in soil. Barban is highly selective to control wild oat in spring wheat, durum wheat, barley, flax, peas, sugarbeet, safflower, mustard and soybean. Phenmedipham and desmedipham are post-emergent herbicides and particularly effective in sugarbeet and used to control annual beets. Asulam is a foliar applied herbicide used for the control of certain perennial grasses such as horsetails (*Equisetum arvenses.*), rushes (*Juncus* spp.), bracken (*Pteridium aluilinum*) and tansy (*Tanacetum* spp).

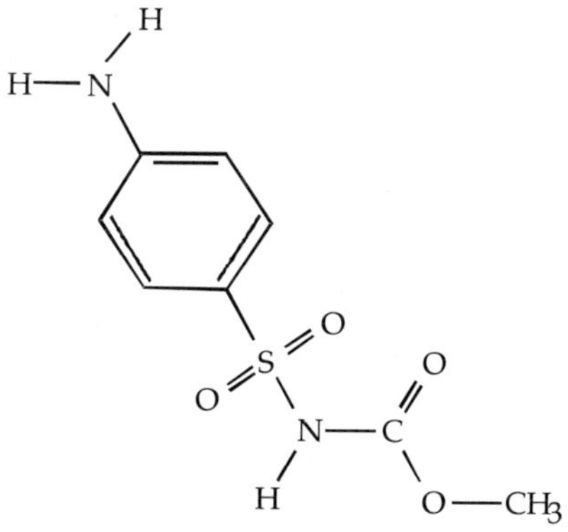

Asulam

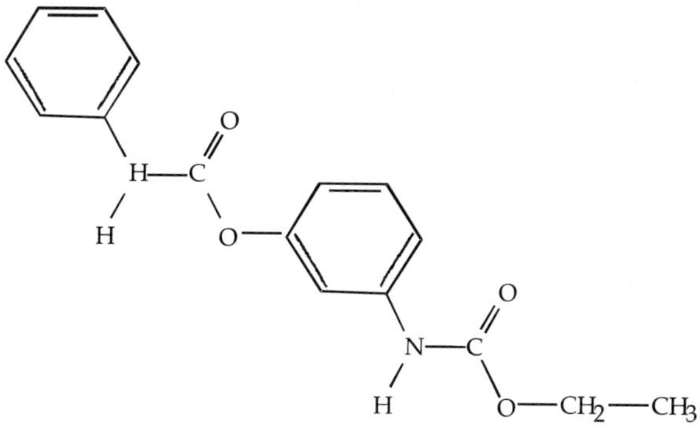

Desmedipham

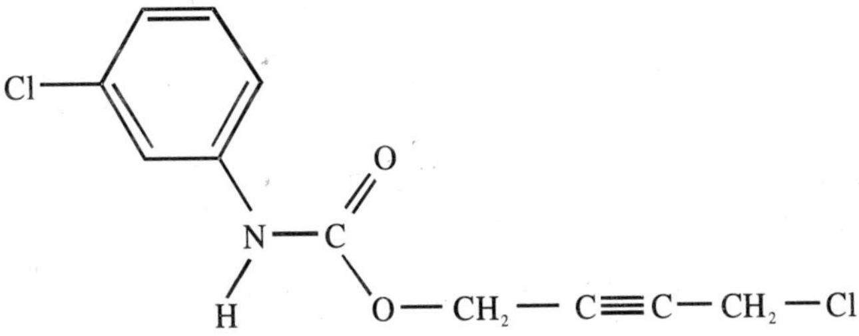

Barban

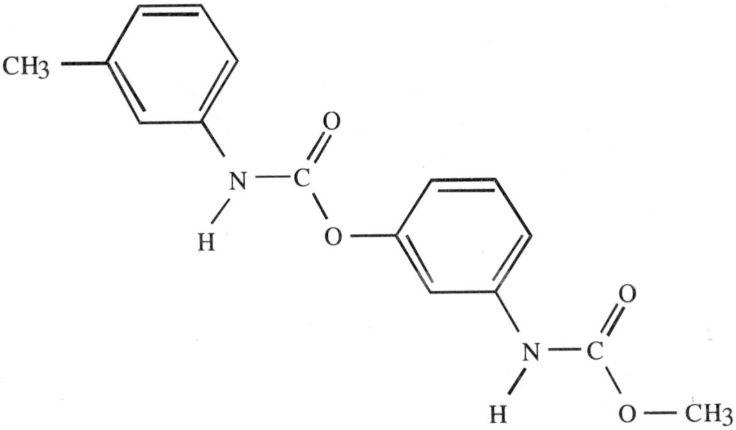

Phenmedipham

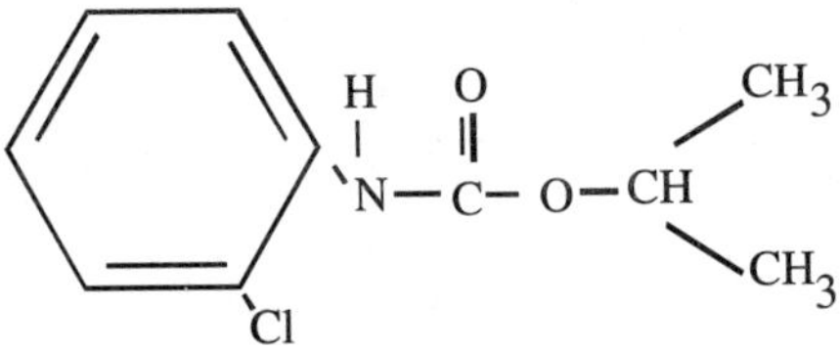

Chlorpropham

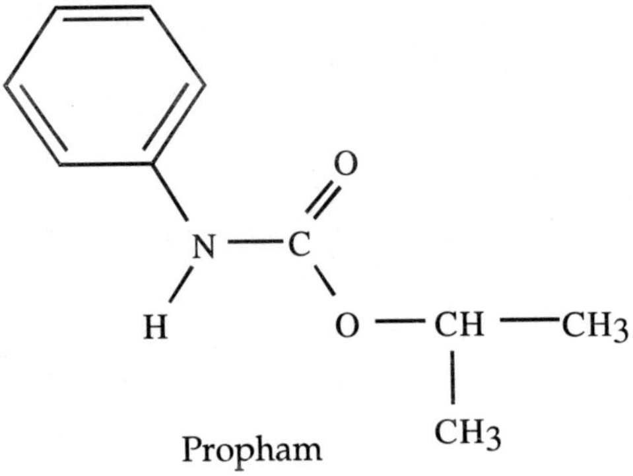

Propham

Dinitroanilines

Dinitroanilines are yellow orange in colour, have low water solubility and are somewhat volatile. They are usually used for selective weed control as preplant soil incorporation treatment prior to weed seed germination. The dinitroanilines herbicides inhibit growth of the entire plant. However, this is apparently largely brought about by initially limiting root growth, especially the development of lateral or secondary roots. The roots which do develop, often only the primary roots are somewhat thickened, stubby and are devoid or have only a limited number of secondary roots. However, when herbicide is absorbed by the shoot it inhibits shoot growth directly.

Dinitroanilines herbicides are strongly adsorbed on soil clay and organic matter and show negligible leaching. Dinitroanilines degrade slowly in soil and alter the chemical composition and several biochemical processes in higher plants. In general, all dinitroanilines persist in soil for more than one season.

NO_2 / F—C(F)(F)— ring —N(CH_2—CH_2—CH_2—CH_3)(CH_2—CH_3) / NO_2

Benfin

NO_2 / CH_3—S(=C)(=C)— ring —N(CH_2—CH_2—CH_3)(CH_2—CH_2—CH_3) / NO_2

Nitralin

NO_2 / CH_3—C(CH_3)(CH_3)— ring —N(H)—CH(CH_3)—CH_2—CH_3 / NO_2

Butralin

Oryzalin

Dinitramine

Pendimethalin

Ethalfluralin

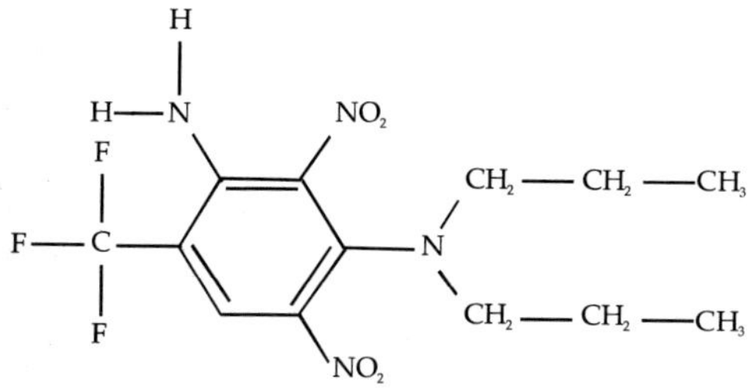

H
H—N NO$_2$
F
F—C— N CH$_2$—CH$_2$—CH$_3$
F CH$_2$—CH$_2$—CH$_3$
NO$_2$

Prodiamine

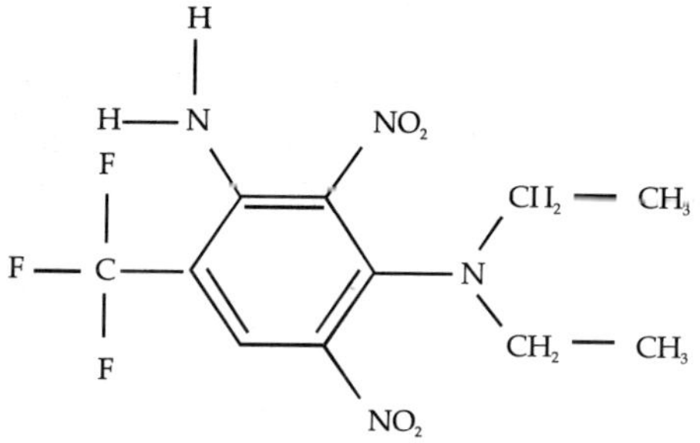

H
H—N NO$_2$
F
F—C— N CH$_2$—CH$_3$
F CH$_2$—CH$_3$
NO$_2$

Fluchloralin

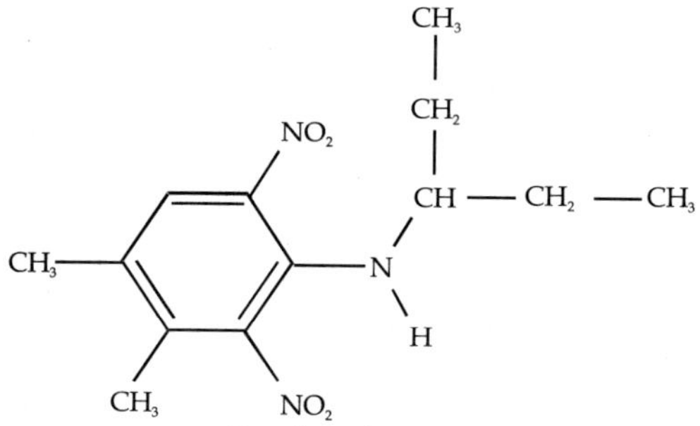

CH$_3$
CH$_2$
NO$_2$
CH—CH$_2$—CH$_3$
CH$_3$— N
H
CH$_3$ NO$_2$

Profluralin

CH$_3$
CH$_2$
NO$_2$
O CH$_2$
H$_3$C
S =N—S— N
H$_3$C
O CH$_2$
NO$_2$ CH$_2$
CH$_3$

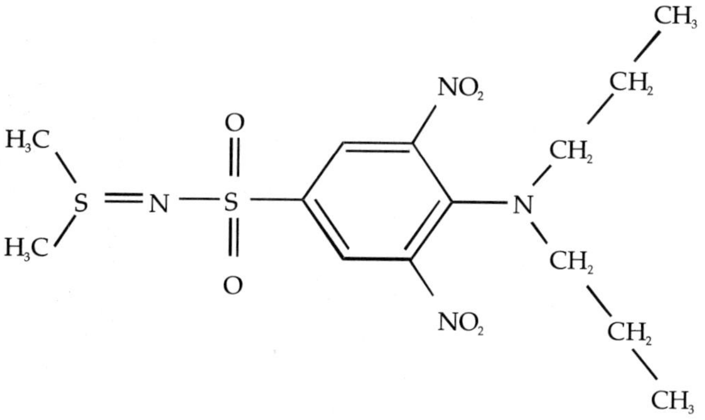

Prosulfalin

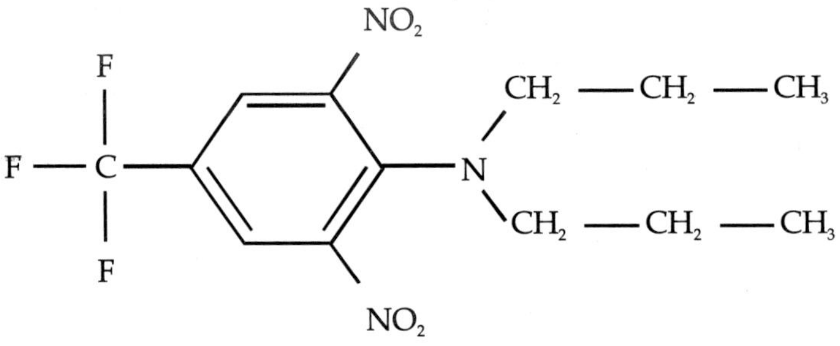

Trifluralin

Diphenyl ethers

The diphenyl ether herbicides are used to control many annual weeds with a pre-emergence or early post -emergence application in several crops. In general, the diphenyl ethers are considered to be contact herbicides, however certain compound also induce growth response. The diphenylether induce chlorosis and necrosis when applied to leaves which have been attributed to loss of membrane integrity.

Some of these compounds also inhibit seed germination, growth and meristematic activity. These herbicides are readily absorbed by the leaves but translocation is limited. These herbicides undergo reduction of the NO_2 group to NH_2 group, cleavage of the ethers linkage and subsequent conjugation with natural plant substances.

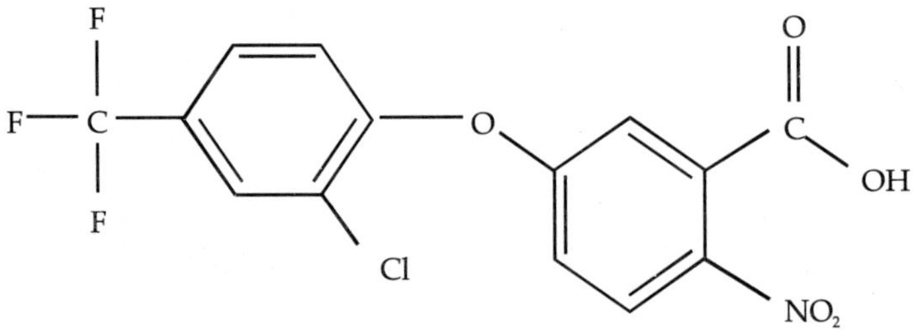

Acifluorfen

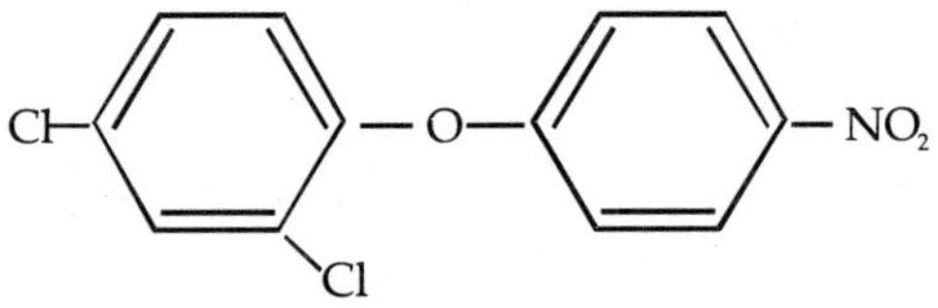

Nitrofen

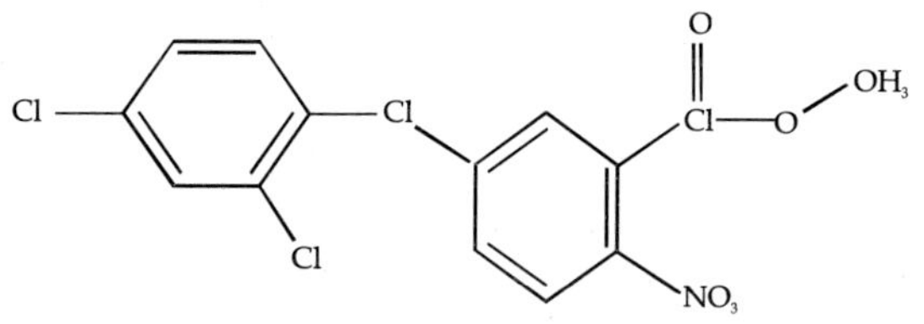

Cl Cl Cl O Cl O OH_3 NO_3

Bifenox

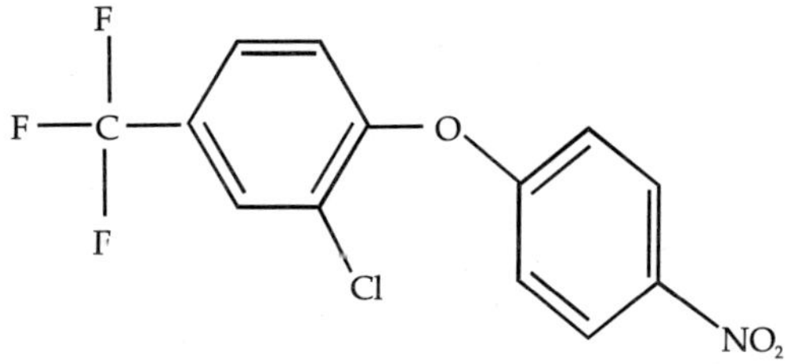

F F C F O Cl NO_2

Nitrofluorfen

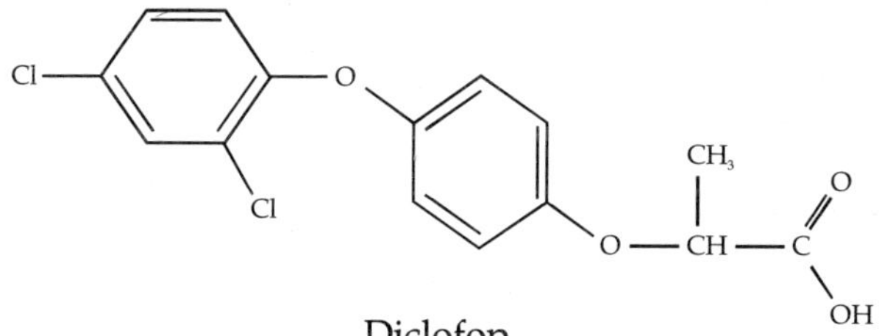

Cl O Cl CH_3 O O CH C OH

Diclofop

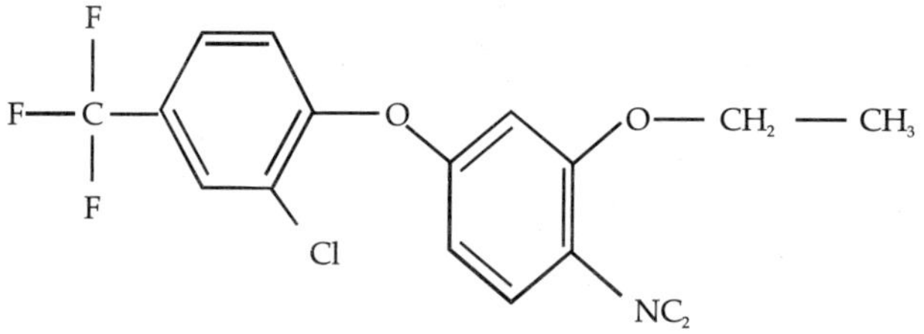

F F C F O Cl O CH_2 CH_3 NC_2

Oxyfluorfen

F F C F O NO_2 NO_2

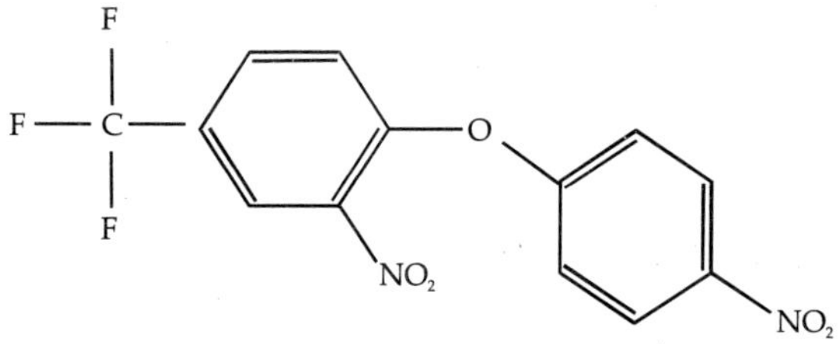

Fluorodifen

Imidazolines

This class of herbicides was discovered by American Cyanamid in 1963 for control of mono and dicotyledonous weeds. The structural formulae for some important herbicides are given variation in the z-aryl substituent on the imidazolines ring which results in considerable variation in the spectrum of herbicidal activity. The imidazolines have low mammalian toxicity (LD_{50} oral to rats >5000 mg/kg). Imazaquin and imazethapyr are effective as pre and post-emergence in soybean and other leguminaceous crops. The herbicides of this class are listed below.

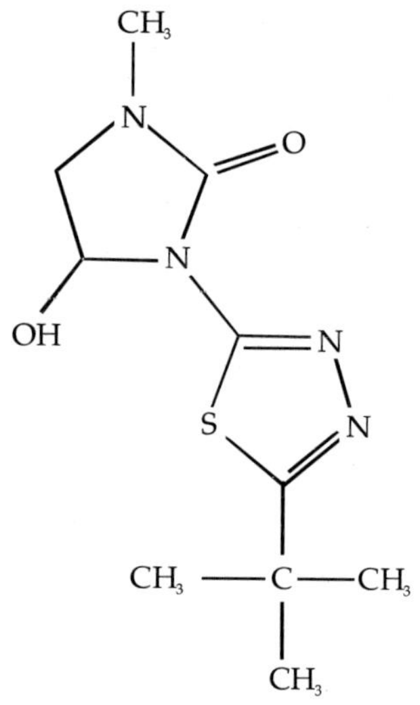

Buthidazole

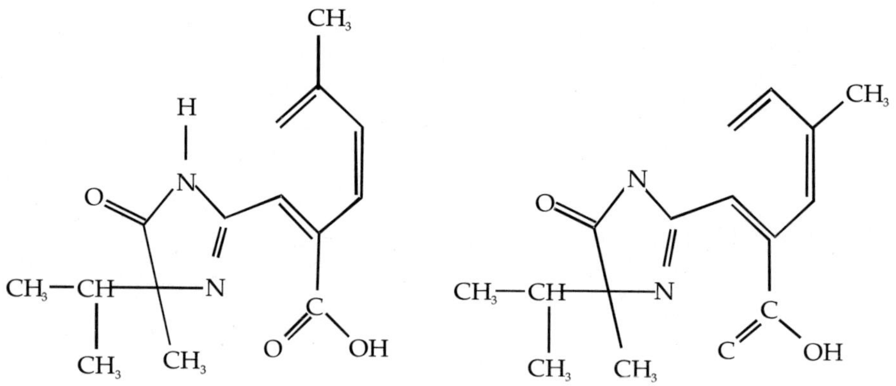

Imazamethabenz

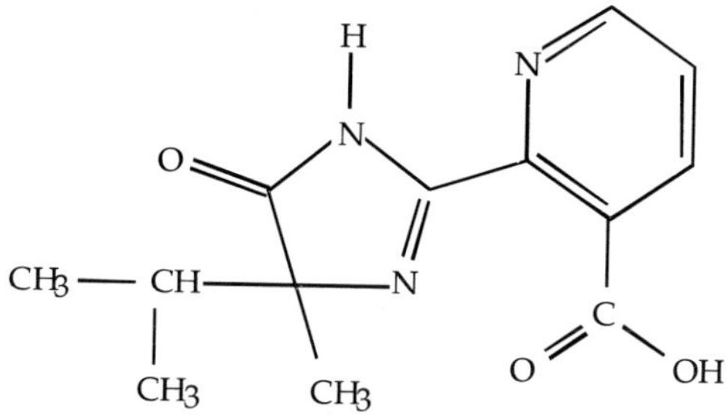

Imazapyr

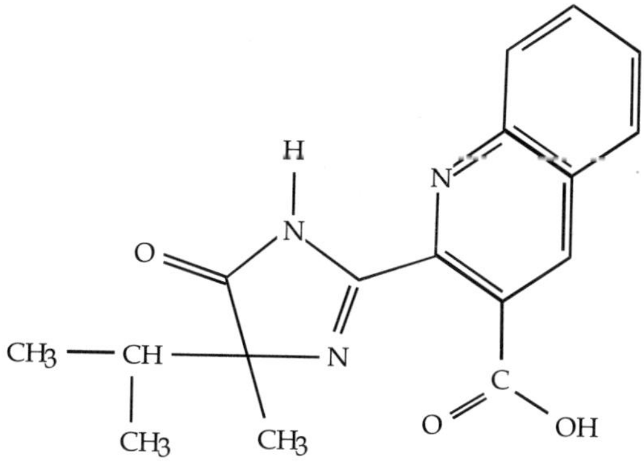

Imazaquin

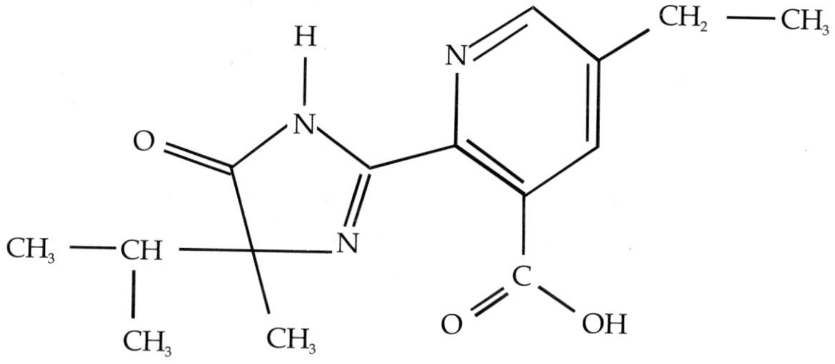

Imazethapyr

Isoxazolidinones

The clomazone a compound of this class was developed in the early 1980 and commercialized by FMC as Command in 1985. It is used as pre plant emergence or pre-emergence. Comazone is used to control several annual broadleaf and grass weeds including *Abutilon*

Isoxazolidinones

The clomazone a compound of this class was developed in the early 1980 and commercialized by FMC as Command in 1985. It is

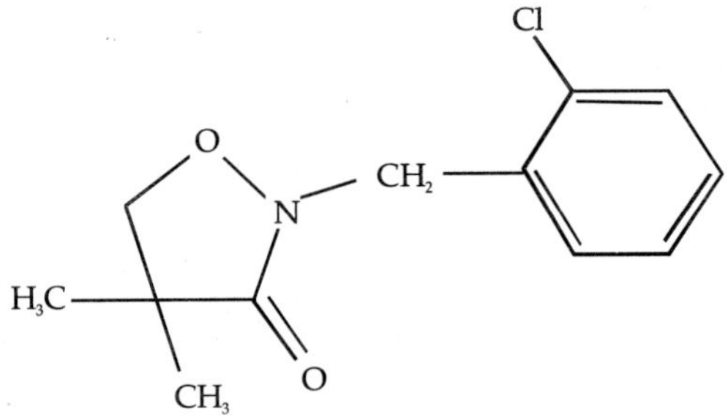

Clomazone

used as pre plant emergence or pre-emergence. Comazone is used to control several annual broadleaf and grass weeds including *Abutilon theophrasti, Ambrosia artemisiifolia, Chenopodium album, Echinochloa crus-galli, Digitaria* spp. etc. Susceptible plants bleached white and become necrotic after several days, however, in post application plants exhibit foliar bleaching in later growth stage.

Nitriles

The nitrile are compound containing –C=N group. Those used as herbicides are benzonitriles with OH-, Cl^- and/or Br^- substitution on the benzene ring. The herbicides of nitrile class are extremely effective inhibitors of nitrification in soils. However, soil microorganisms are believed to have an ability to adapt to these herbicides and degrade them. Dichlobencil was first described as phytotoxic in 1958 by Koopman and Daams (1960). In 1959 personnel of Amchem Products Inc, discovered the herbicidal activity of ioxynil and subsequently they also discovered the phytotoxicity of bromoxynil. Dichlobenil was the first compound of this class, introduced as Casoron by Philip Duphar in 1960. Dichlobencil is a soil applied herbicide for control of mono and dicot weeds by preventing the establishment of seedlings. It is used as pre- or post-emergence for control of annual as well as perennial weeds such as *Pteridium aquillinum, Agropyron* spp., *Astemisia* spp., *Cynodon* spp. It may also be used for aquatic weed control. Bromoxynil and ioxynil were introduced by May and Bekar and

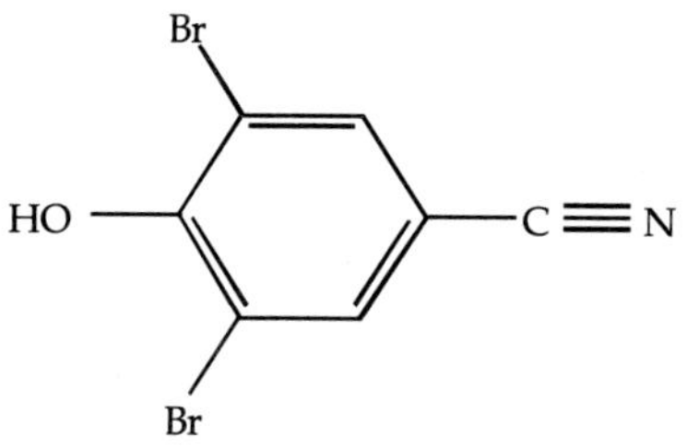

Bromoxynil

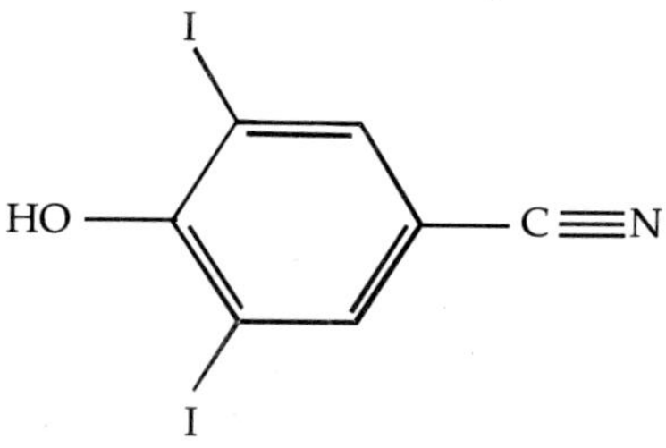

Ioxynil

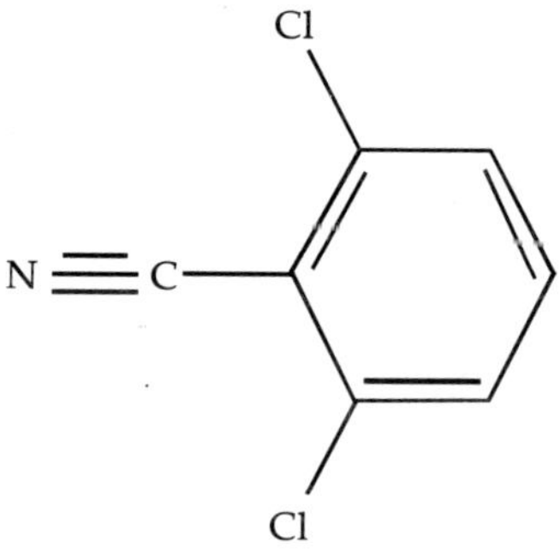

Dichlobenil

Oxadiazoles

The oxadiazon a compound of oxadiazole class was first reported in 1969 and introduced as pre-emergence herbicide by Rhone-Poulenc. It is active against many annual and broad leaf weeds. It is used in groundnut, onion, garlic, potato, sunflower, rice, sugarcane, soybean, vegetables, tea, rubber and banana. It is particularly useful to control weeds that grow from bulbs, rhizomes and other deep rooted plant propagules including *Cynodon dactylus* and *Lolium perenne.*

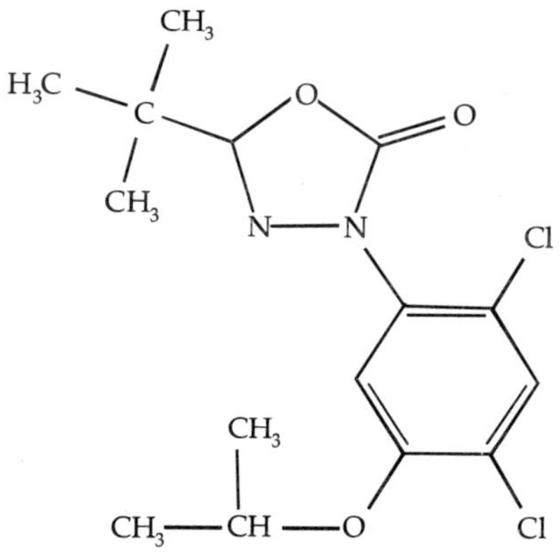

Oxadiazon

Oxadiazoles

The oxadiazon a compound of oxadiazole class was first reported in 1969 and introduced as pre-emergence herbicide by Rhone-Poulenc. It is active against many annual and broad leaf weeds. It is used in groundnut, onion, garlic, potato, sunflower, rice, sugarcane, soybean, vegetables, tea, rubber and banana. It is particularly useful to control weeds that grow from bulbs, rhizomes and other deep rooted plant propagules including *Cynodon dactylus* and *Lolium perenne.*

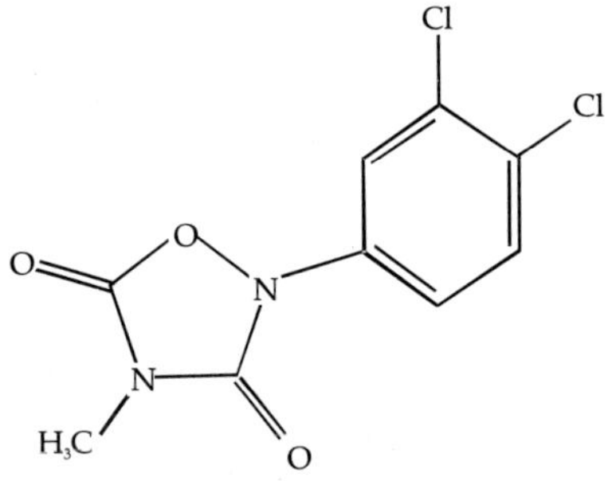

Cl Cl O O N N H_3C O

Methazole

Oxadiazolides

Methazole was synthesized by Sandoz in 1965 and its herbicidal activity was reported in 1970. Foliar application of methazole caused veinal chlorosis followed by general chlorosis and desiccation. It is used as preplant, pre-emergence or post-emergence against wide range of annual and broad leaf weeds in cotton, sorghum, maize, wheat, oats, soybean, onion, grapes and several fruits crops. Methazole do not accumulate in soil and do not injure rotational crops in the following season.

Phenols

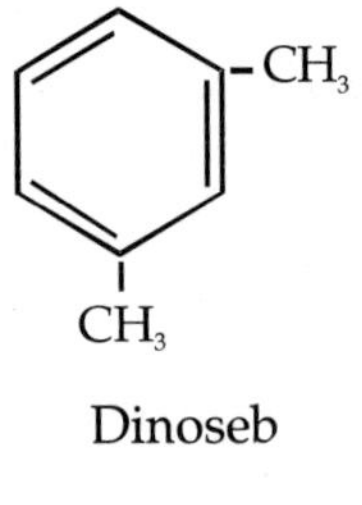

$-CH_3$ CH_3

Dinoseb

Phenoxys

The chlorinated phenoxy acid compounds comprise of phytotoxic substances that are used as herbicides in the form of the parent acids or more commonly as salt and esters, the chlorophenoxy compounds along with certain benzoic and picolinic acid derivatives are growth regulators with hormone like activity. One of the compound of this class was first introduced in 1944. At relatively lower dose, they bring about growth responses in regions distant from the point of application. Practically this property of translocation is responsible for the fact that they can be used by the low volume method and they can be used to kill root system of perennial weeds by foliar application.

CH_2 - COOH, O, Cl, Cl

2, 4-D

CH_2 - CH_2 - CH_2 - COOH, O, CH_3, Cl

MCPB

CH_2 - CH_2 - CH_2 - COOH, O, Cl, Cl

2, 4-D, B

Cl, Cl, O — CH — C, CH_3, O, OH

Dichlorprop

Cl, Cl, Cl, O — CH_2 — C, O, OH

2, 4, 5-T

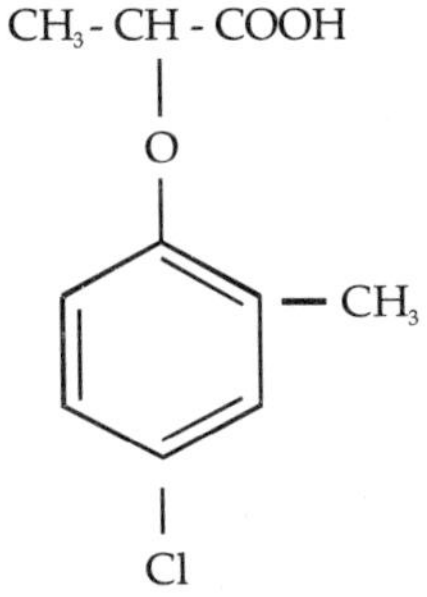

Mecoprop 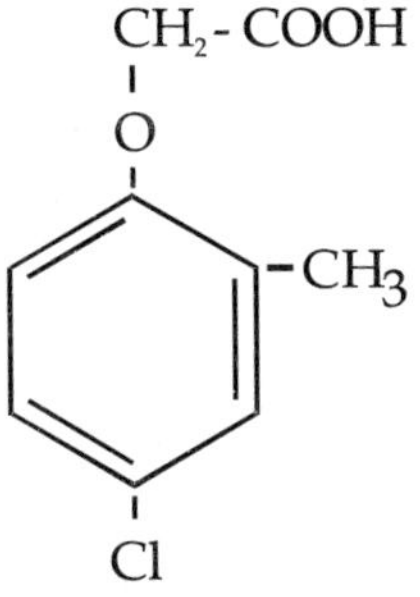 MCPA 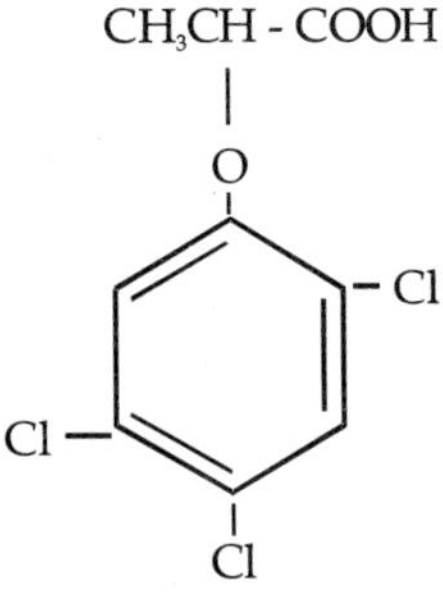 Silvex

The phenoxy herbicides induce abnormalities in growth and plant structure. Epinastic bending is rapid following foliar applications; subsequently tumours, secondary roots and fascinated structures often develop in susceptible species. Absorption by leaves and roots readily occurs, however, the formulation and environmental conditions influences that rate. These herbicides are considered to be primarily translocated from leaves in the symplast; therefore the amount translocated from leaves is acceptable but minimal from roots. Many biochemical and metabolic changes are induced by phenoxy herbicides, many of them are undoubtly secondary and tertiary in nature.

N-phenylphthalamides

Flumiclorac a compound of N-phenylphthalamide chemical class was discovered in 1981 by Sumitomo, Japan. It is commercially available as a phenyl ester foumulation under its trade names *Resource* and *Sumiverde.* It is used as post-emergence in soybean and maize to control *Abutilon theophrasti, Chenopodium album, Ambrosia artemisiifolia, Amaranthus* spp, *Euphorbia maculate* etc. The susceptible plants exhibit injury symptoms within one day under bright sunlight followed by wilting and bleaching. The leaves then become brown, desiccated and necrotic. It is a non persisting herbicide.

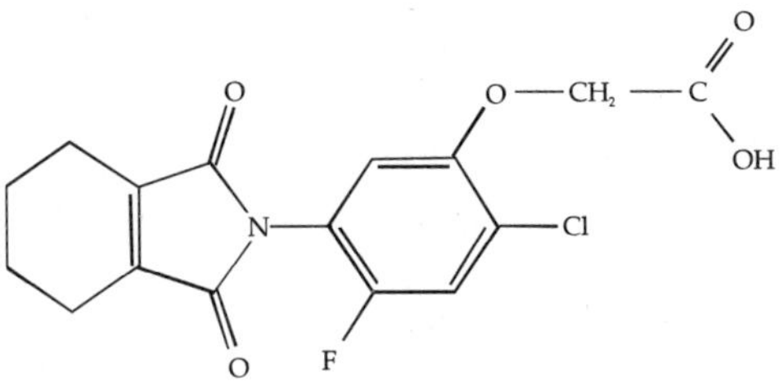

Flumiclorac

Phenylpyridazines

Pyridate was discovered in 1974. It is used as post-emergence in alfalfa, groundnut, orchards, maize, onion, rice, vineyards and some vegetable crops. The susceptible plants show withering at leaf edges and the effect is complete within 5-10 days. It is non-persistent in soil and has short half-life of 7-2 days.

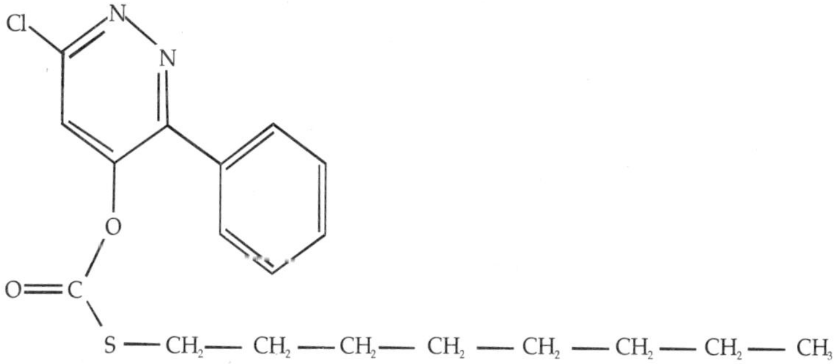

Cl N N O O=C S — CH_2— CH_2 —CH_2— CH_2 — CH_2 — CH_2 — CH_2 — CH_3

Pyridate

Phenyl triazinones (Aryl triazinones)

Sulfentrazone is a new phenyl triazinone herbicide introduced in 1991 and registered in 1998 for soybean by the FMC Corporation. It is used as preplant, pre and post-emergence for control of numerous broadleaf weeds in tobacco, sugarcane and some species of turfgrass. Several sensitive weeds are controlled by foliar application of sulfentrazone at 18 g ai/ha. Its post herbicidal activity can be enhanced by the use of surfactant.

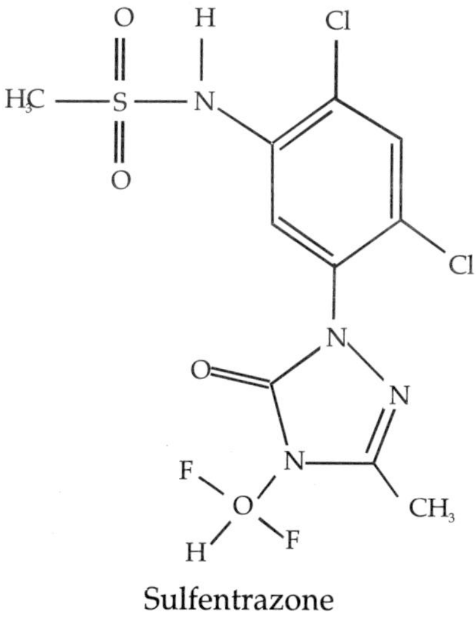

O H Cl H_3C —S—N O Cl N O N N F O F H CH_3

Sulfentrazone

Phthalamates

Naptalam a compound of pthalamate class was discovered by Uniroyal in 1947 and first registered as a herbicide in 1956. It is used as pre or post-emergence to control many annual broadleaf weeds and certain annual grasses in cucumber, melon, watermelon, groundnut, soybean etc. It is a non persistent herbicide with short half life of 14 days and the residues do not injure crops planted in the following season. Susceptible plants show strong epinasty. Naptalam also produces a strong antigeotropic response.

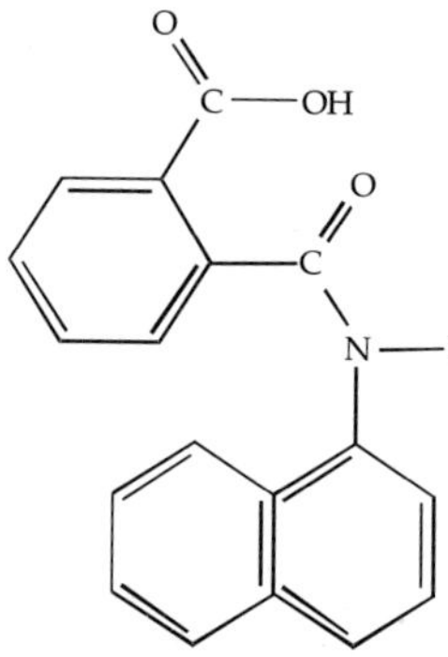

O
C—OH
O
C
N—H

Naptalam

Pyrazoliums

The herbicidal properties of difenzoquat-methyl sulfate were first reported in 1973. It is used as post-emergence herbicide for wild oat control. Efficiency of difenzoquat can be increased by the addition of a non-ionic surfactant. Plant growth ceases soon after application and injury symptoms appear after 3-7 days. Meristematic region become chlorotic, leading to necrosis.

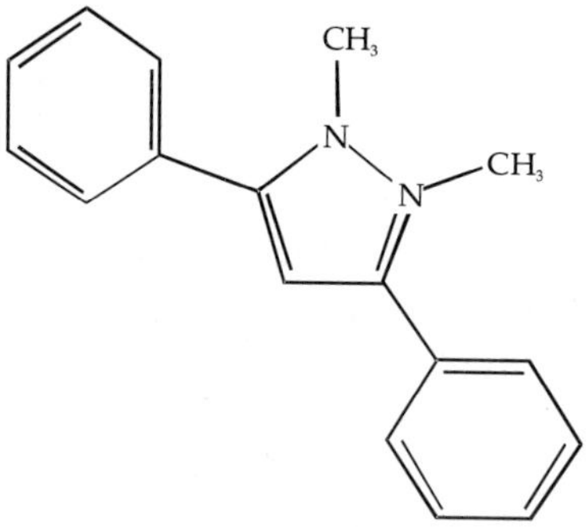

CH_3
N
N
CH_3

Difenzoquat

Pyridazinones

Pyrazon, also known as chloridazon was the first pyridazinone herbicides developed in 1963. It is used as pre or post-mergence to control many annual and broad leaf weeds. It provides 4-8 weeks weed control depending on soil moisture and temperature. It is a temperature sensitive herbicide resulting in crop damage under very unusual temperature extremes. Norflurazon, pyrazon and chloridon belong to this chemical class. Norflurazon was first introduced in 1968 by Sandoz and used as pre-emergence in tree crops, vineries, orchards ornamental and plantation crops to control many grasses and broadleaf weeds. Pyrazon and norflurazon have moderate to long persistence in soil.

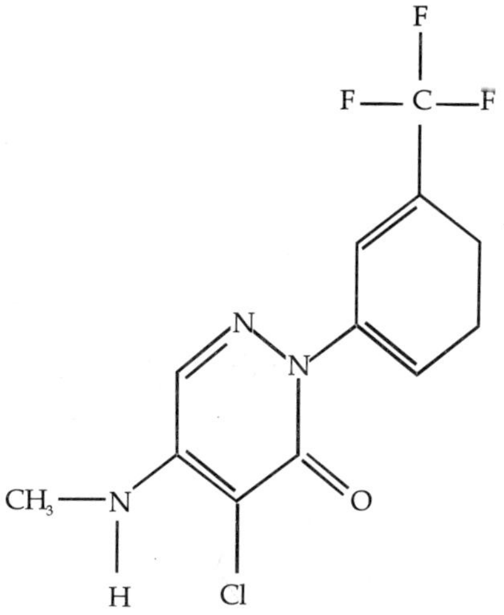

Norflurazon

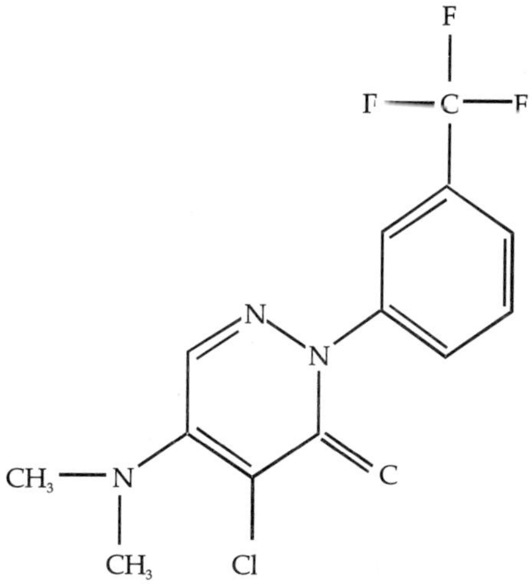

Metflurazon

Pyridines

Pyridines are mostly soil applied herbicides with little or no post activity. These include dithiopyr and thiazopyr, which causes root growth inhibition and swelling in meristematic region such as root tips but do not inhibit germination. Dithiopyr and thiazopyr were first introduced by Monsanto but purchased by Rohn and Haas in 1994. Dithiopyr is used as pre or post-emergence in direct seeded and transplanted rice, orchards and ornamental crops to control several annual and broadleaf weeds. Susceptible plants may show thickened or swallow hypocotyls or internodes. Thiazopyr is used as soil applied in several crops such as maize, potato, groundnut, vegetables, alfalfa, soybean, sugarcane and sunflower etc.

Fluridone belongs to pyridinones group and its activity was first described in 1976 as an aquatic herbicide. Fluridone can be applied on the water surface or subsurface or as a bottom application just above the hydrosol. It controls most submerged and emerged aquatic weeds including *Utricularia* spp., *Myriophyllum* spp., *Najas* spp., *Potatogeton* spp., *Hydrilla verticillata, Brachiaria mutica* etc. The affected weed becomes white, pink, and chloratic appearances in 7-10 days with complete kill requiring more than 30 days.

Pyrimidinylthio-benzoates also referred as benzoates as a new class of herbicides being developed since the early 1990s. Pyrithiobac is the first developed herbicide of this class developed by Dupont. It is used as preplant, pre or post-emergence to control numerous broadleaf weeds. Cotton is very tolerant to this herbicide. Antagonism occurs when pyrithiobac applied with graminicides such as quizalofop-p, fluazifop-p, clethodim and sethoxydim.

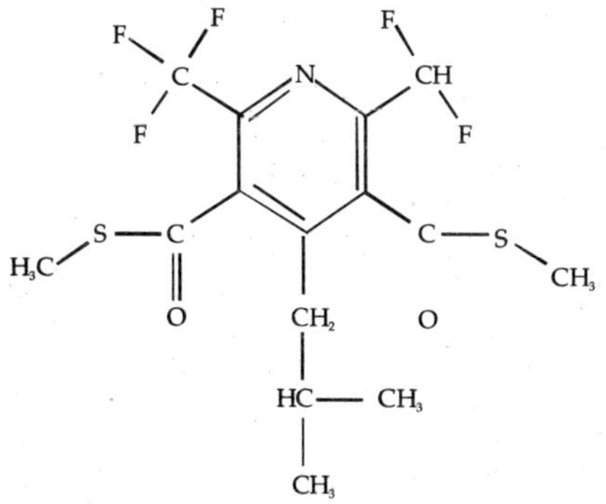

Dithiopyr

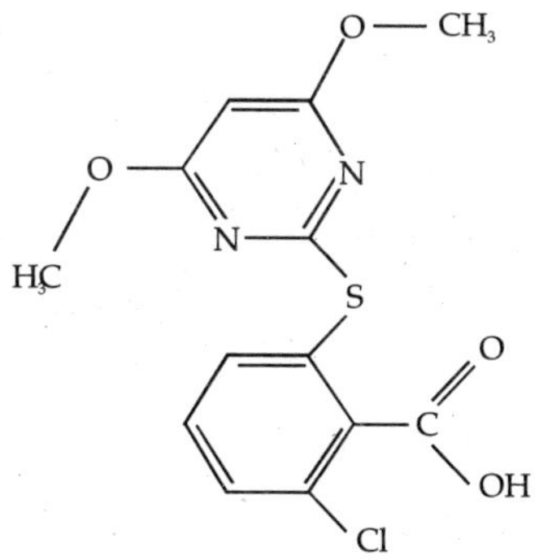

Pyrithiobac

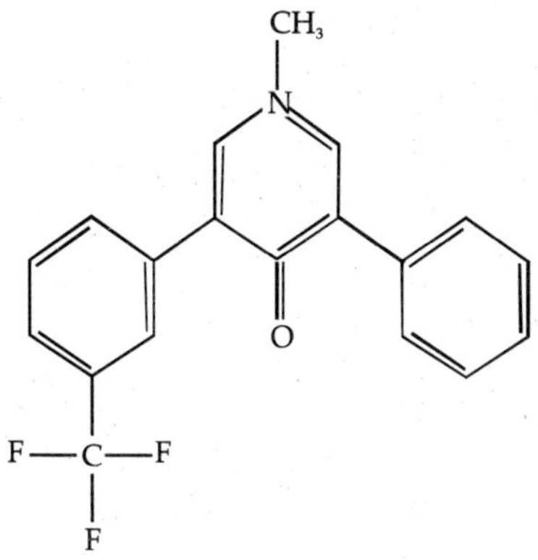

Fluridone

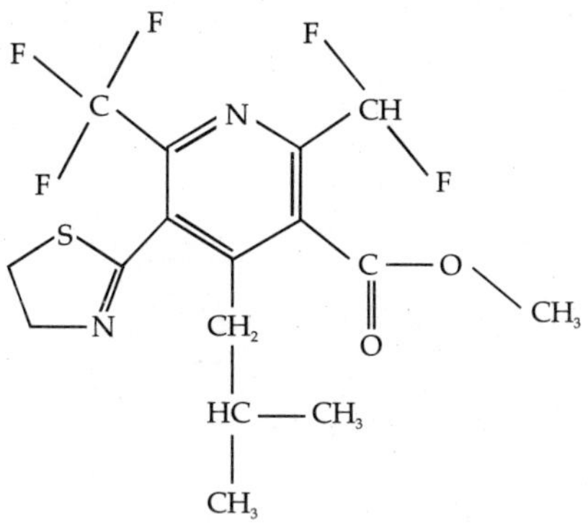

Thiazopyr

Quinolines

The first herbicide of this class was quinclorac developed by BASF. It is an auxin type herbicide used in direct seeded and transplanted rice. Quinclorac is particularly effective against *Echinochloa crus-galli, Setaria* spp. etc. it is used as preplant, pre or post-emergence to control certain annual and perennial broadleaf weeds. It inhibits shoot growth and caused chlorosis and subsequently necrosis of the entire shoot. It is highly persistent in soil and residues can injure subsequently crops planted one year after application.

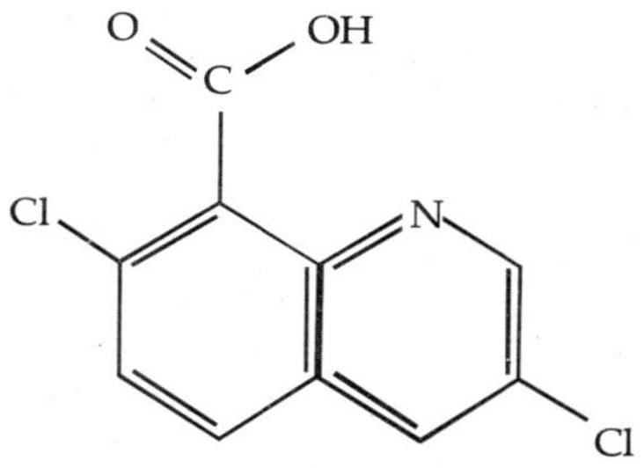

O OH C Cl N Cl

Quinclorac

Sulfonylureas

Sulfonylureas are a class of potent herbicides that are characterized as acetolactate synthase-inhibitors (ALS) and introduced by Dupont in 1982. Sulfonylures within a comparatively short period of time have made a major impact on weed control technology. Due to their high potency, these are typically applied at low rates. Sulfonylureas are thermally labile compounds and are directly metabolized in the environment. While the level of toxicity to animals from these compounds is low, their ability to persist in soil and water can cause damage to crops in later years. All sulfonylureas share the same basic structure, containing an aryl ring linked to a heterocycle (triazine or pyrimidine) through the herbicidally active sulfonylurea 'bridge'. To increase herbicidal activity, the aryl ring is usually substituted to 'ortho' to the sulfonylurea bridge with the substituent typically a carboxylic acid ester (e.g. metsulfuron-methyl) or halogen (e.g. chlorsulfuron). Substituent on the heterocycle ring are typically alkyl or alkoxy, and can be partially halogenated. Certain sulfonylureas posses a thienyl (e.g. thifensulfuron-methyl) or pyridinyl (e.g. nicosulfuron and rimsulfuron) moiety rather than the analogous aryl ring. The basic structure is presented below:

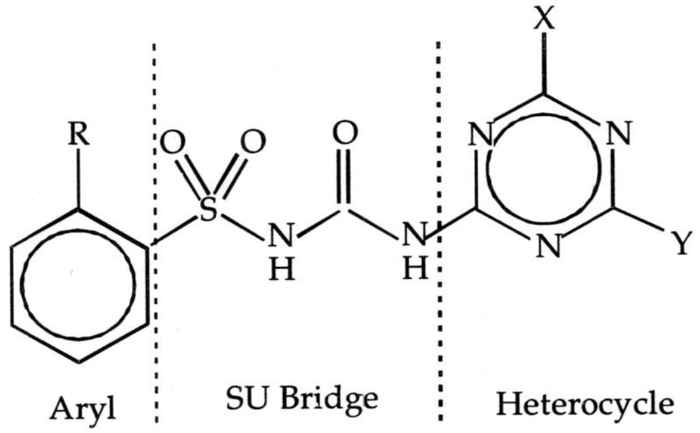

Basic structure of sulfonylurea

Sulfonylureas

Sulfonylureas are a class of potent herbicides that are characterized as acetolactate synthase-inhibitors (ALS) and introduced by Dupont in 1982. Sulfonylures within a comparatively short period of time have made a major impact on weed control technology. Due to their high potency, these are typically applied at low rates. Sulfonylureas are thermally labile compounds and are directly metabolized in the environment. While the level of toxicity to animals from these compounds is low, their ability to persist in soil and water can cause damage to crops in later years. All sulfonylureas share the same basic structure, containing an aryl ring linked to a heterocycle (triazine or pyrimidine) through the herbicidally active sulfonylurea 'bridge'. To increase herbicidal activity, the aryl ring is usually substituted to 'ortho' to the sulfonylurea bridge with the substituent typically a carboxylic acid ester (e.g. metsulfuron-methyl) or halogen (e.g. chlorsulfuron). Substituent on the heterocycle ring are typically alkyl or alkoxy, and can be partially halogenated. Certain sulfonylureas posses a thienyl (e.g. thifensulfuron-methyl) or pyridinyl (e.g. nicosulfuron and rimsulfuron) moiety rather than the analogous aryl ring. The basic structure is presented below:

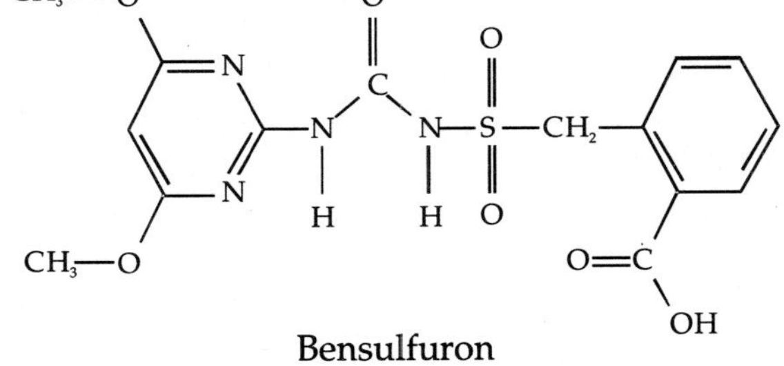

Bensulfuron

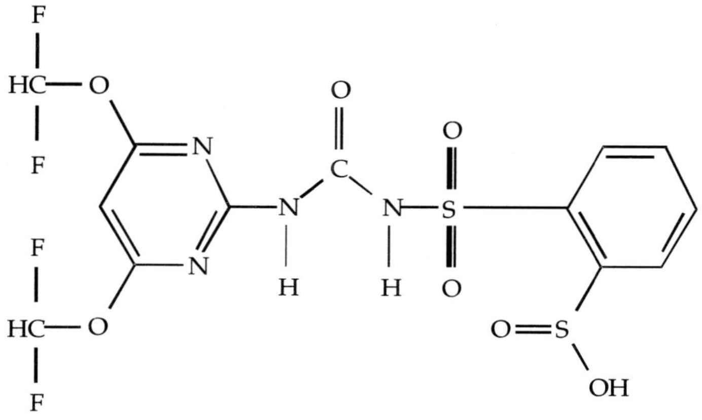

Primisulfuron

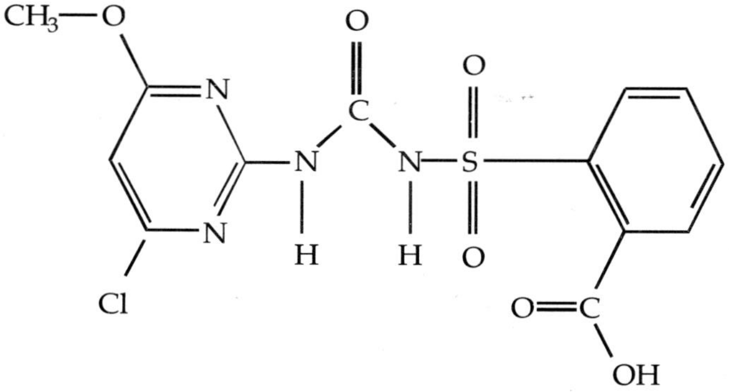

Chlorimuron

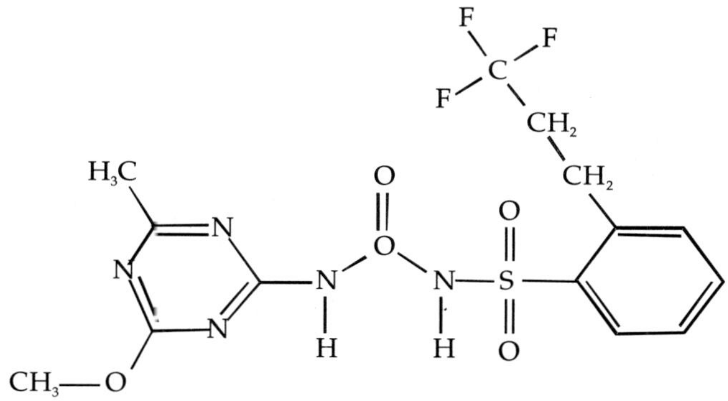

Prosulfuron

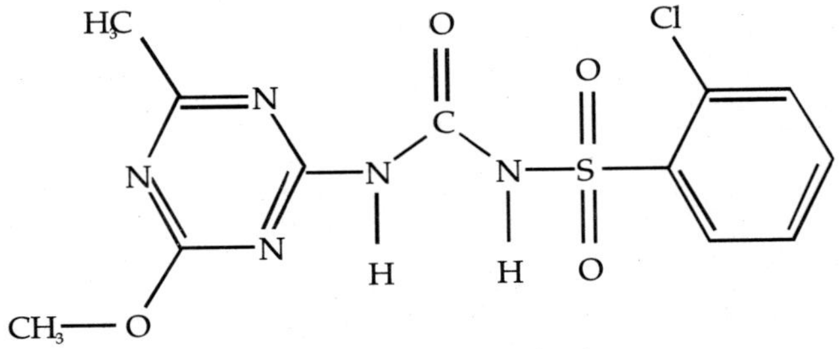

Chlorsulfuron

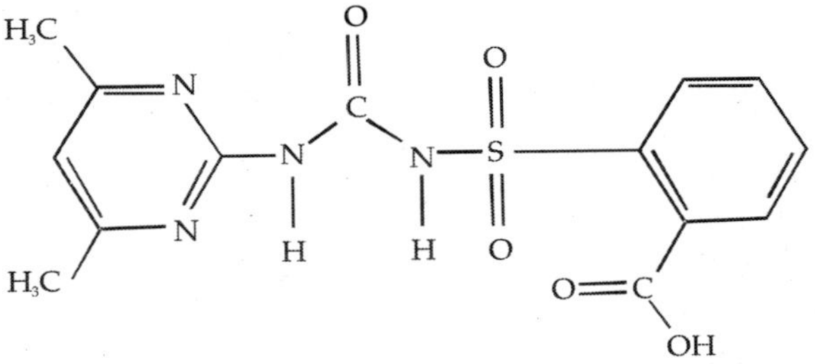

Sulfometuron

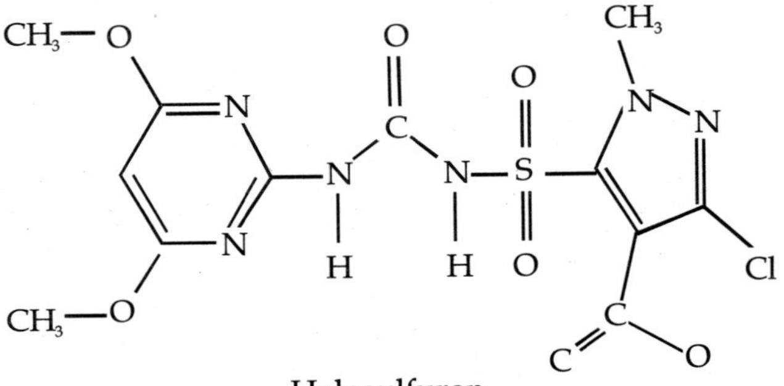

Halosulfuron

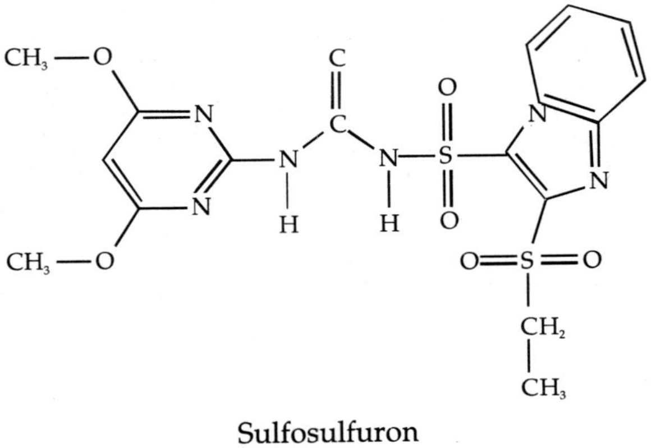

Sulfosulfuron

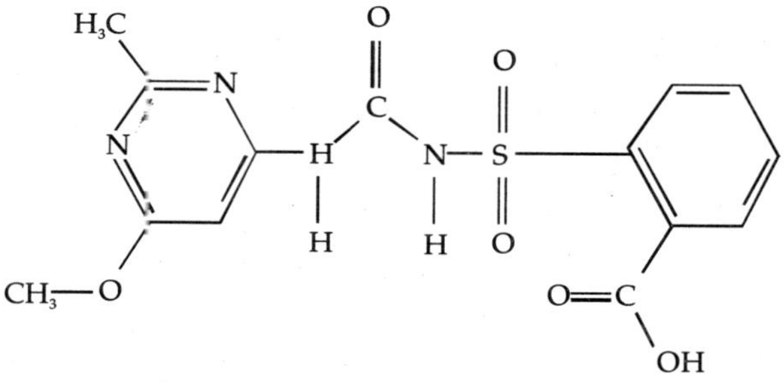

Metsulfuron

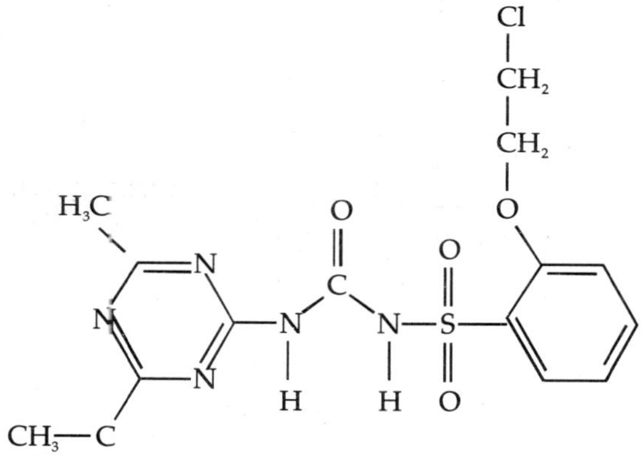

Triasulfuron

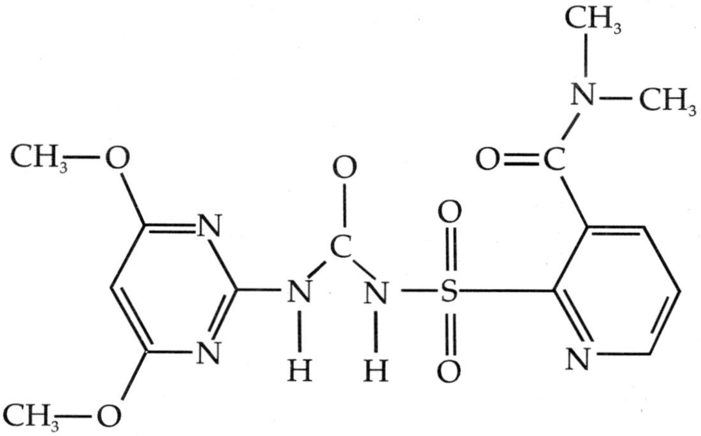

Nicosulfuron

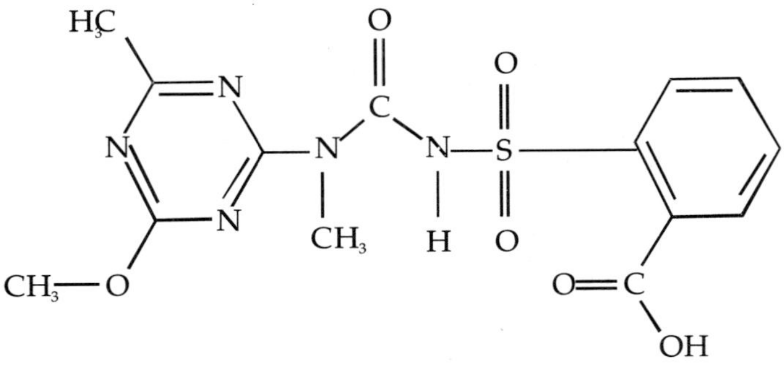

Tribenuron

Thiocarbamates

These are derived from the carbamic acids molecules ($NH_2.COOH$) by the substitution of one or two oxygen atoms by sulphur atom to form thiocarbamic acid or dithiocarbamic acid. The development of the dithiocarbamates was introduced by Stauffer Chemical Company in 1956. Thiocarbamate herbicide preceeded the development of the dithiocarbamate herbicides. The thiocarbamates are usually applied to the soil to selectively control grass and many broadleaf weeds during seed germination or early seedling growth in numerous crops.

Except for thiobencarb, these herbicides are usually mechanically incorporated into soil immediately after application. In the dry soil, incorporation by overhead irrigation following application is required. Certain antidotes have been added to the formulation of some thiocarbamates herbicides to protect the crop from herbicide injury.

In general, thiocarbamate and dithiocarbamate herbicides inhibit the growth of germinating seedling to a greater extent than the roots. The thio and dithiocarbamate herbicides seem to alter normal plant metabolism in a variety of ways such as photosynthesis, respiration, lipid synthesis, protein synthesis and nucleic acid metabolism.

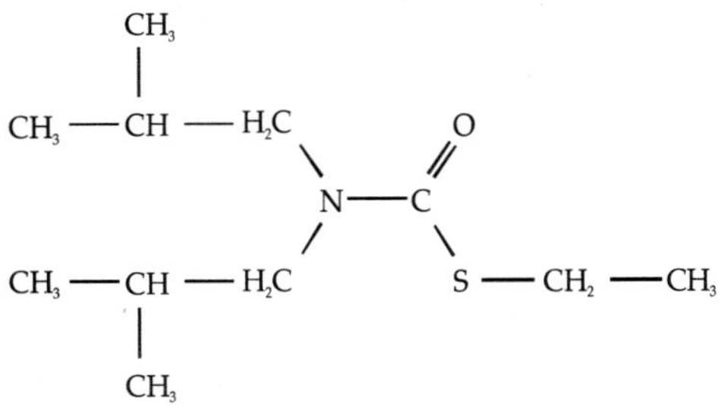

Butylate

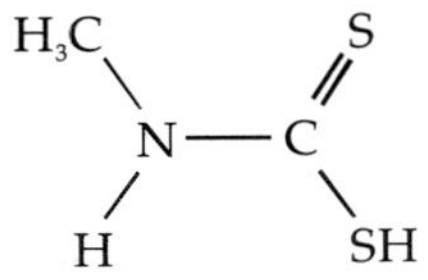

Metham

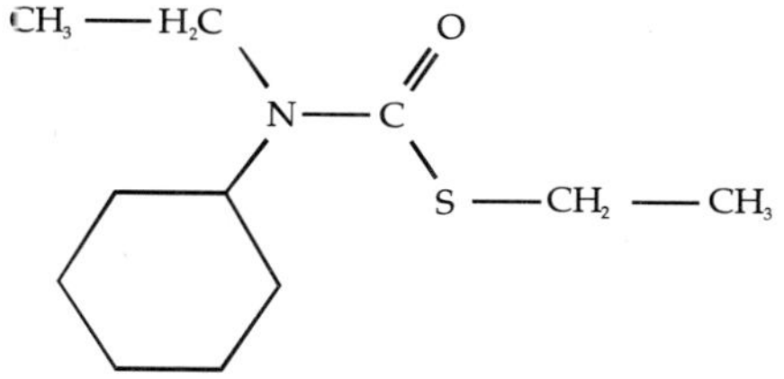

Cyclorte

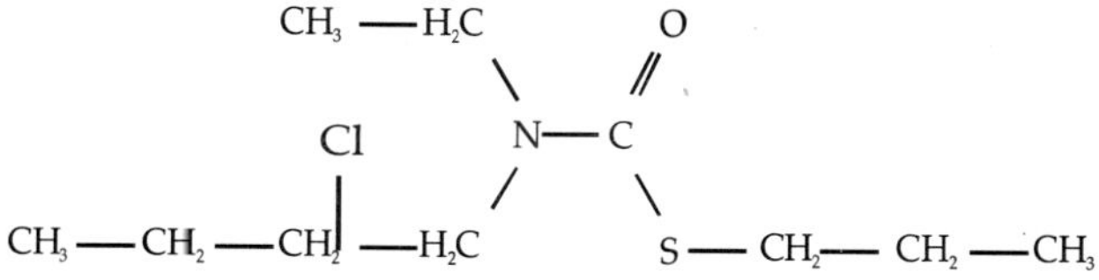

Pebulate

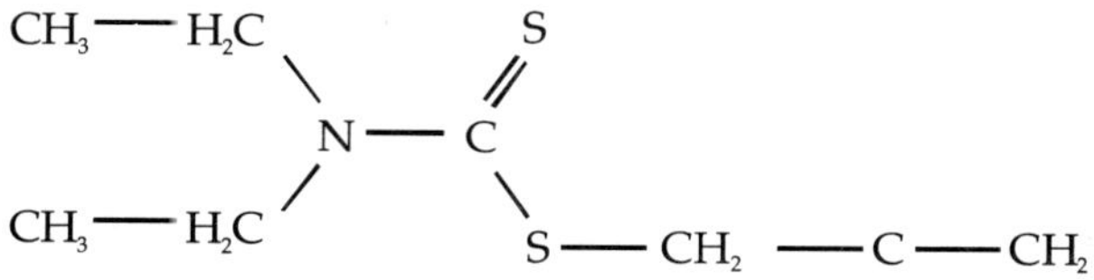

CDEC

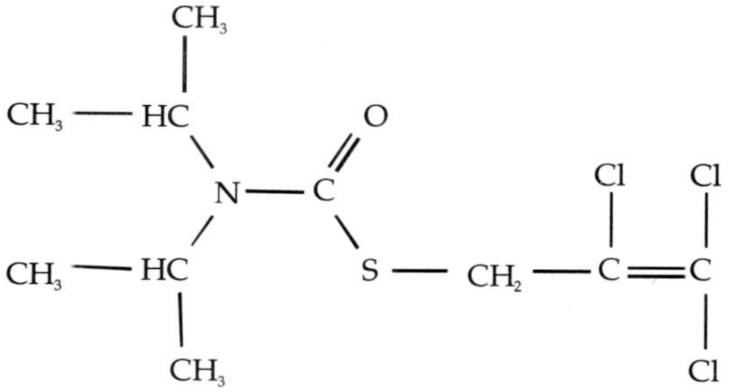

Triallate

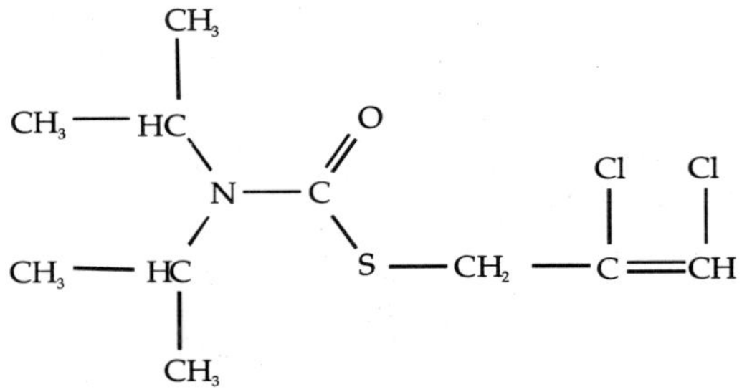

CH_3
CH_3 — HC
O
Cl Cl
N — C
CH_3 — HC
S — CH_2 — C = CH
CH_3

Diallate

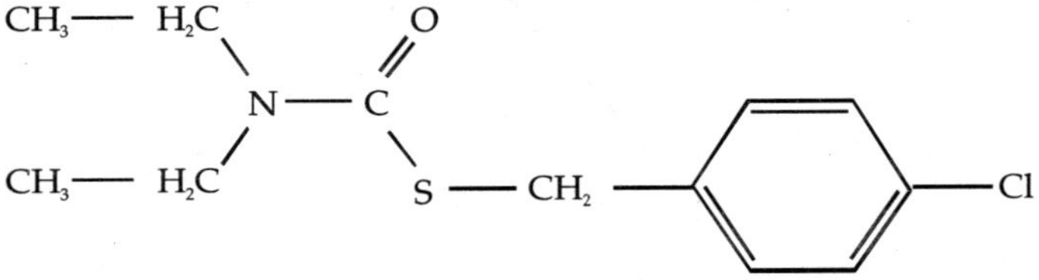

CH_3 — H_2C
O
N — C
CH_3 — H_2C
S — CH_2 — Cl

Thiobencarb

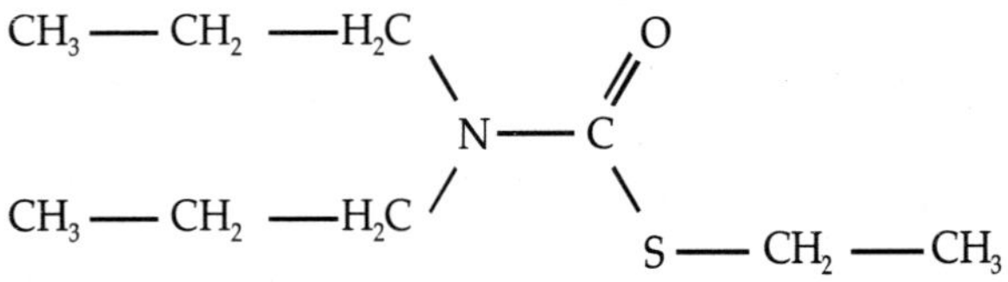

CH_3 — CH_2 — H_2C
O
N — C
CH_3 — CH_2 — H_2C
S — CH_2 — CH_3

EPTC

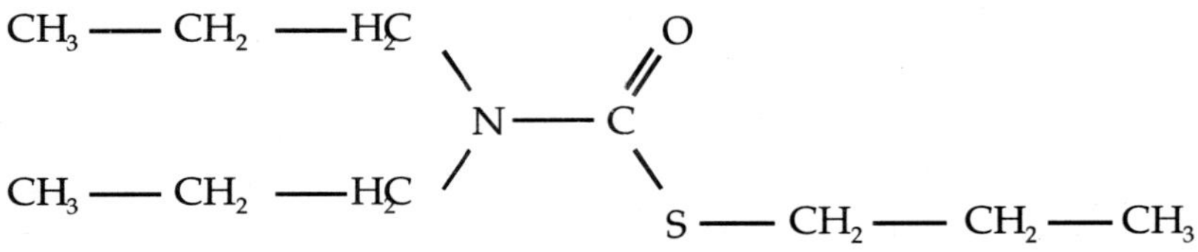

CH_3 — CH_2 — H_2C
O
N — C
CH_3 — CH_2 — H_2C
S — CH_2 — CH_2 — CH_3

Vernolate

O
N — C
S — CH_2 — CH_3

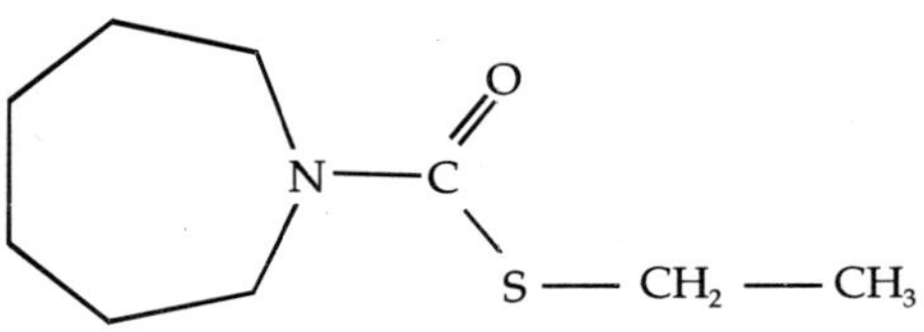

Molinate

(>4000mg/kg and 3000mg/kg).

Thiocarbamates

These are derived from the carbamic acids molecules ($NH_2.COOH$) by the substitution of one or two oxygen atoms by sulphur atom to form thiocarbamic acid or dithiocarbamic acid. The development of the dithiocarbamates was introduced by Stauffer Chemical Company in 1956. Thiocarbamate herbicide preceeded the development of the dithiocarbamate herbicides. The thiocarbamates are usually applied to the soil to selectively control grass and many broadleaf weeds during seed germination or early seedling growth in numerous crops.

Except for thiobencarb, these herbicides are usually mechanically incorporated into soil immediately after application. In the dry soil, incorporation by overhead irrigation following application is required. Certain antidotes have been added to the formulation of some thiocarbamates herbicides to protect the crop from herbicide injury.

In general, thiocarbamate and dithiocarbamate herbicides inhibit the growth of germinating seedling to a greater extent than the roots. The thio and dithiocarbamate herbicides seem to alter normal plant metabolism in a variety of ways such as photosynthesis, respiration, lipid synthesis, protein synthesis and nucleic acid metabolism.

Triazines

Investigations on potential herbicidal activities of triazine derivatives was done by Geigy and Basle (1952). The herbicidal activity of clorazine was reported by Gast *et al.*, (1955) and Antognini and Day (1955). Subsequently numerous triazine derivatives were evaluated for weed control.

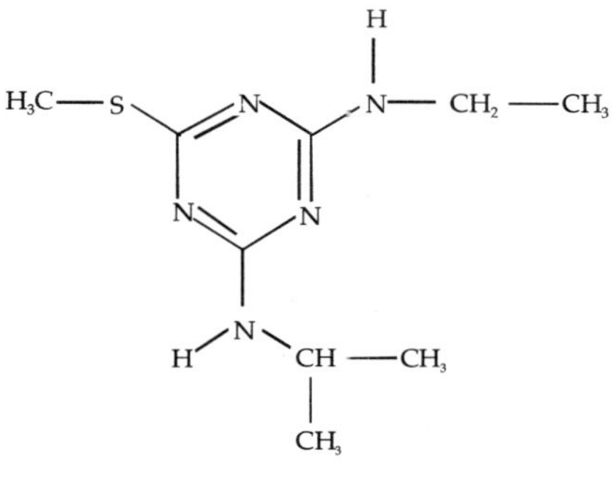

Ametryn

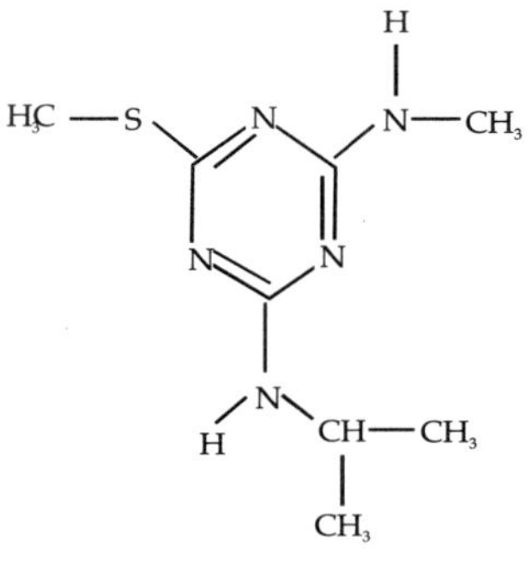

Desmetryn

i.e. simazine was introduced as Gesatop in 1956 for the selective residual pre-emergence control of annual grasses and broadleaved weeds in a variety of deep rooted crops, due to its low solubility in water (3.5 mg/l at 20 ⁰C).

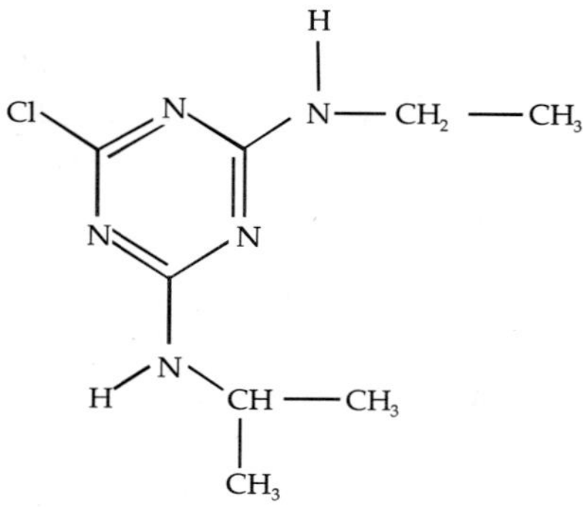

Atrazine

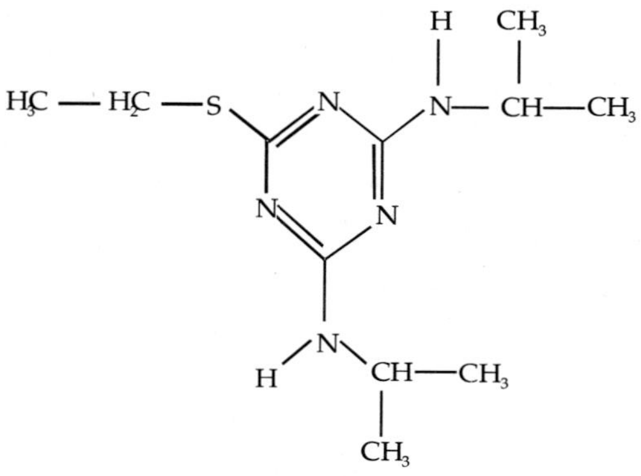

Dipropetryn

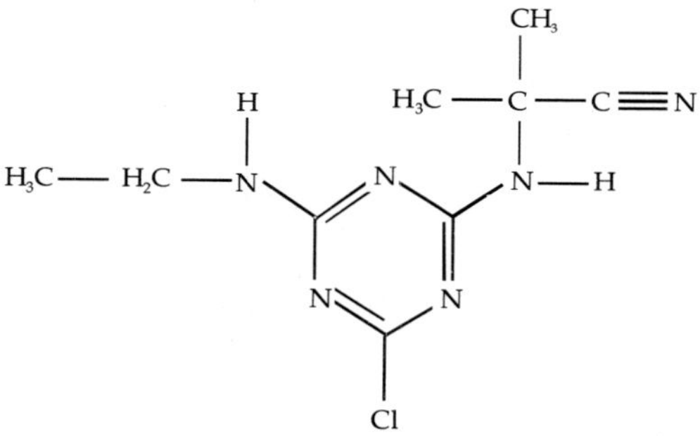

Cyanazine

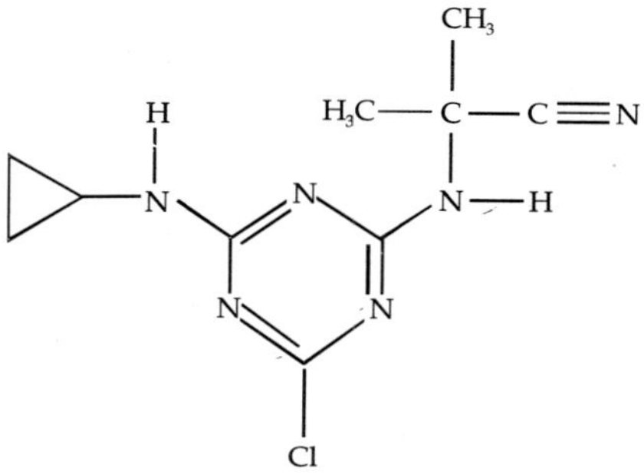

Procyanzine

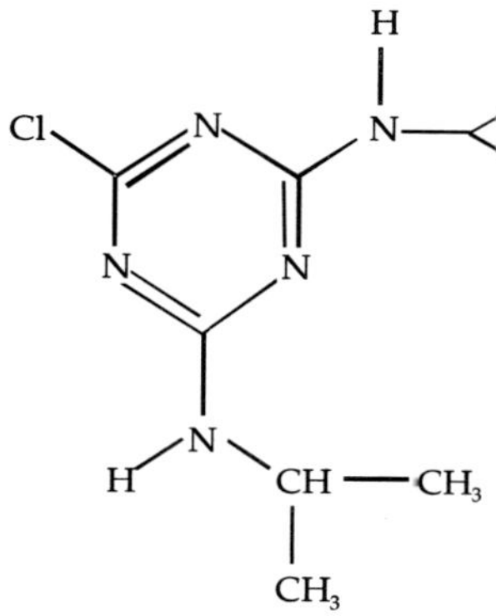

Cyprazine

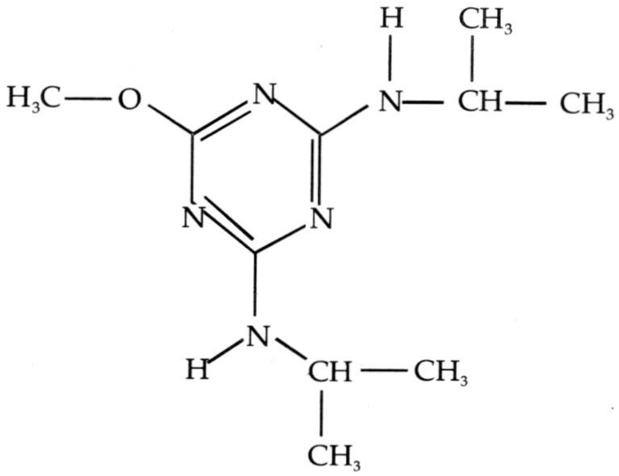

Prometon

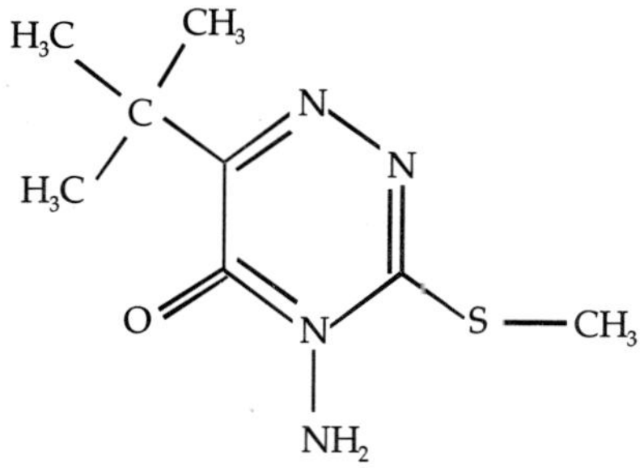

Metribuzin

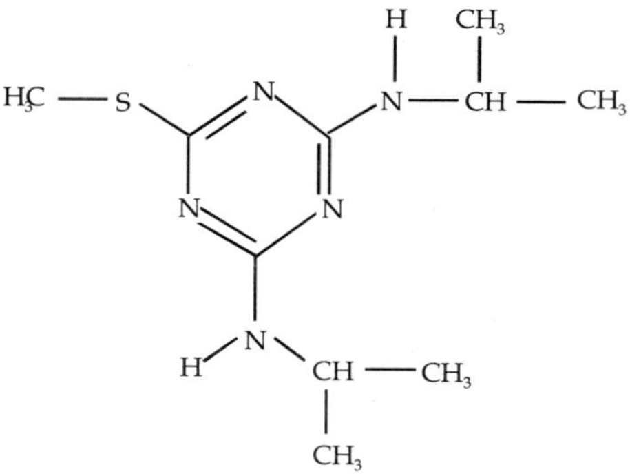

Prometryne

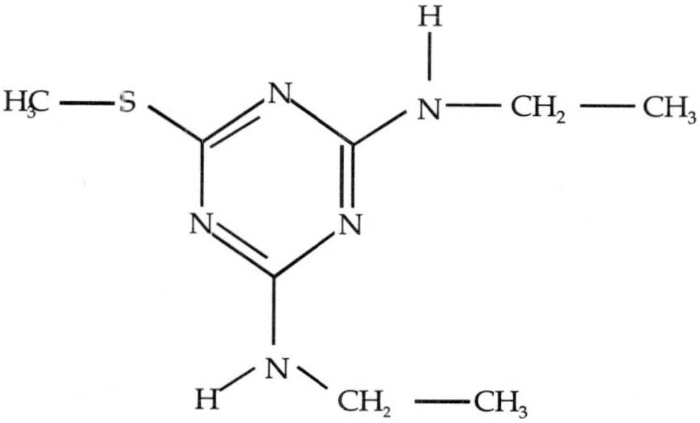

Simetryne

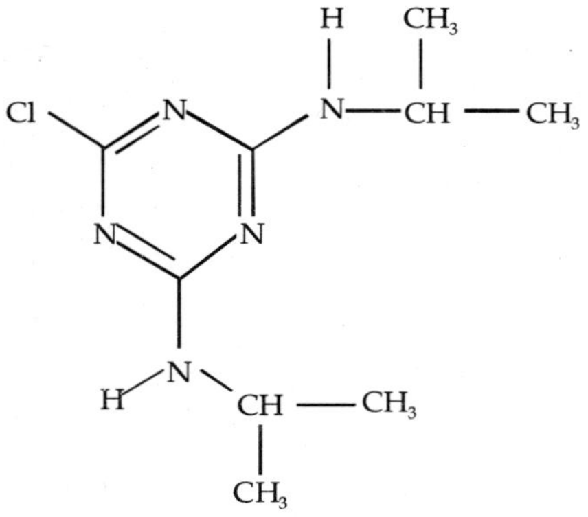

Propazine

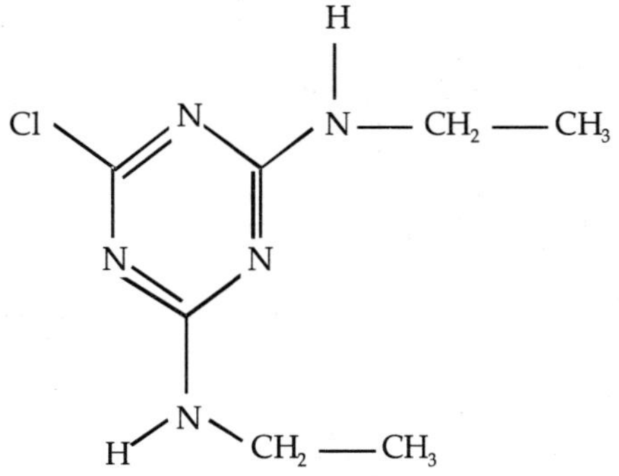

Simazine

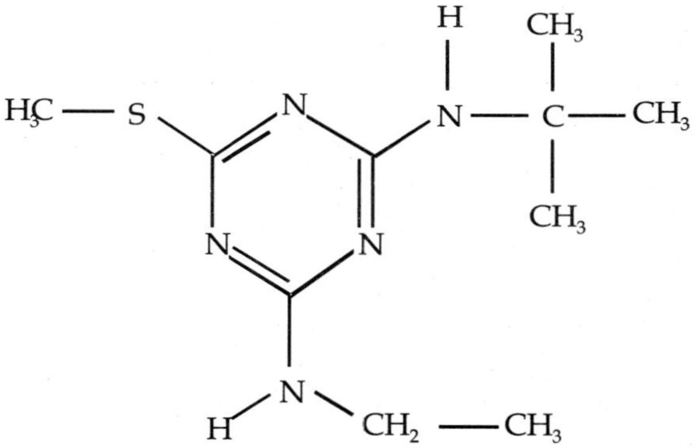

Terbutryne

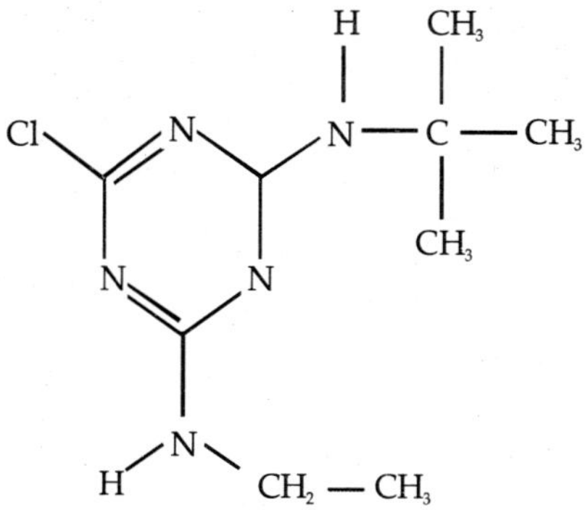

Terbuthylazine

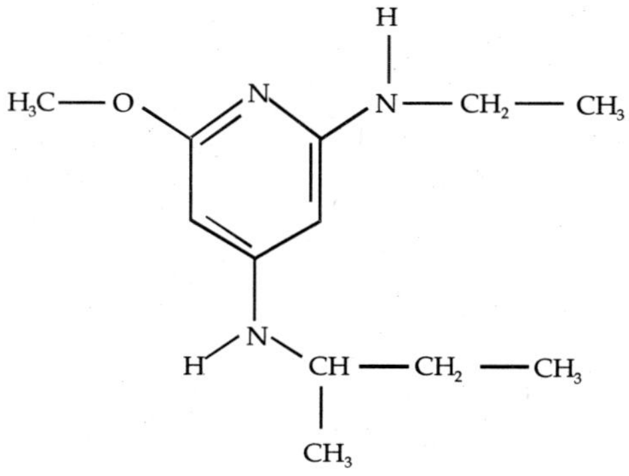

Secbumeton

Pyramidines (Uracil)

These herbicides are derived from uracil applied to the soil, and are absorbed *via* roots but eventually inhibit the light reaction of photosynthesis causing chlorosis and death, therefore resemble the substituted urea herbicides in their mode of action (Hilton *et al.*, 1964) Bromacil is used in non-cropped situations and in citrus orchards to control many annual and perennial weeds as well as seedling weeds in pineapple. Terbacil is used to control many annual and perennial weeds in fruit orchards. Lenacil has been useful for weed control in sugarbeet. Foliar chlorosis is primarily symptoms of the uracil herbicides, where root and shoot growth is also inhibited.

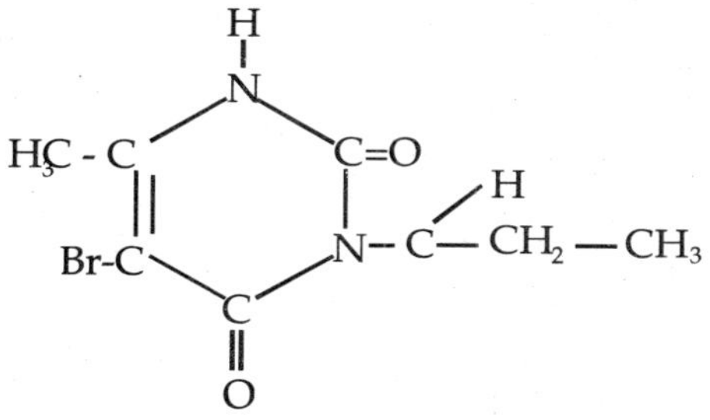

Bromacil

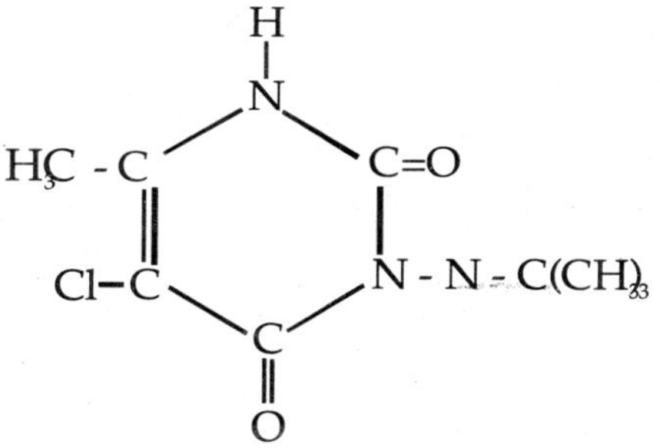

Terbacil

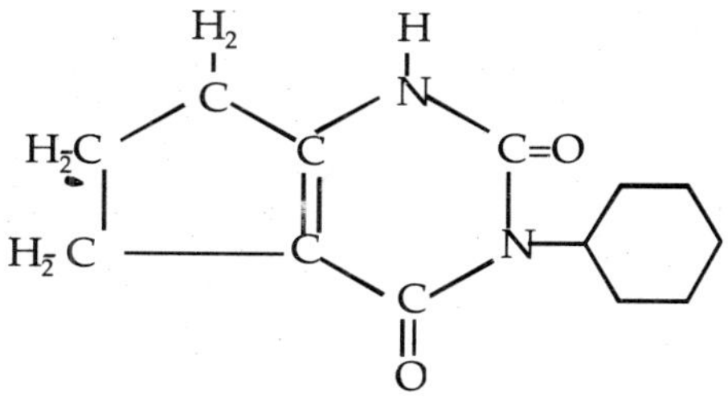

Lenacil

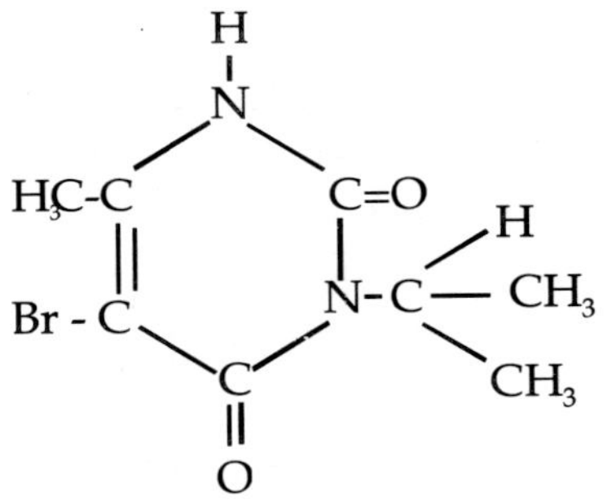

Isocil

Substituted ureas

Several substituted urea compounds were included by Thompson *et al.*, (1946) for growth inhibitory activities. DCU [1, 3-bis (2,2,2-trichloro-1-hydroxyethyl) urea] was the first substituted urea used commercially for weed control as pre-emergence herbicide to control grasses and certain broadleaf weeds.

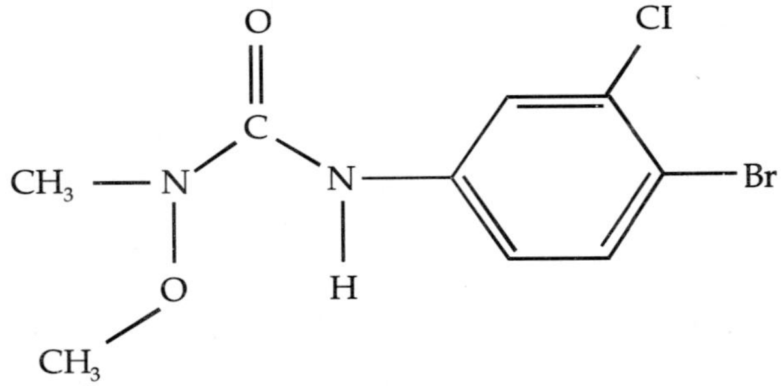

Chlorbromuron

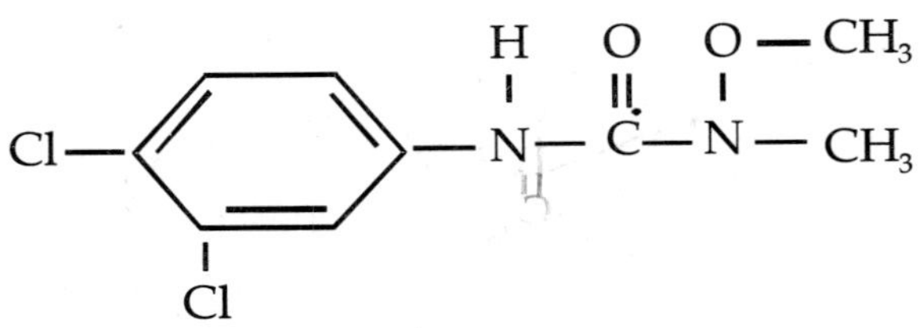

Linuron

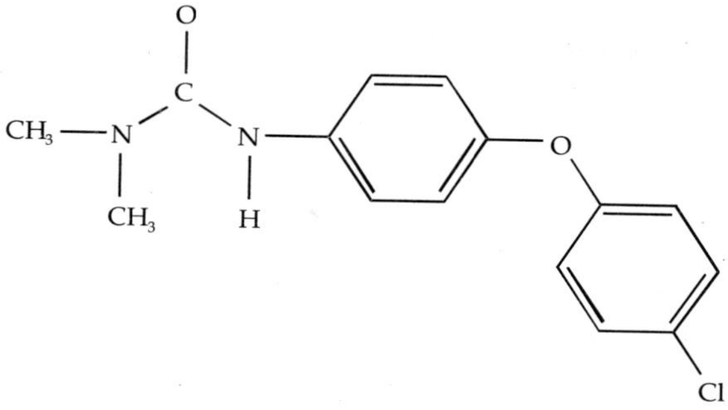

Chloroxuron

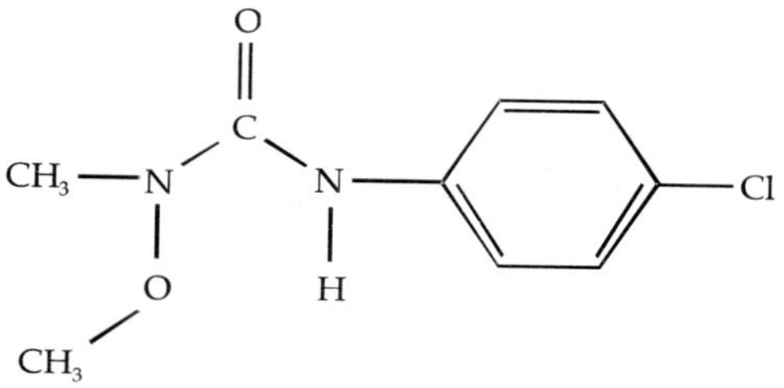

Monolinuron

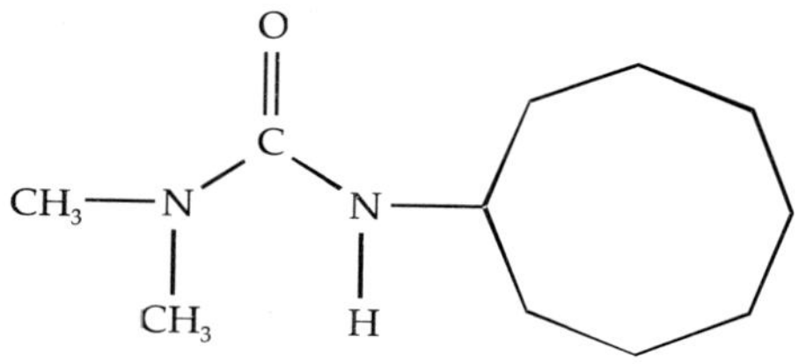

Cycluron

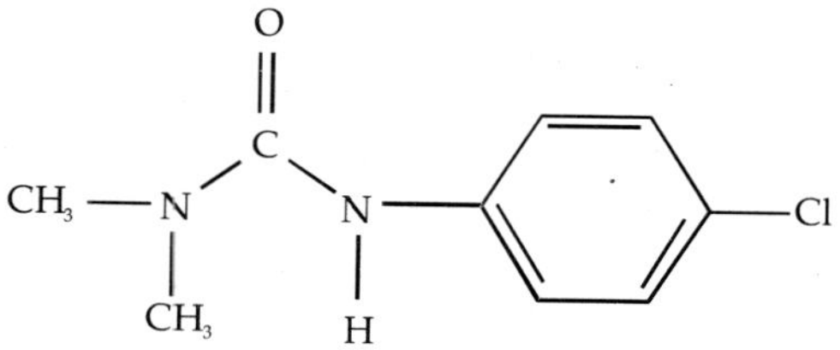

Monuron

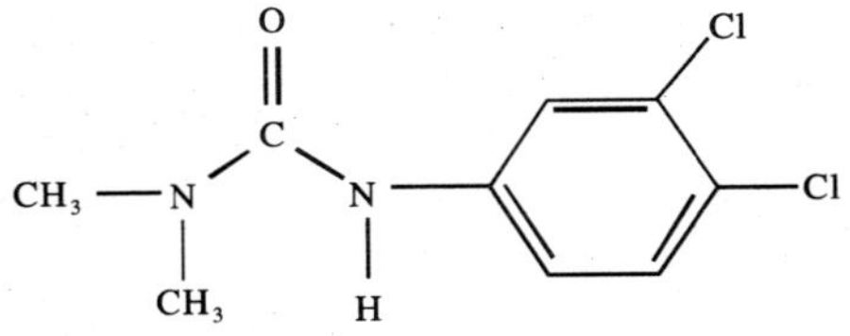

Diuron

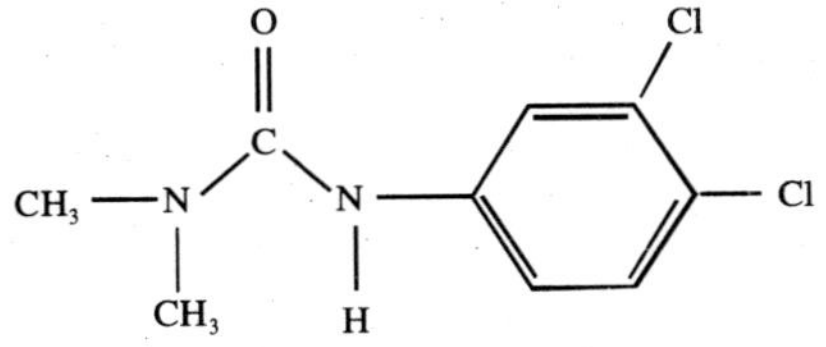

MonuronTCA

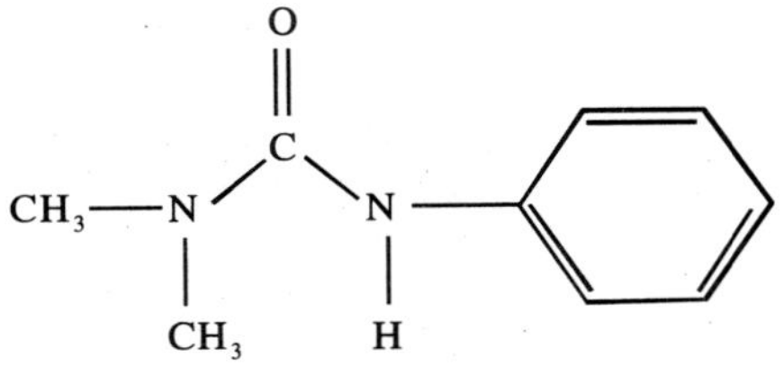

Fenuron

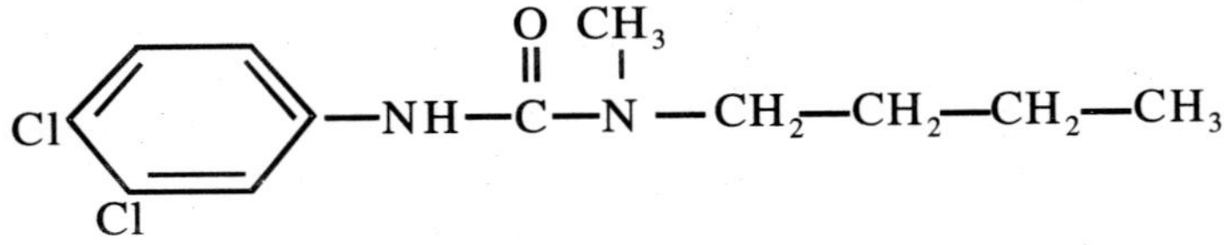

Neburon

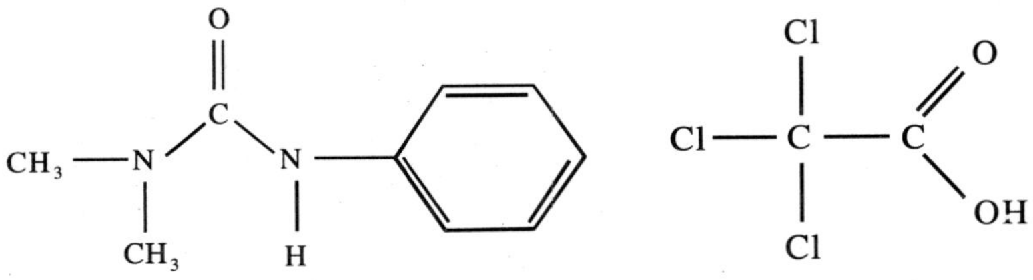

FenuronTCA

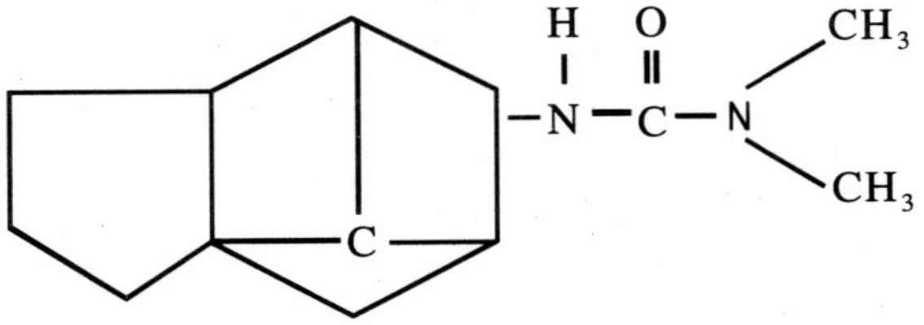

Norea

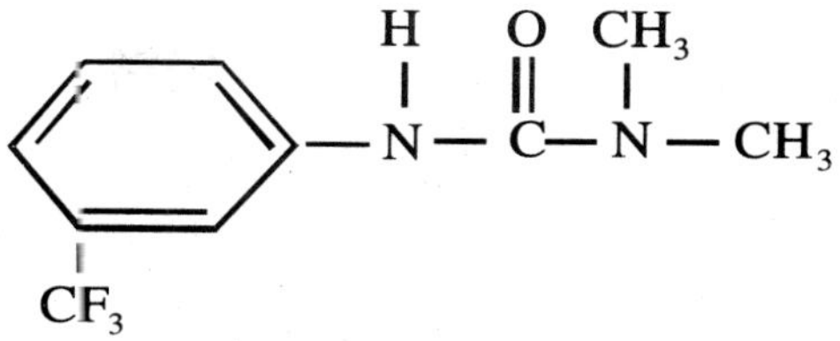

H O CH_3
N — C — N — CH_3
CF_3

Fluometuron

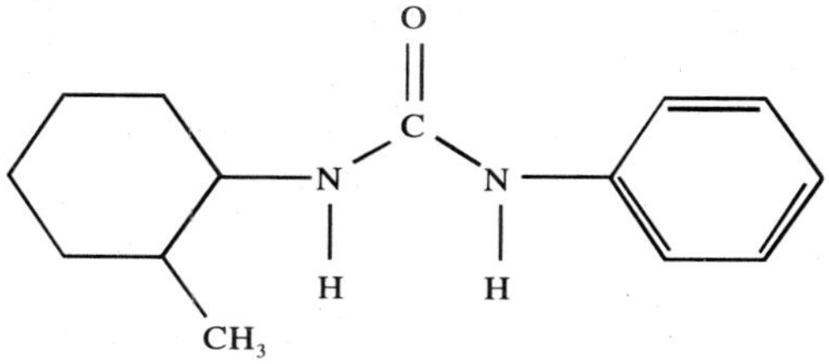

O
C
N N
H H
CH_3

Siduron

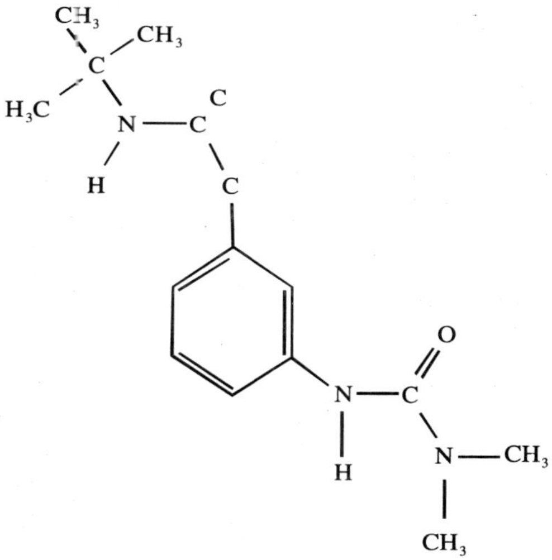

CH_3 CH_3
C
H_3C C
N — C
H C
O
N — C
H N — CH_3
CH_3

Karbutilate

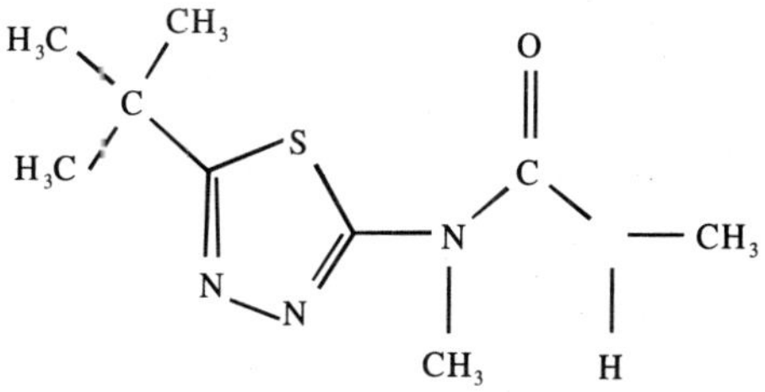

H_3C CH_3 O
C
H_3C S C
N — CH_3
N N
CH_3 H

Tebuthiuron

Unclassified heterocyclic compounds

There are many compounds that do not fall to previously described categories and have been used for many years. Some of them are relatively new. Amitrole is one of the oldest heterocyclic herbicide used for total weed control. It is non-selective and is often used as synergist in many total herbicide mixtures having ureas and triazines and acts by inhibition of carotenoid biosynthesis. Bentazon is a post-emergence contact herbicide for control of dicot weeds in winter and spring cereals. Mechanism of action of bentazon is still unknown. Bensulide and DCPA is used as pre-emergence to control grassy weeds and dichondra lawns. It is also used as pre-emergence or preplant soil incorporation in cotton and several vegetable crops. It is adsorbed on root surface but a small amount is absorbed (Anon. 1979). DCPA is absorbed from the soil but not by foliage and not translocated in the plant (Anon. 1979). Dinoseb was described as potential herbicide in 1945 (Craft 1945). Dinoseb inhibit the incorporation of ^{32}P into ATP and to stimulate O_2 utilization in tomato leaf disk (Wojtaszek *et al.*, 1966). Dinoseb inhibits enzymatic synthesis of glycerides *in vitro* and intact-wheat seedlings (St. John and Hilton 1973) and nitrate reduction in the dark and nitrite reduction in the light (Finke *et al.*, 1977).

Endothall is a contact herbicide used primarily for aquatic weed control and preharvest desiccation in potatoes and cotton as well as several crops grown for seed. It is readily absorbed by roots and translocated to a limited extent to the foliar plant parts *via* xylem. It is not phloem mobile. Fenac is used as pre-emergence to control annual and perennial weeds in sugarcane, as post-emergence against broadleaf perennial weeds on non-cropped area and found very effective against puncture grass (*Tribulus terrestris*) and field bindweed (*Convolvulus arvensis*). Herbicidal activity of fluridone was first described by Waldrep and Taylor (1976). It is effective against annual grasses and many broadleaf weeds as a pre-emergence or preplant incorporation. Fluridone is relatively persistent in soil.

Methazole is a selective herbicide used as preplant, pre-emergence or post-emergence in cotton to control annual weeds. It is commonly used in combination with MSMA as a post-emergence herbicide to control grasses and nutsedges. In methazole mechanism of action and phytotoxicity is similar to urea herbicides. Norflurazon is a pre-emergence herbicide used to control many annual broadleaf weeds and grasses in cotton, cranberries, nut tree crop etc. Perfluidone is used as a pre-emergence treatment for the control of nutsedge

(*Cyperus* spp.), grasses and certain broadleaf weeds in cotton. Picloram is used for general woody plant control as well as for control of most annual and perennial broadleaf weeds. Most grasses are tolerant to picloram. Picloram is considered as relatively persistent in the environment. Picloram intended to accumulate in young growing leaves following either root or foliar uptake. Picloram is translocated both in the symplast and in apoplast.

Oxadiazon is selectively used in rice, wine grape and orchard fruits. Pyrazon is generally a pre-emergence and post-emergence herbicide used principally in sugarbeet. Under irrigated conditions pyrazon may be applied as pre-emergence in bands and sprinkled or as a shallow preplant soil incorporated treatment where furrow irrigation is practiced. Pyrazon is readily absorbed by roots and distributed throughout the plant *via* xylem. Triclopyr is used for control of dicot weeds in cereal and in non-crop land.

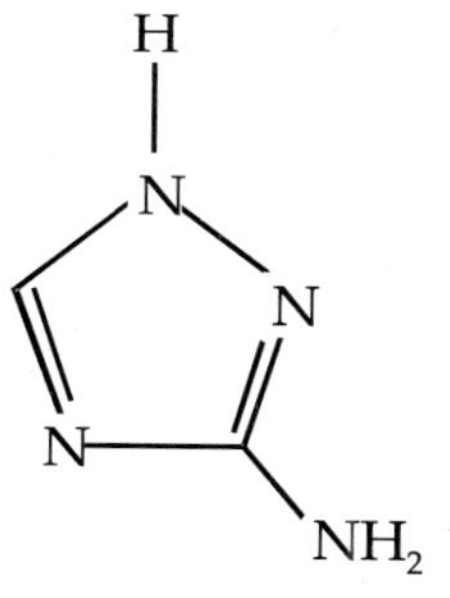

H N N N NH_2

Amitrole

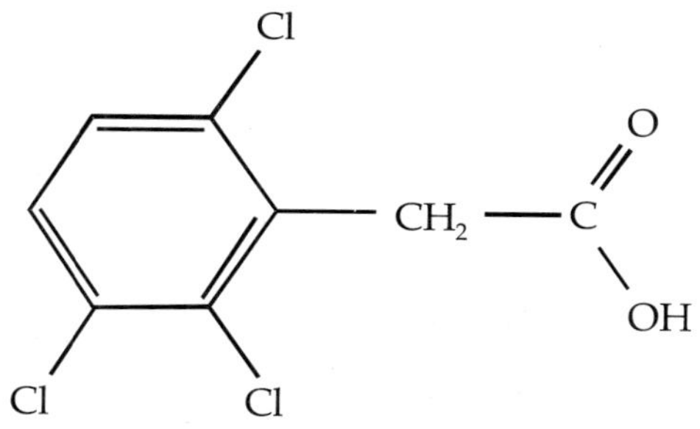

Cl CH_2 C O OH Cl Cl

Fenac

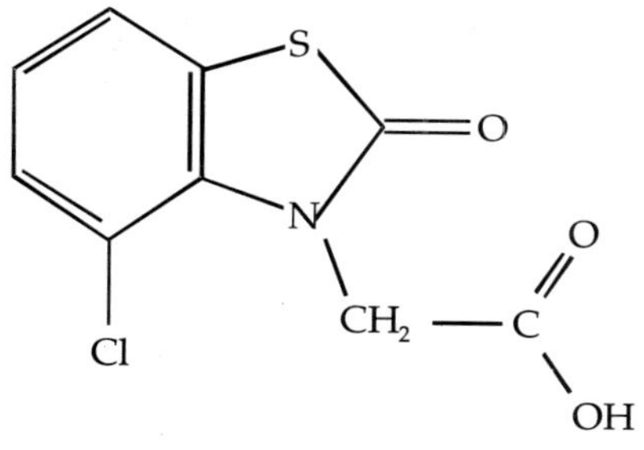

S O N CH_2 C O OH Cl

Benazolin

S O N CH_2 C O OH Cl

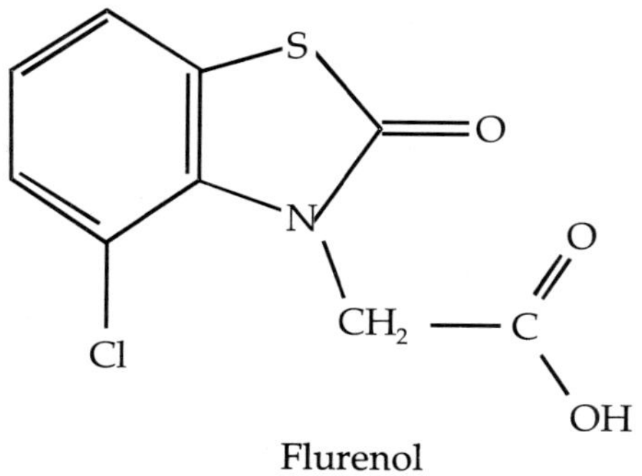

Flurenol

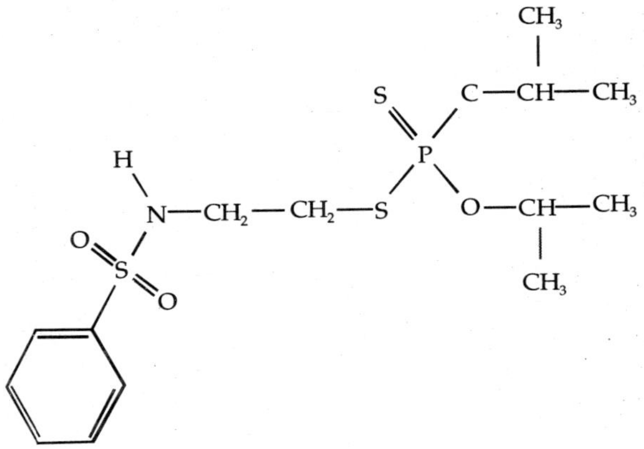

Bensulide

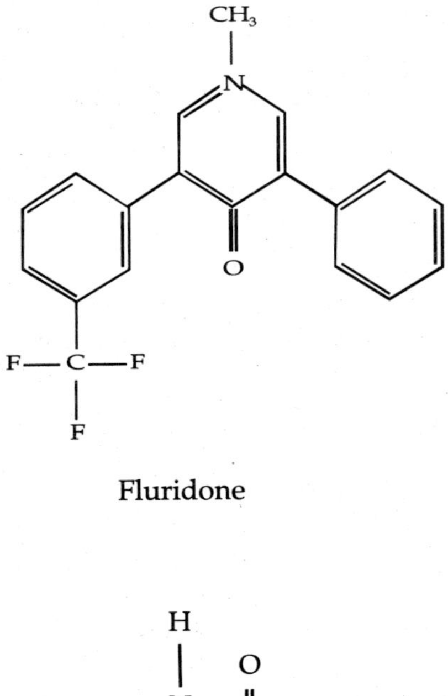

Fluridone

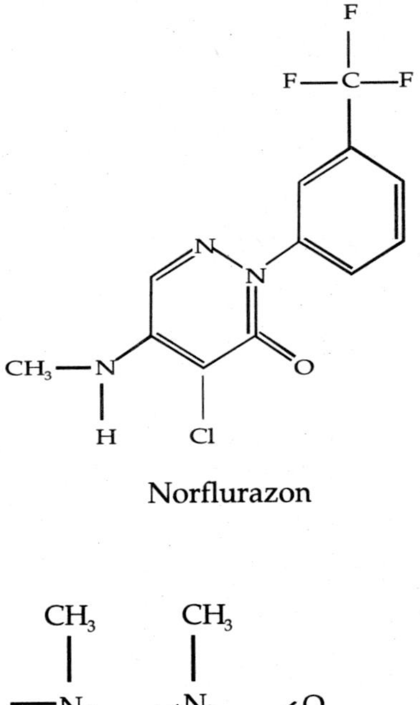

Norflurazon

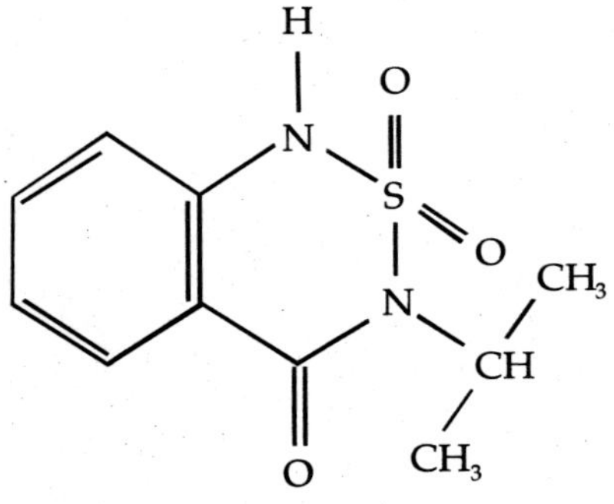

Bentazon

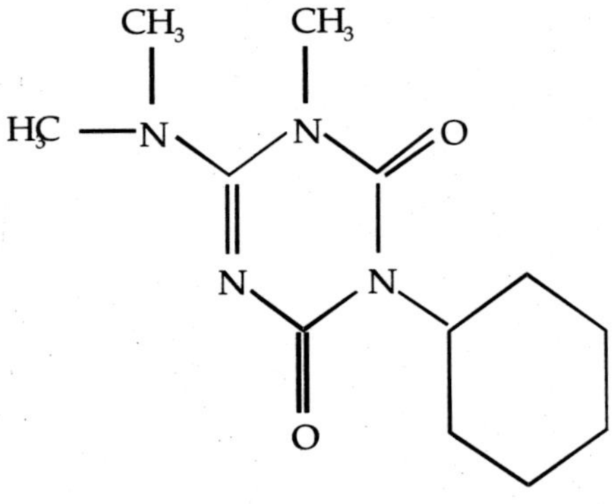

Hexazinone

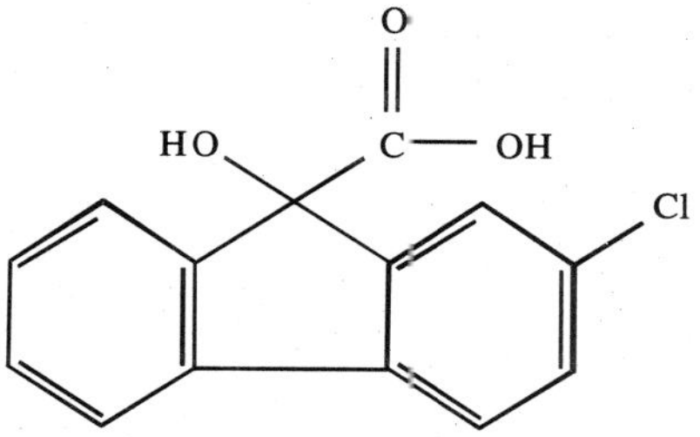

Chlorflurenol

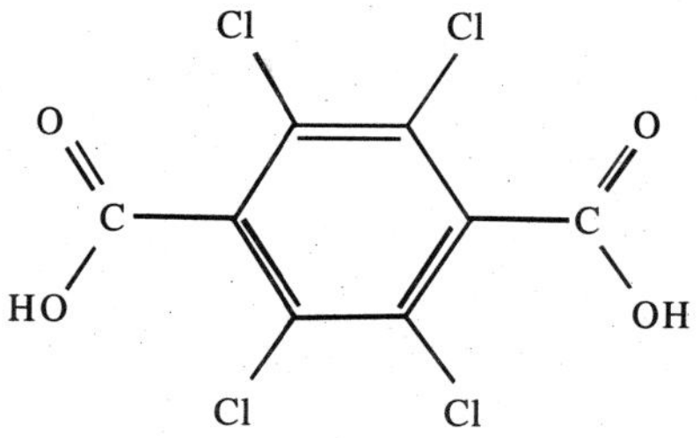

DCPA

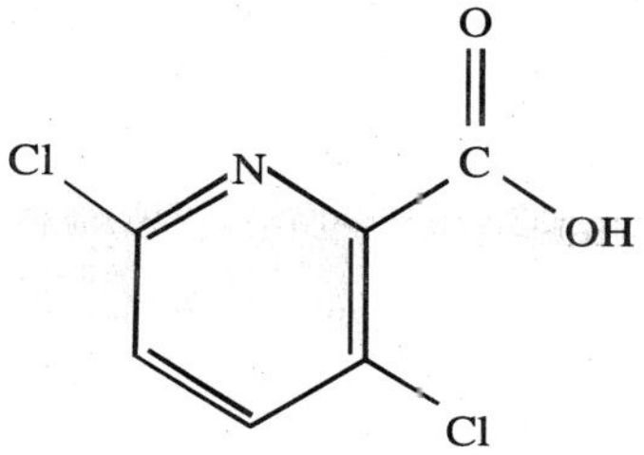

3,6-Dichloropicolinic acid

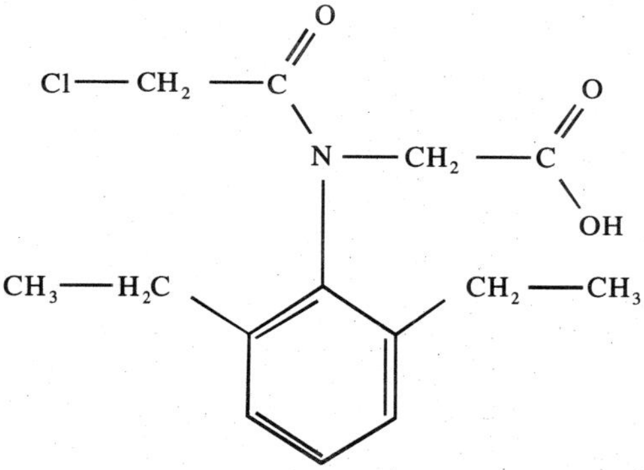

Diethatyl

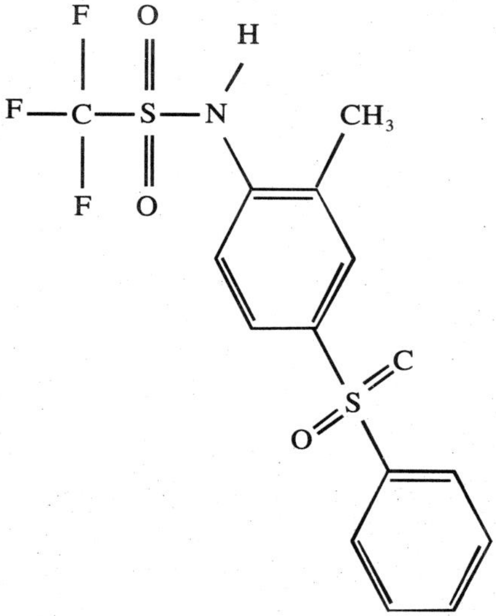

Perfluidone

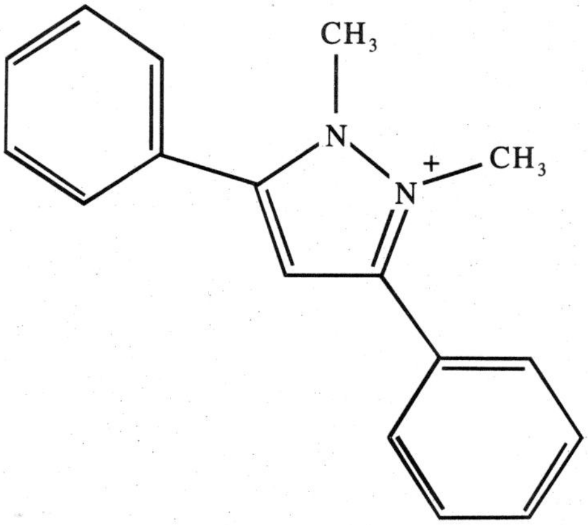

Difenzoquat

Picloram

Dinoseb

Pyrazon

Endothall

Triclopyr

(B) Based on use

On the basis of the effects produced by the herbicides, these are grouped broadly into selective or non-selective herbicides. Selective herbicides are chemicals that suppress or kill certain weeds without significantly injuring an associated crop or other desirable plant species. Usually some weeds are not injured by selective herbicides. Non-selective herbicide, suppress a wide range of vegetation.

Based on the methods of application, herbicides are categorized into two main groups: soil applied and foliage applied herbicides. Soil applied herbicides are used before planting, before crop or weed emergence, or after the plants emerge in specific situation. The time of herbicide application is referred as pre-plant, pre-emergence or post-emergence, respectively. Soil applied herbicides must move into the soil profile by water or mechanical incorporation to be effective because some of the herbicides are volatile or photodecomposible, i.e. trifluralin and fluchloralin. Movement in soil is an important factor that influences herbicide persistence and fate. The physiological activity of soil applied herbicides depends on the degree of inherent plant tolerance, the location of the herbicide in the soil, and depth of plant roots. Some soil applied herbicides are applied as bands, either over or between crop rows to enhance selectivity and decrease costs of application.

Herbicides applied as post-emergence are included in foliage applied herbicides. Some herbicides, either by their rapid action or limited movement, injure only the portion of the plant actually touched or contacted by the chemical or spray solution and are called contact herbicides. Herbicides in this category are usually applied to foliage (McHenry and Norris, 1972). Paraquat, bromoxynil and dinoseb are examples of foliage applied contact herbicides. In some cases, herbicides may be directed away from crops or applied in shields to minimize foliage exposures to these chemicals. Some plant applied and many foliage herbicides move or translocate in treated plants. Herbicides of this type often effectively suppress root, rhizome or shoot growth at a considerable distance from the point of application that is either the soil (root) or the foliage.

Examples of herbicides under various categories based on their uses are summarized in Table 1.

Table 1. Classification of herbicides based on their uses

Chemical class	Soil applied herbicides		Foliage applied herbicides	
	Systemic (Translocated)	Contact	Systemic (Translocated)	Contact
Acetamides	acetochlor, alachlor, acetolachlor, napropamide, propamide, propachlor, cisanilide dipheninamid, naptalam	-	benzadox, cypromid acetochlor, alachlor, aetolachlor, apropamide, propanil	-
Aliphatic Organic arsenicals	TCA	-	dalapon	-
Benzoics	dicamba	-	-	-
Bipyridium	-	-	-	diquat, paraquat
Carbamates	-	-	asulam, barban, profam, phenmedipham desmedipham	-
Cyclohexane-dione			sethoxydim, clethodim tralkoxydim, cycloxydim	
Dinitroanilines	benfin, butralin ethalfluralin, dinitramine, fluchloralin, isoproturon, nitralin, oryzalin, pendimethalin, prodiamine, profluralin, prosulfalin, trifluralin	-		-
Diphenyl Ethers	fluorodifen, oxyfluorfen	nitrofen, diclofop-methyl	acifluorfen, bifenox, fluoroglycofen, fomesafen, lactofen, oxyfluorfen	-
Imidazolines	buthidazole	-	buthidazole	-
Isoxazolidinones	imazapyr, imazaquin, imazethapyr	-	imazapyr, imazaquin, imazethapyr, imazamethabenz, CGA-248757	-
Nitriles	dichlobencil	-	-	bromoxy nil, ioxynil

Contd...

Table 1. Contd...

Chemical class	Soil applied herbicides		Foliage applied herbicides	
	Systemic	Contact	Systemic	Contact
Oxadiazoles	oxadiazon	-	-	-
Oxadiazolides	methazole	-	methazole	-
Phenoxys	-	-	2, 4-D	-
Phenols	-	-		dinoseb
N-phenylphtha lamides	-	-	flumiclorac	-
Phenylpyridazines	-	-	pyridates	-
Phenylpyridazones	sulfentrazone	-	sulfentrazone	-
Phthalamates	naptalam	-	naptalam,	-
Pyrazoliums	norflurazon, metflurazon	-	difenzoquat	-
Pyridines	dithiopyr, thiazopyr fluridone	-	fluridone	-
Picolinic acids			clopyralid, triclopyr	
Quinolines	quinclorac	-	quinclorac	-
Sulfonylureas	bensulfuron, chlorimuron, chlorsulfuron, halosulfuron, sulfometuron	-	bensulfuron, chlorimuron, chlorsulfuron, halosulfuron, sulfometuron, nicosulfuron, primisulfuron, prosulfuron, thifensulfuron, triasulfuron,	-
Thiocarbamates	butylate, diallate, EPTC, molinate, pebulate, thiobencarb, triallate, vernolate	-	-	-
Triazolopyrimidine sulfonamide	cloransulam		Cloransulam, flumetsulam	
Triazines	ametryn, atrazine, cyanazine, hexazinone, propazine, prometon,	-	atrazine, cyanazine, hexazinone, prometon, prometryne	-

Contd...

Table 1. Contd...

	prometryn, simazine, metribuzin			
Uracil	bromacil, terbacil, lenacil	-	-	-
Ureas	diuron, flumeturon, linuron, methabenzth iazuron, metoxuron, isoproturon, monuron, siduron, tebuthiuron	-	diuron, fluometuron, linuron, isoproturon, tebuthiuron	-
Unclassified	bensulide, ethofumesate, DCPA, endothall, fenac, perfluridone pyrazon	-	endothall, ethofumesate, fosamine, glufosinate, glyphosate, perfluridone, pyrazon	-

Summary

Herbicide are named by (i) chemical name (ii) trade and (iii) common name. The chemistry of a compound determines how the herbicide will act in biological and physical systems such as plants, animals, soils, and water. On the basis of the effects produced by the herbicides, these are grouped broadly into selective or non-selective herbicides. Selective herbicides are chemicals that suppress or kill certain weeds without significantly injuring an associated crop or other desirable plant species. Usually some weeds are not injured by selective herbicides. Non-selective herbicide, suppress a wide range of vegetation.

Based on method of application, herbicides are categorized into two main groups; soil applied and foliage applied herbicides. Soil applied herbicides are used before planting, before crop or weed emergence, or after the plants emerge in specific situation. These times of herbicides application are referred to as pre-plant, pre-emergence or post-emergence, respectively. Some soil applied herbicides are applied as bands, either over or between crop rows to enhance selectivity and decrease costs of application.

Herbicides applied as post-emergence are included in foliage applied herbicides. Some herbicides, either by their rapid action or limited movement, injure only the portion of the plant actually touched or contacted by the chemical or spray solution and are called contact herbicides. Herbicides in this category are usually applied to foliage. In some cases, herbicides may be directed away from crops or applied in shields to minimize foliage exposures to these chemicals. Some plant applied and many foliage herbicides move or translocate in treated plants. Herbicides of this type often effectively suppress root, rhizome or shoot growth at a considerable distance from the point of application that is either the soil (root) or the foliage.

□□□

Chapter 3

Absorption and Transport in Plants

The major requirement of herbicide to be effective is that the herbicide come in contact of plant surface, remain there long enough to penetrate or be absorbed into the plant and reach a living cellular site through translocation where it can disrupt a vital process or structure. However, for some herbicides the site of action could be the absorption site by itself when translocation may not be prerequisite to witnessing their action on the plant. The absorption and translocation of a herbicide depends primarily on its molecular configuration, which in turn determines its chemical and physical properties. Differential absorption and translocation which forms the basis of herbicide selectivity, determines the tolerance and the susceptibility of a plant species to a particular herbicide. Plant characteristics and environmental conditions also influence the amount of a compound that enters a plant and its consequent movement. The site of application also has an important bearing on the absorption and translocation of herbicides.

Absorption

Herbicide enters plants through shoots, roots, other below ground organs and seed (Fig. 1). The process of herbicide entry into treated plants is called absorption, which involves contact, penetration and movement of the chemical into the plant, whereas, adsorption is the attraction of ions or molecules to the surface of a solid. After application, many herbicides adsorb (bind) to the clay and organic-matter fractions of soils. However, herbicides adsorb poorly to the sand and silt fractions of soil. Therefore, the extent of herbicide adsorption increases as the percentage of organic matter and clay increases. The dinitroaniline herbicides, dithiopyr, oxadiazon and many other pre-emergence herbicides readily bind to soil particles. There are several ways by which herbicide are absorbed.

Foliar absorption

Many herbicides are applied to plant surfaces as foliar spray. Leaves are the primary means of herbicide entry through shoots, although herbicide absorption can also occur through other aerial organs such as substantial amounts of some herbicides are absorbed through stems or emerging coleoptiles. Foliar herbicide absorption involves the following three steps;

i. Retention of spray droplets on a leaf surface
ii. Penetration of the herbicide into plant cells
iii. Movement into the cytoplasm of the plant cell

If the retention time is too short or if an insufficient amount of herbicide is intercepted by the plant, penetration through the cuticle and eventually the control will not be satisfactory. The type of herbicide carrier and adjuvant used, the amount of spray volume, the amount of shoot growth and the occurrence of rainfall after application, all factors affect the herbicide retention and plant coverage.

Maximum retention occurred when leaves are positioned at 50^0 to 90^0 to the orientation to the incidence of the spray. Similarly, leaf orientation was found to be important for both easy and difficult to wet soybean cultivars. Spray retention also be influenced by the trichome, stomata, veins *etc*. Trichomes are common features on many plants surfaces. They vary markedly in size, morphology, frequency, distribution and function. Challen (1962) showed that water retention was greater on leaves having an open than on those having a closed trichome pattern. The open pattern may enhance the wetting due to capillary action, while the closed pattern depresses it by entrapment of air beneath the water droplets. Surfaces over veins where trichomes are often plentiful over guard cells and around the bases of trichomes often differ in wettability from areas over other epidermal cells. Such differential wettability not only leads to variation in retention over a given surface but may be the basis for selective permeability often associated with specialized structure.

The cuticle is primarily composed of cutin, epicuticular wax, embedded wax, and pectins. A herbicide moving from a leaf surface to the cytoplasm of cell first encounters wax occur on the surface (epicuticular waxes) and within the cuticle itself (embedded waxes), which is composed of long hydrocarbon chains mostly C_{20} to C_{37} that include fatty acid which are nonpolar and lipophilic in nature and form an effective water repellent layer over the shoot surface, then cutin, it is also lipophilic in nature and composed of intermediate length fatty acids (C_{16} to C_{18}), free carboxylic (-COOH) and hydroxyl group (-OH) that can become charged ($-COO^-$ and $-O^-$, respectively) making the cutin more compatible with water. The third layer is pectic layer which is composed of polymers of galacturonic acid and thus has charged carboxylic group. Pectin is the most hydrophilic of the cuticle component and when hydrated can provide aqueous pathway for the uptake of water soluble herbicides, then cell walls and finally the plasmalemma (outer membrane) of the cell. These substances create gradients in polarity and water solubility so that all chemicals must traverse to enter plant leaves. As a general rule, movement of a herbicide into and through the cuticle is inversely related to its solubility in water. The lower the rate of movement through the culticle; the more lipophilic the herbicide, the more likely it is to penetrate. Two pathways have been proposed to explain the absorption of both polar and non-polar herbicides into plant foliage (Ashton and Crafts, 1981; Hull *et al.*, 1982). These pathways are aqueous and lipoidal routes of herbicides entry, respectively (Fig. 1).

The cutin and pectinaceous strands of the cuticle are believed to constitute the aqueous route for herbicide absorption into leaves. After entry into cracks, punctures or fissures in the epicuticular waxes on the leaf surface, the herbicide moves internally along the relatively polar components of the cell wall. Since water soluble herbicides generally do not penetrate leaf surfaces easily, the aqueous route is enhanced by a hydrated atmosphere that cause expansion of the distance between epicuticular and cuticular wax plates and makes herbicide movement through the cuticle easier. Maleic hydrazide and paraquat enter into the plant leaves *via* the aqueous route.

The bulk of the cuticle volume consists of polymerized hydroxylated fatty acids, which gives it a lipophilic character. Lipophilic formulations (Emulsion Concentrations) tend to absorb into the cuticle easier than hydrophilic formulations (salts) due to the lipophilic property of the leaf cuticle. Sethoxydim, dinoseb and 2, 4- D (ester form) enter into plants by this pathway. The fate of a herbicide can also be influenced by its charge. The cuticle has a slightly negative charge, at a physiological pH, which can repel anionic formulations (sodium salts). Cationic formulations tend to readily adsorb to the cuticle. Absorption of both lipophilic and hydrophilic herbicide is done by simple diffusion. Rate of penetration is dependent on the permeability in the cuticle and the driving force (concentration gradient). Hydrophilic herbicides will tend to be slower because of their low permeability within the cuticle.

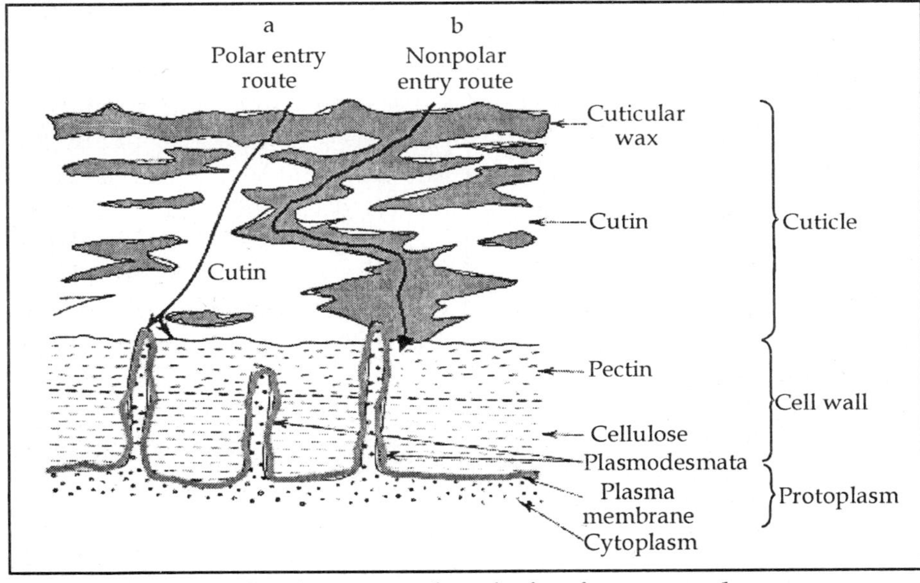

Fig. 1. Absorption route from leaf surface to cytoplasm. (Ashton and Crafts, 1981)

The nature of carrier is very important in determination of herbicide retention by the plant surface. For example, oil readily spread out and adhere to plant surface, granules tend to roll off when applied to foliage and provide a suitable carrier for foliage uptake only in site such as turf, in which the weed leaves are close to the soil surface and are dense enough to keep the granules suspended on the leaves. Since granules do roll off leaves, they can be used for some post-emergence treatments with greater safety to crops than can liquid sprayed broadcast. In contrast to oil, water has a high surface tension and tends to bead or ball up when it hits the waxy surface of leaves and stem cuticle. The subsequent lack of wetting or spread over the plant surface results in lower herbicide penetration. In fact, herbicide often is lost because it can bounce or runoff the cuticle. Modification can be accomplished by the addition of adjuvant such as a wetting agent to the spray solution. Wetting agents act by reducing the surface tension of the water droplets allowing them to spread and make close contact with the plant surface. Wetting agent also minimizes the problem of herbicide retention upright, vertically positioned leaves. Without wetting agents, aqueous spray droplets rapidly roll off the leaves of plants such as grasses, wild oat, wild garlic and carrot.

Retention and uptake of symplistically translocated herbicides can be improved by using lower carrier volumes and finer spray droplets, leading to improved performance. Adding the same amount of herbicides to a reduced volume of carrier increases the concentration of herbicide per droplet. Since penetration of herbicide into leaf tissue is due to diffusion presumably the greater herbicide concentration gradient across the cuticle, thus leading to the potential for incensed diffusion. Decreasing the spray droplet at a given volume of carrier can enhance coverage and retention on hard to wet plant surface, thereby enhancing the performance of translocated herbicides.

Rainfall soon after the herbicide application can wash herbicide from the leaf surface. Retention times of 6 to 24 hour frequently are needed to prevent the loss of water soluble herbicide by heavy rains. High retention time is required for negative charge herbicides such as sodium salt of 2,4-D because these herbicides do not absorb to the cuticle and therefore do not penetrate plant tissue rapidly. Positive charge herbicides such as paraquat are rapidly absorbed by the cuticle and less subjected to removal from leaves by rain. Retention time for oil soluble herbicides, which tend to penetrate rapidly into cuticle, is shorter possibly as short as one hour. Although the required time of retention varies with individual herbicides and environmental conditions it is wasteful and always disappointing to apply a post-emergence herbicides, if rainfall is expected.

In addition to influencing wetting and retention trichomes are more directly involved in foliar absorption. Leaf hair, depending on their morphology may provide a microclimate, which can alter the drying time of aqueous sprays and thus the absorption pattern. Since in most instances leaf hairs are extension of epidermal cells, an increased epidermal area is exposed to the spray solution. Those trichomes containing living protoplasm may be of particular significance.

Stomata appears to play a two fold role in foliar penetration, firstly, under certain conditions the aqueous spray solution may pass over in mass through the stomatal pore and diffuse through the air space of the leaf (Dybing and Currier, 1961; Greene and Bukovac 1974). Secondly the cuticle over the guard and associated accessory cells may be permeable and these structures, *per se* serve as preferred sites of entry (Neumann and Jacob, 1968).

Absorption from soil

Soil applied herbicides are applied directly on the soil surface or incorporated in the soil and roots are the primary absorbing organs for herbicides present in soil, although absorption by other subterranean plant organs has been observed.

Herbicides enter roots by three possible routes: apoplast, symplast and apoplast symplast (Devine *et al.*, 1993). Collectively all living portions of a plant form the symplast, which is the interconnected, continuous, living protoplasm of plants (Fig. 2). The apoplast comprises all the non-living plant tissues and any spaces between cells. Phloem and other living cells are the major component of the symplast, whereas xylem, inter-cellular spaces and cell walls form the apoplast. The apoplastic route allows the herbicides to enter roots freely until they encounter the casparian strip of the root endodermis. After passing through the endodermis they then enter the xylem of the vascular cylinder. The symplastic route involves initial herbicide entry through the cell walls of the roots hairs and subsequent movement into the cytoplasm of the epidermis and cortex. The herbicide then passes through the cytoplasm of the epidermis and cortex. Herbicides then pass through the cytoplasm of the endodermis, avoiding the casparian strip and enter the phloem of the vascular cylinders by means of interconnecting protoplasmic strands. The apoplast-symplast route is identical to the symplastic route, except that after passing through the endodermis the herbicide may re-enter cell walls and then enter the xylem of the vascular cylinder. Some herbicides are restricted to only one route of entry. However, it is more likely that most herbicides enter plant roots through more than a single pathway (Devine *et al.*, 1993). The route or routes of entry is determined by the physico-chemical properties of herbicides.

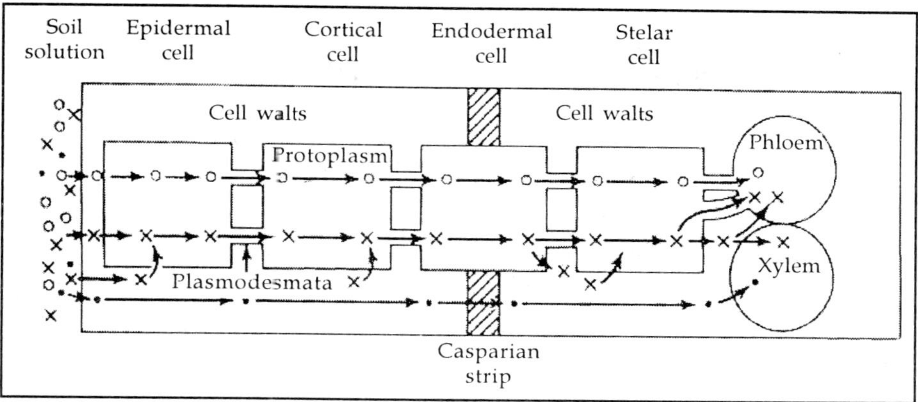

Fig. 2. Hypothetical diagram representing herbicide absorption into roots (Epstein, 1973).

Absorption into the roots does not have as many barriers as does absorption into the foliar surfaces of a plant, the primary reason being the absence of a cuticle where most herbicide absorption occurs. The most important point of entry is passive flow (co-migration with water) through the root hair zone (zone of differentiation) at root tips. The root hairs can increase the surface area available for herbicide uptake by over two-fold. However, the herbicide must be contained within the soil water solution. Once inside the root, either by apoplastic or symplastic movement, the herbicide molecule will travel with water and come to the casparian strip in the cell wall. The strip is highly lipophilic due to the presence of suberin. This layer acts to regulate the passing of ions as water cannot pass. However, this is not a major factor in herbicide translocation to the xylem.

Absorption through plant stem

The direct application of herbicides to plant stems is rare except to control woody plants, generally trees and shrubs. However, the stems of herbaceous crops and weeds usually are exposed to herbicides that are applied primarily to leaves. Herbicides applications directed to soil at the base of plants may also result in some stem exposure. The penetration of herbicides through the bark of tree or shrubs presents a much more difficult problem than penetration into foliage or herbaceous stems. Bark is a suberized covering (which are composed of tannins, fatty acids, lignin, cellulose and terpene) of corks cells that represents a formidable barrier to herbicide penetration. However, the presence of lenicels, which provide radial channel through the bark layer, could aid herbicide penetration. Herbicide entry through the bark of woody plants is normally done by mechanical cuts or by injections. When bark is uniform without cracks or fissures, aqueous sprays of herbicides usually are ineffective. An oil carrier is required for adequate herbicide absorption through woody stems.

The stem has an extensive and highly developed transport system. The degree of penetration through the stem depends on the growth and stage of development of the plant. Penetration through a young and the immature stem is more rapid than in a mature stem. The mechanism of penetration through the stem is similar to that in the foliage.

Absorption by seeds and coleoptiles

Absorption by seeds and shoot is very important with regards to soil-applied herbicides that target newly emerging seedlings. Herbicide entry into seed occurs in a passive manner with the water necessary for germination (Anderson, 1977, 1996; Aldrich, 1984). Several reports have shown that herbicides are absorbed by seeds of many species (Haskell and Rogers, 1960; Aston and Helfgott, 1966; Helfgott, 1969; Scott and Phillips 1971; Phillip *et al.*, 1972). In general, herbicide uptake occurs during inhibition of water but proceeds at rates different than the uptake of water. However, certain volatile herbicides can be absorbed by dry seeds. The amount of herbicides absorbed varies from species to species. Phillips *et al.*, (1972) reported that the total quantity of herbicide absorbed was determined by total oil and percent oil of the seeds. However, this generalization is not supported by Helfgott (1969). EPTC, diallate, CDEC and possibly DAA are dependent on uptake through the shoots prior to emergence. However, certain volatile herbicides and some soil fumigants enter dry seeds as a gas.

Researches since 1963 have shown that some herbicides may be absorbed from the soil by coleoptile and young shoots as they develop and push upward through the soil following germination of seeds. Dawson (1963) studying the response of barnydgrass (*Echinochloa crus-galli*) seedlings to soil applied EPTC found that EPTC in soil gave little or no response, whereas exposure of the young shoot resulted in severe injury and in most instances death. Some soil-applied herbicides are absorbed primarily through young shoots or coleoptiles of emerging seedlings when they contact herbicide treated soil. Absorption occurs with the soil water or as a gas, if the herbicide is volatile. Before emergence, the shoot has a poorly developed cuticle. The shoot does not have a casparian strip. Herbicide uptake is by passive diffusion with the water in contact with the shoot. Triazine and urea herbicides kill the plants *via* the entry in shoot. They allow the plants to emerge and effect shoot growth by inhibiting the 'Hill' reaction of photosynthesis. The plants die shortly after emergence.

Absorption across plant membranes

All biochemical target sites for herbicide action are located within the symplast. In order for herbicides to reach their target sites, they must cross the membrane located at the cell wall (plasma membrane) and often an additional organelle membrane (e.g. chloroplast envelope). Normally this is done by passive diffusion; however, equilibrium will be reached. Most herbicides cross membranes by simple diffusion. The driving force is the concentration gradient across the membrane. The herbicide will move from a region of higher concentration (outside the membrane) to a region of lower concentration (inside the membrane) until the concentration on both sides are equal. Metabolic energy is not required for this movement. Lipophilic herbicides tend to diffuse more freely across the plasma membrane than hydrophilic herbicides. Some herbicides diffuse across the plasma membrane but also are connected within the cell; in other words, they move against a concentration gradient and accumulate at a higher concentration in the cytoplasm than outside the cell.

Regardless of their location within the plant, all membranes have the same basic structure. They consist of a lipid bi-layer with a hydrophilic exterior and a lipophilic interior. The lipophilic interior consists of two chains containing mostly 16-18 carbons attached to a glycerol backbone. The third substituent on the glycerol backbone is a polar (hydrophilic) group. The membranes will also contain proteins that actually may aid in the transfer into the symplast. Uncharged herbicide molecules continue to diffuse into the cell because the concentration of uncharged herbicide is still higher outside the membrane than inside. The most important of these proteins is the ATPase "pump" which acts to create a pH difference between the symplast (pH as high as 8.0) and the cell wall (pH as low as 5.0). This concentration gradient drives the ion trapping process. Many herbicides have an ionizable group such as bentazon, clopyralid, sethoxydim, chlorsulfuron and other sulfonylureas and imidazolinones. When herbicides become exposed to this low pH, they become protonated (increasing lipophilicity) and can readily diffuse across the membrane. Once inside the membrane, the herbicide becomes ionized and can not get back out.

Translocation

Once the herbicide absorption process is complete, translocation of the herbicide to the site of action becomes the primary physiological function involved in the mode of action. Herbicides are translocated within the plant through the symplast and apoplast. Some herbicides are primarily translocated in the symplast, some in the apoplastic system and some in both systems. However, all the herbicides also appear to be able to move readily from one system to the other system (phloem > xylem) during transport.

The apoplastic system constitutes the total continum of intercellular spaces, cell walls and mature xylem, it is considered to be non-living. Apoplastic mobile herbicides, which are absorbed by the roots, follow the same pathway as water. They enter the xylem and are swept upward in the transpiration stream. The driving force for this movement is the removal of water from the leaves by transpiration. When these herbicides are absorbed by the leaves they remain in the treated leaf can occur under conditions that permit reversal of the transpiration stream that is very high humidity and very dry soil.

The symplastic system constitutes the total continum of protoplasm throughout the plant, including the cytoplasm of each cell their interconnecting plasmodesmata and the phloem, it is considered to be living. Symplastic mobile herbicides, which are absorbed by the, leaves move along with the photosynthate *via* the same pathway.

Some herbicides move (translocated) within the plant. Systemic herbicides are translocated once they are taken up by the leaves, stems or roots. Herbicides that do not move after they enter the plant are called contact herbicides. Contact herbicides are sprayed like other herbicide but they reveal their effect at or close to the site of application example, paraquat, a non-selective herbicide and among selective contact herbicide propanil is used for control of barnyard grass control. To be effective, contact herbicides must be applied to the site of action. Most foliar-applied contact herbicides work by disrupting cell membranes. Contact herbicides damage the top growth that the spray solution contacts, but the underground portion of perennial plants remains unaffected and can rapidly initiate new growth.

Contact herbicides cause membrane destruction, burning, and necrosis of leaf and stem tissues, often within hours after application. Since these compounds do not move or redistribute themselves in the plants, complete coverage of the plant surface is required for shoot kill. Contact herbicides often are more effective on broadleaves than on grasses. The growing point of young grasses is located in the crown region of the plant. which is at or below the soil surface, and thus, difficult to contact with the spray. In contrast, the growing point on young broadleaf plants is exposed to the spray treatment. Thus, paraquat may not kill all the growing points of a tiller grass plant, and regrowth can occur. Some soil applied herbicides also show limited movement in plant. They are most effective when used on seedling plants to or at emergence from the soil. There are relatively few barriers to penetration and the herbicide does not have to move far into the plant to cause damage.

Systemic herbicides (tranlocative herbicides) can be translocated to other parts of the plant either in the xylem or the phloem or both. Translocation depends on the chemical and the plant species. Herbicides are translocated only in the xylem are most effective as soil-applied or early post-emergence treatments because translocation is only upward. Atrazine is a good example of a herbicide that is translocated only in the xylem. Phloem translocated herbicides that move downward and suppress root and rhizome growth, as well as top growth, provide the best perennial weed control. 2,4-D and glyphosate are examples of systemic herbicides that will translocate in the phloem and provide good, long-term control of certain perennial weeds.

Once a herbicide has penetrated the leaf or stem cuticle or the root epidermis, there are still many barriers to its movement to the site of action. The chemical can be compartmentalized into portions of the plant away from the site of action and, therefore, it is rendered inactive. Short distance herbicide movement across a few cell layers occurs by simple diffusion, ion trapping, or carrier mediated processes. Assuming the herbicide is not immobilized in the leaf or root, it is available for long distance movement in the plant by utilizing the xylem and phloem transport systems (Fig. 3).

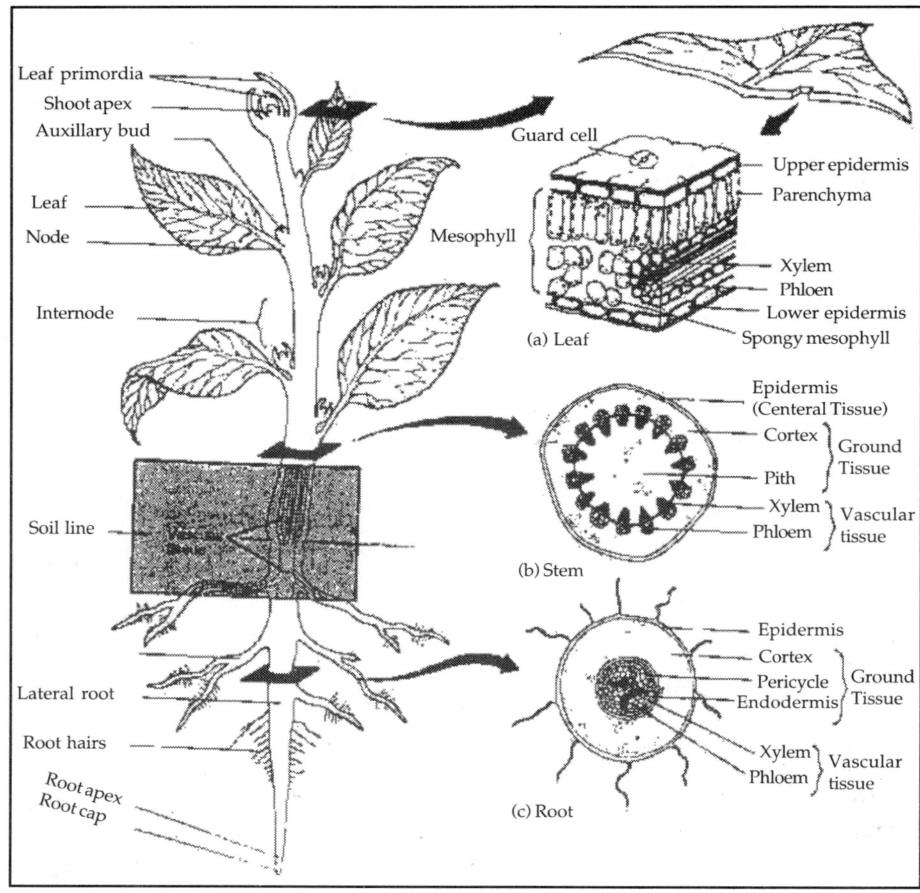

Fig. 3. Body of the seed plant, showing cross section of (a) leaf, (b) stem and (c) root (Taiz and Zeiger, 1991)

Xylem

Herbicides that enter plants *via* the root hairs move upward in the xylem with the mass flow of water. Water and substances dissolved in water move in the xylem forward by transpirational pull and root pressure. The degree of translocation is often associated with the hydrophilic/lipophilic balance of the herbicide. As lipophilicity (affinity for lipophilic substances in the cell) of a herbicide increases this slows movement into the xylem. Most herbicides are xylem mobile, however, there are several reasons that a herbicide may not be:

a. Absorption to apoplastic or symplastic cellular components
b. Com partmentalize within cellular components
c. Conjugate to cellular substrates that are not xylem mobile

The major difference in the translocation behaviours of a phloem mobile and a xylem mobile herbicide is that the later does not bypass mature leaves on its way to the plant tops, whereas the former does so. Also a xylem mobile herbicide translocate only in a direction acropetally to the point of its application while a phloem mobile herbicide translocates both in acropetally and basipetally directions. But picloram when applied to an intact root, translocates acropetally up to its crown level and then move into the phloem. Once picloram has reached the phloem tubes, it can translocate both acropetally and basipetally to its sink by passing the mature leaves (Reid and Hurtt, 1969).

When a xylem mobile herbicide is placed on a plant shoot. It moves through the apoplast and is translocated upward in the xylem. It will not show any basipetally translocation unless it was capable of leaking into the adjoining phloem tubes (Crafts, 1966). For example, atrazine and diuron both xylem mobile herbicides will kill only shoots of establish annual grasses, leaving their crown buds, roots and hidden meristems unhurt. This is because these herbicides do not translocate from the treated leaves, but cause contact injury to the foliage. On the other hand, when atrazine and diuron are applied to soil, these are absorbed by the grass roots and are translocated through xylem to the meristematic tissues, crown buds and blades which may be damaged seriously.

Reverse xylem mobility

Plants, growing under moisture stress, such as prevailing in the arid regions, certain herbicides show reverse xylem mobility i.e. they translocate rapidly from plant shoots to its roots through xylem instead of phloem. One to two percent solution of arsenic trioxide, borax, sodium chloride and ammonium thiocyanate form common reverse xylem mobile herbicide. They are applied as liquids by jar method. In this method the shoots of a weed growing under an intense moisture stress conditions are bent into a jar containing the herbicide which injures the leaf surfaces and renders them so permeable that it is heavily taken in by the moisture stressed plant into its xylem vessels and is translocated to its deep roots. For example, translocation of herbicides absorbed by the cut stems of unwanted trees.

Phloem

In general, the movement in the xylem is faster than the phloem; therefore, one problem can occur: when certain herbicides can move readily between the symplast (phloem) and apoplast (xylem) their movement within the vascular system is in direct correlation with the transpirational stream. This can cause foliar applied herbicides never to reach the root systems of target species. The transport is along a physical turgor or osmotic gradient in the phloem maintained by a source-sink relationship. High concentration of sugar in the phloem causes water to move into the phloem by osmosis, the high turgor pressure then forces the contents of the sieve tubes of the phloem to flow *en masse* to areas of low turgor pressure (sink). For this to occur the osmatic pressure or concentration of sugar of the contents at the source must be greater than that of the surrounding cells. Therefore the movement of sugar into the phloem from the surrounding cells against a concentration gradient must require energy. This process is called vein loading. The mechanism of herbicide entry into the phloem is unknown, however, the pathway of phloem-mobile herbicide follows the source-sink relationship of the photosynthate. Two mechanisms allow herbicides to remain in the phloem long enough to translocate;

pH gradient

The active loading of sucrose into sieve elements, which involves an ATPase pump, moves hydrogen ions into the cell wall, thus creates a pH gradient between the phloem cell and the cell wall (pH 8 and pH 5 respectively). In the acidic cell wall, herbicides having an ionizable group with a pKa of less than 6 will favour a protonated (lipophilic) form, which is soluble in the membrane. Once in the phloem, the high pH (8.0) will favour the ionized form (anion), which is not soluble in the membrane, thus the herbicide will remain trapped in the phloem.

Intermediate membrane permeability coefficients

Once inside the phloem, these compounds can remain long enough for translocation to occur due to their $\log K_{ow}$ ((octanol/water partition coefficients (-1 to +1 is ideal)). The major limiting factors of phloem movement are the ability of the herbicide to enter the phloem, and then remain long enough for transport to sink tissues.

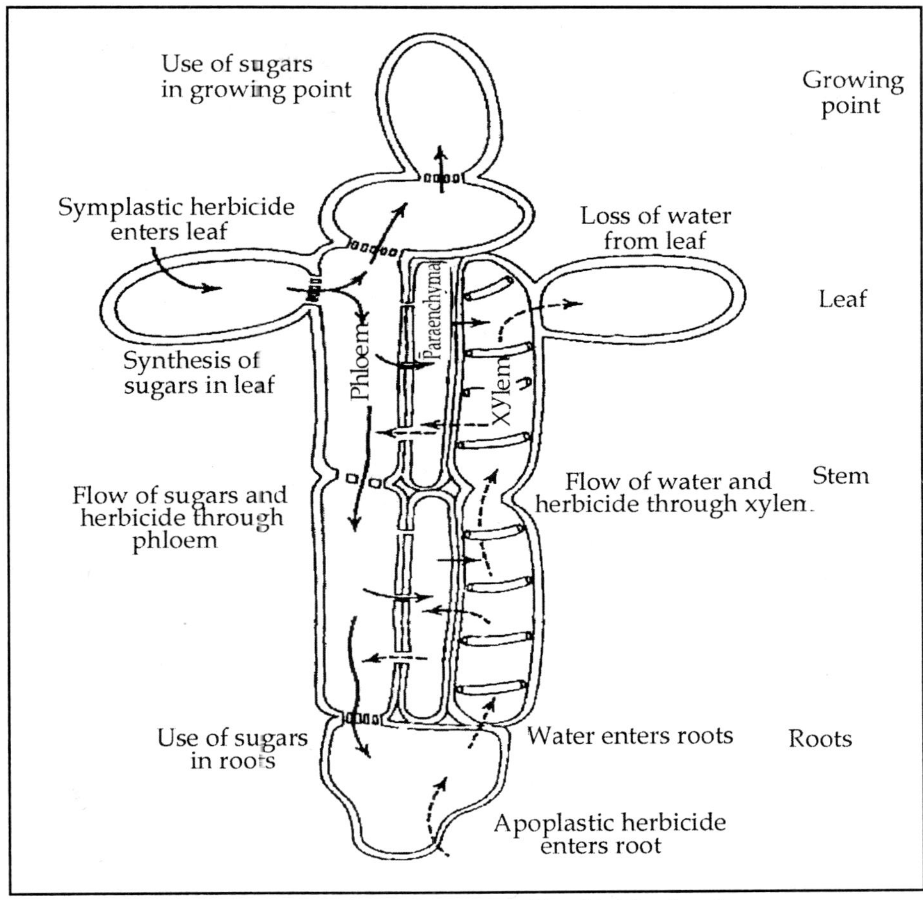

Fig. 4. Routes of translocation of herbicides in plants (Bonner and Galston, 1952).

Stage of the plant growth greatly affects the amount of sugar translocation to various parts of the plant changes during the life cycle of the plant. In general, plants are most susceptible to symplastically translocated herbicide when young or when undergoing periods of rapid growth, vegetative reproductive structure development or food accumulation in storage structure (Fig. 4).

Complete plant coverage with symplastically translocated herbicide is not required, however, it is essential that the herbicide come in contact with and be adequately retained by structures that are actively photosynthesizing and exporting sugars. Young leaves and old senescing leaves do not export sugars and therefore do not transport herbicides well. The application of a symplastic herbicide to photosynthezing and sugar exporting lower leaves results in its movement downward with the sugar to underground structure. Application to fully functioning upper leaves sends the herbicides to the developing shoot tip.

Active and functional sucrose translocation is required for the optimal movement and distribution of symplastically translocated herbicides. Movement of a phloem mobile herbicide to the underground portions of a plant can take several hours to a few days. Thus immediate foliage kill or damage to the symplast, such as the plugging of sieve tubes with callose or death of the cytoplasm of the conducting cell almost always results in poor herbicides transformation and unsatisfactory control of the underground plant part. Best results can be obtained by frequent application of several low dosage of a herbicide that can move underground rather than a single dose that kills the plant too rapidly. Multiple applications also provide time for dormant buds to initiate growth after the initial application and to become susceptible to the second application.

Summary

The absorption and translocation of a herbicide depends primarly on its molecular configuration, which in turn determines its chemical and physical properties. Differential absorption and translocation which form the basis for herbicide selectivity, determine the tolerance and the susceptibility of a plant species to a particular herbicide. Plant characteristics and environmental conditions also influence the amount of a compound that enters a plant and its consequent movement.

Herbicide enters plants through shoots, roots, other below ground organs and seed. The process of herbicide entry into treated plants is called absorption, which involves contact, penetration and movement of the chemical into the plant, whereas, adsorption is the attraction of ions or molecules to the surface of a solid. In foliar absorption leaves are the primary means of herbicide entry through shoots, although herbicide absorption can also occur through other aerial organs such as stems or emerging coleoptiles. Soil applied herbicides are applied directly on the soil surface or incorporated in the soil. Roots are the primary absorbing organs for herbicides present in soil, although absorption by other subterranrean plant organs has also been observed. Herbicide entry through the bark of woody plants is normally done by mechanical cuts or by injections. Absorption by seeds and shoot is very important with regards to soil-aplied herbicides that target newly emeging seedlings. Herbicide entry into seed occurs in a passive manner with the water necessary for germination. However, certain volatile herbicides can be absorbed by dry seeds.

Once the herbicide absorption process is complete, translocation of the herbicide to the site of action becomes the primary physiological function involved in the mode of action. Herbicides are translocated within the plant through the symplast and apoplast. Some herbicides are primarily translocated in the symplast, some in the apoplastic system and some in both systems. However, all the herbicides also appear to be able to move readily from one system to the other system (phloem $\leftrightarrows$ xylem) during transport. Herbicides that enter plants *via* root hairs move upward in the xylem with the mass flow of water. The apoplastic system constitutes the total continum of intercellular spaces, cell walls and mature xylem. It is considered to be nonliving. Apoplastic mobile herbicides, are absorbed by the roots, follow the same pathway as water.

❑❑❑

Chapter 4

Mode of Action

After a herbicide enters the plant and reaches a specific site in plant cells, it inhibits a biochemical process. The specific biochemical process inhibited often depends on the chemistry of the herbicides and in some cases, the plant species involved. Both the site and biochemical reaction inhibited are known for many herbicides groups. For other only the effect of herbicide rather than the cause of the response is known. Mode of action in respect of most of the herbicides fall into the following categories

1. Amino acid synthesis inhibitors
2. Growth regulator herbicides
3. Lipid synthesis inhibitors
4. Pigment inhibitors
5. Photosynthesis inhibitors
6. Seedling growth inhibitors
7. Cell membrane disruptors and organic arsenicals
8. Cellulose biosynthesis inhibitors
9. Unknown mode of action

1. Amino acid synthesis inhibitors

Plants use proteins in functional, storage and structural roles. Functional proteins are called enzymes. Enzymes catalyze thousands of chemical reactions necessary for plant growth and development.

Storage of proteins commonly occur in seeds and supply essential amino acids to young, developing seedlings. Both enzymes and seed proteins consist of long chains of interconnected amino acids. Commonly, about 17 to 20 different amino acids occur in plants. However, the amino-acid composition between different plant proteins varies greatly. In the absence of amino acid and protein synthesis, plants cannot complete the chemical reactions necessary for growth. Amino acid synthesis inhibitors act on a specific enzyme to prevent the production of specific amino acids, key building blocks for normal plant growth and development. There are three classes of amino acid synthesis inhibitors:

Branched chain aminoacid inhibitors (ALS or AHAS inhibitors)

Branched chain amino acid synthesis is a perfect target site for herbicide activity as it only occurs in plants. The main enzyme inhibited by these herbicides is ALS (acetolactate synthase) also known as AHAS (acetohydroxyacid synthase). ALS is an early enzyme in the biosynthesis pathway of valine, leucine and isoleucine. Within plants, the ALS enzyme binds with pyruvate to eventually make the three-branched chain amino acids mentioned above. The two related pathway in which acetolactase is produced from pyruvate and acetohydroxy-butyrate from threonine are catalyzed by a common enzyme acetolactase synthase (acetohydroxy acid synthase) (Fig. 1). This is effectively inhibited by sulfonylureas, imidazolinones, sulfonamides and triazolopyrimidines (Dekker and Duke, 1995).

When ALS inhibitors are able to inhibit the function of the ALS enzyme, the amino acids cannot be produced. Recently, it was shown that pyruvate oxidase might have been an early precursor to ALS. Based on these findings, it was shown that ALS inhibitors have a binding site on ALS that is an "evolutionary vestige" of the quinone-binding site that was part of the pyruvate oxidase enzyme. This inhibits ALS as the shape of the substrate-binding site has been changed. Thus branch chain amino acid production ceases. The question has arisen whether or not depletion of branched chain amino acids is the sole cause of plant death for herbicides that inhibit ALS. There are many other sites of action that have been considered for the ALS inhibitors, however, most of the herbicide activity may well be due to the general disruption in plant metabolism due to a depletion of branched chain amino acids.

Inhibition of amino acid biosynthesis will lead to growth inhibition, in as much as new amino acid production is necessary to sustain the protein synthesis required for plant growth. Analysis of plant tissue (duck weed) after chlorsulfuron treatment showed a large accumulation of alpha-aminobutyrate (Rhodes *et al.*, 1987). Some evidence suggests that singlet oxygen accumulates and involve in the mechanism of action of ALS-inhibitors (Durner *et al.*, 1994). The ALS enzyme has oxygen consuming side reaction, which may generate singlet oxygen if its normal action is inhibited. Fig. 2 exibits structural formulae of some of the brached chain amino acid inhibitors.

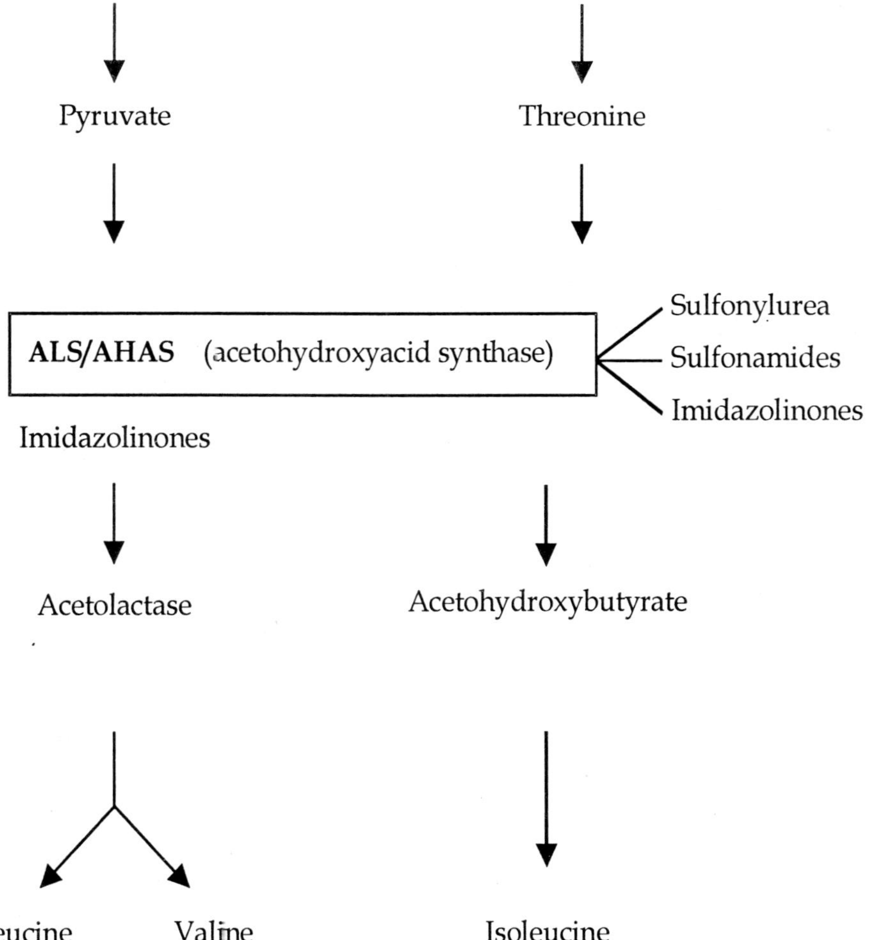

Fig. 1. Part of the biosynthesis pathway of branched chain amino acid synthesis showing the site of action of herbicides at the enzyme acetolactase synthase (acetohydroxy acid synthase), Hawkes *et al.*, 1989.

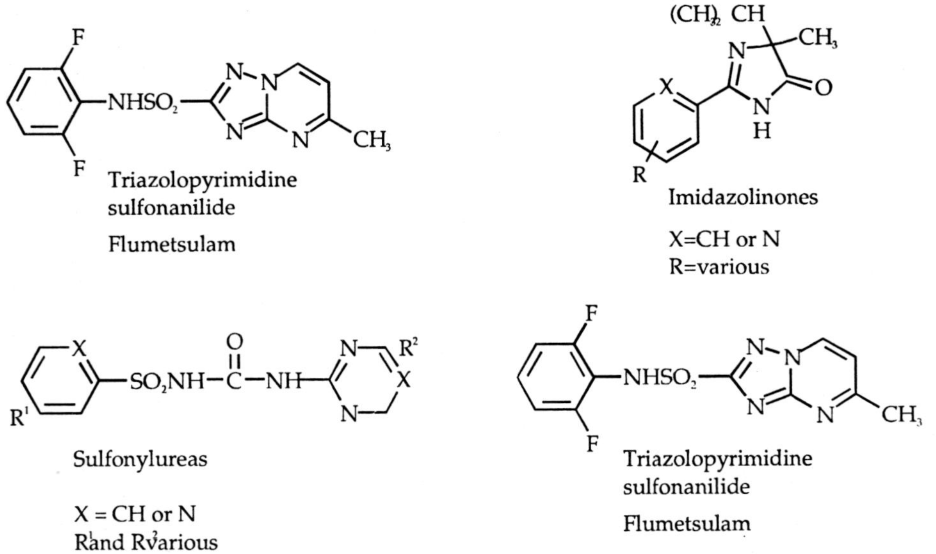

Triazolopyrimidine sulfonanilide

Flumetsulam

Imidazolinones

X=CH or N
R=various

Sulfonylureas

X = CH or N
R^1 and R^2 various

Triazolopyrimidine sulfonanilide

Flumetsulam

Fig. 2. Compounds that inhibit branched chain amino acid biosynthesis

Injury symptoms

The ALS inhibitors are effective at very low concentrations. Action of these herbicides is quite slow, however, ALS inhibitors are potent inhibitors of plant growth. Control of treated plants does not occur until several weeks after application. The meristematic tissue appears to be the first tissue affected. This is followed by chlorosis, necrosis and eventual plant death. Grass plants may be stunted, with interveinal yellowing (chlorosis) or purpling. Corn plants may be stunted and show symptoms of root inhibition such as pruning of lateral roots. Leaves emerging from the corn whorl may not unfurl properly and may be yellow to translucent in appearance. Broadleaf plants may be stunted and chlorotic or purple and sometimes blackened terminals. Symptoms take 1-2 weeks to develop. Branched chain amino acid inhibitors are prone to resistance development. Herbicides in these categories are:

Imidazolinones: imazapyr, imazapic, imazethabenz, imazamox, imazaquin, imazethapyr.

Sulfonylureas: metsulfuron-methyl, sulfometuron, bensulfuron, chlorimuron, chlorsulfuron, halosulfuron, nicosulfuron, primisulfuron, prosulfuron, rimsulfuron, sulfometuron, thifensulfuron, triasulfuron, tribenuron.

Sulfonamides: They are also known as triazolopyrimidine sulfonanilides or TPS: cloransulam- methyl, flumetsulam, diclosulam, lumetsulam, pyrimidinyl thio, pyrithiobac.

Aromatic amino acid synthesis inhibitors (ESPS inhibitors)

Shikimate pathway is one of the important biosynthesis route in plant that leads to the formation of the aromatic amino acids tryptophan, tyrosine and phenylalanine. A large number of secondary compounds such as flavanoids, anthocyanins, auxins and alkaloids arise from these amino acids. The failure of the shikimate pathway would therefore not only have a serious effect upon protein synthesis but also on the synthesis of compounds associated with growth regulation and defense. Glyphosate inhibits the pathway at the enzyme enopyruvyl shikimate-3-phosphate synthase (ESPS) (Fig. 3). Glyphosate has shown to be translocated rapidly to meristematic tissue and stem apices, especially underground rhizomes where growth failure occurs. Glyphosate is xylem and phloem mobile; however, phloem mobility is not as great as with other phloem mobile herbicides. The reason is thought to be due to glyphosate's inefficient ability to remain in the phloem. Therefore, the slow rate of control is an advantage in allowing maximum phloem mobility to roots and other perennial organs prior to death of treated leaves. At a physiological pH, glyphosate is a zwitterion, and is one of the few herbicides that have been shown to cross the plasma membrane *via* a carrier protein (phosphate carrier).

Aromatic amino acid biosynthesis involves a photosynthetic carbon reduction cycle. Within this cycle, PEP (phosphoenolpyruvate) and erythrose-4-P combine to eventually make shikimate. Through another reaction, shikimate is used to make chorismate, which eventually makes the aromatic amino acids (tryptophan, phenylalanine, tyrosine). The glyphosate molecule finds its site of action between shikimate and chorismate. After shikimate is produced, a phosphate group is added to it (now called S3P) and it combines with PEP to make EPSP (Enolpyruvylshikimate-3-P). EPSPS (5- enolpyruvylshikimate-3-phosphate synthase) is responsible for catalyzing the reaction. This is the enzyme that glyphosate inhibits resulting a buildup of shikimate (Lydon and Duke, 1988). This deregulation and enhanced carbon flow into the shikimate pathway drains other biosynthesis pathway of necessary building blocks (Dekker and Duke, 1995). Thus the blockage of shikimate pathway can lead to a large number of potentially damaging physiological effects. Glyphosate is able to bind to an allosteric site near PEP on the EPSPS enzyme. Therefore, PEP cannot bind with S3P to produce EPSP. Without EPSP, chorismate concentrations are reduced and also those of the aromatic amino acids mentioned above. A reduction in amino acid production will reduce protein synthesis and subsequently cause an inhibition in growth.

Injury symptoms

The first noticeable symptom after glyphosate treatment is growth inhibition, followed by chlorosis of treated tissue. After 5-10 days, chlorosis begins to turn into necrosis and the affected plants begin to die. Plant foliage, especially new growth, will first become yellow and then turn brown and die within 10 to 14 days after herbicide application.

Examples: Glyphosate, sulfosate.

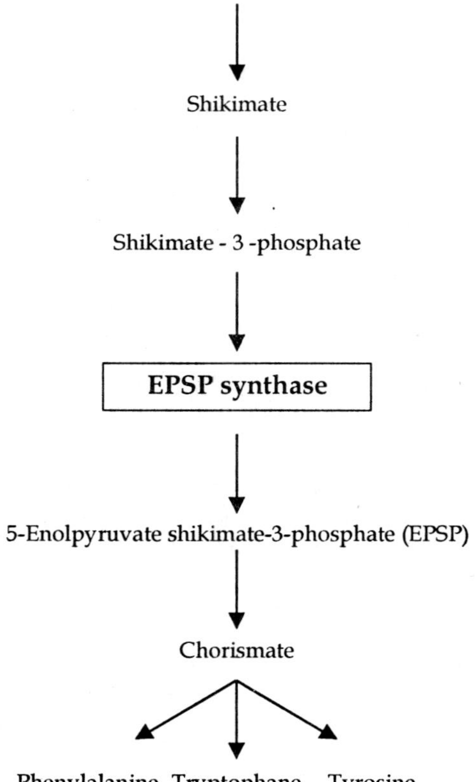

Fig. 3. Part of the biosynthesis pathway of aromatic amino acid synthesis showing the site of action of glyphosate at EPSP synthase

Glutamine synthesis inhibitor

The symptoms of glufosinate injury are chlorosis followed by necrosis. Symptoms are somewhat like membrane disrupting herbicides, however, the speed of membrane disruption is slower than other herbicides having a direct membrane disruption mode of action (e.g. paraquat). After application, ammonia level in leave, which is usually very low, increases dramatically. Within four hours after treatment, the ammonia level is about 10 times greater and after 1 day, level exceed 100 times that of a non-treated leaf.

The accumulation of ammonia in glufosinate treated plants is known to be due to a direct inhibition of the glutamine synthetase (GS) enzyme. This accumulation is accompanied by a significant decrease of glutamine and glutamate, which are required for many nitrogen assimilation reactions in the cell wall, thus their absence causes substantial disruption of plant cell function. GS is responsible for converting glutamate plus ammonia to glutamine. Glutamine is key in the transamination of keto-acids in the synthesis of several amino acids and is also an effective method of maintaining a low level of ammonia within the plant's cells (Fig. 4).

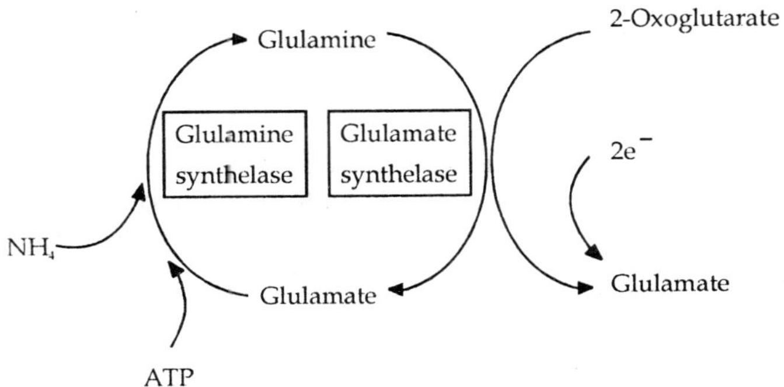

Fig. 4: Biosynthesis of glutamine

Further this is an ATP requiring reaction. Important evidences show that ammonia is not directly responsible for the toxic effects of glufosinate. Glufosinate can be considered a structural analog of glutamate. The inhibition of GS by glufosinate is a two step process.

Glufosinate is an efficient competitor to glutamate for its binding site to GS. At this stage, binding of glufosinate to GS is reversible. In the second step, glufosinate bound to GS can be phosphorylated by ATP. The phosphorylated form of glufosinate is not released from the enzyme (irreversible inhibition), thus effectively blocking the further use of the GS enzyme in forming glutamine. Most conclude that photosynthesis inhibition is indirectly related to a treatment of glufosinate, and primarily related to the disruption of amino acid metabolism as a result of GS inhibition. After glufosinate treatment, the loss of amino donors (i.e. glutamate) for the glycolate pathway (glycolate> glyoxylate> glycine) leads to break down of the transamination reaction of glyoxylate to glycine in the photorespiration cycle. The accumulation of glyoxylate inhibits RUBPcarboxylase, which is the key enzyme for carbon fixation. Inhibition of RUBP carboxylase, in turn, will lead to an inhibition of the light reactions of photosynthesis. In summary, glufosinate disrupts many important nitrogen metabolism (nitrogen assimilation) synthetic reactions in plants by inhibiting glutamine and glutamate formation. The loss of both of these important substrates is due to inhibition of GS. Electron flow in photosynthesis is indirectly inhibited through the decrease in amino donors (from glutamate) for glyoxylate. Glyoxylate accumulates to a level that inhibits RUBP carboxylase and this induces an inhibition of the light reaction in photosynthesis. In the light, inhibition of electron flow in photosynthesis causes induction of lipid peroxidation from triplet chlorophyll.

2. Growth regulator herbicides

Indole acetic acid (IAA) and benzoic acids were first reported to have growth regulating properties in plants in the 1940's (Zimmerman and Hitchcoch, 1942). IAA is responsible for numerous aspects of plant growth and morphogenesis, including elongation growth and cell division in stems. The phenoxy herbicides are often referred to as auxin like herbicides because they induce twisting and curving (epinasty) of the petiole and stems of broadleaf plants, reminiscent of plant response to high application doses of the native auxin. These herbicides are widely used to control broadleaf weeds in grass crops such as wheat, corn, sorghum, forages and turf grasses. One member of this group, 2,4-D, was one of the first selective herbicides developed. Although the 2,4-D binding site in plants is unknown, there is speculation that 2, 4-D competes with the auxin hormones for binding to an unknown auxin binding hormones receptor on plasma membrane (Holt *et al.*, 1993). Growth regulator herbicides upset the normal hormonal balance that regulates processes such as cell division, cell enlargement, protein synthesis and respiration. That is why this group of herbicides is sometimes called the "hormone herbicides" (Fig. 5).

Biochemical and metabolic changes in plants reported to be induced by growth regulator herbicides are numerous. Studies suggest that nucleic acid metabolism and metabolic aspects of cell wall plasticity are most relevant to the mode of action of growth regulator herbicides. Early work showed that IAA stimulated the synthesis of RNA and DNA in tobacco pith cell. Phenoxy herbicides also have similar results. However, IAA has endogenous control mechanism that maintains its concentration within the appropriate physiological range. Growth regulator herbicides have no such mechanism. The level of the growth regulating herbicides in the meristem and developed organs of the intact treated plant increases with time after application. Thus there is first a stimulation of cell metabolic processes, resulting in uncontrolled growth and later a inhibition of these processes and plant death. After application of growth regulator herbicides early plant responses are associated with cell wall acidification and change in gene expression. Auxin and auxin growth regulating herbicides induce proton efflux through the plasma membrane by stimulation of proton pumping ATPase, which leads to the acidification of the cell wall matrix. Low pH increases cell wall extensibility and activates enzymes that degrade cell wall. Together these events weaken the cell wall and enable growth *via* turgor-driven cell expansion. Approximately 25 auxins responsive genes have been identified. However, with the exception of acetyl coenzyme A synthetase (ACC synthase) the precise biochemical action of other auxins responsible gene is unknown. ACC synthase is the key regulatory enzymes in ethylene biosynthesis. Ethylene has been suggested to be involved in the effects induced (epinasty) in susceptible plants by growth regulating herbicides. Tissue proliferation induced by a growth regulator herbicides leads to epinasty, stem swelling and disruption of the phloem, preventing photosynthesis movement from the leaves to the root system. This unproductive growth causes death in several days or weeks.

The effects associated with auxins help set them apart from other downwardly mobile herbicides. Bending and twisting of leaves and stems is evident almost immediately after application. Delayed symptom development includes root formation on dicot stems; misshapened leaves, stems and flowers and abnormal roots. Auxin growth regulator herbicides are organic acids, which take on a negative charge after ionization of acids and salts. Esters are hydrolyzed to acids or salts in both plants and soils.

Injury symptoms

These herbicides produce profound effects upon the growth and structure of plants. The most common symptoms for these herbicides are leaf and stem malformations. Broadleaf plants exhibit stem curl, twisting (epinasty), droop, callus tissue formation and leaf malformations (cupping, crinkling or have a "drawstring" appearance caused by irregular growth at the leaf edges, parallel veins, leaf strapping). Corn plants exhibit rolled leaves (onion-leafing), fused brace roots, stalk bending (goose necking) and brittleness, and missing kernels, small grains or missing grain (blasting), exhibit twisted flag leaves, sterile florets or multiple florets, due to over application or application at the wrong stage of development. Blasting and malformed seed heads are common symptoms of ill-timed applications in wheat. Resistance may occur but is not yet a problem. Herbicides in this category are:

Phenoxy acetic acids: 2, 4-D, 2, 4-DB, MCPA, dichlorprop, MCPB, mecoprop; MCPP.

Benzoic acids: dicamba

Pyridines: triclopyr, clopyralid, picloram, quinclorac.

Others: diglycolomine; 2,4-DP.

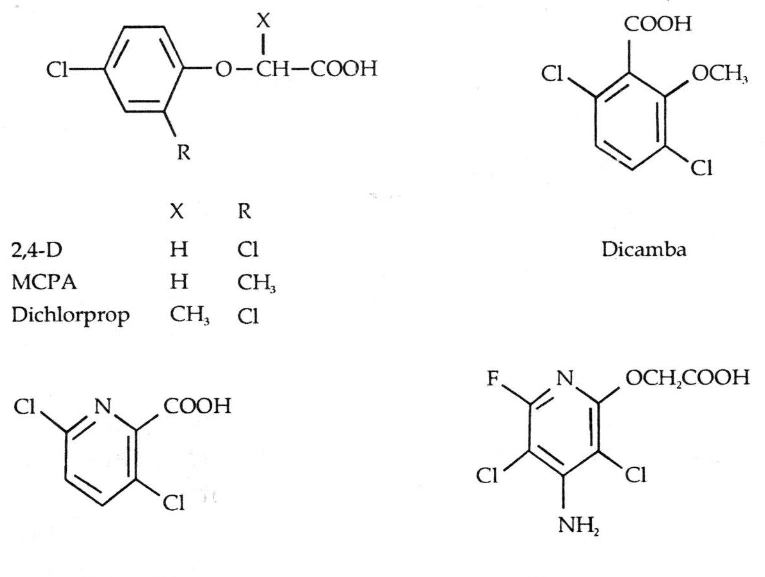

Fig. 5. Some hormone or auxin herbicides

3. Lipid biosynthesis inhibitors (ACCase inhibitors)

Lipid synthesis inhibitors act only on annual and perennial grasses, not on broadleaf plants. Plastid contains acetyl coenzyme A synthetase, acetyl coenzyme A carboxylase and fatty acid synthetase. These enzymes are the key component of fatty acid synthesis in plants (Kleczkowski, 1994). Herbicides that affect any of the steps involving these enzymes can block glycerolipid and phospholipid synthesis, which results in inhibition of membrane formation. The most affected are young developing leaves and meristematic tissues, depending upon the efficient fatty acid supply.

Acetyl coenzyme-A carboxylase (ACCase) is a key enzyme in the lipid biosynthesis pathway. The step catalyzed by ACCase is thought to be the rate limiting step in lipid biosynthesis. ACCase inhibitors of the anyloxy phenoxy type are readily absorbed through the cuticle and into leaf cells where it is deesterified by esterase enzymes. The other herbicides class having this mode of action are the cyclohexanediones. The acid is then translocated to meristematic areas of the shoots and roots. Roots and shoots of grass plants cease growth within a few hours after treatment, long before visual signs of injury appear. Plants then develop red coloration in the leaves (anthocyanin formation) and necrotic areas in the meristematic tissue at the growing point and the nodes, then extending throughout the plant. Respiratory activity declines with a resulting increase in sugar and anthocyanin levels. Mitosis is affected as a result of inhibition of cell wall formation. Inhibition of lipid biosynthesis could explain the reduction in growth (a lipid requiring process), the reported increase in membrane permeability and the ultrastructural effects observed after treatment. ACCase, is responsible for converting acetyl-Coenzyme-A (acetyl-CoA) to malonyl Coenzyme-A (malonyl-CoA) by adding CO_2 (in the form of bicarbonate- HCO_3). This is a key early biosynthetic reaction in the formation of lipid biosynthesis. Malonyl CoA is required for the synthesis and elongation of fatty acids and the synthesis of a number of secondary products (Walker *et al.*, 1989). ACCase consists of three reaction regions:

1. A biotin carboxylase, which catalyzes an ATP dependent carboxylation of biotin with CO_2.
2. A biotin carrier protein (BCP), which is covalently linked to the biotin group and allows the biotin to move between the two catalytic centers.

3. A carboxytransferase, which transfers the activated CO_2 from the biotin to acetyl-CoA (Fig. 6).

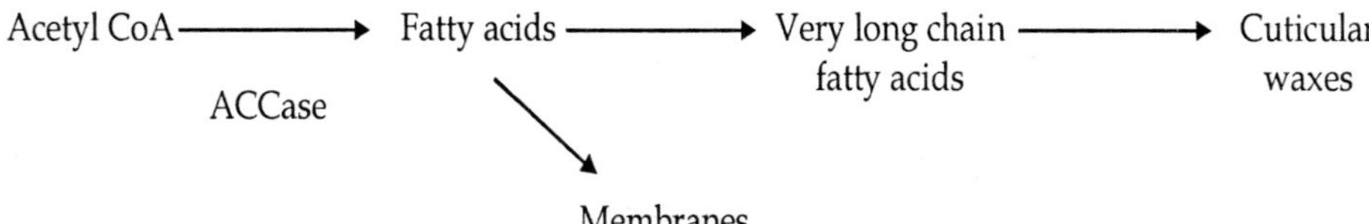

Fig. 6. Biosynthetic pathway to fatty acids, lipids and waxes

Since the acetyl-CoA/malonyl-CoA exchange is so strongly inhibited, these herbicides are thought to react mostly on the carboxytransferase region. As this is the case, CO_2 is unavailable to the acetyl-CoA, therefore, malonyl-CoA cannot be manufactured, thus, lipid synthesis is inhibited. In the mid 1990's, information emerged regarding the forms of ACCase in plants. Two types of ACCase enzymes were discovered in higher plants: one, a prokaryote form consisting of four protein subunits; the second, a eukaryote form consisting the three functional domains in a single large protein. Only the eukaryote form is affected by ACCase inhibitors. In dicots, the prokaryote form is located in the chloroplast where all of the components for lipid synthesis are housed, and is not inhibited by these herbicides. In grass species, the eukaryote form is located in the chloroplast and functions in lipid biosynthesis.

Burton *et al.*, (1987) found that sethoxydim and haloxyfop inhibited ACCase isolated from chloroplast of corn seedling with IC_{50} concentration of 2.9 and 0.5 μM respectively. Lichtenthaler *et al.*, (1987) reported that cycloxydim, sethoxydim and clethodim inhibited fatty acid biosynthesis in a chloroplast enzyme preparation from barley. ACCase was found to be the site of action for acid forms of diclofop, fenoxaprop, fluazifop and haloxyfop (Koblek *et al.*, 1988), (Fig. 7). A direct effect on membrane function has also been proposed as a site of action of these herbicides. It has been suggested that diclofop acts as a proton ionophore or that it binds to a membrane bound receptor (Shimabukuro and Hoffer, 1992). However, the action at the site does not appear to kill plants at herbicidal doses (Dekker and Duke, 1995). The membrane effects appear to be antagonism of auxin stimulated proton flux at the receptor level (Barnewell and Cobb, 1993). Auxin antagonism of these compounds has been known since they were first discovered (Duke and Kenyon, 1988).

Injury symptoms

Symptoms develop slowly on plants and may not appear for 7 to 14 days, which include; discoloration and disintegration of meristematic tissue at and above the nodes, including nodes of rhizomes, leaves turn yellow, reddish and sometimes wilt. Seedling grasses tend to lodge by breaking over at the soil. Injury is seen on grass plants only. Newer leaf tissue will be yellow (chlorotic) or brown (necrotic) and the leaves in the leaf whorl can be easily separated from the rest of the plant. Reddish-blue pigmentation may also be observed on the stem sheath, leaf margins, and/or leaf blade.

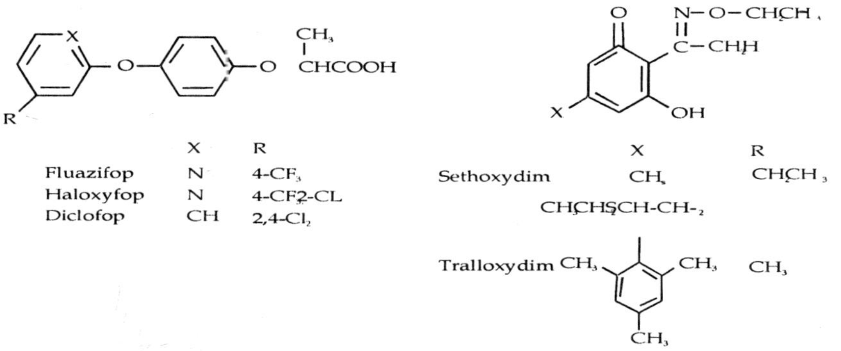

Fig. 7. Aryloxyphenoxypropionates and cyclohexanedione oximes

There are two families of herbicides that disrupt lipid biosynthesis.

Cyclohexanediones: clethodim, sethoxydim, tralkoxydim.

Aryloxyphenoxypropionates: fenoxaprop, fluazifop, diclofop, quizalofop.

The conversion of fatty acids into very long chain fatty acids is specially inhibited by the thiocarbamate herbicides such as EPTC and triallate (Fig. 8). These compounds are applied as pre-plant incorporated for the control of grass and some small seeded broad-leaved weeds in crops such as maize and small grain cereals.

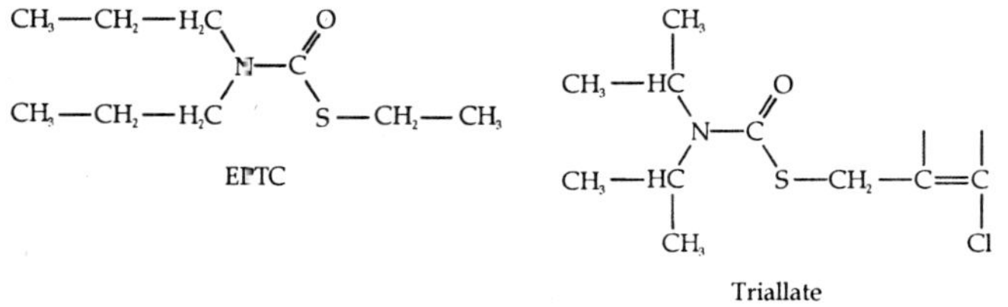

Fig. 8. Thiocarbamate herbicides

4. Pigment inhibitors

Herbicides classified as pigment inhibitors destroy the green pigment (chlorophyll) in leaf tissue. These herbicides are often described as "bleaching herbicides" because they cause new leaves to appear yellow or white. These herbicides are absorbed by roots and translocate to the shoot tissue where they inhibit the production of carotenoids substances, which protect the chlorophyll molecules. Without carotenoids, chlorophyll is destroyed. These herbicides do not destroy carotenoids already formed, but prevent the formation of new ones. Norflurazon initially causes bleaching within veins, while clomazone initially cause bleaching between veins. Clomazone is taken up by plant roots and shoots and can move in the xylem to plant leaves. The newly developed foliage of many plant species is so sensitive to clomazone that very small amounts can whiten new plant growth. Norflurazon is taken up by plant roots and moves to the growing points of susceptible plants. Susceptible weeds will emerge as white plants before dying.

Carotenoids and chlorophyll are plant pigments located in the chloroplasts of leaf cells. Both carotenoids and chlorophyll absorb light during photosynthesis. These herbicides do not directly inhibit green pigment biosynthesis. The loss of chlorophyll is the result of destruction of chlorophyll by photo-oxidation (damage from excessive light) and also due to the absence of carotenoids indirectly disrupting normal chlorophyll synthesis and formation of pigment-protein complexes required for assembly of PS II in the thylakoids of chloroplasts. Carotenoid biosynthesis inhibitors also disrupt the normal development sequence of chloroplasts. When developing shoot tissues (ex. cotyledons) are grown in the dark in the presence of carotenoids biosynthesis inhibitors, the number of thylakoids particularly the granna thylakoids is significantly reduced (Wrischer *et al.,* 1998). The pigment protein complexes required for assembly of photosystem II (PS II) in the thylakoids of chloroplast are also reported to be inhibited (Moskallenko and Karapetyan, 1996).

After chlorophyll is synthesized and is fully functional, some of the chlorophyll, which has been electronically excited by absorbing light photons, is transformed from the singlet state to the longer lived, but more reactive, triplet form. Carotenoids are able to quench the excitation energy of triplet chlorophyll, which is formed in cells under high light intensity. When carotenoids are not present, these triplet chlorophyll states initiate degrading reactions, such as chlorophyll and membrane destruction. Thus, without carotenoids, plant cells cannot survive under high light intensity. Herbicides that prevent carotenoid biosynthesis act at following sites:

Phytoene desaturase (Desaturation pathway)

This is the area where the most research has been carried out. Isopentyl pyrophosphate (IPP) is an important carotenoid and is a component of chlorophyll and plastoquinone. After geranylgernyl pyrophosphate (GGPP) is formed, two GGPP molecules combine to form the 40 carbon intermediate phytoene, which undergoes a series of desaturation reactions (hydrogen) then cyclisation to form a alpha caraotene and eventually lutein, beta carotene and zeaxanthin (Fig. 9, 10). The site of action of norflurazon and floridone is the inhibition of phytoene desaturase so that phytofluene is not produced. Inhibition cause a large accumulation of the enzyme inhibited i.e. phytoene desaturase, which is responsible for desaturating the carbon backbone of phytoene, which in turn forms two additional double bonds along the desaturation pathway. With an interruption along this pathway, functional carotenoids are no longer produced. Inhibition of this enzyme causes a buildup of phytoene and phytofluene.

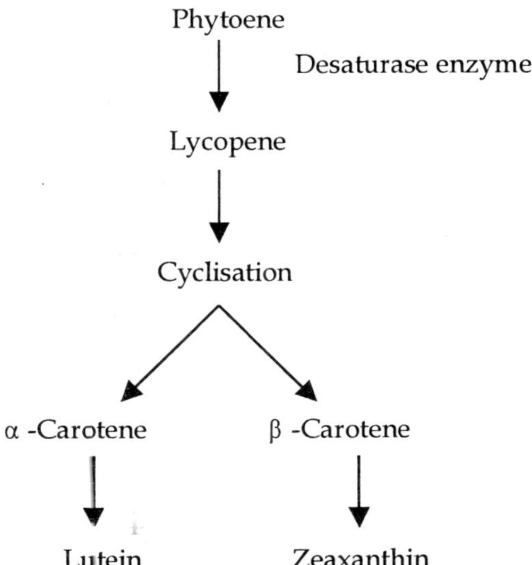

Fig. 9. Part of biosynthesis pathway to carotenoids (Eastin, 1971).

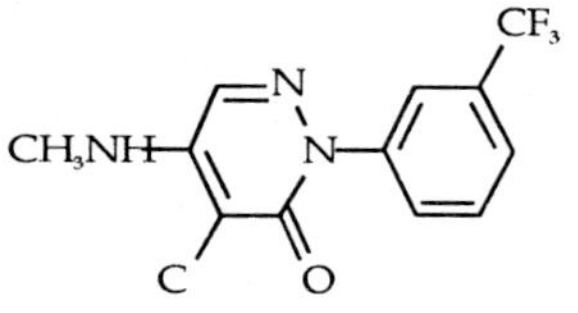

Norfluazon

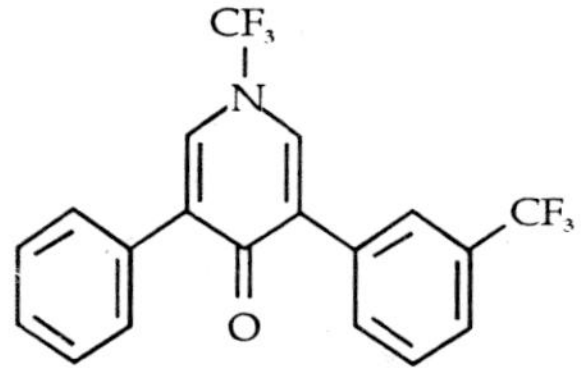

Fluridone

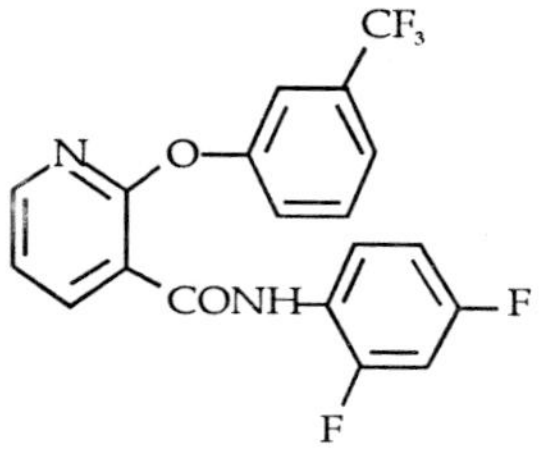

Dituferican

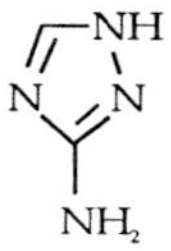

Amitrote

Fig. 10. Herbicides that interfere with carotenoid biosynthesis

Zeta carotene desaturase

This carotene desaturase enzyme, which forms an additional two double bonds, is located along the desaturation pathway between zeta carotene and lycopene, thus, an inhibition of carotene synthesis is realized if inhibited by a herbicide (Börge 1993; Börge and Santteman 1996).

IPP Isomerase (Mevalonic acid to phytoene isoprenoid pathway)

It is responsible for adding carbon in between three reactions within the isoprenoid pathway. Without the addition of these carbons, subsequent reactions cannot take place thus phytoene cannot be made. The herbicide clomozone is thought to act somewhere in this pathway. Vivid white new growth, sometimes tinged with pink or purple, characterizes the symptoms associated with the pigment inhibitors. New growth initially appears normal except for the conspicuous lack of green and yellow pigments. Amitrole is the only compound of this group, which moves well in the symplast, however other compounds in the group show initial movement into shoot tips causing new growth to be devoid of green and yellow pigments.

4-hydroxyphenylpyruvate dioxygenase (4-HPPD inhibitors)

An additional new site of action for carotenoid synthesis inhibitors was discovered in 1983. Syclotrione and isoxaflutole was shown to inhibit the enzyme 4-hydroxyphenylpyruvate dioxygenase (HPPD) that converts 4-hydroxyphenylpyruvate to homogenatisate 4-hydroxy-phenylpyruvate (Shulz *et al.*, 1993; Pallett *et al.*, 1997). The effect of this enzyme inhibition results in a depletion of plastoquiniones, which are needed for proper functioning of the phytoene desaturase enzyme. Clomazone (site of action currently unknown) does not interfere with this enzyme. The new herbicide isoxaflutole is a pro-herbicide that rapidly undergoes ring opening at the isoxazole ring to form at the diketonitrile derivative (Pallett *et al.*, 1997, 1998); this diketonitrile is thought to be active for herbicides as HPPD, is strongly inhibited by this derivative. There is no instance of resistance development in any weed following repeated use of the carotenoid inhibitors herbicides except for amitrole.

Injury symptoms

Pigment inhibitors prevent plants from forming photosynthetic pigments. As a result, the affected plant parts become white to translucent. Plants (all or portions of the true leaves) turn white, often becoming translucent at the leaf tips. Plant death occurs slowly due to the eventual depletion of stored food reserves and the inability of the plant to manufacture new sugars. Those herbicides that inhibit carotenoid biosynthesis do not affect pre-existing carotenoids. Thus, plant tissues formed before treatments do not show typical albino symptoms, only new growth in treated plants is white. There are two classes of pigment inhibitors that bleach plant tissue.

Diterpene inhibitors: isoxazolidinone, amitrole, clomazone.

4-hydroxyphenylpyruvate dioxygenase (4-HPPD inhibitors): isoxazole, isoxaflutole, mesotrione, sulcotrione.

Pyridazinones: norflurazon

5. Photosynthesis inhibitors

About half of the commercially important herbicides act by interrupting photosynthetic electron flow. The precise sites of action of many of these agents have been found to lie either at the reducing side of Photosystem I or in the quinone acceptor complex in the electron transport chain between the two photosystems. Photosynthesis inhibitors are broadleaf herbicides, but also control annual grasses to some extent. Photosynthesis inhibitors shut down the photosynthetic (food producing) process in susceptible plants by binding to specific sites within the chloroplast. Inhibition of photosynthesis could result in a slow starvation of the plant; however, in many situations rapid death occurs perhaps from the production of secondary toxic substances. Herbicides that directly inhibit photosynthesis interfere with or block electron transport and prevent ATP and $NADPH_2$ production. This leads to decreased sugar or food formation. However, the visual injury symptoms (chlorosis, desiccation or browning of plant tissue) occur too rapidly to be the result of starvation of the plant. Instead, chlorosis of leaf tissue is due to the photo-destruction (damage from excessive light) of chlorophyll and other plant pigments. When herbicides block electron transport, chlorophyll continues to absorb light energy but cannot pass this energy on to make ATP and $NADPH_2$. Thus, chlorophyll either self-destructs from the energy it is absorbing or passes this absorbed energy on to oxygen. This forms radical oxygen, which is highly destructive to cell membranes and other cell structures. Cell-membrane destruction causes leakage of the cellular contents and results in the desiccation of plant tissue. When chlorophyll accepts light energy it changes from a ground energy state to a singlet energy state. This singlet energy is normally transferred to the P_{680} reaction center and the chlorophyll molecule returns to the ground state. When electron transfer is blocked due to the herbicide binding to the Q_B site, the singlet chlorophyll energy cannot be converted back to its ground state. The singlet energy state then can be transformed to a more reactive triplet energy state. Some triplet state chlorophyll is produced during normal photosynthesis and is dissipated by carotenoids (Fig. 11). However, because of the large buildup caused by electron flow inhibition in PS II, the carotenoid quenching system gets overloaded.

The pigments, proteins and other chemical entities involved in photosynthesis reactions are located within the chloroplasts of plant cells. The light reaction of photosynthesis occurs within two reaction centers, Photosystem I and Photosystem II. PS II is composed of at least a dozen proteins and is the site of action for photosynthesis inhibitors. In this system, light captured by chlorophyll and carotenoids is transferred to a special reaction center, P_{680}, creating an excited state

Table 1. Herbicides which interrupt electron flow in non-cyclic photophosphorylation (Photo system II)

Herbicide class	*Example*	*Date of introduction*			
1,3,5-Triazine	Atrazine	1958	$R^1 = CH_2CH_3$	$R^2 = CH\ (CH_3)_2$	
	Simazine	1956	$R^1 = CH_2CH_3$	$R^2 = CH_2CH_3$	
	Cyanazine	1970	$R^1 = CH_2CH_3$	R^2=-2-methylpropio mtrile	
	Terbethylazine and many more	1967	$R^1 = CH_2CH_3$	$R2 = C(CH_3)_3$	
	Diuron Linuran Chlorotoluron			Diuron Linuran Chlorotoluron	
			R^1	R^2	X
Phenylurea	Diuron	1959	CH_3	CH_3	3,4-Cl_2-Phenyl
	Linuran	1962	CH_3	OCH_3	3,4-Cl_2-Phenyl
	Chlorotoluron	1971	CH_3	CH_3	3-Cl,4-CH_3-Phenyl
	Fluoenturon				
	Isoproturon	1965	CH_3	CH_3	3-CF3-Phenyl
	Methabenzthiazuron	1975	CH_3	CH_3	4-$CH(CH_3)_2$-Phenyl
	and many more	1968	CH_3	H	Henzothinzole
Acylanilide	Propanil	1961			
			X		
Hydroxy-benzonitrile	Ioxynil	1970	I		
	Bromoxynil	1971	Br		
			R		X
Phenyl carbamate	Phenmedipham	1968	CH_3		CH_3
	Desmedipham	1970	CH_2CH_3		H
Uracil	Bromacil	1965			
	Lenacil	1965			
			R^1		R^2
Triazinone	Methribuzin	1972	$C(CH_3)_3$		SCH_3
	Metamitron	1977	Phenyl		CH_3
Miscellaneous	Bentazone	1972			
	Chloridazon	1962			

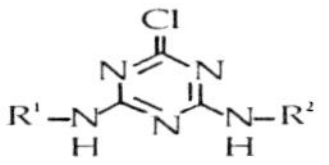

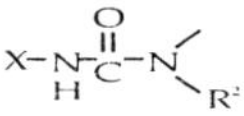

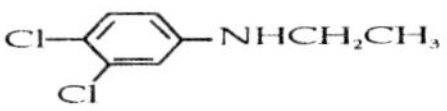

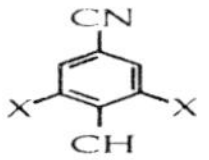

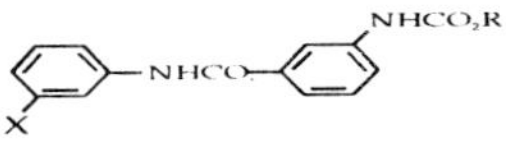

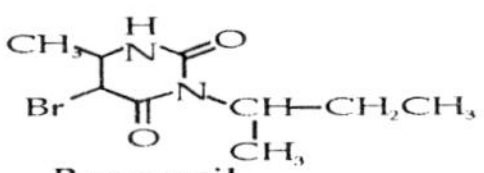

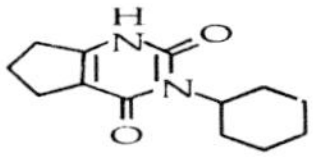

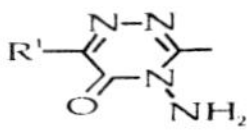

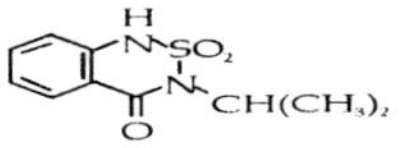

Bromacil

Lenacil

Bentazone

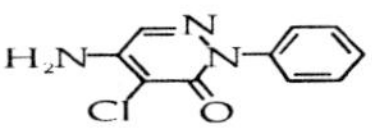

Chloridazon

6. Photosystem I inhibitors (PS I inhibitors)

Unlike photosynthetic inhibitors, these herbicides do not block electron flow, but rather they accept the electron given off in PS I and form radicals. Auto oxidation is a normal defense mechanism to curb the buildup of superoxide radicals (O_2°). Auto oxidation incorporates water and oxygen to turn superoxide into H_2O_2 (hydrogen peroxide). This process is termed as enzymatic dismutation. H_2O_2 is not harmful to plants due to the fact that it can be broken down by enzymes. However, the mass loading of superoxide overcomes the amount of enzymes present. Due to this build up, H_2O_2 and O_2- can further react to form unbearable amounts of a hydroxyl radical (OH) (Fig. 12). This reaction is catalyzed by transition metals (Fe or Cu). Hydroxyl radicals are the most potent biological oxidants known which quickly and effectively initiate lipid peroxidation. This allows cellular components to leak into the intercellular space and changes the refractive index of the cell. Within a few hours after treatment with these products, plant foliage becomes severely damaged. The first visual damage is the appearance of many dark green spots on the leaf (termed water soaking) followed by wilting and necrosis of the tissues. Paraquat and diquat are both in the form of dications, therefore, cannot readily diffuse across the lipid bilayer of the plasma membrane. Thus the movement into cells requires the polyamine carrier system. Once inside the plant, they move to the photosynthesis reaction centers in the chloroplasts, which is their site of action.

Injury symptoms

Plant leaves will have a limp, water-soaked appearance, which is followed by browning (necrosis) of the plant tissue. Drift injury will appear as speckling on leaf tissue. They are strong cations and show rapid absorption and adsorption by the leaf decreases rain sensitivity and their action is much more rapid in the light than in the dark and show very little translocation because of rapid death of cell.

Herbicides in this category are:

Bipyridyliums: paraquat, diquat.

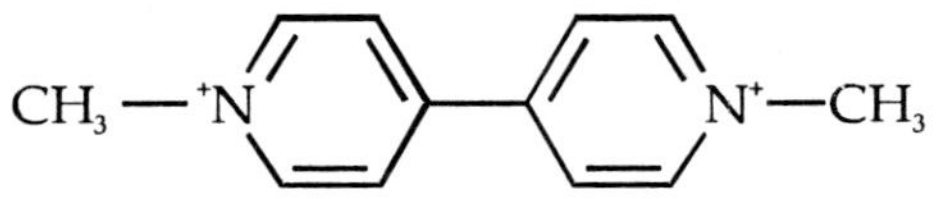

Paraquat

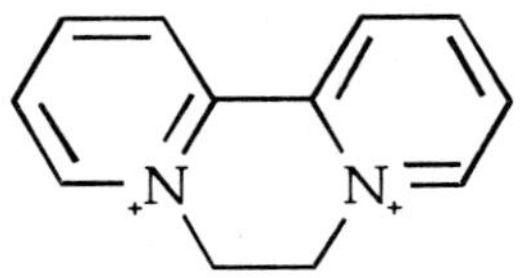

Diquat

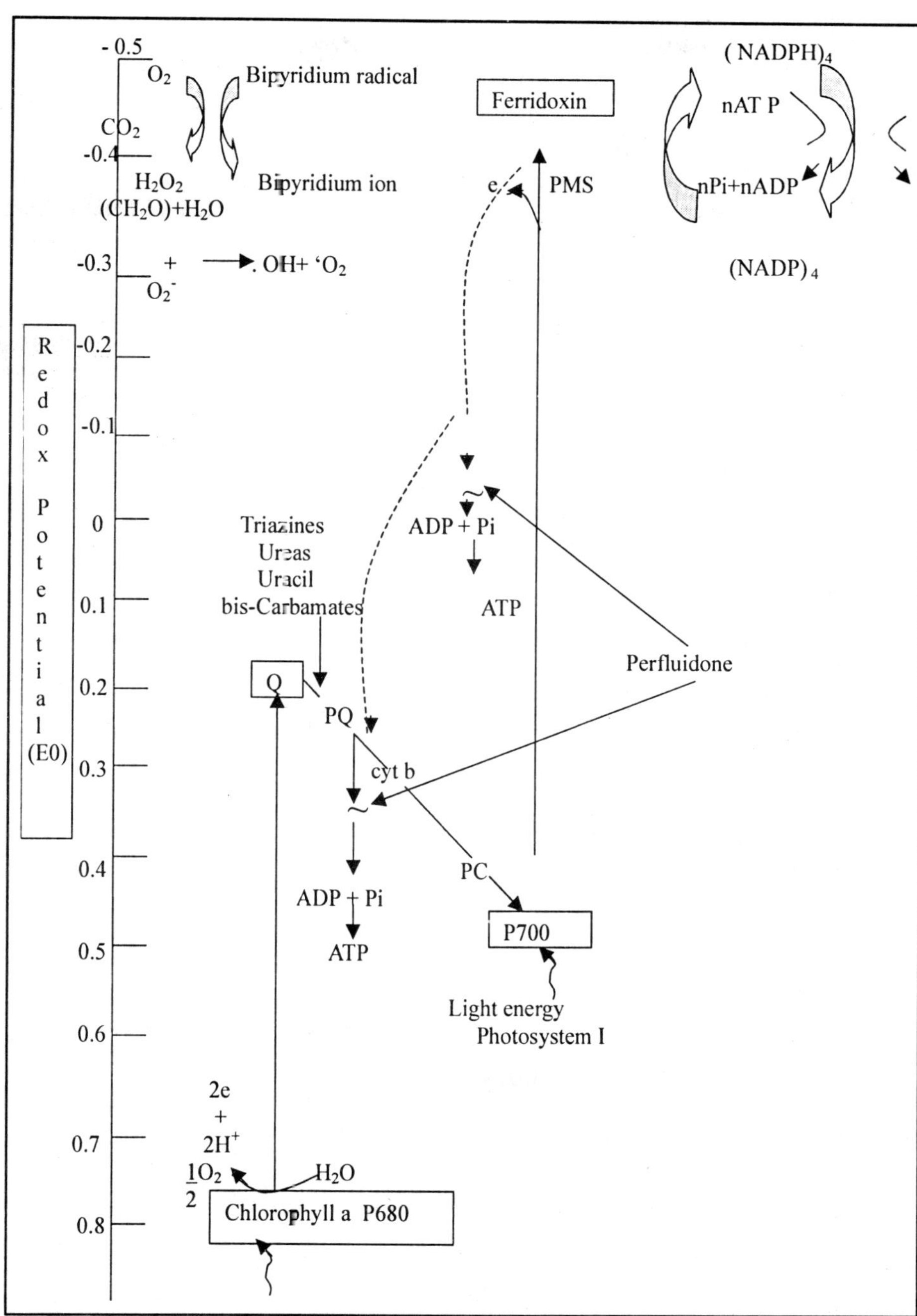

Fig. 12. Schematic presentation of electron transport and phosphorylation reactions considered to occur in chloroplast lamella and site of action of several herbicides (Roger, 1971).

There are a number of other herbicides that affect photosynthesis indirectly. Pyrazole herbicides such as benzofenap, pyrazolynate and pyrazoxyfen interfere with chlorophyll biosynthesis and have found commercial application for the control of annual and perennial weeds in paddy rice and maize (Fig. 13).

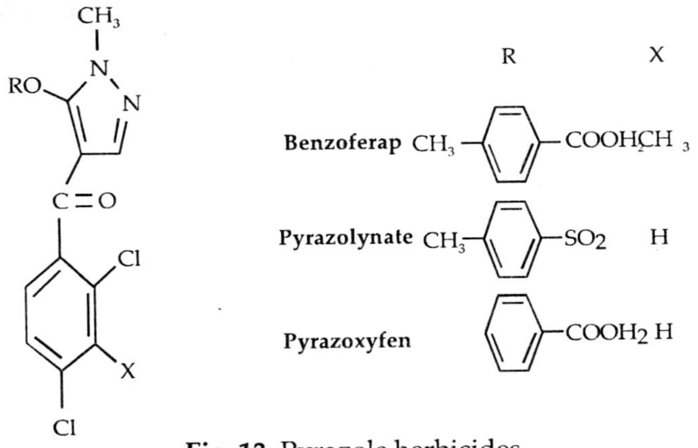

Fig. 13. Pyrazole herbicides

7. Seedling growth inhibitors

Seedling growth inhibitors interfere with new growth, thereby reducing the ability of seedlings to develop normally. Plants can take up these herbicides after germinating, until the seedling emerges from the soil. Plants that have emerged from the soil uninjured are likely to remain unaffected. Seedling growth inhibitors are active at two main sites, i.e. the developing shoot and the root. Much more is known about the action of seedling root inhibiting herbicides than seedling shoot inhibitor herbicides. The root inhibitors stop cells from dividing, which inhibits shoot elongation and lateral root formation. Uptake is through developing roots and shoots. Because herbicide movement within the plant is limited, herbicide injury is confined primarily to plant roots and shoots. Shoot inhibiting herbicides are taken up by developing roots and shoots and can move *via* the xylem to areas of new growth. Present evidence suggests that these herbicides can affect multiple sites within a plant, primarily interfering with lipid and protein synthesis.

Many pre-emergence herbicides inhibit growth shortly after seed germination, thus preventing emergence of the weed above the soil surface. Growth of plant roots and shoots is a combination of cell division and cell enlargement that results in an irreversible increase in size. An inhibition of either or both of these processes will disrupt growth. These herbicides do not inhibit the onset of mitosis, but rather disrupt the mitotic sequence once initiated. Cell division during seedling growth takes place primarily in the meristems located at the root and shoot tips. Within these areas, specialized cells go through a continuous cycle of events, which includes mitosis (cell division) and cell elongation. Processes such as RNA synthesis, amino acid synthesis, protein synthesis, lipid synthesis and ATP synthesis are fully functional and required for the cell cycle to progress. Mitosis is a sequence of events that forms two cells, each containing one complement of chromosomes, from a single cell. Which has four different steps *viz.* prophase, metaphase, anaphase and telophase (Fig. 14). During prophase the nuclear envelope breaks down and the chromosomes are more or less randomly distributed around the cell. During metaphase, the chromosomes become aligned in the middle of the cell, during anaphase, one set of chromosomes is moved to each end of the cell and during telophase a new cell wall is formed between the two sets of chromosomes and the nuclear envelope reforms around each chromosome set. The spindle apparatus, which is composed of protein structures called microtubules, is the framework responsible for moving chromosomes during each stage of mitosis. The structural protein in the microtubules is tubulin. During mitosis, these microtubules rapidly depolymerize to tubulin and repolymerize to microtubules several times in order to accomplish the needed chromosome movement. This assembly-disassembly equilibrium is termed dynamic instability.

Herbicides known to interfere with chromosome movement do so by interfering with the spindle apparatus. These products work by slowing or preventing the assembly portion of the equilibrium in microtubule formation by herbicide-tubulin complexes binding to the assembly site of tubulin protein. This inhibits further polymerization without stopping disassembly, therefore, causing a disappearance of the spindle apparatus. The prophase sequence appears normal, however, without the presence of microtubules, the chromosomes are unable to move to the metaphase configuration, the daughter chromosomes cannot migrate to their respective poles (anaphase), and cell wall formation does not occur (telophase). After a time in the prophase state, the chromosomes coalesce in the middle of the cell and the nuclear envelope reforms, causing a polyploid nucleus. Without the production of new cells, growth will eventually stop. Dinitroaniline cause characteristic root tip swelling. This swelling is due to abnormal cell enlargement in the zone of elongation. In this zone, cells of treated roots enlarge to a spherical shape rather than a normal cylindrical shape. This is caused by the loss of microtubules. Microtubules are responsible for the orientation of the cellulose microfibrils that provide support to the cell wall. Without microtubules, the microfibril orientation is random and the cell wall has no preferential support in any direction (Holmsen and Hess, 1984). Dithiopyr and thiazopyr inhibit the root growth and the tip becomes abnormally enlarged, cell division is disrupted at metaphase because of the absence of microtubule. In protein binding studies dithiopyr was found to bind to protein other than tubulin (Armbruster *et al.*, 1991).

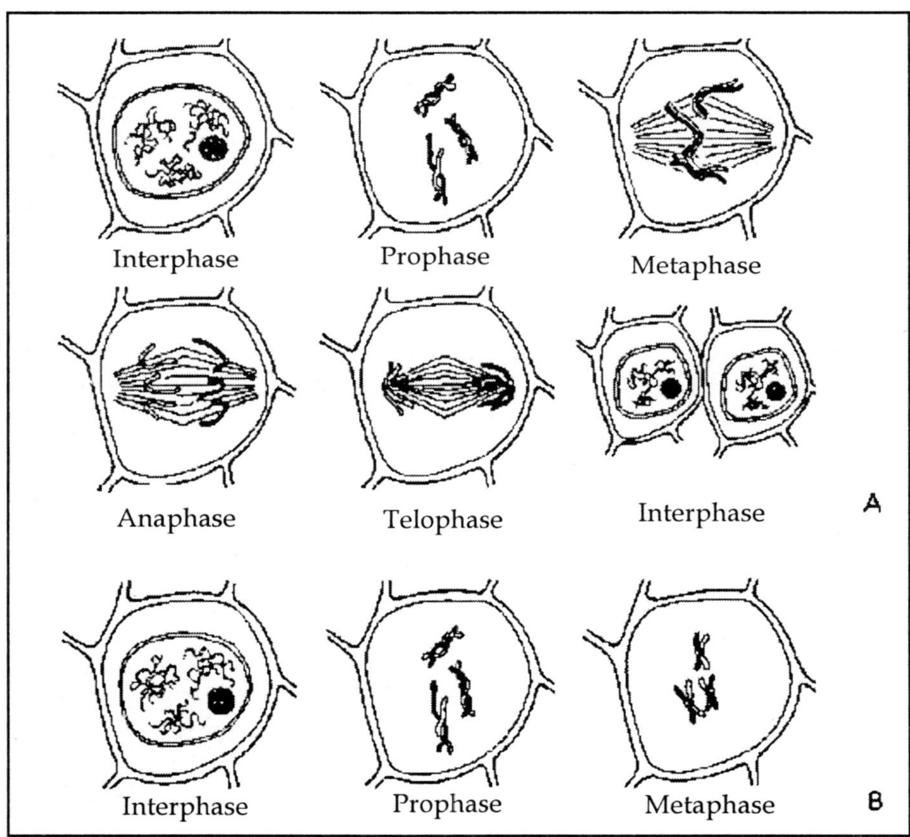

Fig. 14. (a). Diagram of normal mitotic sequences. (b). Bottom row shows the effect of dinitroanilines herbicides on mitotic sequence in root cell (Ross and Lembi, 1985)

Injury symptoms

Injury to tolerant plants is caused by root damage. Grass crops may be stunted and have a purple discoloration because roots cannot take up enough phosphorus. Root systems appear stubby and thick, especially the lateral roots. Broadleaf plants may have swollen and cracked hypocotyls. If these herbicides are incorporated shallowly or applied pre-emergence they sometimes cause callus tissue (tumors) to form on the plant stem at the soil surface. This weakens the stem and causes lodging.

Symptoms caused by the shoot inhibitors are much different than those of the root inhibitors. Over application or extended periods of cool, wet weather shortly after planting may sometimes cause injury to tolerant crops such as corn or sorghum. Symptoms include leafing out underground, improper leaf unfurling (shepherd's crooking), buggy whipping (tightly rolled leaves) and leaf crinkling, malformed (twisted) dark green shoots and leaves on injured young plants. In broadleaf plants, the center vein (mid-rib) may draw in the leaf edge in a drawstring effect. Leaf puckering is also a symptom on broadleaf plants. There may also be stunting that persists until the soil warms enough to promote plant growth. Thiocarbamates reduce the formation of epicuticular wax on leaves, which can cause leaves to stick together rather than unfold normally. Affected plants may be stunted, leaves may be shortened and thickened, or true leaf development may be inhibited. Some severely stunted plants may die while others will start producing new leaves and will produce a nearly normal-sized roots at harvest. Severely stunted plants may grow very little for two or more weeks after emergence and then make a complete or nearly complete recovery.

Root inhibitors: Dinitroanilines (Dinitrobenzenamines): pendimethalin, prodiamine, oryzalin, trifluralin, benefin, ethalfluralin.

Shoot inhibitors: Thiocarbamates (Carbamothioates): EPTC, butylate, pebulate, cycloate.

Substituted amides (chloroacetamides and acetanilides) : acetochlor, alachlor, butachlor, metolachlor, pronamide, propachlor, dimethenamid, napropamide, propachlor, dimethenamid.

8. Cell membrane disruptors and organic arsenicals

The cell membrane disruptor post-emergence herbicides control both grasses and broadleaf weeds by destroying cell membranes and causing rapid desiccation of the plant. There are two types of cell membrane disruptor herbicides: the bipyridiliums and the diphenylethers. These herbicides are post-emergence contact herbicides that are activated by exposure to sunlight to form oxygen compounds such as hydrogen peroxide. These oxygen compounds destroy plant tissue by rupturing plant cell membranes. Destruction of cell membranes results in a rapid browning (necrosis) of plant tissue. On a bright and sunny day, herbicide injury symptoms can occur in 1 to 2 hours. Because these are contact herbicides, they are excellent for burn down of existing foliage and post-emergence control of annual weeds. Perennial weeds usually re-grow because there is no herbicide movement to underground root or shoot systems.

The bipyridilium herbicides require thorough plant coverage to be effective, and they have no soil activity. The diphenylether herbicides act in a similar way but more slowly. Some of them are more selective between crops and weeds. The herbicides oxyfluorfen and formesafen have significant soil activity. Compounds in this group result in rapid disruption of cell membranes and very rapid kill. The bipyridyliums and the diphenylethers penetrate into the cytoplasm, cause the formation of peroxides and free electrons (light is required), which destroy the cell membranes almost immediately. Herbicidal oils dissolve membranes directly. Rapid destruction of cell membranes prevent translocation to other regions of the plant. There are two classes of cell membrane disruptors:

Protex inhibitors (Protoporphyrinogen oxidase inhibitors)

Diphenyl ether activity is expressed as foliar necrosis after 4-6 hours of sunlight following post-emergence application. Diphenyl ether activity does not involve photosynthesis, however, the substance involved in the light requirement for herbicidal activity is some sort of light absorbing plant pigment other than chlorophyll itself. Membrane damage induced by diphenyl ether herbicides is the result of lipid peroxidation. This is due to the accumulation of protoporphyrin IX. Protoporphyrin IX is a precursor to chlorophyll biosynthesis. Protoporphyrinogen oxidase (Protox) is a protein located in the chloroplast, where it is involved in chlorophyll and heme synthesis and in mitochondria where it is involved in nonplastic heme synthesis. Oxygen and light are known to interact with protoporphyrin IX to produce singlet oxygen, which is known to be an efficient initiating factor of lipid peroxidation. This accumulation is due to the inhibition of Protox. This enzyme converts protoporphyrinogen IX to protoporphyrin IX. The very rapid accumulation of protoporphyrin IX in herbicide treated plants is due to deregulation of the entire pathway resulting from a decrease in heme levels (Lydon and Duke 1989; Matringe *et al.*, 1989; Witkowski and Halling 1989). In the absence of a herbicide, the heme levels would keep the entire pathway in a balanced production of products. Heme is an iron containing porphyrin, which regulates chlorophyll formation. However, in the presence of a herbicide heme levels decrease and are a signal for increased pathway activity and result in high levels of proporphyrinogen IX formation in the chloroplasts. This protoporphyrinogen IX then leaks into the cytoplasm where it is oxidized to protoporphyrin IX, which forms free radicals that lead to membrane disruption (Lehnen *et al.*, 1988).

Injury symptoms

The first symptoms are a water soaked appearance (dark green spots on the foliage) followed by the water soaked areas. Water soaking is indicative of membrane damage and cellular substances leaking into intercellular spaces, which changes the refractive index of the tissue. With pre-emergence applications, tissue is damaged by contact with the herbicide as the plant emerges above the soil surface. As with post-emergence applications, the damage is tissue necrosis. Following absorption and movement to the site of action in chloroplasts, light is an obligate requirement for herbicidal activity. Severe injury is evident hours after application, first as water-soaked areas, which later turn yellow or brown. Maximum kill is attained in a week or less. Partial coverage of a plant with spray results in spotting and/or partial shoot kill. New growth on surviving plants will be normal in appearance. Foliar activity alone can provide only shoot kill. Plant leaves become yellow and then turn brown and die. Reddish-colored spotting on the leaf surface may appear shortly after the herbicide is applied. Plants that do not die may be stunted for a week or more. Crop oils and other additives, as well as extremely cool or warm temperatures, may increase plant injury. Roots of perennial weeds are seldom killed because these herbicides do not usually translocated to the roots. Herbicides in this category are:

Diphenylethers (nitrophenyl ethers): acifluorfen, oxyfluorfen, formesafen, lactofen. N-phenyl-phthalimides; flumiclorac.

Triazolinones: carfentrazone, sulfentrazone.

Organic arsenicals

The organic arsenical herbicides disodium methyl arsenate (DSMA) and monosodium methyl arsenate (MSMA) are contact herbicides and are cell membrane disruptors. However, their true mode of action is unknown. These organic arsenical accumulate in root and leaf tips and symptoms are first seen on leaf tips. They rapidly kill leaf and stem tissue. MSMA and DSMA are more effective on grass weeds than broadleaf weeds, except for common ragweed (*Ambrosia artemisiifolia*) and cocklebur. When applied over cotton to control grasses or cocklebur, they sometimes cause speckled leaf burn and red stems on the cotton plants, however, this has little effect on overall growth. They are used to selectively control wide-leaved grasses such as crabgrass (*Digitaria sanguinalis*) or dallis grass (*Paspalum dilatatum*) in narrow-leafed grasses such as Bermuda grass (*Cynodon dactylon*) lawns. These herbicides bind tightly to soil clay and organic matter so they have no residual, pre-emergence activity.

Examples: cacodylic acid, DSMA, MSMA

9. Cellulose biosynthesis inhibitors

The cellulose biosynthesis inhibitors herbicides are a diverse group of chemically unrelated compounds. The common herbicidal effect is either a direct or indirect inhibition of cellulose biosynthesis, this effect leads to a lack of cell structure integrity.

No definitive experimental data exist to describe a single specific site of inhibition for these herbicides in the cellulose biosynthesis pathway. However, recent evidence suggest that these herbicides act at different points in the biosynthesis pathway (Sabba and Vaughn 1999). Therefore there is a lack of cellular integrity, which leads to arrested or abnormal growth resulting in plant death. Dichlobencil is known to have a site in the cellulose synthase enzyme complex and it aparantly acts to inhibit the stunting of glucose molecules necessary for building the cellulose molecules. (Vaughn *et al.*, 1996). These workers show that during telophase in onion root tips, dichlobenil inhibited the stiffening and straightening of the plate stage of cell wall formation, which is associated with the accumulation of cellulose in the newly forming cell wall.

Injury symptoms

The most characteristic symptoms of cellulose biosynthesis inhibiting herbicide are a swelling of the apical regions of the germinating seedlings. Histological analyses of root tip meristems show a progressive disappearance of the meristematic zone. Cells in the meristematic zone become enlarged and sometimes are almost devoid of cytoplasm (Lefebvre *et al.*, 1987).

Examples: Quinclorac, dichlobenil, isoxaben, dyclomec.

10. Unknown mode of action

Herbicides in the amide family (also referred to as acetanilides or acetamides) inhibit root and shoot growth causing stunted, malformed seedlings. The specific site of action and mode of action of this herbicide family is unknown. Normal cell division, cell elongation, and protein synthesis are potentially inhibited. The herbicides must be present in early stages of germination and growth of weeds for effective control. These herbicides are most effective on annual grass weeds, although some small-seeded annual broadleaf weeds are also sensitive. Injury symptoms to corn from these herbicides include leafing out underground and failure of leaves to properly unfurl. Soybean injury from these herbicides occurs in the form of a shortened mid-vein in the leaflets resulting in crinkling and a heart shaped appearance. Dimethenamid and flufenacet have slightly different chemical structures than the amide herbicides, but it is believed they kill plants in the same manner as the amides.

Examples: Flufenacet, DCPA, siduron.

The herbicide dichlobenil (Fig. 15) is believed to exert its effect through the inhibition of cellulose biosynthesis of actively growing plant tissue, leading to a cessation of cell division and death.

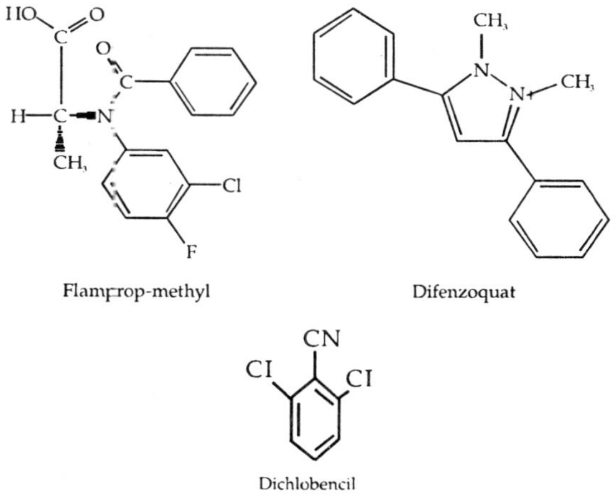

Flamprop-methyl

Difenzoquat

Dichlobencil

Fig. 15. N-arylalanine ester herbicides and difenzoquat.

There are a number of herbicides whose mode of action is not known or that is uncertain. These include compounds such as the organo-phosphorous herbicides anilofos, bensulide, butamifos, fosamine and piperophos (Fig. 16, 17) and the benzofuranyl alkanesulfonates, ethofumesate and benfuresate.

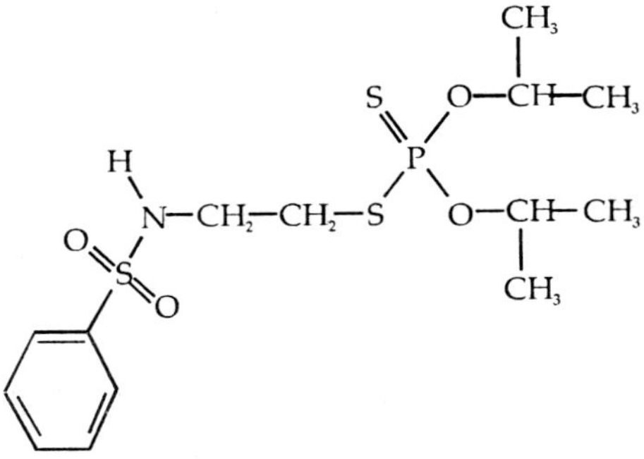

Bensulide

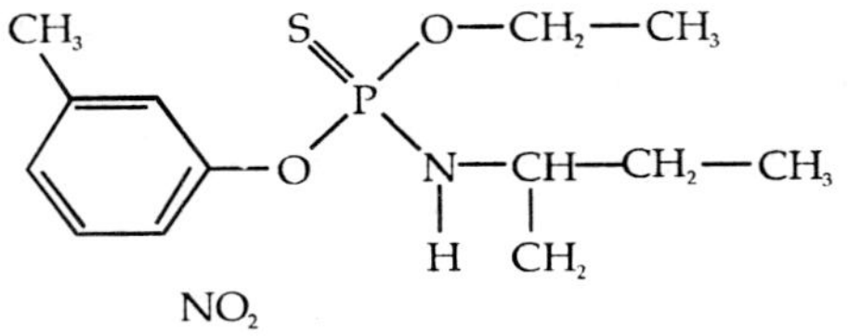

Butaimifos

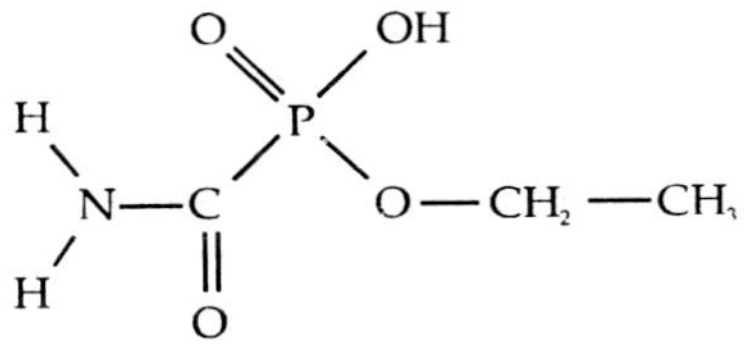

Fosamine

Fig. 16. Organophophorus herbicides

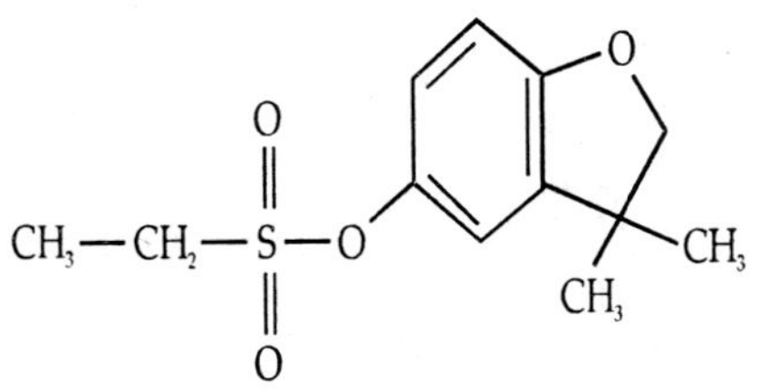

Benfuresate

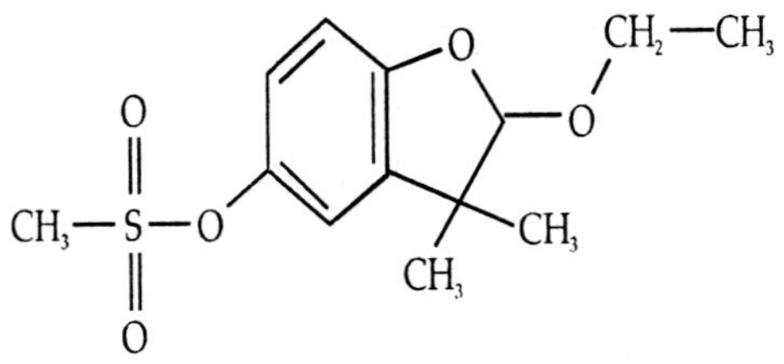

Ethofumesate

Fig. 17. Benzofuranyl alkane sulfonate herbicides

Metolachlor, inhibits the synthesis of fatty acids and lipids, proteins and gibberellins. It also inhibits both shoot or root growth of susceptible weeds.

Table 2. Mode of action of herbicides and chemical families. (Retzinger and Smith, 1997).

HRAC Group	Mode of Action	Chemical Family	Active Ingredient	WSSA Group
A	Inhibition of acetyl CoA carboxylase (ACCase)	Aryloxyphenoxy-propionates ('FOPs')	clodinafop-propargyl, cyhalofop-butyl, diclofop-methyl, fenoxaprop-p-ethyl, fluazifop-P-butyl, haloxyfop-R-methyl, propaquizafop, quizalofop-p-ethyl, diclofop	1
		Cyclohexanediones ('DIMs')	alloxydim, butroxydim (clefoxydim proposed), clethodim, cycloxydim, sethoxydim, tepraloxydin, tralkoxydim	
B	Inhibition of acetolactate synthase ALS (acetohydroxyacid synthase AHAS)	Sulfonylureas	Amidosulfuron, azimsulfuron bensulfuron-methyl, chlorimuron-ethyl, chlorsulfuron, cinosulfuron, cyclosulfamuron, ethametsulfuron-methyl ethoxysulfuron, flazasulfuron flupyrsulfuron-methyl-Na, foramsulfuron, halosulfuron-methyl, imazosulfuron, iodosulfuron, mesosulfuron, metsulfuron-methyl, nicosulfuron, oxasulfuron primisulfuron-methyl, prosulfuron, pyrazosulfuron-ethyl, rimsulfuron, sulfometuron-methyl, sulfosulfuron, thifensulfuron-methyl, triasulfuron, tribenuron-methyl, trifloxysulfuron triflusulfuron-methyl, tritosulfuron	5
		Imidazolinone	Imazapic, imazamethabenz-methyl imazamox, imazapyr, imazaquin, imazethapyr	
		Triazolopyrimidine	cloransulam-methyl, diclosulam, florasulam, flumetsulam, metosulam, penoxsulam	
		Pyrimidinyl (thio) benzoate	Bispyribac-Na, pyribenzoxim pyriftalid, pyrithiobac-Na, pyriminobac-methyl	
		Sulfonylaminocarbonyl-triazolinone	Flucarbazone-Na, propoxycarbazone-Na	
		Pyridazinones	Chloridazon	
		Phenyl-carbamates	Desmedipham, phenmedipham	

Contd...

C1	Inhibition of photosynthesis at photosystem II	Triazine	Ametryn, atrazine, cyanazine, desmetryne, dimethametryne, prometon, prometryne, propazine, simazine, simetryne, terbumeton terbuthylazine, terbutryne, trietazine	5
C2	Inhibition of photosynthesis at photosystem II	Ureas	Chlorobromuron, chlorotoluron chloroxuron, dimefuron, diuron, ethidimuron, fenuron, fluometuron (see F3), isoproturon, isouron, linuron, methabenzthiazuron, metobromuron, metoxuron, monolinuron, neburon, siduron, tebuthiuron	7
		Amides	Propanil, pentanochlor	
C3	Inhibition of photosynthesis at photosystem II	Nitriles	Bromofenoxim bromoxynil ioxynil (also group M)	6
		Benzothiadiazinone	Bentazon	
		Phenyl-pyridazines	Pyridate, pyridafol	
D	Photosystem-I-electron diversion	Bipyridyliums	Diquat, paraquat	22
E	Inhibition of protoporphyrinogen oxidase (PPO)	Diphenylethers	Acifluorfen-Na, bifenox, chlomethoxyfen, fluoroglycofen-ethyl, fomesafen, halosafen, lactofen, oxyfluorfen	14
		Phenylpyrazoles	Fluazolate, pyraflufen-ethyl	
		N-phenylphthalimides	Cinidon-ethyl, flumioxazin flumiclorac-pentyl	
		Thiadiazoles	Fluthiacet-methyl, thidiazimin	
		Oxadiazoles	Oxadiazon, oxadiargyl	
		Triazolinones	Azafenidin, carfentrazone-ethyl sulfentrazone	
		Oxazolidinediones	Pentoxazone	
		Pyrimidindiones	Benzfendizone, butafenacil	
		Others	Pyrazogyl, profluazol	
F1	Bleaching: Inhibition of carotenoid biosynthesis at the phytoene desaturase step (PDS)	Pyridazinones	Norflurazon	12
		Pyridinecarboxamides	Diflufenican, picolinafen	
		Others	Beflubutamid, fluridone, flurochloridone, flurtamone	
F2	Bleaching: Inhibition of 4-hydroxyphenyl-pyruvate-dioxygenase (4-HPPD)	Triketones	Mesotrione, sulcotrione	28
		Isoxazoles	Isoxachlortole, isoxaflutole	
		Pyrazoles	Benzofenap, pyrazolynate, pyrazoxyfen	
		Others	Benzobicyclon	

Contd...

F3	Bleaching: Inhibition of carotenoid biosynthesis (unknown target)	Triazoles	Amitrole (*in vivo* inhibition of lycopene cyclase)	11
		Isoxazolidinones	Clomazone	13
		Ureas	Fluometuron (see C2)	
		Diphenylether	Aclonifen	
G	Inhibition of EPSP synthase	Glycines	Glyphosate, sulfosate	9
H	Inhibition of glutamine synthetase	Phosphinic acids	Glufosinate-ammonium, bilanaphos	10
I	Inhibition of DHP (dihydropteroate) synthase	Carbamates	Asulam	18
K1	Microtubule assembly inhibition	Dinitroanilines	Benfluralin, butralin, dinitramine ethalfluralin, oryzalin, pendimethalin trifluralin	3
		Phosphoroamidates	Amiprophos-methyl, butamiphos	
		Pyridines	Dithiopyr, thiazopyr	
		Benzamides	Propyzamide, tebutam	
		Benzenedicarboxylic acids	DCPA	3
K2	Inhibition of mitosis / microtubule organisation	Carbamates	Chlorpropham, propham carbetamide	23
K3	Inhibition of cell division (Inhibition of VLCFAs; see Remarks)	Chloroacetamides	Acetochlor, alachlor, butachlor	15
			Dimethachlor, dimethanamid metazachlor, Metolachlor, pethoxamid	
			Pretilachlor, propachlor, propisochlor, thenylchlor	
		Acetamides	Diphenamid, napropamide naproanilide	
		Oxyacetamides	Flufenacet, mefenacet	
		Tetrazolinones	Fentrazamide	
		Others	Anilofos, cafenstrole, indanofan piperophos	
L	Inhibition of cell wall (cellulose) synthesis	Nitriles	Dichlobenil, chlorthiamid	20
		Benzamides	Isoxaben	21
		Triazolocarboxamides	Flupoxam	
M	Uncoupling (Membrane disruption)	Dinitrophenols	DNOC, dinoseb, dinoterb	24

Contd...

N	Inhibition of lipid synthesis - not ACCase inhibition	Thiocarbamates	Butylate, cycloate, dimepiperate, EPTC, esprocarb, molinate, orbencarb, pebulate, prosulfocarb, thiobencarb, tiocarbazil, triallate, vernolate	8
		Phosphorodithioates	Bensulide	
		Benzofuranes	Benfuresate, ethofumesate	
		Chloro-Carbonic-acids	TCA, dalapon, flupropanate	26
O	Action like indole acetic acid (synthetic auxins)	Phenoxy-carboxylic-acids	Clomeprop, 2,4-D, 2,4-DB, dichlorprop, 2,4-DP, MCPA, MCPB, mecoprop	4
		Benzoic acids	Chloramben, dicamba, TBA	
		Pyridine carboxylic acids	Clopyralid, fluroxypyr, picloram triclopyr	
		Quinoline carboxylic acids	Quinclorac (also group L), quinmerac	
		Others	Benazolin-ethyl	
P	Inhibition of auxin transport	Phthalamates Semicarbazones	Naptalam, diflufenzopyr-Na	19
	...			
R	...	...	...	
S	...	...	...	
		...	...	
Z	Unknown	Arylaminopropionic acids	Flamprop-M-methyl /-isopropyl	25
		Pyrazolium	Difenzoquat	8
		Organoarsenicals	DSMA, MSMA	17
		Others	Bromobutide, (chloro)-flurenol	27
			Cinmethylin	
			Cumyluron	
			Dazomet	
			Daimuron, methyl-dimuron, etobenzanid, fosamine, metam, oxaziclomefone, oleic acid	
			Pelargonic acid, pyributicarb	

Summary

After a herbicide enters the plant and reaches a specific site in plant cells, it inhibits a biochemical process which depends on the chemistry of the herbicides and in some cases, the plant species involved. Herbicides inhibit biochemical process by different mode of action. Amino acid synthesis inhibitors act on a specific enzyme prevent the production of specific amino acids. There are three classes of amino acid synthesis inhibitors: (a) branched chain amino acid inhibitors inhibit the function of the ALS enzymes, thus amino acids cannot be produced. In general, branched chain amino acid inhibitors are prone to resistance development (b) aromatic amino acid synthesis inhibitors (ESPS inhibitors) inhibit shikimate pathway which is responsible for the formation of the aromatic amino acids *viz.* tryptophan, tyrosine and phenylalanine and can lead to a large number of potentially damaging physiological effects (c) glutamine synthesis inhibitor caused chlorosis followed by necrosis.

Biochemical and metabolic changes in plants reported to be induced by growth regulator herbicides are numerous. Nucleic acid metabolism and metabolic aspects of cell wall plasticity are most relevant to the mode of action of growth regulator herbicides. These herbicides produce profound effects upon the growth and structure of plants. Lipid synthesis inhibitors inhibit Acetyl coenzyme-A carboxylase (ACCase) which is a key enzyme in the lipid biosynthesis pathway. These herbicides act only on annual and perennial grasses, not on broadleaf plants. Pigment inhibitors destroy the green pigment in leaf tissue and often described as "bleaching herbicides". Pigment inhibitors prevent carotenoid biosynthesis act on Phytoene desaturase (Desaturation pathway), Zeta carotene desaturase, IPP Isomerase (Mevalonic acid to phytoene isoprenoid pathway) and 4-hydroxyphenylpyruvate dioxygenase (4-HPPD inhibitors) sites.

Photosynthesis inhibitors shut down the photosynthetic process in susceptible plants by binding to specific sites within the chloroplast. The precise sites of action of photosynthesis inhibitors lie either at the reducing side of photosystem I or in the quinone acceptor complex in the electron transport chain between the two photosystems. inhibition of photosynthesis could result in a slow starvation of the plant; however, in many situations rapid death occurs perhaps from the production of secondary toxic substances. Seedling growth inhibitors interfere with new growth, thereby reducing the ability of seedlings to develop normally. The root inhibitors stop cells from dividing, which inhibit shoot elongation and lateral root formation. The cell membrane disruptor herbicides control both grasses and broadleaf weeds by destroying cell membranes and causing rapid desiccation of the plant. These are activated by exposure to sunlight to form oxygen compounds such as hydrogen peroxide. The organic arsenical herbicides disodium methyl arsenate (DSMA) and monosodium methyl arsenate (MSMA) are contact herbicides and are also cell membrane disruptors. However, their true mode of action is unknown.

The common herbicidal effect of cellulose biosynthesis inhibitors herbicides is either a direct or indirect inhibition of cellulose biosynthesis, which in effect leads to a lack of cell structure integrity. Herbicides in the amide family inhibit root and shoot growth causing stunted, malformed seedlings. The specific site of action and mode of action of this herbicide family is unknown.

❑❑❑

Chapter 5

Metabolism in Plants

Metabolism is one of the most important ways a plant can escape the toxic effects of a herbicide. Herbicide-tolerant plants often have the ability to metabolize or break down the chemical to non-active compounds before it can build up to toxic levels at the site of action. Susceptible plants are unable to detoxify herbicides. Selectivity of many herbicides is based on differing rates of metabolism. Some plants have the ability to render certain herbicides inactive through metabolizing them.

Metabolism of a herbicide may result in inactivation of the herbicide or deactivation where phytotoxicity is reduced. This is the most common result of herbicide metabolism since plants vary in their ability to modify chemical structures, this determines tolerance e.g. atrazine is rapidly degraded by corn but not by weeds. Hence the corn is tolerant and the weeds are susceptible. This makes atrazine a selective herbicide. Plants metabolize sulfonylureas such as chlorimuron-ethyl by oxidative mechanisms (Sweetser *et al.*, 1982; Hutchinson *et al.*, 1984) with subsequent conjugation to glucose or by direct conjugation to glutathione (Hutchinson *et al.*, 1984).

Degradation experiments of simazine in pine seedlings have shown the presence of three unknown metabolites. In wheat crop, a plant species was found sensitive to simazine, known as hydroxy simazine which is one of its metabolites (Hamilton, 1964). Resistant species such as maize degrade simazine to hydroxysimazine non-enzymatically. A compound named benzoxazinone was isolated from maize extracts, and appears to be responsible for apparent catalytic degradation of simazine. It is possible that red pine seedlings contain very little benzoxazinone. This was evident from the fact that the amount of simazine degraded by red pine seedlings was small and there was no hydroxysimazine (Dhillion *et al.*, 1968).

Example: Inactivation of simazine

Cl

N N

C_2H_5NH N C_2H_5NH

Simazine

→

OH

N N

C_2H_5NH N C_2H_5NH

Hydroxysimazine

The simazine molecule is de-chlorinated (dehalogenation reaction) and the chlorine is replaced with a hydroxyl group. The end product of the inactivation reaction is hydroxysimazine, which is 1000 times less toxic than simazine. In most cases, crops are able to metabolize the herbicide, but weeds are not. A summary of the types of metabolic reactions involved in the molecular fate of herbicide in higher plants is given in this chapter.

(A) Inactivation of herbicides

Herbicides are made less toxic by altering the molecular configuration through biochemical and chemical reactions. Biochemical reactions are primarily enzymatic reactions that occur within a living plant. There are a number of key plant enzymes that are used in the inactivation of herbicides. Microsomal mixed function oxidases are able to hydroxylate a wide range of herbicides such as bentazone and diclofop-methyl (Fig. 2). It is often the case that these hydroxylated metabolites are subsequently glucosylated by sugars in the tissue and these conjugants can be stored in the cell vacuole where they can have no phytotoxic effects.

Chemical reactions are breakdown of herbicide to low molecular weight chemical products, but do not necessarily require enzymes. Clearly, if a plant is able to metabolize a herbicide more quickly than the herbicide can accumulate at the site of action within the plant, then that plant will be tolerant of that herbicide. This is detoxification.

The metabolism or molecular fate of herbicides can be described using a 3-phase process (Shimabukuro, 1985).

Phase I reactions

Dechlorination, hydroxylation, decarboxylation, dealkylation, oxidation reduction, and hydrolysis are the initial chemical reactions that occur. These reactions detoxify the herbicide and predispose the resultant metabolites to conjugation. Many of the products from these reactions are acted on immediately by Phase II reactions as well.

Phase I reactions are not synthetic reactions in that the herbicide does not increase in size, but rather this phase is needed to introduce a reactive substituent group not already present. Some functional or reactive groups and their names follow (the hyphen indicates where the group attaches to the remainder of the molecule):

Amide	= $-CONH_2$
Amino	= $-NH_2$
Carboxylic acid	= -COOH,
Cyano	= $-C \equiv N$,
Halogens (chloride, bromide, iodide, fluoride)	= -Cl, -Br, -I, -F,
Hydroxyl	= -OH,
Methoxy	= $-OCH_3$,
Methyl	= $-CH_3$,
Methylthio	= $-SCH_3$,
Sulfhydryl	= -SH

Some specific phase I reactions are as follows:

Decarboxylation

Removal of -COOH group from the herbicide is called decarboxylation. It is a part of the degradation sequence for many phenoxy, benzoic acid, and substituted urea herbicides. Several herbicides including the phenoxy, benzoic and urea derivatives have been shown to undergo decarboxylation (Fig. 1). With a urea herbicides decarboxylation is accomplished by a simulation deaminization, which follows demethylation reaction (Fig. 1). This reaction also requires one molecule of water and therefore could be considered a hydrolysis.

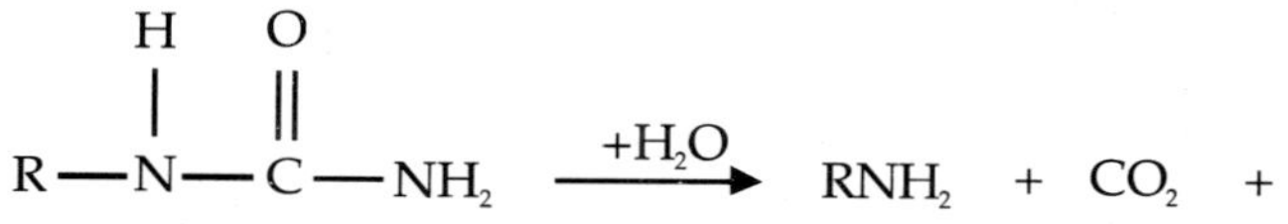

$$R-\overset{H}{\overset{|}{N}}-\overset{O}{\overset{\|}{C}}-NH_2 \xrightarrow{+H_2O} RNH_2 + CO_2 +$$

Decarboxylation-deamination-hydrolysis of an aniline degradation product of a urea herbicide

Fig. 1. Some common reactions involved in herbicide degradation (Ashton and Crafts, 1981)

Hydroxylation

The addition of a -OH group to the molecule is called hydroxylation. It is often accompanied by removal or shifting of another atom such as chlorine. Hydroxylation of the ring of herbicide molecules has been demonstrated in higher plants. Phenoxy, benzoic and triazines herbicides undergo hydroxylation (Fig. 2). With the phenoxy herbicides, ring hydroxylation may be accomplished by a shift in the position of a chlorine atom on the ring. However, with the benzoic acids hydroxylation occurs without a chlorine atom shift. Whereas in triazine herbicides, ring hydroxylation involves dechlorination, demethoxylation or demethylation.

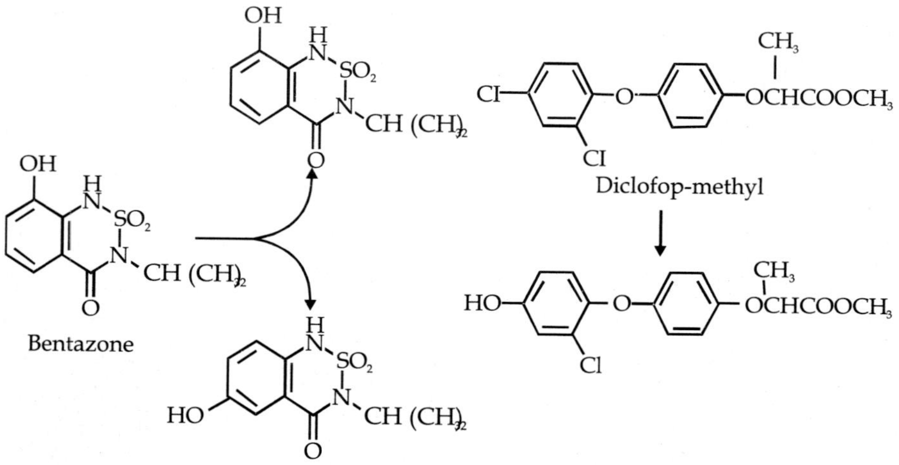

Fig. 2. Hydroxylation reactions

Hydrolysis

Splitting of a molecule through the addition of water is called hydrolysis. Hydrolysis is an important selective mechanism for herbicides in the carbamates, thiocarbamates, substituted ureas, sulfonylureas and triazines group of chemicals. Hydrolysis may result in either activation or deactivation of the herbicides. Shimabukuro (1985) describes that proponil is hydrolytically degraded by an aryl acryamidase. This enzymatic reaction results in differential tolerance between rice (tolerant) and barnyard grass (susceptible) to propanil.

The degradation of herbicides in higher plants by hydrolysis is a common phenomenon. It is involved in the degradation of various formulations such as phenoxy esters as well as the basic molecules of several classes of herbicides. Hydrolysis of a herbicide molecule usually causes a major split in the molecules yielding two relatively large fragments that are non-phytotoxic. These two fragments are also often subject to additional degradation.

Some herbicides are applied in a form that is inactive as a herbicide and it is the plant's metabolic processes that converts the applied compound into a herbicidal compound. Many acids are applied as esters, amides or salts and these have to be hydrolyzed to the free acid before they become effective. Examples of this are found with the aryloxyalkanoic acids, the aryloxyphenoxypropionates and the N-arylalanine esters (Fig. 3). If these compounds are hydroxylated or detoxified in other ways before the free acid is released the compound will be ineffective as a herbicide on that weed or selective as a herbicide in that crop. (Dusky *et al.*, 1980; Hill *et al.*, 1978; Hagland *et al.*, 1974; Hogland and Grab, 1972).

Degradation of sulfonylureas typically occurs through hydrolysis of the aryl (or pyridyl or thienyl) ring, facilitating a cleavage of the sulfonylurea bridge. This results in formation of an aryl (or pyridyl or thienyl) sulfonamide and 'heterocyclic amine' with a net loss of CO_2, as portrayed in Fig. 3.

Amide hydrolysis is probably most important for the acylanilide group of herbicides (Fig. 4). The major mechanism of selectivity for barnyardgrass control in rice using propanil is that rice hydrolyzes propanil using the enzyme, acyl arylamidase. Barnyardgrass has about sixty-fold less acyl arylamidase in its tissue than rice does. Interestingly, some insecticides (carbamates and organic phosphates) act as synergists, by binding to the enzyme and inhibiting propanil hydrolysis resulting in more rice injury (Fig. 5,6).

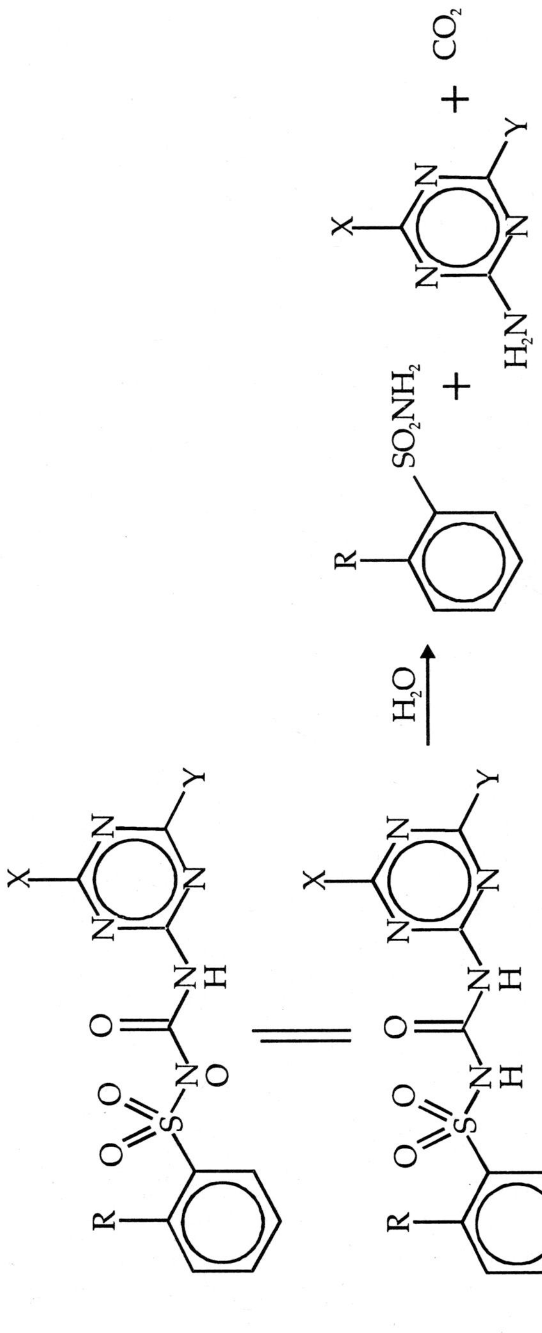

Fig. 3. Hydrolysis of a sulfonylurea

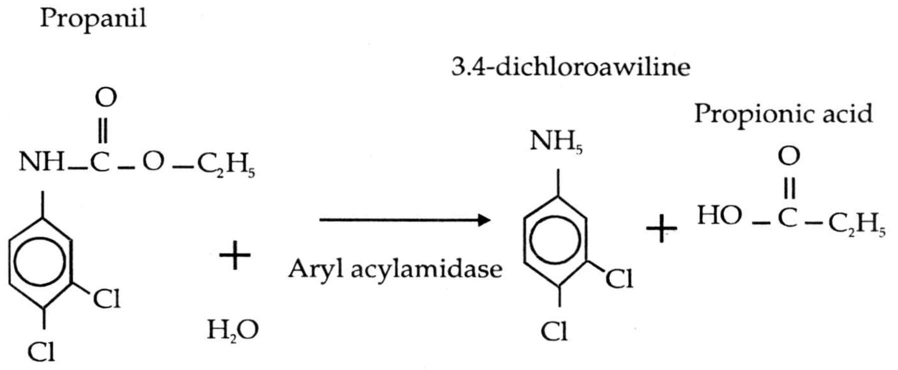

Propanil

3.4-dichloroawiline

Propionic acid

$NH - \overset{O}{\overset{\|}{C}} - O - C_2H_5$ + H_2O —Aryl acylamidase→ NH_5 + $HO - \overset{O}{\overset{\|}{C}} - C_2H_5$

Cl Cl Cl Cl

Fig. 4. Amide hydrolysis

OH_3 CH_3

Cl—O—OCH–COOCH$_3$ → Cl—O—OCHCOOH

Cl Cl

Diclofop - m:thyl Diclofop acid

CH_3 CH_3

CO N CHCOOCHCH$_3$ → CO N CHCOOH

Cl Cl Cl Cl

Benzoylprop-ethyl Benzoylprop acid

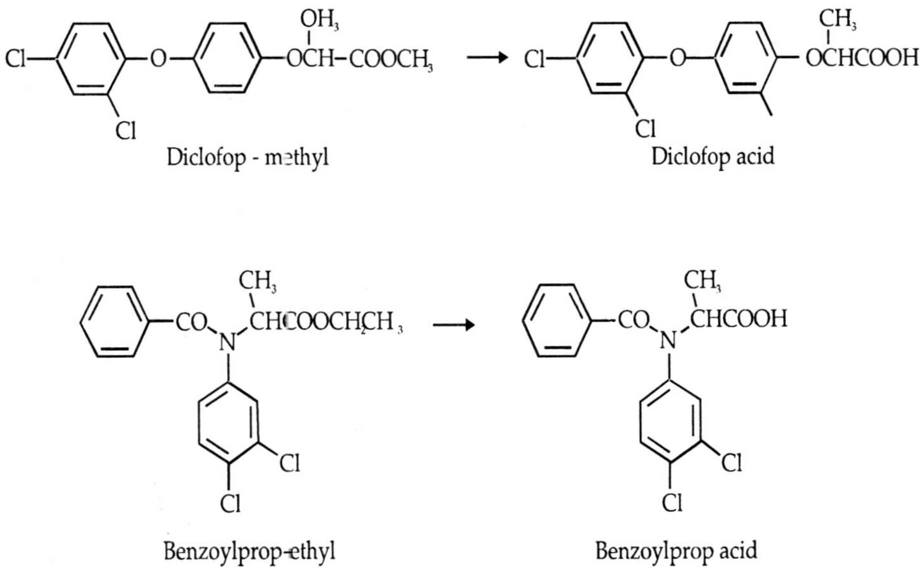

Fig. 5. Desertification reaction to release active free acids

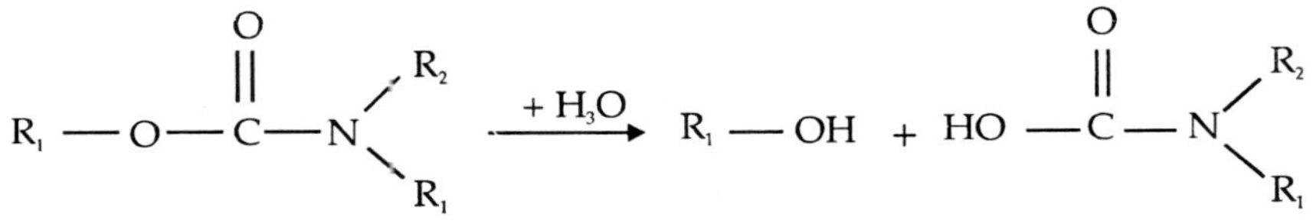

$$R_1 - O - \overset{O}{\overset{\|}{C}} - N \begin{matrix} R_2 \\ R_1 \end{matrix} \xrightarrow{+ H_3O} R_1 - OH + HO - \overset{O}{\overset{\|}{C}} - N \begin{matrix} R_2 \\ R_1 \end{matrix}$$

Hydrolysis of a carbamate herbicide

Fig. 6. Hydrolysis of a carbamates herbicide

Dealkylation

Dealkylation is another common inactivating process. Removal of alkyl side chains from the herbicide is called dealkylation; e.g. triazine, substituted urea, carbamate, thiocarbamate, and dinitroaniline herbicides. The substitution of various alkyl group on a basic structure of a given class of herbicide molecules has not only altered their absolute toxicity but it has also resulted in varied selectivity of different plant species such as triazine and urea herbicides. Certain herbicides molecules have alkyloxy substitution rather than merely alkyl substitution. They seem to undergo dealkyloxylation almost as readily as dealkylation occurs.

The phenyl urea herbicides and the 1,3,5-triazines are dealkylated to inactive metabolites almost certainly through the action of the microsomal mixed function oxidases discussed above (Fig. 7). The 1,3, 5- triazines also conjugate with glutathione to form inactive conjugates and a similar process is seen with thiocarbamates although the conjugation of these compounds demands the formation of the sulfoxide (Fig. 7).

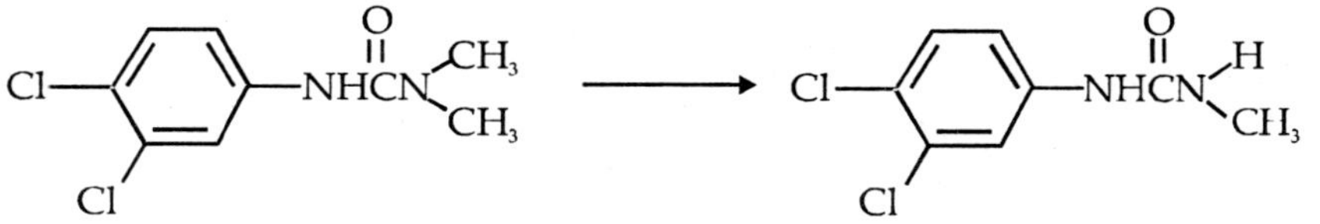

Diuron

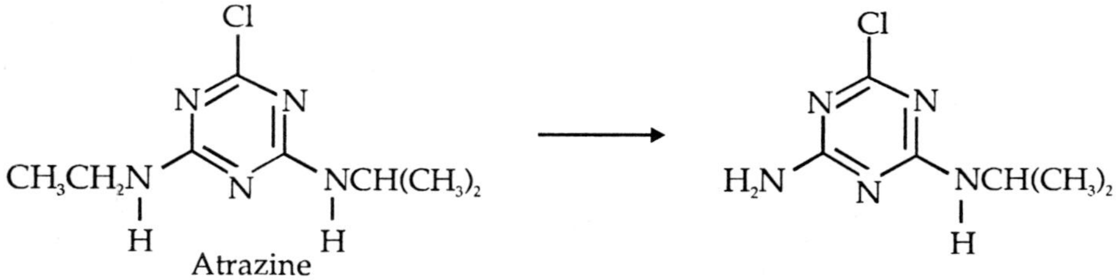

Atrazine

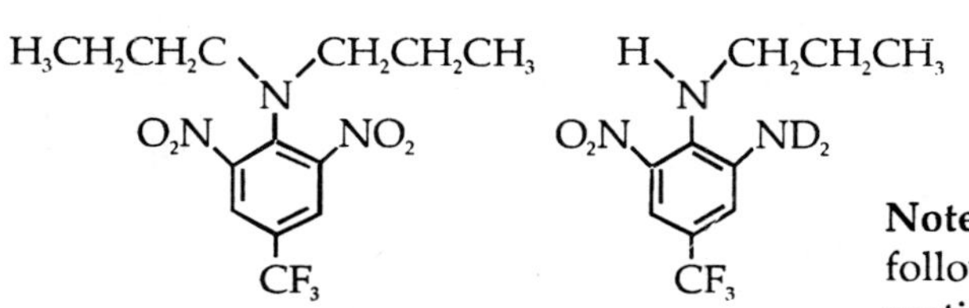

Trifluralin

Note : This step usually is followed by a nitroreduction reation ($NO_2 \rightarrow NH_2$)

Contd...

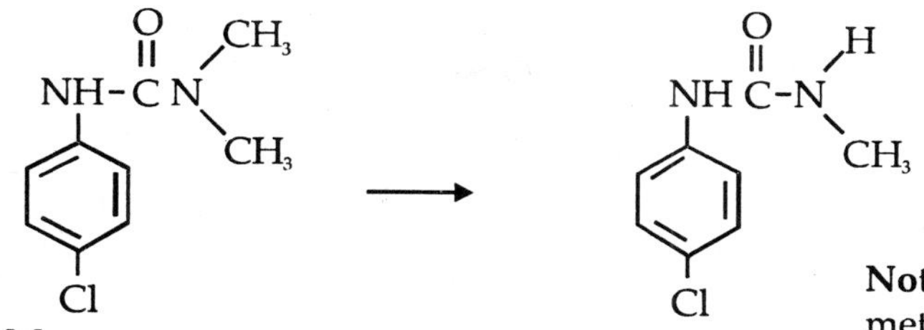

Monuron

Note : Removal of both methyl groups is required for detoxification

Fig. 7. N-dealkylation reactions

Deamination

Deamination is another detoxifying metabolic process. Removal of an amine group (NH_2) from the herbicide is called deamination. Herbicides such as metribuzin and metamitron are deaminated almost certainly by a peroxisome-based deaminase to inactive intermediates (Fig. 8).

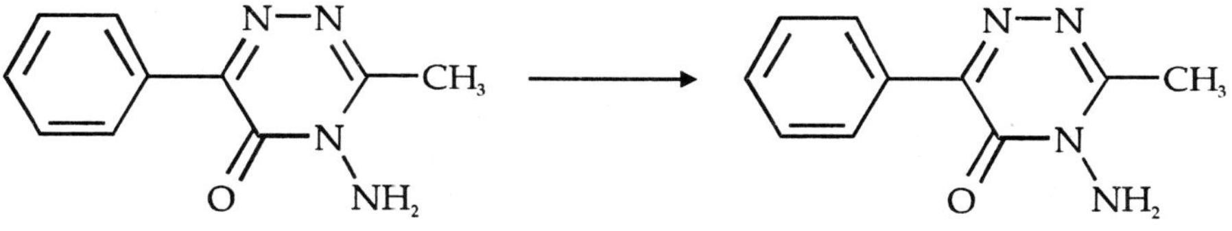

Metamitron

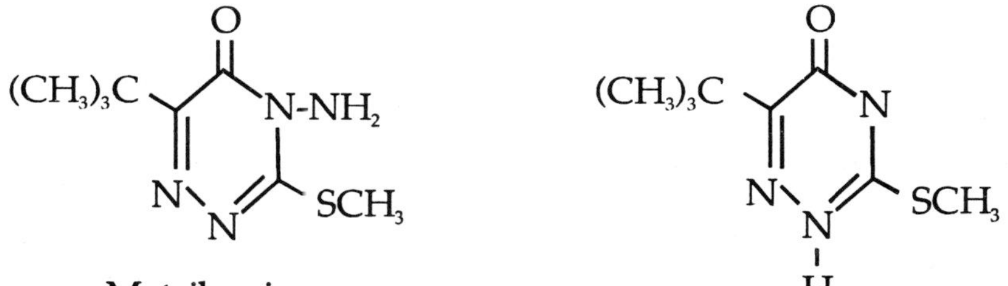

Metribuzin

Fig. 8. Deamination of metamitron and metribuzin

Ring hydroxylation

The addition of a -OH group to the ring structure of the molecule is called ring hydroxylation. The hydroxylation of the ring of some herbicides is an important mechanism of detoxification. Imidazolinone herbicides are taken up by root and leaf tissue and detoxified by hydroxylation and conjugation with glucose (Fig. 9, 11).

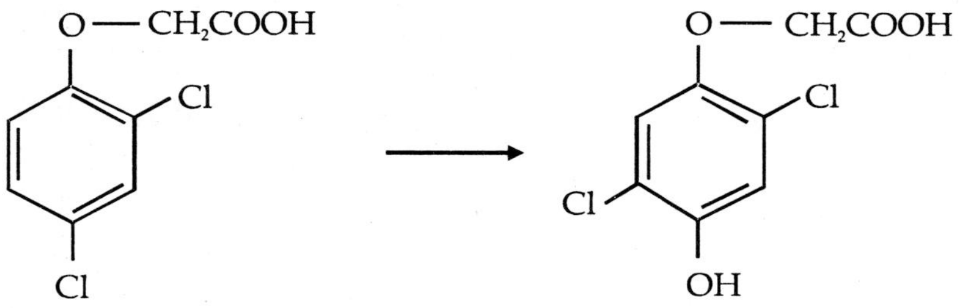

Fig. 9. Ring hydroxylation of 2, 4-D

Dealkyloxylation

Removal of an alkyl group with an attached 'O' is called deakyloxylation. Urea herbicides are generally metabolized by dealkylation, followed by dealkoxylation (Fig. 10).

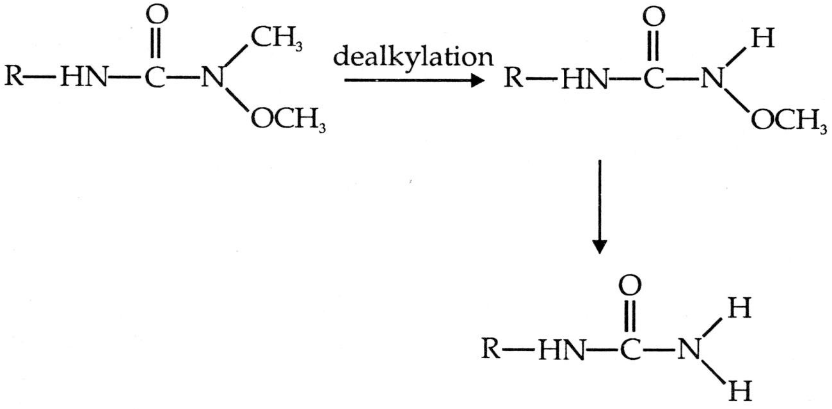

Fig. 10. Stepwise dealkylation and dealkyloxylation of a urea herbicide

Oxidation

For most herbicides, this process is driven by the metabolizing enzyme, cytochrome P_{450} mono-oxygenase, which is located inside the plasma membrane. P_{450} is able to inactivate herbicides by various oxidation reactions, such as hydroxylation. After the attachment of a oxygen to the herbicide, conjugation to glucose or other natural plant

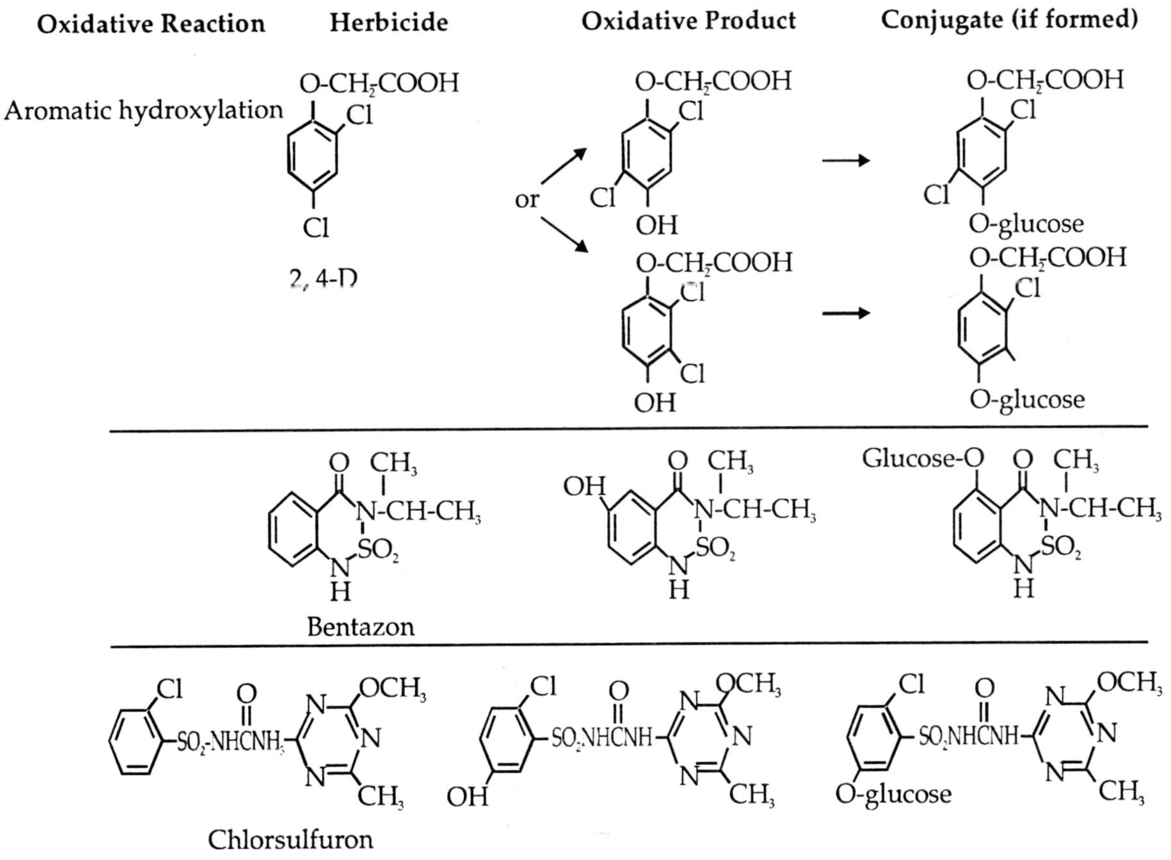

Oxidative Reaction	Herbicide	Oxidative Product	Conjugate (if formed)
Aromatic hydroxylation	O-CH_2-COOH, Cl, Cl 2, 4-D	or: O-CH_2-COOH, Cl, Cl, OH O-CH_2-COOH, Cl, Cl, OH	→ O-CH_2-COOH, Cl, Cl, O-glucose → O-CH_2-COOH, Cl, O-glucose
	O, CH_3, N-CH-CH_3, SO_2, N, H Bentazon	OH, O, CH_3, N-CH-CH_3, SO_2, N, H	Glucose-O, O, CH_3, N-CH-CH_3, SO_2, N, H
	Cl, O, SO_2-NHCNH-, N, OCH_3, N, N, CH_3 Chlorsulfuron	Cl, O, SO_2NHCNH-, N, OCH_3, N, N, CH_3, OH	Cl, O, SO_2NHCNH-, N, OCH_3, N, N, CH_3, O-glucose

Fig. 11. Some examples of aromatic hydroxylation

constituents occurs due to another enzyme transfer system. Following conjugation, there is evidence that conjugates are then removed from the cytoplasm *via* active transport into the vacuole or into the cell wall. Plants metabolize sulfonylureas by oxidative mechanisms (Sweetser *et al.*, 1982; Hutchinson *et al.*, 1984) with subsequent conjugation to glucose or by direct conjugation to glutathione (chlorimuron-ethyl: Hutchinson *et al.*, 1984).

There are three type of oxidation α (alpha), β (beta) and *w* (omega) which involve oxidation at three different sites on the side chain. The omega 2,4-dichlorophenoxy alkane nitirles undergo alpha oxidation.

α-oxidation of a model phenoxy alkane nitrile is described below,

$$\text{R-O-(CH}_2)_n\text{-CN} \rightarrow \text{R-O—(CH}_2)_{n-1}\text{ –COOH}$$

Beta-oxidation is a pathway in plants whereby long chain carbon segments are degraded by removing two carbons for each cycle of the pathway (Fig. 1).

The oxidation of the thiocarbamates to form sulfoxides is another activating process as it has been shown that the sulfoxides are more effective herbicides than the thio-compounds (Fig. 10) but the sulfoxide also conjugates with glutathione, rendering it herbicidally inactive. The classical activation process is the β-oxidation of the herbicidally inactive MCPB into the herbicidally active MCPA (Fig. 12), a process that takes place slowly in legumes and that, thereby, allows the use of compounds such as MCPB for the control of broad- leaved weeds in cereals under sown with clover.

Cl– –$OCH_2CH_2CH_2COOH$ —β - oxidation→ Cl– –OCH_2COOH

CH_3 CH_3

MCPB MCPB

Fig. 12. β- oxidation of MCPB to MCPA.

The omega-oxidation of 10-phenoxy-n-decanoic acid was suggested by Fawcett *et al.*, (1954) to explain the large amounts of phenol found as a degradation product in flax (Fig. 13).

$$R{-}O{-}CH_2{-}(CH_2)_8 - COOH \longrightarrow R - O - \underset{\underset{O}{\|}}{C} - (CH_2)_8 - COOH$$

$$\downarrow$$

$$R{-}OH + CH_3 - (CH_2)_8{-}COOH$$

Fig. 13. ω-oxidation of 10-phenoxy-n-decanoic acid

Ring cleavage

Some evidence exists for the partial cleavage of the aromatic ring found in most organic herbicides. However, the process appears to be slow and relatively unimportant in higher plants. It is more likely for the ring to be incorporated into insoluble, non-phytotoxic residues following biological substitution or removal of the various substitutions on it. There is a little evidence for the complete degradation of any herbicides ring structure to CO_2 at a biologically important rate. During ring cleavage, the aromatic and heterocyclic ring structure is spilt, resulting in formation of monochloroacetic acid. The cleaved products are further degraded readily. Ring cleavage transform intrinsically inactive herbicides to herbicidally active products. The cleaved products are further degraded readily. For example, methazole is converted by the ring cleavage mechanism to the Photosystem II inhibitors DCMU [1-(3,4-dichlorophenyl)-3-methyl urea]. Similarly Buthidazole with an imidazolinone ring, also undergoes ring cleavage to yield an active N-methyl urea.

Apparently the splitting of the aromatic and hetrocyclic ring structure contained in many herbicides proceeds very slowly in higher plants. Although the various substitutions on the ring are usually cleaved off, the ring itself may persist as a nontoxic compound throughout the life of the plant. Frequently the rings are bound to insoluble residues, which complicate the reach on them.

Phase II reactions

Conjugation with sugars, amino acids or natural plant constituents: Metabolites are strongly bound as compared to parent compounds. Conjugations with sugars, amino acids or natural plant constituents (metabolites are strongly bound) are common phase II reactions.

Conjugation

Phase II reactions are also known as conjugation reactions. Conjugation is the reaction of a herbicide with another substance in the plant resulting in a new compound of higher molecular weight. Generally conjugation is a process in which plants convert herbicide molecules into less toxic and more water soluble metabolites. Conjugation reactions also function in translocation of some herbicides. Conjugation with glucose happens readily in most plant species. The conjugation of herbicides with amino acids such as glutamic and aspartic acid are also well known biological reactions. These conjugation reactions have little effect on the ultimate activity of herbicides although they too are important for translocation. The glutathione s-transferase enzyme within a cell will covalently bind the herbicide to glutathione through the sulfur atom of glutathione. The herbicide is bound (adsorbed) to another compound, e.g., glucose, amino acids. The 1,3,5- triazines also conjugate with glutathione to form inactive conjugates and a similar process is seen with thiocarbamates although the conjugation of these compounds demands the formation of the sulfoxide (Fig. 14).

The susceptibility of a plant species to herbicide damage is often correlated with its ability to degrade the herbicide to less active or immobile metabolites. Most plant species detoxify herbicide molecules in three phases. Detoxification begins with oxidation of the parent molecule, during Phase I, followed by conjugation with natural substrates, such as glucose, amino acids or, more commonly, glutathione, during phase II. In some cases, parent molecules bypass Phase I and enter Phase II directly. Although these reactions usually reduce the phytotoxicity and mobility of the herbicide, phase I metabolism is required for the activation of some herbicides, such as fenoxaprop-ethyl. In this case, the parent compound is de-ethylated during Phase I to produce the active compound, fenoxaprop (Fig. 5). During Phase III, phase II metabolites may undergo further conjugation to produce insoluble residues, which are sequestered in vacuoles or bound in lignin biopolymers. Alternatively, Phase II conjugates can be directly sequestered in the vacuole by the activity of glutathione or glucoside-conjugate pumps located in the vacuolar membrane (Cole, 1994).

The majority of oxidative Phase I reactions are catalyzed by cytochrome P_{450} dependent mono-oxygenases. These enzymes are haemoproteins associated with the smooth endoplasmic reticulum and are characterized by their ability to bind carbon monoxide, producing an absorption maximum at 450 nm. Over 50 different systems are believed to exist in plants, including biosynthetic systems involved in synthesis of natural substrates such as sterol, terpene, isoflavonoid, gibberellin, abscisic acid, cytokinin and lignins, as well as detoxification systems involved in the metabolism of herbicides. This class of enzymes has been demonstrated to be responsible for the degradation of the herbicides monuron, chlorotoluron, metolachlor, alachlor, bentazone, chlorotoluron, 2, 4-D, diclofop, flumetsulam, primisulfuron, prosulfuron and triasulfuron, (Barrett, 1995; Schuler, 1996).

Conjugation with glutathione

Glutathione conjugation involves the attachment of the small tripeptide, glutathione (γ-glutamylcysteinyl-β-glycine) (Fig. 14), to herbicides containing halogen, phenolate or alkyl sulfoxide groups (Fig. 14). On a herbicide molecule, these groups can serve as leaving groups when reacted with the nucleophilic glutathione anion (GS^-). In the reaction (equation 1), X is the herbicide and Z is the leaving group. These reactions are catalyzed by glutathione-*S*-transferases (GST). Conjugation proceeds as a nucleophilic displacement reaction with the glutathione anion (GS^-) serving as the nucleophile. The sulfhydryl group (-SH) on the cysteine residue of the glutathione molecule is the reactive component required for this reaction. A simplified form of the reaction follows:

Equation 1

$$X\text{-}Z + GS^- \rightarrow GS\text{-}X + Z^-$$

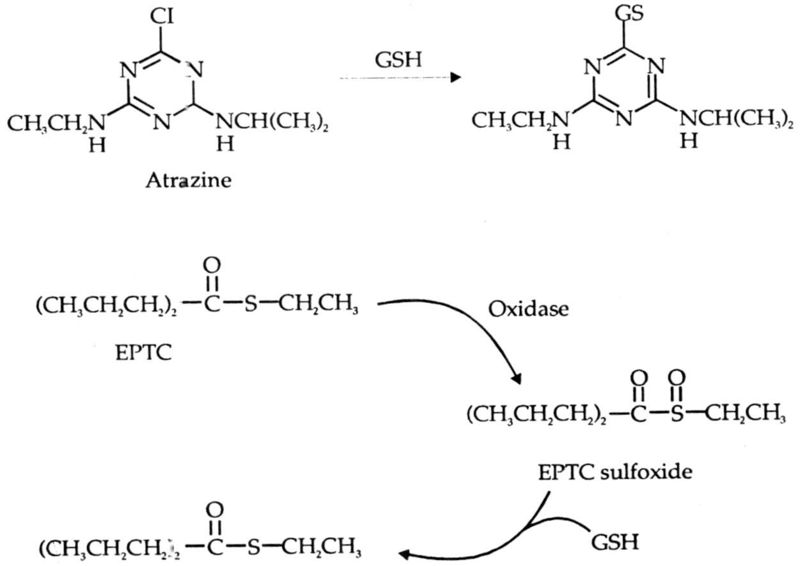

Cl

N N

CH_3CH_2N N $NCH(CH_3)_2$

H H

Atrazine

GSH →

GS

N N

CH_3CH_2N N $NCH(CH_3)_2$

H H

$(CH_3CH_2CH_2)_2-\overset{O}{\overset{||}{C}}-S-CH_2CH_3$

EPTC

Oxidase

$(CH_3CH_2CH_2)_2-\overset{O}{\overset{||}{C}}-\overset{O}{\overset{||}{S}}-CH_2CH_3$

EPTC sulfoxide

GSH

$(CH_3CH_2CH_2)_2-\overset{O}{\overset{||}{C}}-S-CH_2CH_3$

Fig. 14. Conjugation with glutathione

All plants are able to perform this conjugation, although some use homoglutathione instead, and it is the speed with which the conjugation takes place that determines the damage to the plant. Some compounds, that themselves have no observable biological effect, have been shown to increase the levels of glutathione S-transferase in treated crops. The application of such compounds in conjunction with a herbicide will reduce the damage from the herbicide to the crop.

Amino acid conjugation

Conjugation with amino acids is particularly common with the phenoxyacetic acid herbicides like 2,4-D. A peptide bond is formed between the herbicide's carboxylic acid residue and the amino acid's amino group. Some of the first evidence of amino acid conjugation with a herbicide was 2,4-D conjugation to aspartic acid (Fig. 15). Other amino acids identified include glutamate, valine, leucine, phenylalanine and tryptophan. The most common amino acids conjugating with xenobiotics in plants are aspartate or glutamate. These conjugates can still be biologically active, but are relatively immobile. There is evidence that amino acid conjugates are excreted to the cell wall.

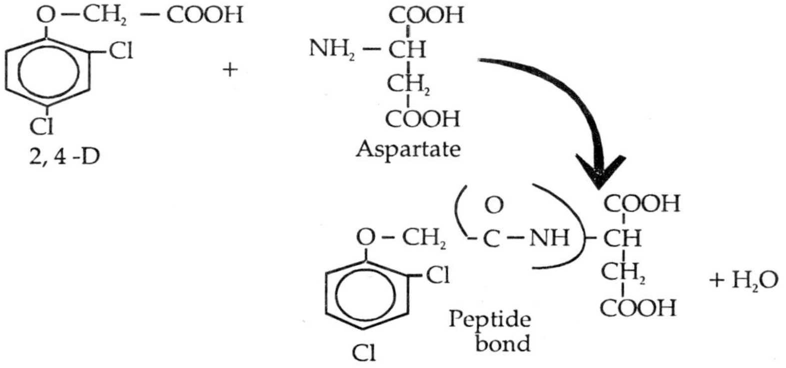

Fig. 15. 2, 4-D conjugation to aspartic acid

Phase III reactions

These are unique to plants and consist of secondary conjugation reactions or the formation of insoluble bound residues. In general, plants "immobilize" these metabolites whereas animals excrete them. In animals, toxins are degraded by the liver and excreted through the kidneys (Fig. 16).

Phase III reactions are unique to plants because plants do not excrete herbicides as animals do. Plants therefore, need to somehow remove the herbicides within their own system. Plants do this by either conjugating Phase II products further by moving metabolites to the vacuole for storage and incorporating the metabolite into the cell wall region.

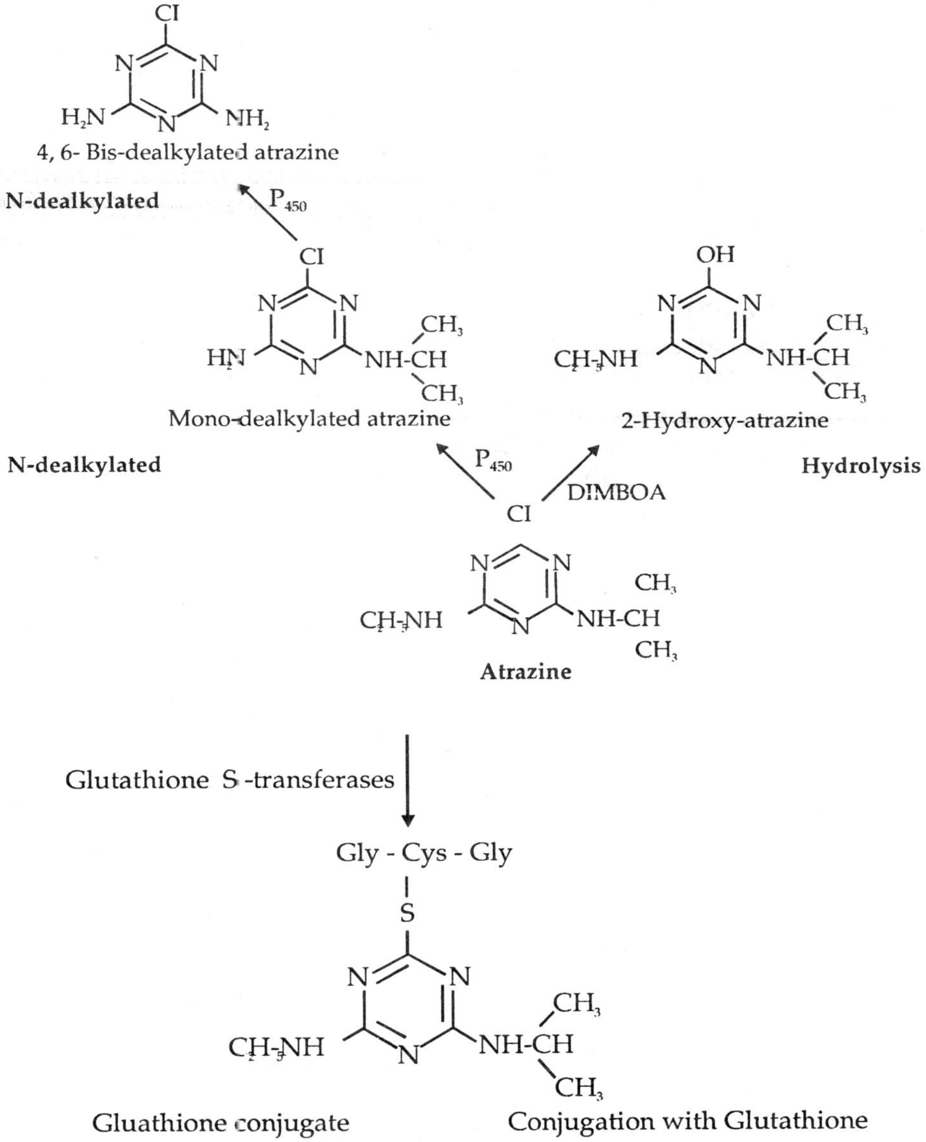

4, 6- Bis-dealkylated atrazine

N-dealkylated P_{450}

Mono-dealkylated atrazine

2-Hydroxy-atrazine

N-dealkylated P_{450} DIMBOA Hydrolysis

Atrazine

Glutathione S -transferases

Gly - Cys - Gly

Gluathione conjugate Conjugation with Glutathione

Fig. 16. Major detoxification pathway for atrazine (Ross and Lembi, 2000)

It is assumed that Phase III products are no longer toxic; however, this area of herbicide fate in plants is poorly understood, especially with reference to the identity of sequestered products and any subsequent fate in herbivores who consume the plants.

Scondary conjugation

Herbicide glucosides can be further processed by the addition of a malonic acid residue to the sugar conjugate. In secondary conjugation or malonyl CoA conjugation, a malonic acid residue is attached to a product from Phase II sugar conjugation by an ester linkage. Malonyl CoA is the acyl donor (3 carbons will be added to the sugar conjugate) and the reaction is catalyzed by malonyl-CoA-transferases. Because malonyl conjugates are labile during extraction and purification, there may be more examples of secondary conjugation than currently identified in the literature. Malonylation may enhance vacuolar sequestration as well. When a herbicide is conjugated to glutathione, it can be catabolized to cysteine by the removal of the glutathione's glutamate and glycine groups, prior to conjugation with malonate (Fig 17).

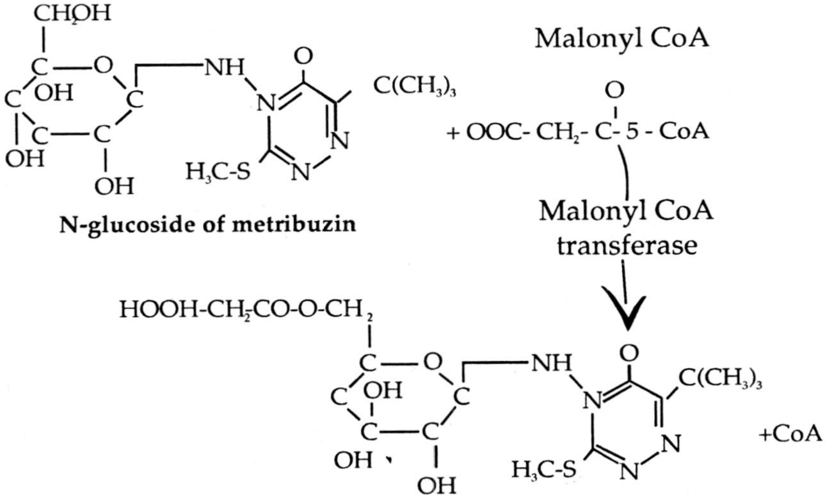

CH_2OH — C—O — C — NH — N — O — $C(CH_3)_3$ — N — N — H_3C-S — OH — OH — OH

N-glucoside of metribuzin

Malonyl CoA

$+ OOC- CH_2- C(O)- S - CoA$

Malonyl CoA transferase

$HOOH-CH_2-CO-O-CH_2$ — C—O — C — NH — O — N — $C(CH_3)_3$ — N — N — H_3C-S — OH — OH — OH

+CoA

Fig. 17. Secondary conjugation of metribuzin

Insoluble residues : Herbicide metabolites may also be incorporated into cell wall material. These are known as insoluble or bound residues because they cannot be removed using conventional solvent extraction techniques. In general, terminal conjugation to insoluble or bound residues is not well understood, but studies have shown that between

1 and 70 % of the herbicide metabolite can be incorporated into structural components of the plant. This wide variation is thought to be due to differences in the herbicides, the plant species or tissue studied and the time of harvest. Herbicides sensitive to this fate usually contain aromatic or heterocyclic rings. Lignin is a polymer of phenolic groups. Peroxidases in the cell wall region create a free radical on the lignin molecule (at the highlighted -OH,) to which the OH-PCP can attach. The PCP is essentially bound irreversibly.

Vacuolar sequestration : Because plants store many toxic metabolites in their vacuoles and these toxic metabolites are released upon herbivore attack, it has long been thought that herbicides metabolites are moved into the vacuole. Recent evidence suggests there is a carrier on the vacuole membrane which recognizes glutathione conjugates (GS-X), transporting them into the vacuole. The carrier is dependent on ATP and apparently transports many glutathione conjugates (herbicides and natural products) into the vacuole. Glutathione and its conjugates are predominately anionic at the high pH found in the cytoplasm of plant cells and therefore, rely on a transporter to move across the lipophilic membrane. To date, only a few glutathione conjugates have been shown to accumulate in plant vacuoles. No transporters of amino acids or glucose conjugates have been identified on vacuolar membranes. Malonylation may aid in vacuolar sequestration as well.

(B) Activation of the herbicide

In some cases a herbicide is metabolized to a more active form. This is less common than inactivation. Example, herbicide activation through conversion of 2, 4-DB to 2, 4-D (Fig. 18)

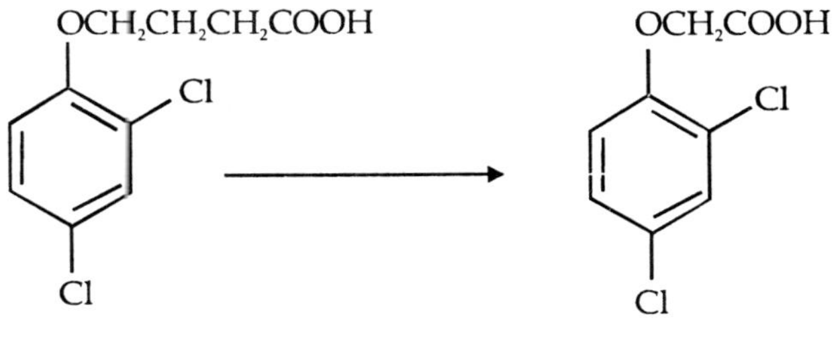

$OCH_2CH_2CH_2COOH$ — Cl — Cl → OCH_2COOH — Cl — Cl

2,4 - DB → 2,4 - D

Fig. 18: Activation reaction of 2, 4-DB

2, 4-DB (2, 4-dichlorophenoxy butyric acid) is a phenoxy herbicide used in soybeans and alfalfa, which are both legume crops. β-oxidation is a pathway in plants whereby long chain carbon segments are degraded by removing 2 carbons for each cycle of the pathway. By removing 2 carbons from 2, 4-DB (butyric acid - 4C's) we are left with the more active form, 2, 4-D (acetic acid - 2C's) (Fig 17).

This selectivity mechanism protects legumes from herbicidal injury as they do not catalyze this reaction; broadleaf weeds cleaving off the two carbons convert the phenoxybutyric acid into the herbicidal phenoxyacetic acid, such as 2,4-D. Another example of bioactivation is the photochemical reduction of paraquat (Fig 19). This reaction is nonenzymatic as an electron from the electron transport system in the chloroplast during the light reactions of photosynthesis reduces paraquat, creating a free radical.

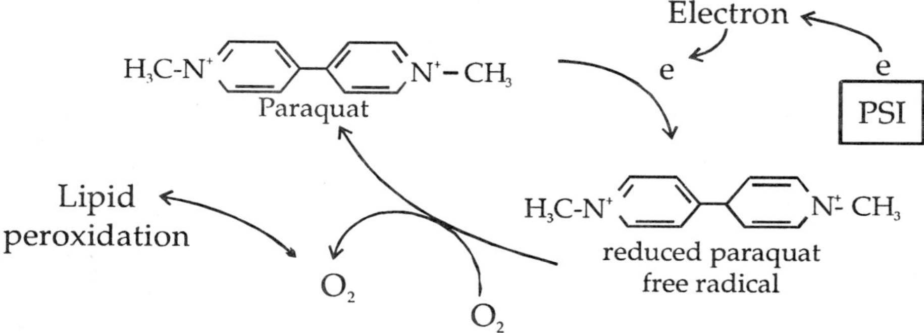

Fig. 19. Photochemical reduction

Summary

Metabolism is one of the most important ways a plant can escape the toxic effects of a herbicide. Herbicide-tolerant plants often have the ability to metabolize or break down the chemical to non-active compounds before it can build up to toxic levels at the site of action. Susceptible plants are unable to detoxify herbicides. Metabolism of a herbicide may result in inactivation of the herbicide or deactivation where phytotoxicity is reduced. The metabolism or molecular fate of herbicides can be described using a 3-phase process Shimabukuro (1985). Phase I reactions involve de-chlorination, hydroxylation, de-carboxylation, dealkylation, oxidation/reduction, ring hydroxylation, deamination and hydrolysis etc. These reactions detoxify the herbicide and predispose the resultant metabolites to conjugation.

Many of the products from these reactions are acted on immediately by Phase II reactions as well. Whereas conjugation with sugars, amino acids or natural plant constituents (metabolites are strongly bound) are common phase II reactions. Phase III reactions are unique to plants because plants do not excrete herbicides as animals do. Plants therefore, need to somehow remove the herbicides within their own system. plants do this by either conjugating Phase II products further by moving metabolites to the vacuole for storage and incorporating the metabolite into the cell wall region. However, this area of herbicide fate in plants is poorly understood, especially with reference to the identity of sequestered products and any subsequent fate in herbivores who consume the plants.

❑❑❑

Chapter 6

Selectivity and Resistance

Herbicide selectivity refers to the differential effect of a herbicide when applied to a mixed population of plants. Selective herbicides kill weeds and leave desirable plants undamaged when applied to both. These mechanisms are characteristics of the plant rather than the herbicide. It is the result of a chemical reaching and disrupting a vital plant function/ process at some biochemical and biophysical site in one plant but not in another and is largely determined by the amount of herbicide that ultimately reaches to the site of action in plants. It is an important factor towards weed control through herbicides and crop safety. Selectivity results from the differential response of plant species to herbicides and is a complex phenomenon achieved by complex interaction between plant, herbicide and environment (climate and soil).The site of uptake of soil active herbicides is a characteristic of the herbicide rather than the plant. In other words for a given herbicide the site of uptake can be the same across plant species. Herbicides considered selective can be tolerated within a wide range of concentration by some species but not by others. The basic principle of herbicide selectivity is that more toxicant reaches the site of action in active form inside target plants than in the non-target species. Beyond a certain limit, however all species will likewise be susceptible.

Plants differ in many ways; their surfaces vary in form and chemical composition; some are easily wet by aqueous sprays; others are difficult to wet. Surface coatings vary in permeability, and protoplasm responds specifically to different chemicals. The gross structure of plants varies in such a way that the meristematic tissues of some are exposed and vulnerable to sprays, others have their growing points protected. Because of these differences certain weeds may be destroyed in growing crops with little or no injury to the latter. Therefore, selectivity is a relative property of a herbicide and is dependent largely on the time, method, rate and formulation of herbicide application, environmental conditions such as temperature, humidity, rainfall soil moisture, soil texture, plant species and the stage of growth of crop and weeds.

Selectivity Index (SI)

Safety of crop plant is described by Selectivity Index (SI) which is the ratio of maximum dose of herbicide tolerated by the crop to the minimum dose of herbicide required to control weeds. The higher the selectivity index, the greater is the safety to a crop and *vice-versa.* If SI is greater than 2, the herbicide is safer for use without much fear of crop damage. If SI is less than 2, the proper method of herbicide application is to be adopted and care must be taken to apply the herbicide strictly at the recommended rate and not beyond that in any case. Otherwise, considerable crop damage may occur from such application.

$$\text{Selectivity Index (SI)} = \frac{\text{Maximum dose of herbicide tolerated by a crop}}{\text{Minimum dose of herbicide required to control weeds}}$$

Mechanism governing the selectivity of herbicides under field conditions can be classified either as physical or physiological. In physical selectivity contact between the herbicides and the plant is minimized or prevented. This is best illustrated by the selectivity of a soil applied herbicide intended to control perennial weeds or deep rooted plants. Highly water soluble herbicides are easily leached to the deeper in the soil profile. The leaching process is fast enough to prevent or limit the absorption of the herbicides by shallow rooted weeds. The herbicides molecules, which accumulate at the root zone of the deep rooted perennial weeds, are then absorbed in sufficient amounts to cause phytotoxicity. Incase of non mobile herbicides that stay at the upper soil layer, absorption by the shallow-rooted species thereby causing phytotoxicity. The deep rooted species are then unharmed. If both the shallow rooted and the deep rooted species are however made to be in full contact with both mobile and non mobile herbicide, they may manifest phytotoxicity symptoms equally. Coating seeds of susceptible species with activated charcoal to prevent direct herbicide seed contact presents another case of physical selectivity.

In physiological selectivity, uninhibited contact between the herbicide and the plant occurs. The tolerant plant can counteract herbicide action through a built in physiological or biochemical mechanism. Under field condition selectivity mechanism may altogether operate these mechanisms.

Selectivity can be based on any or a combination of the following mechanisms/factors such as

1. Interception and uptake of herbicides
2. Metabolic pathways and rates
3. Tolerating product phytotoxicity
4. Differential plant metabolism
5. Chemical characteristics of herbicides,
6. Mode of application;
7. Use of antidotes or safeners and
8. Genetic engineering/biotechnology.

Herbicide selectivity should not be evaluated only by the chemical itself, but considering the interaction between other factors, including environment, rate, and crop aspects (Oliveira Júnior, 2001). Therefore, selectivity must be seen as the joint consequence of several plant protection factors to minimize or exclude injuries. Hence these mechanisms and factors affecting herbicide selectivity are discusses here.

(1) Differential absorption or interception and uptake of the herbicide

Physical selectivity based on differential herbicide absorption of foliage applied herbicides can be controlled by the difference in morphology between plant species or group. In many laboratory studies, molecular peculiarities of herbicides have been found responsible for differences in their time-bound rates of absorption by different plant species. In an attempt to determine the basis of the selectivity of chlorsulfuron, Sweetser *et al.,* (1982) examined uptake, translocation and metabolism of chlorsulfuron in a variety of tolerant and susceptible plants. The results of this short-term study (24 h) indicated that there was no correlation between selectivity, herbicide uptake and translocation. By contrast, there was a clear relationship between phytotoxicity and the rate of metabolism of chlorsulfuron. In young plants of susceptible cotton, soybean, mustard and sugarbeet, between 80 and 97% of the applied herbicide was recovered unchanged. In resistant wheat, barley, *Avena fatua* and *Poa annua,* less than 10% of parent chlorsulfuron was recovered. A half-life of 2-3 days in wheat leaves suggested that the rate of metabolism was sufficient to account for the tolerance of this species (Sweetser *et al.*, 1982). Slife *et al.*, (1962) studied tolerance of wild cucumber *(Sicyos angulatus)* to 2,4-D in comparison to susceptible cultivated cucumber *(Cucumis sativus)* and found slow absorption of 2,4-D in wild cucumber that it kept pace with its metabolism easily, thus the plant proved tolerant (Fig. 1). Carter and Naylor (1960) found similar reasons for the tolerance of honeysuckle and bigleaf maple to amitrole as against its phytotoxicity to bean and lucerne plants. However, there are only a few examples of real inherent differences among plant species for the absorption of a given herbicide.

In field conditions differential absorption of herbicides by plants may occur for many other more important reasons.

1. The plant species may differ in their morphology and growth habits,
2. The herbicides may be applied at different times and by different methods,
3. Adsorbents and antidotes may be used to prevent herbicide absorption by the non-target plants,
4. Herbicides and their formulations may differ in their ability to contact non-target plants. These points are discussed here;

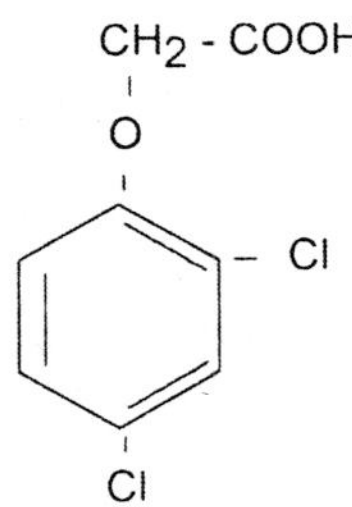

Fig. 1. Wild cucumber easily metabolized 2,4-D due to slow absorption in comparison to susceptible cultivated cucumber.

Differences in morphology of plants

Certain morphological features of crop plants will allow limited retention of aqueous sprays of herbicides on their foliage. This offers an opportunity to obtain selective control of weed species of variable morphology that may allow easy leaf wetting with the aqueous sprays (Fig 2).

Important limited spray retention characteristics found in certain plants are:

(i) Narrow upright leaves,

(ii) Corrugated or finely ridged leaf surfaces

(iii) Waxy leaf surfaces

(iv) Pubescent leaves

(v) Presence or abundance of trichmoes

(vi) Thickness of leaf

(vii) Exposure of the growing point

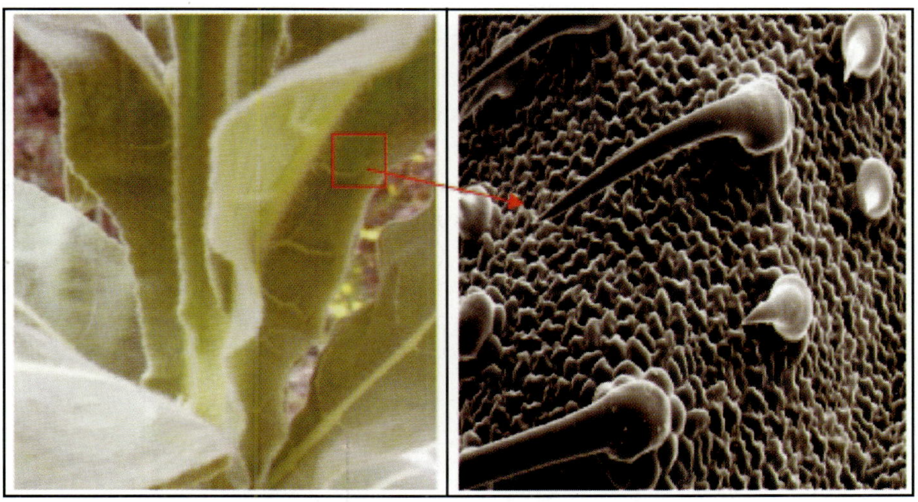

Fig. 2. Leaf surface showing trichomes

Wheat, cabbage, sugarcane, colocasia and some other crop plants possess these characteristics. Aqueous sprays either bounce off as droplets from their foliage or wet them only in small spots. Crops like wheat and sugarcane are further protected against herbicide sprays by their covered growing points. The growing point in wheat plant is located at the base and initially below the soil surface while, in sugarcane, it is hidden by the upper leaf sheaths. Spray droplets may land on trichome tips and prevent a direct leaf herbicides contact. This would lesson the amount of herbicide absorbed by the plant. Rice has been observed to absorb less propanil than barnyard grass because of its ticker leaf cell slowing down the penetration of herbicide into the mesophyll where primary phytotoxic action takes place (Sierra and Vega, 1967).

The limited spray retention provides plants protection only against contact selective herbicides that are sprayed in carrier water without any wetting agent. Schafter and Chilcote (1970) found that limited spray retention was the first line of defense of wheat against post emergent ioxynil and bromoxynil which brought about selective control of many broad-leaf weeds (Fig 3). Likewise, Zawierucha and Penner (2000) evaluated spray retention of quinclorac-sensitive *Digitaria sanguinalis* and quinclorac-tolerant *Eleusine indica* using ^{14}C quinclorac at the one- to two-tiller growth stage cultured under hydroponic conditions. They observed that spray retention *E. indica* (tolerant) retained more applied quinclorac than *D. sanguinalis* (sensitive). Pereira (1970) found that selectivity of nitrofen to many *Brassica* crops and rice was due to differential wetting. With translocated herbicides limited spray retention is of not much help in protecting the non-target plants from herbicide injury. It is because the translocated type of herbicides, even when they wet plant surfaces in small areas, can be absorbed and translocated to other plant parts and thus injure the whole plant. Hernández *et al.,* (2004) evaluated selectivity of trifloxisulfuron + ametryn in sugarcane and recommended lower rate of application in a younger weed growth (5 to 15 cm) stage, and the higher rate in a more advanced (20-40 cm) stage.

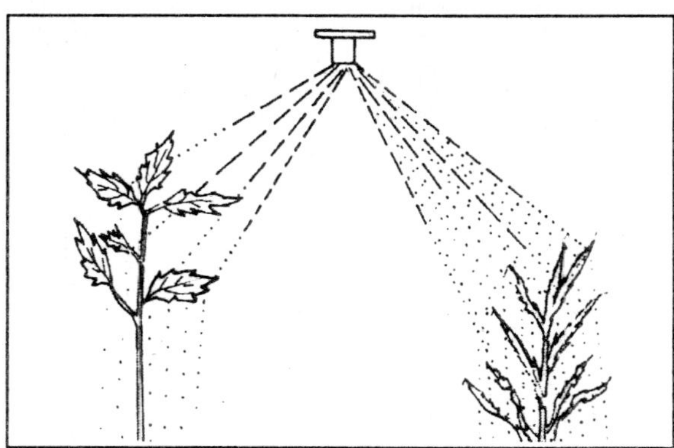

Fig. 3. Differences in spray droplet retention on grass and broadleaf weeds due to leaf orientation

Differences in growth habits of plants

Plant height

When crop rows have a clear advantage in height over the inter-row weeds, directed spraying of herbicides is a common method of achieving selective control of weeds. Mulches are used in standing crop rows for affecting selective control of germinating weeds.

In slow germinating crops like onion and sugarcane, weeds often establish themselves even before crop emergence. These weeds can be controlled selectively by spraying a contact type of herbicide before more than 10 per cent of the crop plants are seen over the ground.

In more advanced stages of crop growth, sometimes specific weed species may grow much above the crop height. In crops like Egyptian clover and spinach, these tall weeds may often completely hide the crop plants. In closely sown crops a low volume application of a contact herbicide can be used wisely to rid them of the crowding, tall weeds. In lawns and gardens, shoots of nutsedges and other erect weeds can be selectively wiped with herbicides from either a herbicide-wax bar or a clothed-stick dipped in concentrated herbicide solution. A wick applicator has been developed at Directorate of Weed Science Research, Jabalpur, India in which herbicides are soaked in wick made-up of cotton (Fig. 4).

Fig. 4. Wick applicator developed at Directorate of Weed Science Research, Jabalpur, India

Root growth

When herbicides are applied to soil, differences in growth habits of underground parts of weeds and crop plants become important in determining their selective absorption. Roots of many annual and shallow-rooted perennial weeds are located in the top soil layer; those of deep-rooted plants may extend many feet into the subsoil such as foxtail (*Hordeum* spp.), shepherd's purse (*Capsella bursa-pastoris*), and chicory (*Cichorium intybus*) as weeds in alfalfa. In general, weed seeds germinate from top 1.25 to 1.5 cm of soil, whereas many crop seeds are planted 5.0 to 7.5 cm deep. When a recommended preemergence herbicide is applied on the soil surface, and the soil moisture conditions are suitable to leach it to about 2.5 to 3.0 cm soil depth, it is readily available for absorption to the germinating weeds. Crop plants that grow their roots beyond 5 cm depth obviously avoid herbicide absorption and escape phytotoxicity. This is the basic principle of selectivity of most of the preemergence herbicides, and it is a function of herbicide structure, formulation and rate, besides the soil texture, organic and inorganic colloids, and the rainfall. When anyone of these factors is unfavourable, the herbicide may either injure crop plants or result in poor weed control. Sometimes, both the adverse effects may occur together when the phenomenon is called 'reverse selectivity'.

Rahman and Ashford (1970) studied selectivity between wheat *(Triticum aestivum)* and *Setaria viridis* with soil incorporated trifluralin and found that *S. viridis* located its coleoptilar node within one cm of soil surface irrespective of the depth from which it germinated while, wheat developed the coleoptilar node much deeper. Since trifluralin is incorporated shallow in soil and the coleoptilar nodes of above plants are of major importance in the absorption of this herbicide, wheat got off safely without absorbing the toxicant. *Setaria viridis,* on the other hand, absorbed trifluralin in phytotoxic amounts and was killed (Fig. 5).

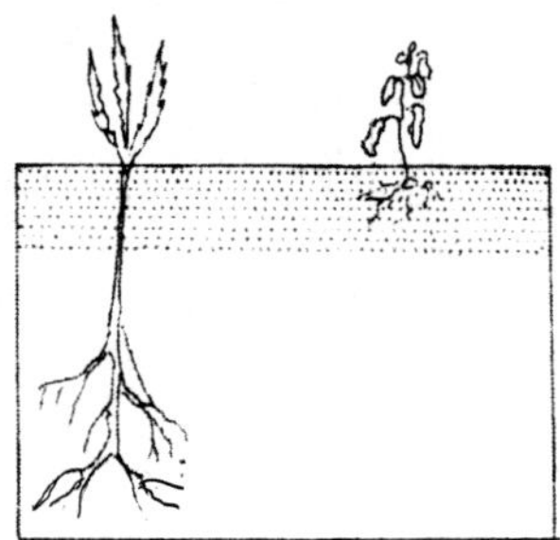

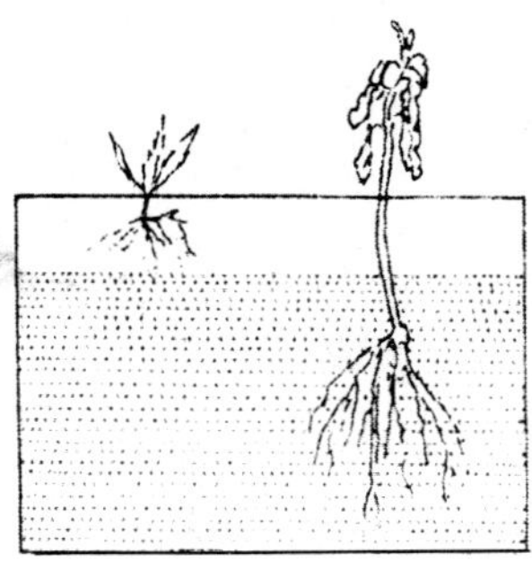

Fig. 5. Placement of an immobile herbicide to the surface will control shallow weeds and leave deep rooted crops unaffected (left). Mobile herbicides can be used to control deep rooted weeds (right)

(2) *Differential translocation of the herbicides to the site of action*

Once the herbicide absorption process is complete, translocation of the herbicide to the site of action becomes the primary physiological function involved in the mode of action. Differential absorption of herbicides by plants results differences in the translocation of herbicides inside plants (for detail see chapter 3). Susceptible plants translocate more amount of herbicide compared to resistant or tolerant ones. Translocation may be apoplastic, symplastic or both based on herbicide. Translocated/systemic herbicides may move from site of application to site of action *via* the symplast, apoplast or both, but contact herbicides have very little or no movement from the site of application. For example, substituted phenylureas are mostly apoplastic except isoproturon which has both apoplastic as well as symplastic movement. Their target site photosynthesis may be affected in varying degrees because of differences in translocation of the same herbicide across plants. As a result, differential selectivity is achieved among plants. Conjugation of herbicides as applicable apart from differential absorption also affects mobility and translocation of herbicides inside the plants.

Slife *et al.*, (1962) found that 2, 4, 5-T was considerably more toxic to *Cucumis trigonus* than 2,4-D because it was translocated much more readily than the later compound inside plants. Ashton (1958) also explained difference in the selectivity between sugarcane (tolerant) and beans (susceptible) to 2,4-D on the basis of its slow translocation in sugarcane and rapid translocation in beans. Better mobility of herbicides in plants, however, does not always mean a quicker plant kills. Actually, in certain instances, it may aid the plant in escaping specific herbicide action. For instance, diphenamid is found selective to *Convolvulus arvensis* because this plant translocates the herbicide rapidly from the roots to the shoots where it is metabolized. Whereas, *Avena sativa* was found susceptible to diphenamid because it failed to translocate diphenamid rapidly away from its roots, which is always the site of phytotoxic action of this herbicide.

(3) *Differential metabolism of herbicides*

Among selectivity factors associated with plant traits, the differential metabolism of herbicides between weeds and crop species, by which susceptible weeds are less able to metabolize selective herbicides, is one of the most important factors (Devine *et al.*, 1993; Cole, 1994; Hatton *et al.*, 1996). In plants, differential metabolism of herbicides is characterized by the conversion of toxic molecules to less toxic compounds, to be stored where these would not affect plant cell survival. One plant species containing such mechanisms can alter or dissipate the herbicide molecule by biochemical reactions producing non-toxic products (Oliveira Júnior, 2001). Herbicide metabolism can be caused by a natural metabolic process of plant detoxification, usually classified into four main phases (Yuan et al., 2007): Phase I, also known as conversion; Phase II or conjugation; Phase III with secondary conversion and transport into vacuole; and Phase IV with deposition of final metabolite (Devine *et al.*, 1993; Cole, 1994; Eerd *et al.*, 2003; Yuan *et al.*, 2007) (see Chapter 5).

Selectivity of metolachlor between maize and yellow nutsedge was found due to chiefly differential metabolism of the herbicide. In addition, yellow nutsedge shoots readily absorbed the herbicide, while in maize it was slow (Dixon and Stoller, 1982). Luckwill and Lloyd Jones (1960) noticed that *Ribes nigrum* (susceptible to 2,4-D), metabolized the herbicide at a rate of 2% of the applied amount in 96 hours whereas the *R. sativum* (tolerant to 2,4-D) metabolized 50% of it during the same period. Selectivity of terbacil between tolerant *Mentha piperita* and susceptible *Ipomoea hederacea* was also found to be due to differential rates of metabolism in the two plants (Barrentine and Warren, 1970). *M. piperita* metabolized terbacil rapidly enough to show only temporary fall in its photosynthetic rates, whereas in *I. hederacea* the herbicide persisted long enough to inhibit photosynthesis to the point of plant mortality. Similar was found true for bentazon and desmedipham selectivities, as well as for many other herbicides.

Atrazine is selective to maize and sorghum. In maize, benzoxaginone, a chemical found in its roots is held responsible for its detoxification. Sorghum, however, lacks benzoxaginone. The possible detoxification mechanism involves an enzyme "glutathione-s-transferase (called triazinyl transferase particularly for this purpose), which forms conjugation with atrazine, increases its molecular weight and thereby affects the mobility of herbicide. Thus atrazine gets prevented from reaching to its site of action, photosystem II of photosynthesis and offers selectivity to sorghum. Another important reaction N-dealkylation is also operative in both sorghum and maize. Therefore, benzoxaginone-mediated non-enzymatic conversion, glutathione-s-transferase-mediated conjugation and N-dealkylation are the possible mechanisms imparting atrazine's selectivity to maize, whereas glutathiones-transferase-mediated conjugation and N-dealkylation (Fig. 6) are likely the mechanisms offering atrazine's selectivity to sorghum (Shimabukuro, 1967). These reactions occur individually and may or may not be simultaneously. Sometimes an individual reaction may impart complete selectivity although a combination of these reactions is responsible.

In some metabolic reactions of herbicides the intermediate chemical structures prove more phytotoxic than the parent compound. Grace in 1939 and Wain in 1955 found that metabolism of 2, 4-DB and MCPB in plants yielded phytotoxic acetic acid intermediates namely 2,4-D and MCPA, respectively though β-oxidation process. It occurred in many non-leguminous plants. Certain legumes like pea, lucerne, berseem, and some other clovers, on the other hand, either lacked the necessary β-oxidation system or operated it sluggishly. Consequently, these legumes proved comparatively tolerent to 2, 4-DB and MCPB (Wain 1957; Norris and Freed 1966; Matsunaka 1969).

Conjugation takes the toxic herbicide concentrations out of the main stream of plant, and plants capable of rapidly doing so obviously prove tolerant to it. Stoller (1969 and 1970) found that HC chloramben conjugated rapidly inside the roots of soybean *(Glycene max)* with glucose molecules, forming N-glucosyl chloramben and an unknown compound chloroamiben-X. In plants susceptible to chloramben the conjugation was found to be so slow that the herbicide translocated from roots to the shoots where it proved phytotoxic. The tolerance of grasses and certain ecotypes of *Convolvulus arvensis* to the foliar application of 2,4-D has also been explained, at least in part, on the basis of its conjugation with glucose and certain amino acids in these plants. With glucose, 2,4-D formed glucoside) D-glucose ester of 2,4-D. Binding of 2,4-D on certain protein films in a tolerant graminaceous plant like sugarcane has also been found an important mode of its selectivity (Brian, 1964). The adsorbed 2,4-D is more rapidly metabolized in plants to the non-phytotoxic compounds than the free 2,4-D molecules. This conjugate, being a volatile substance, escaped from the plant cells into the atmosphere. The pyrazon susceptible test plants like tomato and millets failed to withhold pyrazon by conjugation, and therefore, succumbed to its phytotoxic action.

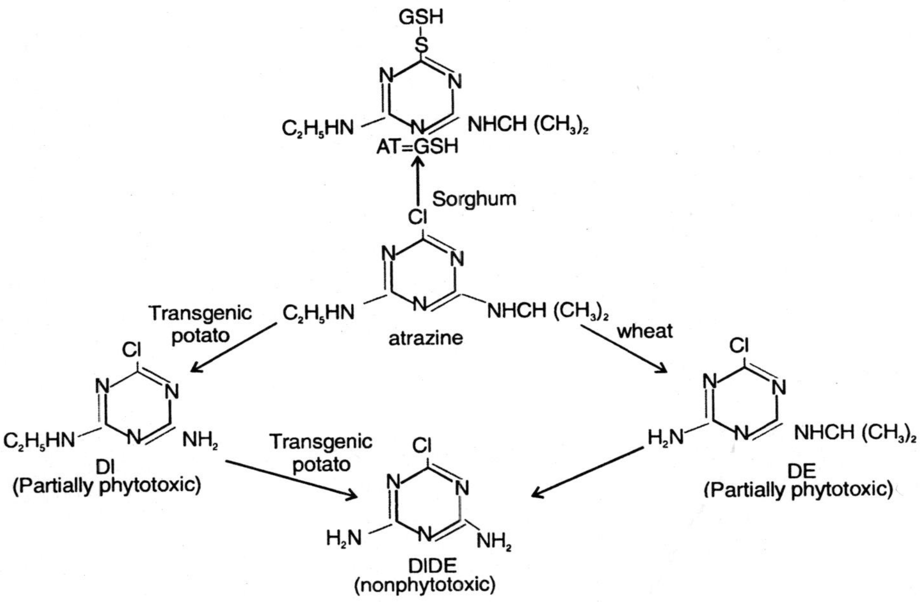

Fig. 6. Metabolism of atrazine (Inui *et al.*, 1999)

It is establised that simazine and atrazine are also deactivated in maize, apple, sugarcane and certain millets by the process of conjugation rather than by earlier believed mechanism of metabolism. Fear and Swanson (1970) purified an enzyme named Glutathione-S-transferase from the leaves of maize. Sorghum, sugarcane, Johnsongrass *(S. halepense)*, and Sundangrass *(S. sudanensis)*, which all are tolerant to simazine. This enzyme was found to catalyze the conjugation of simazine (and probably of other S-triazines) with reduced glutathione in foliage, forming S-(4-ethyl amino-6-isopropylamino-2-s-triazino)-glutathione. Chloride ions were released during the process. Smith (1983) found rapid ^{14}C-atrazine peptide conjugate formation in certain cultivar of *Paspatum notatum.* The role of metabolism in herbicide selectivity is probably the most important one and is dealt with in numerous textbooks (Cobb, 1992; Devine *et al.*, 1993). Fedtke (1982) points out that for the xylem-mobile herbicides the selectivity of a compound can be described by the formula given below,

$$\text{Selectivity} = \frac{\text{Detoxification}}{\text{Uptake} \times \text{Translocation}}$$

As long as the detoxification keeps pace with the incoming uptake and translocation in the denominator, the treated plants will exert tolerance towards this compound. Thus, a low or high capacity of detoxification in itself is not sufficient for the determination of sensitivity or tolerance, but should be associated with the uptake and translocation.

(4) Differential plant metabolism

Different species may follow different pathway for the same process. An example of this is preference of some species of the C_4 pathway over the C_3 or vice versa. In respiration, the hexose monophospahte shunt may be more preferred to the Krebs cycle. Hence when the herbicide acts especially on a reaction involved in the C_3 pathway of photosynthesis, those species following the C_4 pathway predominantly will be tolerant to the herbicide. Likewise, a herbicide attacking the Krebs cycle will hardly affect a species with greater preference for the hexose monophophate shunt.

(5) Tolerating product phytotoxicity

When a plant is exposed to a stressful condition, in which the tolerance threshold is reached, latent physiological or metabolic distresses are revealed as irreversible injuries or chronic symptoms (Larcher, 2000). In the case of selectivity, the amount of visible injuries presented by a plant in consequence of herbicide application is named phytotoxicity. Therefore, phytotoxicity can be interpreted as the overcome of the maximum protection capacity offered by the group of selectivity factors, whereas considering metabolism as the main factor, as the overcome of the species intrinsic ability to detoxify an unfamiliar molecule to the natural metabolism of the plant. In this way, in tolerant plants, part of the herbicide is taken up, quickly metabolized and inactivated before exerting its phytotoxic effects (Oliveira Júnior 2001). However, part of the herbicide reaches the target site of action, leading the plant to express symptoms.

In phytotoxic cases, energy disposal of a plant treated with an herbicide with partial selectivity occurs through two processes. The first process is due to the rapid metabolization of part of the herbicide molecules, and the second is related with the recovering from damages caused by molecules that reached the herbicide site of action. Even considering that the energy used in metabolic processes of detoxification is naturally available and does not cause losses in crop yield, the energy for recovering damaged structures (phytotoxicity) may not be considered as a natural physiological response, in a way that plant recovery can result in a larger or smaller yield losses. Some plants during certain stages of development present a considerable tolerance to defoliation (e.g. soybean) or to the loss of tillers (e. g. sugarcane), with significant compensatory effects. In these cases, the presence of only structural damages looses importance for the physiological damages, mainly when related to effects on growth (Fig 7). Some agrochemicals can affect Cytochrome P_{450} systems, modifying product metabolization (Eerd *et al.*, 2003). In this case, enzyme activity is stimulated (safeners) or repressed (decrease in selectivity), which could affect plant growth and development.

The application of an herbicide might result in a case of temporary or permanent stress (Larcher, 2000), depending on the characteristics of the product (mode of application and rate), of the crop (cultivar or hybrid; stage of development; nutrition; water balance); and of the environment (Oliveira Júnior, 2001; López-Ovejero *et al.*, 2003). There are several articles reporting yield losses or changes in growth caused by serious phytotoxicity. For example, Arruda *et al.*, (1999), observed that the pre-emergence application of sulfentrazone reduced growth of soybean cultivars with low tolerance to the product, leading to reduced leaf area, total dry matter, and mainly dry root matter. Similar results were described by Young *et al.*, (2003), caused by the application of imazethapyr and acifluorfen (Fig. 8).

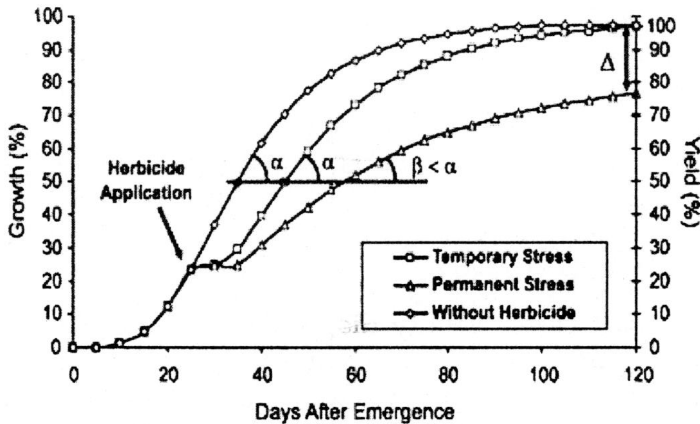

Fig. 7. Mathematic model of phytotoxic effect of herbicides on crop growth and yield. Temporary stress: plant recovers damage without definitive changes on growth rate (α); permanent stress: after partial recovery, the growth rate is permanently damages (β<α). Δ represents the estimative of yield final reduction (Larcher, 2000)

For maize, the combined application of organo-phosphorate insecticides and sulfonylurea herbicides can reduce maize herbicide tolerance, resulting in smaller yields (Porpiglia *et al.*, 1990; Morton *et al.* 1991). Application of trifloxysulfuron-sodium in early stages of cotton development reduced the accumulation of dry matter (Freitas *et al.*, 2005a), whereas application under recommended rate and stage of development presented signs of phytotoxicity that affected crop cycle, and just temporarily the plant growth rate, which did not reduce cotton yield (Freitas *et al.*, 2005b; Freitas *et al.*, 2006).

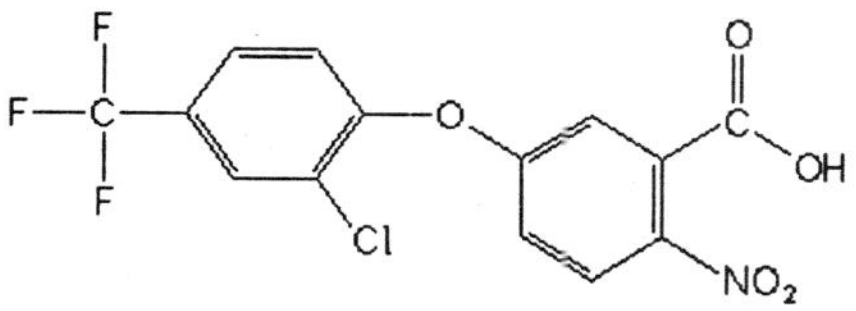

Acifluorfen 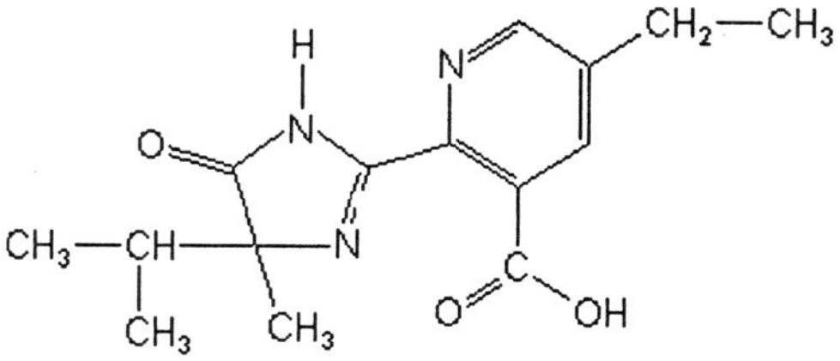   Imazethapyr

Fig. 8. Acifluorfen and imazethapyr caused phyotoxicity to soybean

(6) Chemical characteristics of herbicides

Herbicides are quite specific in their structures as to whether or not herbicidal activity is possible. Slight changes in conformation or structure will alter herbicidal activity. A particular molecular structure required for a herbicide to become selective to crops. Diuron (Fig. 9), a substituted phenylurea herbicide, is selective to cotton and sugarcane, but phytotoxic to wheat. When the "para chlorine atom" (p-CI) of the phenyl ring of diuron is changed by a "methyl" ($-CH_3$) group, the compound formed is chlortoluron, another herbicide of the substituted phenylureas. Chlortoluron is not phytotoxic to wheat like diuron, rather selective to wheat. Similar example of 2, 4-D, MCPA against 2, 4-DB, MCPB can be cited. These 2, 4-D, MCPA, 2, 4-DB and MCPB are broad-leaved killers and belong to phenoxyalkanoic acids (Fig 9). However, 2, 4-DB and MCPB are selective to pulses, whereas 2, 4-D and MCPA not. This may be attributed to structural changes in the side chain of those herbicides on subsequent reactions inside the plants.

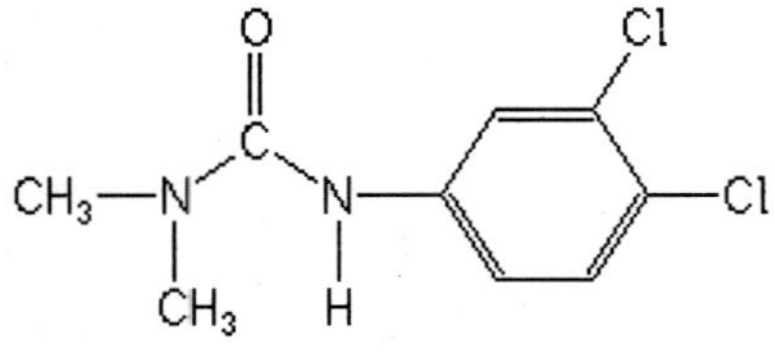

Diuron

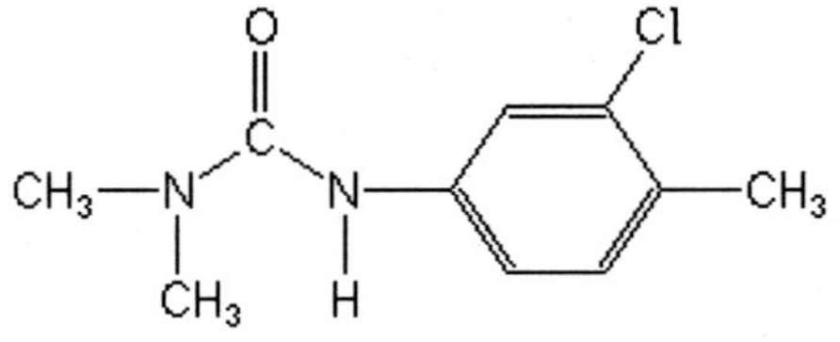

Chlorotoluron

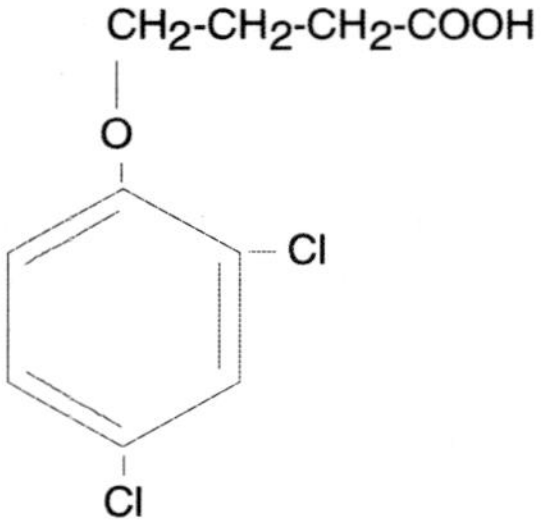

2,4-DB

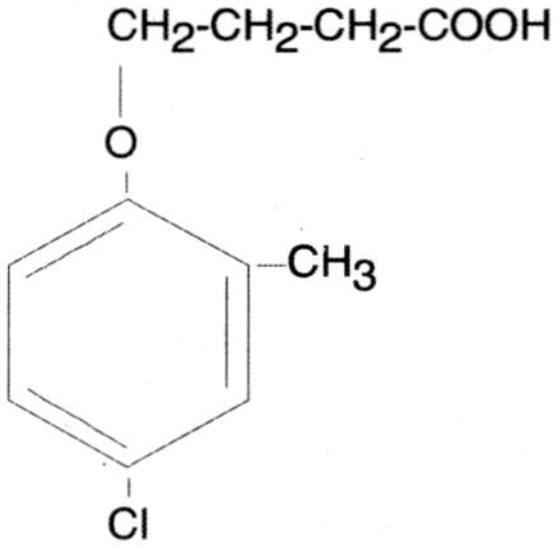

MCPB

Fig. 9. Structural similarities between 2,4-DB and MCPB and diuron and chlorotoluron

Trifluralin and benefin differ in only a methyl group moved from one side of the molecule to the other, yet trifluralin is about twice as active as benefin (Fig. 10). Esters of phenoxy (MCPP etc.) acids are usually much more active than are amines. The manner of formulation of an herbicide can affect its selectivity. The most extreme case of this might be granular formulations which bounce off desirable plants to reach the soil where they then limit germinating weeds. Other substances known as adjuvants or surfactants are often added to improve the application properties of a liquid formulation and increase activity.

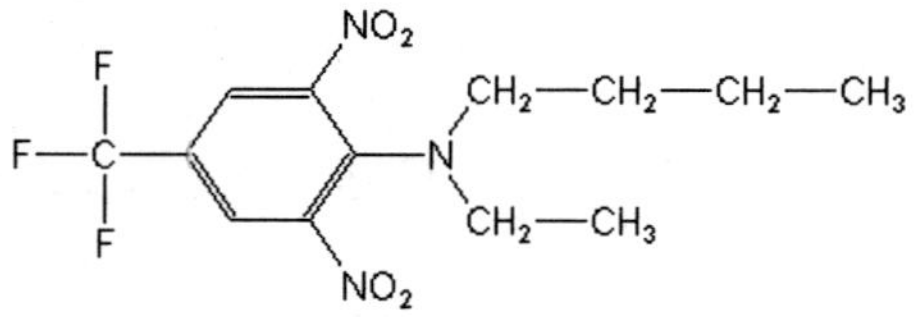

Trifluralin

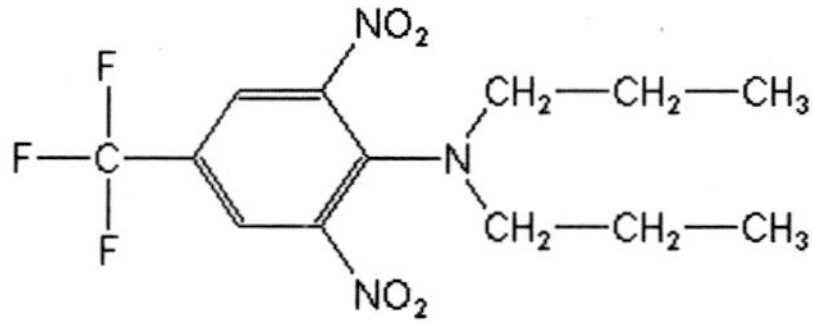

Benefin

Fig. 10. Trifluralin and benefin differ in only a methyl group, yet trifluralin is about twice as active as benefin.

The manner in which an herbicide is applied can also affect its selectivity. When a broad-spectrum postemergence herbicide like glyphosate is applied as a shielded, directed, or wicked application within a susceptible crop, susceptible foliage is avoided and selectivity is achieved with this normally non-selective herbicide. Herbicides can be grouped into families based on the type of action that they have within affected plants (their mode of action). Herbicides which affect similar sites or processes within affected plants produce similar injury symptoms.

(7) Use of granular formulations

Selectivity between established crop plants and associated germinating weeds can be successfully achieved by application of granular formulations which gives opportunity to weeds for selective absorption of the toxicant. Granules filter through the crop foliage, leaving little herbicide for absorption by it and then settle on the moist soil where germinating weeds absorb the herbicide released slowly by the granules. Good selectivity has been achievement with granular formulations of dinoseb, diuron chlorpropham and nitrofen in crops that would otherwise be injured by their spray liquids. An important desirable character of herbicides embodied in the granules is their low leachability in soils, else the toxicant may move down to the crop roots.

(8) Use of adsorbents and antidotes (safeners)

Activated charcoal has been established as a strong adsorbent of many herbicides such as 2, 4-D, EPTC, 2, 4, 5-T, propham, pyrazon, propachlor, trifluralin, nitralin, chlorambean, diuron, neburon, atarzine, simazine, and butachlor. Activated carbon was first used as an adsorbent of 2, 4-D in the form of a uniform 2.5 cm wide band over partially covered crop seed rows. (Linscott and Hagin, 1967). If germinating crop seeds and seedlings are surrounded by a layer of activated charcoal within the soil, they can be made to escape absorption of many soil-applied herbicides. It offers an excellent opportunity for stretching selectivity of herbicides to a wide variety of crops.

In dibbled vegetables activated carbon can be placed over crop seeds in seed holes. Pre emergence herbicides sprayed later are retained by carbon, thus avoiding their absorption by the germinating crops seeds. Robinson (1965) reported success with dipping roots of horticultural transplants in activated carbon before transplanting them in herbicide treated soils. Both, 'seed hole' and 'root dip' methods of applying activated carbon for 'improving herbicide' selectivity require much less adsorbent than the earlier mentioned 'furrow method'.

In seed pelleting with activated carbon method, carbon is coated over crop seeds using gum or PVA (polyvinyl acetate). This technique has proven successful in improving selectivity of chloramben, butachlor, and EPTC to rice and EPTC to maize and cowpea, (Gupta and Niranwal, 1976). Various chemicals have been evaluated as herbicide antidotes, also known as safeners, which protect crops from herbicide damage without protecting weeds (Hatzios, 1983). The mode of action of herbicide antidotes in protected species remains unclear, particularly in cases where the mechanism of herbicide. Hoffman (1948) observed antagonism of 2, 4-D to 2, 4, 5-T on tomato plants. NA seed coating has been found to provide protection to cultivated oat *(Avena sativa)* against preemergence alachlor, and to maize and sorghum against perfluidone and diclofop.

Seed-dress treatment with NA (1, 8 naphthalic anhydride) and EPTC are reported as highly successful safeners of EPTC and butylate in maize (Fig. 11). R-25788 (=*N*, *N*-dially 1-2, 2-dichloroacetamide) was successfully applied to soil as a tank-mix with EPTC without fear of protecting any EPTC susceptible weeds. (Burnside *et al.*, 1971; Chang *et al.*, 1972 and 1972a). R-25788 had been found an antidote of alachlor and metolachlor in protecting maize seedlings. CGA-43089 [oc-(cyanomethoximino)-benzacetonitrile] provides safety to sorghum against metolachlor by seed treatment at 1-1.5 kg seed. It is likely that NA, and possibly other safeners, enter the target plants and compete there with herbicide molecules for a binding site on some native enzyme. Thus safeners have proved successful only against herbicides that affect cell division in one way or the other, and not against the photosynthetic inhibiting herbicides.

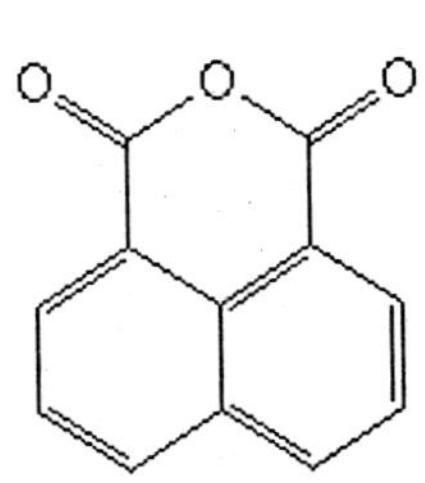

Fig. 11. NA (1, 8naphthalic anhydride) is reported as highly successful safeners of EPTC and butylate in maize

(9) Genetic engineering

An exciting recent development in the field of herbicide selectivity has been the recognition of the potential for genomic modifications of the plant resulting in altered herbicide sensitivity. Using recombinant DNA technology involving gene isolation, construction of a vector and transfer of the gene into the recipient plant species, some very encouraging results have already been reported with modifications to the target site of action of glyphosate (Comai *et al.,* 1985). Alternative methods of the transfer of genes determining herbicide selectivity have included plant breeding techniques (Faulkner, 1982) and protoplast fusion (Gressel *et al.,* 1982).

Several genes which code for pesticide detoxifying enzymes have already been considered as candidates for gene transfer (Hahlbrock *et al.,* 1984; Sandermann, 1985). Ideally, such genes code for a single step detoxification process and a list of potential plant enzymes and herbicide substrates is presented in Table 3. An alternative source of detoxifying genes is microorganisms. Gene transfer of a mutant *aro A* gene coding for a modified 5-enolpyruvylshikimate 3-phosphate synthase, which was less sensitive to glyphosate inhibition than the plant enzyme, has already been achieved using *Salmonella typhimurium* as the source and tobacco plants as the recipient (Comai *et al.,* 1985).

Recently, the transformation of plants with the monooxygenase gene from *Alcaligenes eutrophus* which cleaves the acetate side-chain from 2, 4-D to give the less phytotoxic 2, 4-dichlorophenol has been suggested (Sandermann, 1985).

Factors affecting selectivity

Plant age and stage of development

Since plants have definite periods or stage of sensitivity to herbicides, selectively can be obtained if herbicides are applied at times when the crop plant is most tolerant and the weed is most susceptible. For example, corn is tolerant to many postemergence herbicides if treated before tasseling and after the dough stage of grain development. Many soil applied herbicide kill germinating and newly emerged seedling but fail to control beyond later flushes of emerging weeds in crops that are established. Similarly herbicides can be applied safely to conifers prior to bud break in the spring and after rapid terminal growth has ceased in the fall.

Height difference between weeds and crops can be exploited by directing the herbicide spray away from the crop and onto the weeds and herbicides with limited selectivity can be used in number of cropping situation to control weeds not effectively or safely controlled using other means. These includes 2,4-DB, linuron or paraquat as directed sprays in soybeans, glyphosate with the rope wick or spongewick applicator in soybeans and cotton and a number of foliage applied herbicides used as directed sprays to control herbaceous weeds in established woody species.

Growth of plant

Plants tolerance to a herbicide varies across growth stages of a crop. Therefore, to avoid crop injury and, to achieve greater selectivity, herbicides are to be sprayed during a 'tolerant' phase of a crop particularly when post-emergence application is sought. For example, wheat is tolerant to 2,4-D in the tillering to late tillering stage, but sensitive to injury from germination to four-leaf stage. Some herbicides are also recommended when the crop is dormant. The time of application is again very important for perennial weed control. A herbicide needs to be transported to the underground parts for killing them. If the plant is newly emerged, the sugars being produced will be translocated to areas of new growth. When the plant is older, the sugars will move into storage tissues. It is at this time when herbicide application will be most effective.

Time and dose of herbicide application

Pre-plant incorporation (PPI) and pre-emergence applications are likely to offer more selectivity to a crop than a post-emergence application. Pre-emergence herbicides having lower residual action or quickly deactivated in soil should be applied to kill geminating weeds in crops (Fig. 12). It is always advisable that non-residual, selective, but systemic herbicides are applied as post-emergence so that they can act through foliage. A good example of a non-residual herbicide is clodinafop-propargyl, which is systemic and selective to wheat, barley and cultivated oats. This also holds true for the non-selective herbicides, paraquat and glyphosate, which have very less or no residual activity in soil. Paraquat is applied to potato at 0-5% sprouting (not beyond that) to control the emerged weeds. The sprouted potato plants may be damaged, but selectivity is maintained for rest of the potato plants, which are yet to sprout. This is chronological selectivity.

A herbicide is selective to a crop at a particular dose and, therefore, selectivity is dose-dependent. For example, atrazine is selective to maize when applied at the recommended dose (1.0-2.5 kg/ha). At higher doses, the conducting tissues of plants may be damaged and further translocation of herbicide may be cut-off. As a result, detoxification of herbicide at the required site(s) may be affected. Even though translocation is not affected, normal metabolism in plants may not be able to detoxify herbicide applied at higher rate than recommended and herbicide residue will build up inside the plants. All these will affect selectivity. In non-crop situations, where selectivity not required, herbicide may be applied at a high rate for indiscriminate killing of

Fig. 12. Manipulation in herbicides application can bring about selectivity

Method/device of herbicide application

Some susceptible crop plants can be kept physically away from the contact of herbicides, which are non-selective to them through manipulation of application technique. This can be accomplished by using spray hood/shield, special applicators like rope-wick applicator (RWA), herbicide glove or by directing spray away from the crops. Paraquat. glyphosate, glufosinate-ammonium, 2,4-D are some of the herbicides where manipulation of the spraying method render their use in some crops safe.

Vigor and level of nutrition of the plant

Well nourished and vigorously growing plants of susceptible species will be rendered more susceptible to herbicides action than those suffering from mineral deficiencies. Some elements are necessary for food translocation and indirectly affect herbicide translocation to the site of action. In tolerant species wherein resistance is based on conjugation or metabolism of the herbicide, a vigorous metabolic activity can further increase resistance through faster generation of conjugating materials and more active degrading agents. Sustained enzyme synthesis requires adequate nutrition for the plant.

Retention

Variation in plant shape, leaf orientation and leaf surface characteristic can account for differences in foliar retention. Plants with broad horizontal leaves are more likely to intercept and retain spray than the vertically oriented leaves such as grasses and sedges. Leaves with a moderate degree of hairiness or surface roughness may be better able to tap and retain the runoff of spray than leaves that are smooth and non-hairy, however an excessive amount of hairs on the other hand may prevent the penetration of the herbicide to the cuticular surface.

Penetration

Differential penetration is probably most important selectivity factor among plants of different age since cuticle thickness tends to increase and cuticle composition may change as the plant matures. This is probably one of the reason that seedling plant are much more susceptible to herbicides than mature plants. The addition of wetting agent or alteration in herbicide formulation can decrease both retention and penetration related selectivity by increasing the amount of herbicide entering the plant. This is often the cause of injury among crops that normally show a high degree of inherent tolerance to a particular herbicide. For example, the addition of corn oil to 2,4-D amine can result in injury to corn when 2,4-D amine alone would be nontoxic. Ester formulation of 2,4-D penetrate plant tissue more readily than amine and other salt formulations and are therefore recommended at lower dosages in order to be used safely for broadleaf weed control in corn and small grains.

Role of other chemicals in toxic mixture

The addition of other chemical can alter the activity and selectivity of herbicides. Combining diclofop with phenoxy herbicide results in reduced effectiveness of the diclofop for wild oat control. Some arylophenoxy propionate herbicides show antagonism when tank mixed with other post emergence herbicides. Safeners or protectants have been shown to improve the tolerance of partially tolerant grass species to herbicides in the thiocarbamates and chloroacetamide class without increasing the tolerance of the susceptible species. Some protectants are packed with herbicide. The safener dichlormid is packaged with butylate or acetochlor and R-29148 is packed with EPTC or butylate. Both safeners increase corn tolerance to the herbicides. In most cases the safeners appear to protect the crop by increasing the metabolism of the herbicides. They do this by increasing the level of either glutathione or glutathione-S-transferase, which increases glutathione conjugation or the P_{450} enzymes, which enhance the oxidation of herbicides. The use of safeners can result in a marginally safe compound becoming an important and effective herbicide Fig.13.

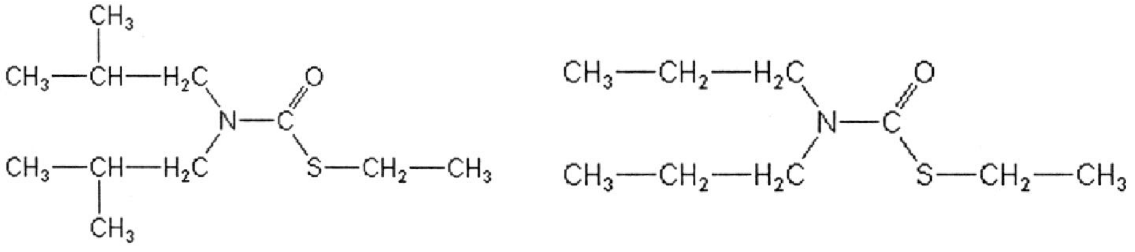

$$(CH_3-CH(CH_3)-H_2C)_2N-C(=O)-S-CH_2-CH_3$$

Butylate

$$(CH_3-CH_2-H_2C)_2N-C(=O)-S-CH_2-CH_3$$

EPTC

Fig. 13. Safener (R-29148) is packed with EPTC or butylate

Adjuvants

Adjuvants increase the potency of a herbicide by increasing the availability of herbicide at the site of action. They may also bring about perceptible changes in the selectivity of herbicides. For example, fenoxaprop-p-ethyl (Fig. 14) is a phenoxyphenoxyalkanoic acid derivative and an exclusively grass killer herbicide. It shows differential selectivity to soybean and wheat mainly due to changes in the surfactant or wetting agent of the formulation.

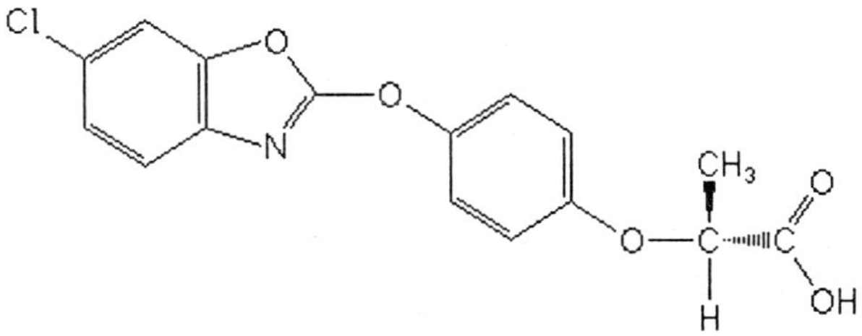

Fig. 14. Fenoxaprop-p-ethyl shows differential selectivity to crops due to changes in the surfactant or wetting agent

Environmental Factors

The environment governs plant's response to any herbicide. Environmental factors such as temperature, sunlight, relative humidity, rainfall/moisture, soil texture etc. affect physiological and biochemical processes for plant growth and development in one hand and the absorption, translocation and metabolism of herbicides on the other. Plant species respond differently to environmental factors and therefore, have differences in growth and development, which may lead to differential absorption and translocation of herbicides across plants. Under severe stress condition, plants may not function properly, metabolism of herbicide may be insignificant, and as a result, a normally tolerant plant may fall susceptible to a herbicide.

At high temperature, greater uptake, translocation and metabolism of herbicides are expected to occur because of higher transpiration. For example, isoproturon and barban have faster metabolism at higher temperatures and thereby have less activity. Atrazine although selective may become phytotoxic to maize in wet and cool conditions due to excessive uptake, but not under hot and dry conditions. However, under conditions of high temperature and high moisture stress, the rate of penetration of herbicide molecules will be definitely reduced to a great extent.

At higher humidities, the leaf cuticle may become hydrated, the herbicide droplets will be retained for more time on the leaf/canopy due to slower evaporation and hence more penetration and absorption of herbicide through leaf are expected. These may likely increase the activity of a foliage-applied herbicide. High humidity also facilitates penetration of water-soluble chemicals through cuticle. In case of soil -applied herbicides, if the humidity is more, the uptake of herbicide is likely to be less due to less transpiration.

Light helps herbicide penetration by stimulating stomatal opening and also activates photosynthesis leading to greater movement of organic solutes and concurrently herbicides from the leaf to other parts of the plants. Light is a pre-requisite for the activity of almost all herbicides since nearly 80% of the herbicides are affecting photosynthesis directly or indirectly. Phytotoxic symptoms (chlorosis, necrosis, burning, yellowing, die-up, wilting etc.) in plants due to herbicide (e.g. paraquat, diquat) also appear quickly in presence of light (Fig. 15). The uptake and translocation particularly of the foliage-applied systemic herbicides are likely to be greater at higher light intensity. Increased light intensity generally increases the rate of photosynthesis which, in turn, increases the translocation of photosynthates inside the plants and hence the translocation of herbicide through photosynthates. Different translocation may lead to differences in selectivity across plant species.

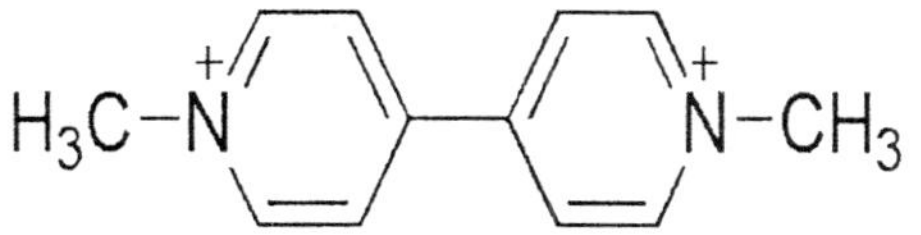

Paraquat

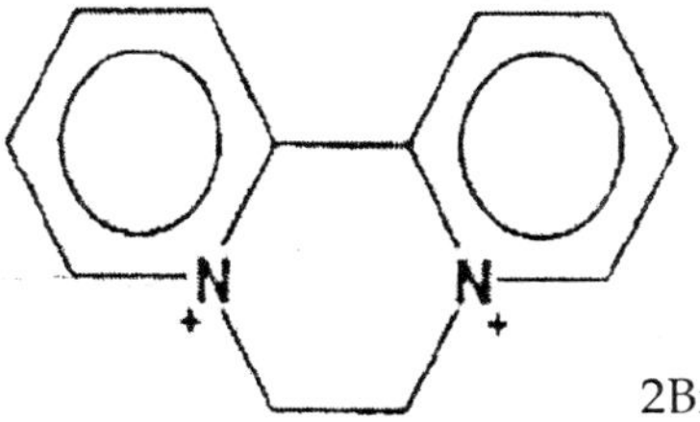

Diquat

$2Br^-$

Fig. 15. Diquat and paraquat caused phytotoxic symptoms in plants in presence of light

Immediate rains after application reduces effectiveness of foliage-applied herbicides due to washing of the spray deposits off the foliage. Whereas drizzling for a very short period after herbicide application may be equally useful to both soil- and foliage-applied herbicides. In case of herbicides applied to the foliage, drizzling may wash the rough cuticle of leaf/foliage, redistribute the herbicide droplets and increase their intactness on the plant/leaf surface and thus may increase the availability, absorption and overall efficacy of a herbicide. In case of soil-applied herbicides, little drizzling may incorporate surface-applied herbicides into soil and may increase their activity and selectivity. However, heavy rains may wash away both foliage-active as well as soil-active herbicides and enhance their loss and reverse selectivity may occur.

The mechanism of selectivity of pre-emergence and pre-plant incorporated herbicides through "depth protection" may be affected since herbicides will percolate to deeper layers of soil and may 'reach to crop roots or seeds. Polar herbicides such as dalapon, 2,4-D amine salts, etc., which are highly water soluble, cannot penetrate the leaf wax rapidly and there is chance that they are washed off by rain occurring within few hours of application. On the contrary, the absorption of paraquat, 2,4-DEE (ester), which penetrate very rapidly, is not severely affected by rainfall even within few hours of application. Generally, herbicides formulated in oil or oil emulsions are less affected by rain then the aqueous solutions.

Soil Factors

Organic matter having huge surface area and negative charges influence herbicide availability to plants. Higher the organic matter content, higher is the dose of herbicide required to achieve the same level of efficiency of a herbicide. Weak basic, cationic, non-ionic polar and nonpolar herbicides undergo sorption with soil colloids and become unavailable to plants. This results in decreased activity of the herbicides.

Heavily-textured clayey soil requires more dose of herbicide compared to light-textured sandy soil for the same degree of weed control. Leaching of herbicide is more in light soil, whereas adsorption is more in heavy soil. In light soil, leaching, therefore, may cause phytotoxicity and affect selectivity by reaching to the crop root zone even though the dose of herbicide is less. On the contrary, in heavy soil due to sorption of herbicides with soil colloids, crops may escape phytotoxicity or selectivity remains unaffected, but there is a chance that weeds may not be controlled effectively (Fig. 4).

Soil temperature affects microbial growth and activity, photodecomposition and chemical decomposition of herbicides in soil and thereby plays an important role in herbicide metabolism. Metabolism of herbicide increases in warmer soil than in cold soil. The activity of microorganisms as well as the degradation of herbicide is higher at higher soil moisture. This results in the lower persistence of herbicide in soil. pH affects adsorption of ionic/polar herbicides but not of non-polar/apolar herbicides.

Herbicide Resistance

Resistance is the ability of a formerly susceptible plant population to survive herbicide doses above that were used to control the original plant population. The resistant population is referred as resistant biotype. The development of resistances become apparent only after several years of repeated application of the same herbicide and is usually characterized by plants that survive a dose several to many times i.e. 20 to 1000 times or even more greater than normally needed for commercially acceptable weed control. These high levels of resistance are most frequent caused by a single gene mutation that modifies the herbicide binding site on an enzyme or protein. If the herbicide can no longer bind to its site of action, it no longer has an inhibitory effect. Occasionally, differential metabolism, uptake, availability or movement will be responsible for the development of resistance (Fig 16). Once a plant population develops resistance to a single herbicide it is likely that it will be resistant to other, closely related herbicides as well. For example atrazine and simazine belong to same chemical family but atrazine and terbacil have the same mode of action.

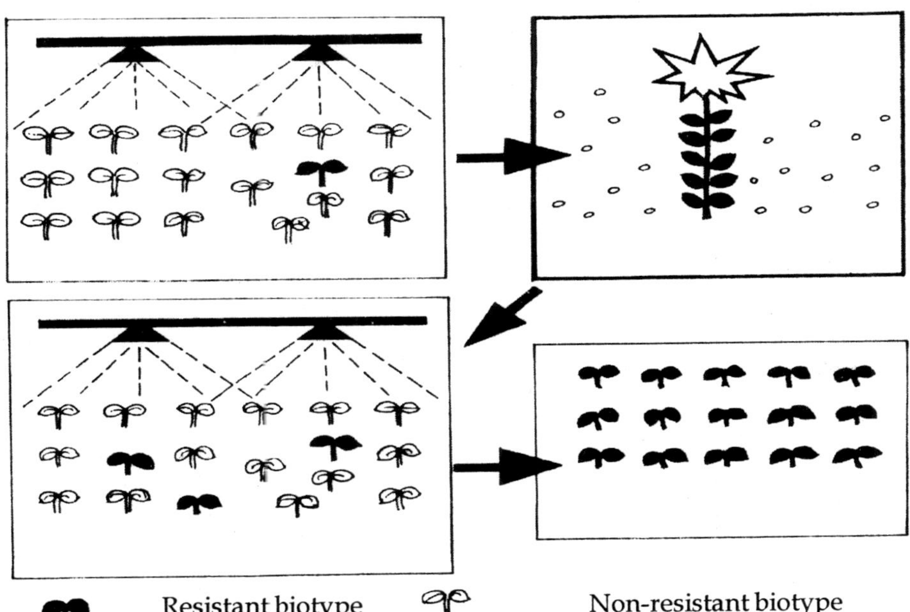

Fig. 16. Evolution of herbicide resistance in weeds

Most recently, in 2003-2006, a very potent broad spectrum herbicide glyphosate (Fig. 17) has been found resistant against some weeds such as grassy weed Johnson grass, *Sorghum helepense* in Argentina (Valverde and Gressel, 2006) and Rigid rye grass bio-type *Lolium rigidum* has exhibited resistance in almond orchar of northern California.(Vargas, 2001). Both hairy fleabane (*Conyza bonariensis)* and buckhorn plantain (*Plantago lanceolata)* have been reported glyphosate resistant in South Africa. Hairy fleabane has been difficult to control with glyphosate in California production system indicating possible resistance. Similarly, reports of poor or ineffective control of *Chenopodium* Sp. in Roundup Ready cotton system have surfaced in the last two years. Recently, Roundup resistance horseweed (marestail) *Conyza canadensis* has been confirmed in the eastern US (Vargas and Wright, 2004). Until 2006, WSSA updated worldwide Roundup (glyphosate) resistant species enlarged to 11. (Table 1).

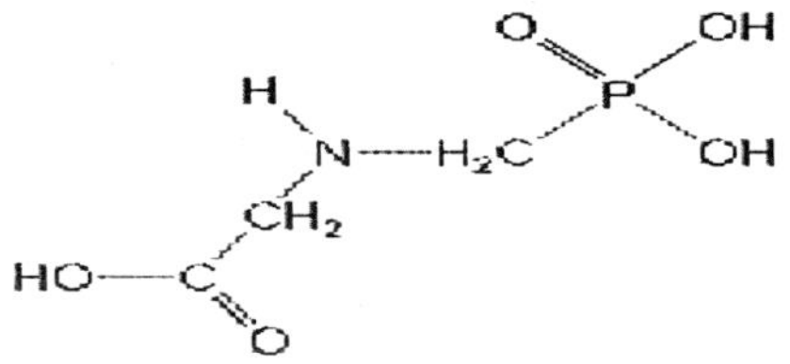

O OH
H P
N—H_2C OH
CH_2
HO—C
O

Fig. 17. Chemical structure of glyphosate

Farmers are being challenged to control glyphosate resistant weeds like *Palmer amaranth, Amaranthus palmeri* and giant ragweed (*Ambrosia trifida*) in certain crops. Monsanto commercialized Glyphosate as RR (Roundup-Ready) with their respective genetically modified (GM) transgenic crop seed. This RR- seed package in soybean, corn, canola and cotton where offered a weed free high farm yield simultaneously, did evolve herbicide resistant biotypes, a challenge that has now become a key issue for all stakeholders. GM transgenic herbicide resistant crops are becoming volunteer weeds are also associated with segregation and introgression of herbicide resistant traits in weed population that has ecological impact on plant communities (Owen and Zelaya, 2004). Most recent world wide updated WSSA inventory has provided a comprehensive information on herbicide resistant weeds by country and by mode of action as compiled by WSSA.

Table 1. Worldwide herbicide resistance against widely used types of herbicide groups. Source: (LeBaron, 1991., Valverde and Gressel, 2006)

Herbicide	Year of resistance found	Year of reporting
2,4-D	1945	1963
Dalapon	1953	1962
Atrazine	1958	1988
Picloram	1963	1973
Trifluralin	1963	1982
Diclofop	1977	1982
Trialate	1962	1987
Chlorsufuron	1982	1987
Glyphosate	2003	2006

In recent years, the appearance of herbicide resistance in plants is increasing exponentially as compared with the case of other pests such as insects and fungi. Besides 61 species of triazine resistance, there are more than 200 species surfaced resistant to 16 other classes of herbicides (Fig. 18). At least one weed species has emerged resistant to herbicide in each country of the world where herbicide is used as farm input. Total number of resistant weed species against site of action based classified herbicides is presented in Table 3. Resistance however, evolves rapidly where monoculture cropping is prevalent. Weeds belonged to one dozen genera have shown developed resistance at almost a dozen or more locations worldwide (Table 2, Fig. 11)

Table 2. Most common genera of weeds developing resistance to herbicides worldwide. Source: (Vargas and Wright, 2004)

Genus	Common name	Number of documented occurrence of herbicide resistance
Amaranthus	Pigweed	42
Chenopodium	Lambsquarters	25
Conyza	Fleabane or horseweed	22
Lolium	Ryegrass	21
Setaria	Foxtail	17
Avena	Wild oat	15
Echinochloa	Barnyardgrass or watergrass	15
Alopecurus	Blackgrass	13
Senecio	Groundsel	12
Polygonum	Knotweed or smartweed	12
Solanum	Nightshade	11

Similarly, little seeded canary grass *Phalaris minor* and Jungle rice, *Echinochloa colona* have reportedly developed resistance against isoproturon and propanil, respectively, within 3 years 1990-93. As a result, crop yield losses are reported in the tune of 40-60 percent in the affected areas of India (Malik and Singh, 1995). *P. minor* is steadily developing resistance against alternate used herbicides such as clodinofop and sulfosulfuron (Mahajan and Brar, 2001).

Cross resistance is a specific term used to describe weed population that are resistant to two or more herbicide that have the same mode of action. In some cases the herbicide may be of the same chemical family such as, atrazine and simazine and in other case they are in different chemical family but having same mode of action such as atrazine and terbacil or the sulfonylurea and imidazolinones herbicide. Cross resistant to two such herbicide initially occurs when a plant is exposed to and develop resistance to the first herbicide, because the modes of action are the same, the plant also is resistant to the second herbicide even though it may never been exposed to that herbicide.

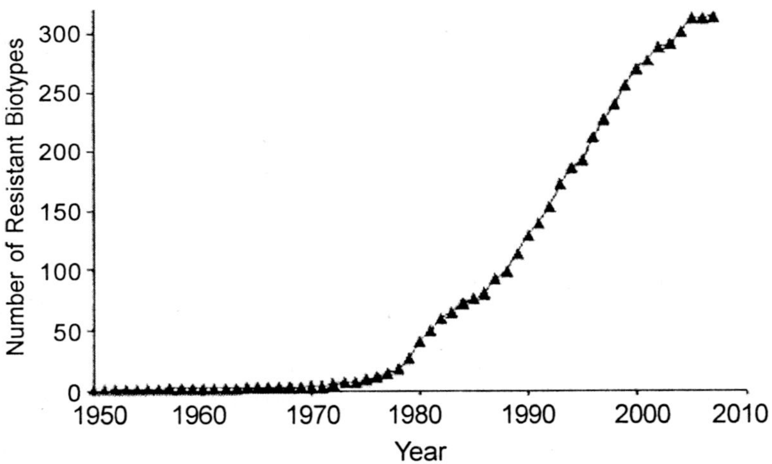

Fig. 18. The Chronological increase in the number of herbicide resistant weeds for several herbicide classes. [Source: Heap 2007]

Multiple resistant is used to describe weed population from different chemical families and with different modes of action. Multiple resistant occurs when weeds are expressed to and develop resistant to different chemistries; it is much less frequent than cross resistance. The most common example of multiple resistant is that of rigid ryegrass (*Lolium rigidum*). One biotype of this species is resistant to at least nine chemical families with five different mode of action (Burner *et al.*,1994). Multiple resistant is attributed to mechanism not associated with the sites of action but to differential uptake, retention, translocation or metabolism. Weeds have been much slower to exhibit resistance to chemical than disease organism and insects. Weeds have fewer generation per growing season and maintain a reservoir of the original population due to buried and dormant seed. In addition, resistant plants that do develop can be less competitive than nonresistant plants. These less competitive plants may produce low numbers of seeds or may never even live long enough to set seed. Not all herbicides have been shown to be resistant to 15 different herbicide modes of action groups somewhere in the world.

Herbicide classes which are frequently associated with resistant belong to different mode of action are; sulfonylureas, imidazolinones and triazoloyrimidines (branched chain amino acid inhibitors oo ALS/ AHAS inhibitors), arylophenoxy propionate and cyclohexanediones (Lipid biosynthesis inhibitors or ACCase inhibitors) (Table 3).

Table 3. Total number of herbicide resistant biotypes of world (Heap, 2002)

Herbicide Group	Mode of Action HRAC	HRAC Group	Example Herbicide	Total
ALS inhibitors	Inhibition of acetolactate synthase ALS (acetohydroxyacid synthase AHAS)	B	Chlorsulfuron	70
Photosystem II inhibitors	Inhibition of photosynthesis at photosystem II	C1	Atrazine	63
ACCase inhibitors	Inhibition of acetyl CoA carboxylase (ACCase) A	A	Diclofop methyl	27
Bipyridiliums	Photosystem-I-electron diversion	D	Paraquat	21
Synthetic auxins	Synthetic auxins (action like indoleacetic acid)	O	2,4 D	21
Ureas and amides	Inhibition of photosynthesis at photosystem II	C2	Chlorotoluron	20
Dinitroanilines and others	Microtubule assembly inhibition	K1	Trifluralin	10
Thiocarbamates and others	Inhibition of lipid synthesis - not ACCase inhibition	N	Triallate	6

contd...

Triazoles, ureas, isoxazolidiones	Bleaching: Inhibition of carotenoid biosynthesis (unknown target)	F3	Amitrole	4
Glycines	Inhibition of EPSP synthase	G	Glyphosate	4
Chloroacetamide s and others	Inhibition of cell division (Inhibition of very long chain fatty acids)	K3	Butachlor	2
Nitriles and others	Inhibition of photosynthesis at photosystem II	C3	Bromoxynil	1
Carotenoid biosynthesis inhibitors	Bleaching: Inhibition of carotenoid biosynthesis at the phytoene desaturase step (PDS)	F1	Flurtamone	1
Mitosis inhibitors	Inhibition of mitosis / Microtubule polymerization	K2	Propham	1

Herbicides belong to these classes have developed resistance after three to five years of continuous use. Approximately 60 species are resistant mostly to the symmetrical triazines such as atrazine, simazine and prometon (photosynthesis inhibitor) within a decade or more of continuous use. Bipyridiliums such as paraquat and diquat (photosystem I energized cell membrane destroyers) and dinitroanilines (microtubular/spindle apparatus inhibitors) class of herbicide showed occasional resistance (Fig 19).

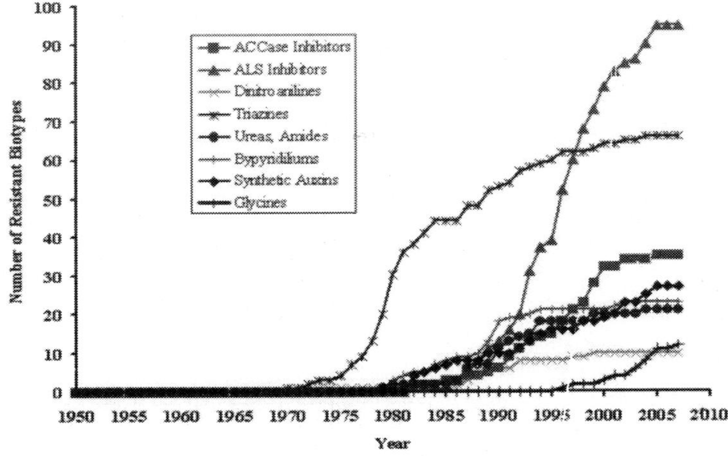

Fig. 19. The Chronological increase in the number of herbicide resistant weeds for several herbicide classes. (Heap, 2007)

Mechanism of herbicide resistance

Mechanisms of herbicide resistance can be broadly grouped into two categories (Dekker and Duke, 1995)

1. Exclusionary resistance

Those that exclude the herbicide molecule from the site in plants where they induce toxic response. In exclusionary resistance mechanism the herbicide is excluded from the site of action in many ways,

(a) Differential herbicide uptake: In resistant biotypes the herbicides are not taken up readily due to morphological uniqueness like overproduction of waxes, reduced leaf area etc.

(b) Differential translocation: In resistant biotypes the apoplastic (cell wall, xylem) and symplastic (plasma lemma, phloem) transport of herbicide is reduced due to different modifications.

(c) Compartmentation: Herbicides are sequestered in many locations before it reaches the site of action. e.g. some lipophilic herbicide may become immobilized by partitioning into lipid rich glands or oil bodies (Stegink and Vaughn, 1988).

(d) Metabolic detoxification: Herbicide is detoxified before it reaches the site of action at a rate sufficiently rapid that the plant is not killed. The biochemical that detoxifies herbicides can be grouped into four major categories: oxidation, reduction, hydrolysis, and conjugation.

Three enzyme systems are known to be involved in resistance due to increased herbicide detoxification. Resistance to atrazine in some population of *Abutilion theophrasti* is due to increased activity of glutathione-s-transferase that detoxifies atrazine. Resistance to propanil in *Echinochloa colona* is due to the increased activity of enzyme aryl-acylamidase that detoxifies propanil. Increased herbicide metabolism due to cytochrome P_{450} monoxygenase is responsible for resistance to inhibitors of ACCase, ALS and PSII in a number of grass weed species.

2. Site of action resistance

Those that render specific site of herbicide action resistance due to,

(a) Altered site of action: Site of action is altered in such a way that it is no longer susceptible to the herbicide e.g. In *Lactuca sativa* biotypes which are resistant to sulfonylurea herbicides, the ALS enzyme is the site of action of herbicide is modified in such a way that herbicide can no longer bind with the enzyme and inactivate it (Eberlein *et al.*, 1999) This target site based resistance is usually associated with resistance involving altered binding of herbicide to their target protein. This results from a single nucleotide change (mutation) in the gene encoding the protein to which the herbicide normally binds. This changes the amino acid sequence of the protein and reduces or destroys the ability of the herbicide to interact with the protein and at the same time do not incapacitate the normal functioning of the enzyme so that the enzyme functions normally in the presence of the herbicide.

However these mutations conferring herbicide resistance may cause changes in other seemingly unrelated physiological processes. Usually these changes adversely affect the biological fitness of the resistant biotypes. For example, in triazine resistant biotypes the mutation in the plastoquinone binding D1 protein of PSII results in reduced photosynthetic efficiency (Radosevich and Holt, 1982); also seeds of some of these resistant weeds biotypes exhibit poor germination as compared to susceptible biotypes. But in some resistant biotypes, as found in *Kochia scoparia*, the mutation conferring resistance to sulfonylurea herbicides will concomitantly reduce or abolish acetolactate synthase sensitivity to normal feedback inhibition patterns, resulting in elevated levels of branch chain amino acids available for cell division and growth during early germination.

Hence sulfonylurea resistant biotypes of this species exhibit a rapid germination even at lower temperature compared to their susceptible counterpart.

(b) Site of action over production: This causes the dilution effect of the herbicide. In the case the site of action is overproduced so that the herbicide at its normal rate of application will not be able to inactivate the entire enzyme produced. Thus the enzyme spared by the herbicide will carry on the normal plant metabolic activities.

Factors responsible for development of herbicide resistance

Herbicides that have a single site specific mode of action are more likely to result in resistance than those that have more general herbicidal properties and affect several processes. Most mutation that result in herbicide resistance are single gene mutation thus the effect is to produce an extremely small change at the site of action, such as in the enzyme or other protein to which the herbicide normally binds and exerts its effect. For example branched chain amino acid inhibitor herbicides bind with and inactivate acetolactate for the formation of the amino acids valine, leucine and isoleucine. A mutation in the gene that codes for ALS changes a single amino acid in the enzymes so that the herbicide can no longer bind to it.

Soil applied herbicides that are persistent and provide control for most of the growing season such as atrazine and simazine further increase the selection pressure for resistance. Susceptible plants of the same and other species do not have an opportunity to grow and provide competition. A similar selection procedure can be obtained with multiple applications over a growing season of a short lived herbicide used for several consecutives years. Other factors that favour the development of resistant plants include a high level of initial control typically 96 to 99 % of susceptible plants and high dosages that eliminates both the susceptible portion of the population and other potential plant competitors.

Herbicide resistance appears most frequently in plants with annual life cycle pigweed (*Amaranthus* spp.), Lambsquaters (*Chenopodium* spp.), kochia (*Kochia scoparia*), barnyardgrass (*Eleucine indica*) than perennial plants. Since annual reproduce from seeds only the potential to transfer resistance to the next generation of plants is relatively high. This is in contrast to perennial plants which can persist for several years, thus the frequency with which resistance appears in the population may be greatly reduced. The selection pressure from consecutive herbicide applications on annual plants is greater than on perennial since annual plants tend to be relatively susceptible to herbicide, in contrast to perennial which because of their underground structures, often escape total destruction.

If the soil is tilled, long lived seeds from previous susceptible populations will dilute the population of resistant seeds and resistance can be delayed. On the other hand seeds that stay on the soil surface and have a short life (little or no dormancy) will either die or germinate within one to several years. Thus management system that leave weed seeds on the soil surface leave mostly resistant seeds since the susceptible ones have germinated and have been killed by the herbicide application.

Factors that stimulate the development of herbicide resistance are many folds; however, the key factors include weed characteristics, chemical properties of the herbicide and cultural practices are discussed below (Vargas and Wright 2004).

If in a particular population some species has relatively more individuals that have specific inherent resistance, then the chances for the resistance to surface are more. Relatively rapid turnover of seed bank to high percentage of seed germination each year or in other words a little or low seed dormancy will lead to death of most susceptible plants by herbicide. Consequently the remainder resistant though fewer, will have brighter chance to grow with vigor and reproduce with increased resistant trait. It is also called hypersensitivity of weeds to a particular herbicide. Due to hypersensitivity, a single application of herbicide can eradicate most (90-95 %) population consequently, high selection pressure will allow the resistant biotypes to prevail and thrive best to stand fit in the field. ?Some weeds have high frequency of resistant gene that develop higher and rapid rate of resistance e.g. *Lolium rigidum.*

Herbicides that are subjected to enhanced metabolism in weeds have least chance to endure resistance in plants than weeds expressing resistance due to change at site of action. However, metabolism-based resistance to ACC-inhibiting herbicides is much less known although this type of resistance seems to be wide-spread (Delye, 2005).

Farm cultural practices can also shift selection pressure on weeds. Because persistently using similar cultural practices such as below increase chances to develop herbicide resistant weed biotype are:

1. A move from old multi-cropping pattern to monocrop cultures where single crop is grown every season or where a fixed horticultural orchard is in place receiving similar input every season.
2. Using reduced soil cultivation, or zero tillage system.
3. Using single herbicide or combination (tank mix) that have same mode of action in every season persistently. This example refers to situations where transgenic herbicide resistant seeds are planted with a package of specific tolerant herbicide.
4. Dose violation i.e. use of low or very high doses in relation to the optimum rates prescribed for a specific crop in a situation.

Management of herbicide resistance in weeds

Alternating non-chemical control measures with herbicides help reduce selection pressure. However alternative non-chemical weed control practices may be limited in number that perform inconsistently. Mixing herbicides with different modes of action will delay the development of resistance provided that both herbicides have activity on the target weed. What one herbicide misses, the other will kill. Therefore, very few resistant survives will remain. Mixtures with one herbicide that control one weed and a second that controls a second weed but not the first will not delay the buildup of resistance. Assessment of risk of developing herbicide resistance based on management options is given Table 4.

Table 4. Assessment of risk of developing herbicide resistance based on management options flowed (Velverde *et al.*, 2000)

Management option		Risk of resistance	
	Low	Moderate	High
Herbicide mix or rotation in cropping system	>2 modes of action	2 modes of action	1 modes of action
Weed control in cropping system	cultural mechanical and chemical	cultural and chemical	chemical only
Use of same mode of action per season	once	more than once	Many times
Cropping system	full rotation	limited rotation	no rotation
Resistance status to mode of action	unknown	limited	common
Weed infestation	low	moderate	high
Control in last three years	good	declining	poor

Herbicide rotation with respect to mode of action from year to year can slow the onset of resistance. Crop rotation where herbicide chemistry and other control methods differ for each crop can delay the buildup of resistance. Controlling the previous crop weeds prevents resistant weeds from reproducing. A resistant weed presents the same type of threat as a serious, new, invading weed, and failure to control it will contribute to its spread. Control measure that vary from standard herbicide applications example hand pulling, limited tillage, spot treatment with other herbicides should help eliminates escaped plants. Control of spreading of resistant weeds at initial stage with a good preventive program will delay the spreading of resistance. Accept less than excellent weed control levels where weeds are already well established. One approach to managing pest is to accept economic threshold levels of weeds control example 60 to 80 % rather than 90 to 95 % control.

Summary

Herbicide selectivity refers to the differential effect of a herbicide when applied to mixed population of plant. Selective herbicides kill weeds and leave desirable plants undamaged when applied to both. A number of factors determining selectivity are plant related/ age and height difference among plants can be used to successfully remove weeds from some crops. Other plant related factor includes difference in herbicides retention, penetration, absorption, translocation, site of action and metabolism. It should always be remembered however that selectivity is dose dependent. At high enough concentration, most herbicides kill all plants. At sufficiently low dosage, herbicide will not kill any plant.

Resistance is the ability of a formerly susceptible plant population to survive doses above that were used to control the original plant population. The resistant population is referred as resistant biotype. The development of resistant become apparent only after several years of repeated application of the same herbicide and is usually characterized by plant that survive a dose several to many times viz 20 to 1000 times or even more greater than normally needed for commercially acceptable weed control. These high levels of resistance are most frequent caused by a single gene mutation that modifies the herbicide site on an enzyme or protein. Control of spreading of resistant weeds at initial stage with a good preventive program, crop and herbicides rotation can delay the buildup of resistance.

□□□

Chapter 7

Fate and Persistence in Soil

The fate of herbicides in the soil is a concern of human welfare. The soil acts as an important buffer governing the persistence and fate of most herbicides in the environment. As long as soil system remains healthy, possible adverse effect of herbicides in the environment probably can be minimized. Herbicides persistence and their fate in the environment are both influenced by the soil processes of adsorption, movement and decomposition. Herbicides are readily adsorbed by the dry soil. Adsorption is also influenced by the chemical charge and water solubility of each herbicide. The amount of organic matter and clay in soil also influences herbicides adsorption drastically. Herbicides that are adsorbed to soil colloids are not readily available for absorption by plants, movement or degradation. Degradation by microbes is probable the most important mechanism for herbicides decomposition in soil.

Most herbicides are organic compounds and are therefore basically unstable in the environment. Inherent instability is essential to prevent these materials from accumulating in the environment as compounds are repeatedly used. Accumulation of pesticides not only poses environmental hazards, but prohibits rotation to sensitive crops. Unfortunately not all herbicides exhibit optimal stability properties and therefore numerous examples of insufficient control or carryover have been reported. Furthermore persistence in the environment prolongs exposure of the material to forces that can cause movement of the herbicides away from the application site. The extent of transport, with the exception of spray drift, is a function of both the stability and the physico-chemical characteristics of the herbicides in the environment to which it is applied.

Most of the soil-applied herbicides are absorbed by plant roots and underground vegetative parts. The duration of activity of herbicides in the soil is the net effect of absorption, volatilization, leaching and degradation whether biological or non-biological (Fig. 1). Herbicides that are chemically bound, or adsorbed, to solids are not available for leaching, plant uptake, or microbial degradation. Such herbicides persist until they are released from the soil surface. Some herbicides, such as paraquat and diquat, bind to soil so tightly that they persist nearly indefinitely and are not typically available for plant uptake.

As soon as a herbicide is applied a number of processes immediately begin to remove the compound from the original site of application. This is the process of environmental dissipation. For the herbicide which is intercepted by plants, the chemicals may be taken up by the plant itself or may be either washed off by precipitation onto the soil, that may undergo photodegradation on plant surface or may volatile back into the air. Herbicides persistence in the soil is expressed as half life or time required to degrade fifty percent of the original molecule. However, the half life is not absolute because it depends on the soil type, temperature, and concentration of the herbicide applied. The persistence varies with the nature of a chemical, soil and climatic conditions. The herbicide should persist long enough to check weeds until the end of critical period of weed competition but should not persist beyond the crop harvest, as it would be injurious to the sensitive crops grown in rotation. Very rapid loss of herbicides from soil will cause insufficient weed control, which is considered as unsatisfactory as their unduly long persistence within soil. Beside herbicides structure, soil conditions prevailing during and after the application of a herbicide as well as herbicide application methods influence the fate of the herbicides in the soil. Heavy rainfall in monsoon will cause greater leaching and runoff. Higher humidity enhances the soil microflora proliferation.

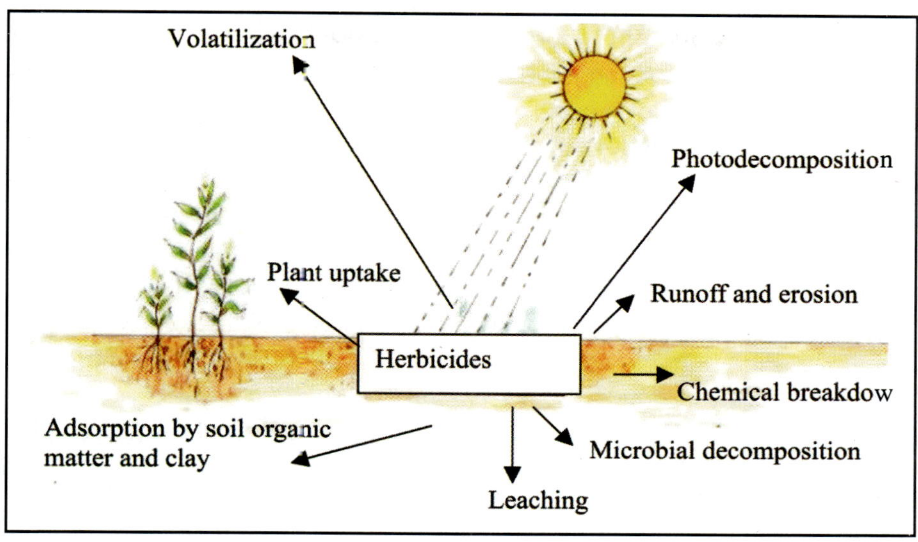

Fig. 1. Fate of herbicide in the environment

Similarly the persistence in dry soil is greater than in wet soil. Under average conditions, triazine herbicides persist in the soil upto 90 days after application. For substituted ureas, particularly isoproturon, 60 to 100 days are required for complete degradation; depending upon the application rate. The persistence of phenoxy carboxylic acid, amide and dinitroaniline groups in soil is short, *viz.* 20 to 60 days (Randhawa *et al.*, 2000; Sondhia, 2008). Field experiments under direct seeded puddled conditions conducted at Coimbatore, showed anilofos at 0.4 kg/ha persisted in the soil for 56 days and residues after two successive rice crops were below the level of detection. At higher doses, both anilofos and 2, 4-D residues were recorded in rice grains and straw, but the residue level was far below the maximum residue limit (Radhamani *et al.*, 1997). The field trials with various rates of application of tralkoxydim 0.25, 0.40 and 0.80 kg/ha for three years (1989-92) revealed no detectable amount of tralkoxydim in soil, wheat grains and straw at harvest (Srivastava *et al.*, 1994). Clodinofop and sulfosulfuron were rapidly dissipated in the soil as compared to isoproturon in rice-wheat cropping system (Sondhia 2006e). Soil pH and water content strongly influence the fate of sulfonylurea herbicides in agricultural fields (Hultgren *et al.*, 2002). Table 1 gives approximate lengths of persistence for various herbicides under Indian tropical conditions. This table is based on experimental work under field and laboratory conditions. Those herbicides persisting one month or less may be used to control weeds present at the time of treatment and can be considered as fast degradating herbicides. Those persisting one to three months may protect the crop only during short period early in the growing season. Those providing three to more than twelve months of persistance may provide more than twelve months of controls are used primarily for total vegetation control or where persistence is desirable and considered as most persisting herbicides.

Table 1. Half-lives of some herbicides in soil (Sondhia 2008)

Herbicides name	Half lives (Days)	Herbicides name	Half lives (Days)
Alachlor	6-8	Metribuzin	23-49
Atrazine	13-58	Metolachlor	8-27
Anilofos	12-78	Metsulfuron-methyl	70-147
Butachlor	5-24	Oxyfluorfen	19-29
Chlorosulfuron	31-93	Pendimethalin	15-77
Chlorimuron	60	Prometryne	43
Dithiopyr	17-25	Pretilachlor	10-11
EPTC	25	Propanil	5-9
Fluazifop-p-ethyl	8-24	Sulfosulfuron	3-27
Fluchloralin	12-13	Triasulfuron	3-9.4
Imazethapyr	57-71	Thiobencarb	19-24
Isoproturon	13-21	Triallate	38
		2,4-D	7-22

Under Indian tropical conditions imidazolines group of herbicides show high persistence in soil and half life varies from 57-71 days followed by substituted urea group herbicides with half-life period ranging from 13-49 days. Chloroacetamides herbicides have the shortest half-life of 6-27 days.

Soil properties such as organic matter content, soil texture, and soil pH also play an important role in the carryover potential of residual herbicides. Soil factors that affect the persistence of residual herbicides are summarized in Table 2.

The fate of herbicide in soil can be classified as follows;

Table 2. Relative behaviours of different herbicides in soil

Herbicides class	Water solubility	Volatility	Soil retention	Persistence
Chloroacetamide	Moderate	Low-moderate	Low	Very short - moderate
Dinitroanilines	Very low	Moderate-high	Moderate -high	Short - moderate
Dipyridiums	Very high	Very low	Very high	N/A
Imidazolines	Moderate- high (pH dependent)	Very low	Very low-low	Very short - short
Organic phosphorus	Very high-Low-moderate	Very low	Very high	N/A
Phenoxys/benzoics/picolinates	Low-high	Very low-Moderate	Very low-low	Very short – short (pH dependent)
Substituted ureas	Very low-low	Very low-Moderate	Low-high	Short - moderate
Sulfonylurea ureas	Very low-high	Very low-low	Very low-low (pH dependent)	Short-long
Thiocarbamate	Low-moderate	High-very high	Low-moderate	Very short - short
Triazines	Very high	Very low	Very high	Moderate-long

1. Transport by adsorption and movement (volatilization, leaching, runoff etc.)

Transport of herbicides within the soil compartment can occur downward into the soil profile (leaching), across the soil surface (runoff), or into the air (volatilization). Each can be a combination of more fundamental processes including adsorption, convection and diffusion.

Adsorption

Adsorption is an important factor that makes herbicides unavailable for uptake by plants. Adsorption is the attraction or adhesion of molecules or ions to the surface of soil particles (colloids). After application, many herbicides adsorb (bind) to the clay and organic-matter fractions of soils. However, herbicides adsorb poorly to the sand and silt fractions of soil. Therefore, the extent of herbicide adsorption increases as the percentage of organic matter and clay increases. The dinitroaniline herbicides, dithiopyr, oxadiazon and most other pre-emergence herbicides readily bind to soils. To be effective and safe, a pre-emergence herbicide must have properties that result in the majority of herbicide being bound to soil colloids with only a small amount remaining in solution. If the majority of herbicide remained in solution the herbicide would rapidly leach through the soil profile or leave the field with runoff. Adsorption can slow or prevent degradation mechanisms that permanently degrade the herbicide. Almost all soil-applied herbicides are adsorbed to some extent. Weed control is inversely proportional to the amount of herbicide adsorbed to the soil.

Adsorption is often dependent on the soil or water pH, which then determines the chemical structure of the herbicide in the environment. Adsorption generally increases with increasing soil organic content, clay content, and cation exchange capacity, and it decreases with increasing pH and temperature.

The availability of an herbicide for transport through the environment or for degradation is determined primarily by the adsorption/desorption process (WHO, 1984). Adsorption to soil particles can stop or slow the rate of microbial metabolism significantly. In other cases, adsorption can facilitate chemical or biological degradation (Farmer and Aochi 1987). Adsorption can change with time and, in most cases, is reversible (i.e. the herbicide can desorb from the soil or sediments and return to the soil solution or water column). The amount of adsorption in a particular soil depends primarily on the nature of the soil mineral and organic component the nature of the herbicides and the soil water status. The nature of soil water on herbicides availability to plants is illustrated in Fig. 2. Normally, the adsorption of herbicides occurs on both the clay and humus fractions. In general, small increase in the organic matter content of a soil greatly increases its ability to adsorb herbicides. A soil high in organic matter content will generally require a higher herbicide rate than a soil with less organic matter. Adsorbed herbicide molecules are unavailable for biological, physical, and chemical processes until released from the soil into the soil solution or vapour phase. Herbicides generally are more tightly adsorbed in dry soils than in wet soils. Water molecules compete and displace herbicide molecules from adsorption sites, making the herbicides available for plant uptake.

Soil organic matter has a wide range of physical, chemical, and biological characteristics. It has been demonstrated that humified matter is more chemically reactive with pesticides than nonhumified matter (Kozak *et al.*, 1983; Gaillardon *et al.*, 1983; Payá-Pérez *et al.*, 1992; Senesi *et al.*, 1997; 2001). Humified matter is composed of humic substances such as humic acids (HAs), fulvic acids (FAs), and humin.

Most soils are relatively rich in humic acids (Martin and Haider, 1971; Stevenson, 1972), and these humic acids are more chemically reactive with pesticides than fulic acids or humin (Hayes, 1970a,b; Khan, 1972; 1974; Stevenson, 1972; Senesi, 1992; Senesi *et al.*, 2001).

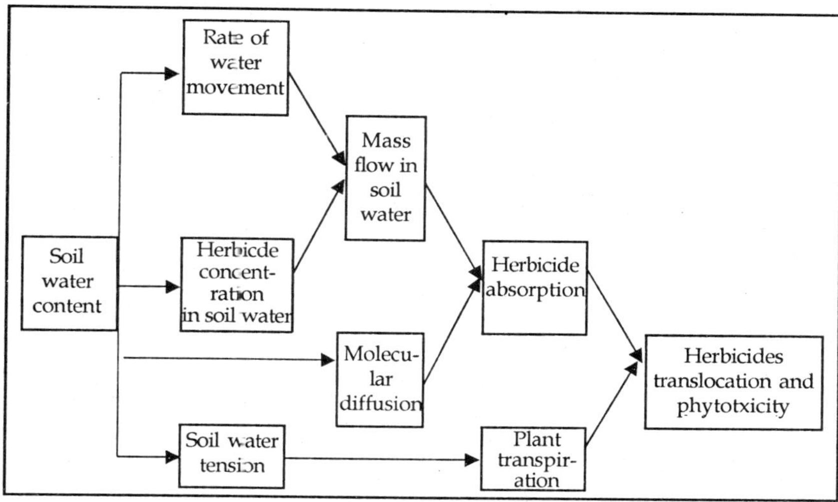

Fig. 2. Influence of soil water on herbicides availability to plants (Green and Obein 1969)

In most situations, the charges are relatively weak and thus the process is reversible. An equilibrium is reached between the amount of herbicide bound to colloids and that found in solution. A plot of quantity of herbicide adsorbed plotted against variable concentration of free herbicide in soil at any given time is called an adsorption isotherm. The adsorption isotherms of many herbicides have been described empirically by the use of Freundlich equation

$$x/m= Kf.\ Ce.\ 1/n,$$

whereas x is the quantity of the adsorbate (herbicide) reversibly adsorbed by a given quantity of adsorbent, in equilibrium with the adsorbed solution of concentration Ce; Kf and n are constants and may be determined by the best line drawn through a plot of log (x/m) versus log Ce. Kf is called Adsorption (or Equilibrium) Coefficient. It depicts relative adsorption capacity of a soil for any given herbicide. An alternate adsorption isotherm equation, which has shown better fit for certain herbicides as suggested by Lambert (1967) is

$$x/m = K_1\ Ce + K_2 Ce^2$$

Where K_1 and K_2 are two equilibria coefficient; K_1 is the adsorption coefficient at infinite dilution (i.e., Ce= O), equivalent to Kf of Frendlich equation when n = 1, K_2 is adsorption coefficient when Ce has some value other than zero.

A adsorption isotherm may vary in different soil at any given application of a herbicides, more of it is adsorbed in a high Kf value soil than in a low Kf value soil, leaving less quantity of the toxicant in soil solution for weed control in the former type of soil. Therefore higher rates of herbicides must be applied to higher Kf value soil and *vice versa* to obtain similar weed control. The ratio of bound to free herbicide is influenced by several factors, including chemical properties of the herbicide, soil characteristics and soil water content. Herbicides are more active under conditions that favour movement into the soil solution. With most herbicides, adsorptivity and solubility are inversely related. Thus, as solubility increases, binding capacity to soil decreases. There are a few exceptions to this rule, including paraquat and glyphosate. Both are highly water soluble yet they bind extremely tightly to soil colloids. With most herbicides, the adsorptivity coefficient (K) is more closely associated with its availability to plants than the water solubility, but it is important to consider both characteristics.

The tendency of herbicides for adsorption in soil is an interplay of their electrical state and properties of soil colloid particulates (Barriuso and Calvet 1992). The soil particulate of colloidal dimension (<1 μ) alone act as adsorbent of the added pesticides. Both soil inorganic (clay) and organic colloids (humic materials) participate in adsorption of pesticides, although the later are generally considered functionally more important. The clay minerals constitute two major types such as the crystalline Si-Al (or Mg) clays and the amorphous Fe-AL hydrous oxide metallic clays, besides some allophanes. The crystalline Si-Al clays are further differentiated into expanding types e.g. montmorillonite and vermiculate and nonexpanding types e.g. kaolinite, illite and chlorite. The expanding types offer much large surface area for adsorption of relevant ions than the non-expanding types. The clay micelles are surrounded by isomorphic substitution and pH dependent negative charge which assign these their cation exchange capacity (CEC). The cation exchange capacity of the expandable micelles is as high as 80-150 meq/100 g soil as compared to 3-40 for the non-expanding clays. Under certain circumstances, Fe-Al hydruous oxide clays, allophanes, and even kalonite clays, also exhibit limited anion exchange capacity (AEC).

The negative charges on humic materials are pH dependent, arising from $COOH^-$ (carboxylic group) and –OH (phenolic groups). In neutral to alkanoic soils, hydrogen of these groups tends to be held rather loosely, and is easily exchanged for other cations. The CEC of humus ranges from 150-300 meq/100 g substrate. Pantani *et al.*, (1994) reported the influence of clay mineral on adsorption of cinosulfuron.

Mechanism of adsorption of organic herbicides

Depending upon their ionizing characteristics, herbicides can be divided into four main groups as below.

(a) Cationic or strong base herbicides
(b) Weak base herbicides
(c) Acidic herbicides and,
(d) Non-ionic herbicides

(a) Cationic or strong base herbicides

Cationic herbicides readily dissolve and dissociate in aqueous systems to form cationic species. Paraquat and diquat are common representatives of strong base herbicides. These are adsorbed on soil colloids by combination of ionic and physical forces. The ion exchange is for cations like Na^+, K^+ and Ca^+ in the case of inorganic colloid adsorbent and for H+ in the case of organic adsorbent in soil.

$Paraquat^{++}$ + 2 Na-clay □ paraquat-clay + 2(Na^+) .... Equation 1

This reaction goes strongly to the right but weakly to the left.

(b) Weak base herbicides

Herbicides such as s-triazines and triazoles are weak basic compounds which under sufficiently acid conditions can accept a proton H^+ at each of their N atom in the ring to become cationic. The process is called protonation. Protons (H^+ ions) are donated by molecular layer of water present on the clay particles in soil (Che *et al.*, 1992).

R + H^+ □ RH .... Equation 2

Where, R= molecular species of a weak basic herbicides, H= hydrogen ions: RH =protonated species.

The protonated basic herbicide molecules (= conjugate acid) dissociate in aqueous solutions to form cationic species which are adsorbed by cation exchange adsorbents according to Equation 3

RH^{+} + X-solid □ RH-solid + X^{+} .... Equation 3

Where X-solid = particular adsorbent matter with X^{+} as exchangeable cation other than hydrogen example Na^{+} and K^{+}. the proton donating capacity of a solution increases with lowering of pH of the media; the maximum protonation occurs at pH near the pKa value of the conjugate acid, which differs markedly with the herbicide structure. The pKa of conjugate acids of atrazine, ametryn and amitrole are 1.68, 3.98 and 4.17 respectively. Maximum adsorption of these compounds has been found to occur at pH near these pKa values and the amount of adsorption was also directly related to their pKa values. A decreased adsorption at pH lower than the pH of the conjugate acid presumably represented increased from solution H^{+} (Wed and weber, 1974). On the contrary, decreased adsorption of s-triazine herbicide at higher pH was attributed to decrease in the number of protonated molecules.

Like cationic herbicides, weak herbicides can also be adsorbed by cation exchange in the interlayer spacing of the expanding clay mineral but they remain biologically available. Besides ionic adsorption, basic herbicides can be adsorbed in molecular form to cation exchange adsorbents through various physical mechanisms, including physical adsorption, hydrogen bonding and complexing with hydrogen ions on the exchange surface (equation 5)

R + X-solid □ RX-solid .... Equation 4

And R + H solid □ RH-solid .... Equation 5

Where X= exchangeable cations other than hydrogen; solid= particulate matter; and H-solid = particulate matter on which hydrogen is the exchangeable cation.

Other physical adsorption processes suggested to occur between basic herbicides and particulate matter surfaces include hydrophobic bonding through Van der Waals forces and perhaps legend exchange. Physical adsorption may make greatest contribution at higher pH when the basic herbicide molecules are not protonated.

(c) Acidic herbicides

Acidic herbicides possess carboxyl or phenolic functional groups and they ionize in aqueous system to yield anionic species. The tendency to ionize decreases at low pH. Various phenoxyalkanoic acids, benzoic acids, pthalamic acids, TCA, dalapon, phenols and nitriles are examples of acidic herbicides. Adsorption of an acidic herbicide by basic anion exchange resins in the laboratory is rapid and is easily reversible (equation 6)

R + Cl-Resin ⇌ R-Resins + Cl .... Equation 6

Where R= molecular species of acidic herbicide.

In soils, ion exchange adsorption of acidic herbicides is limited mainly to non-silicate clay minerals. The herbicide anions exchange sites on these clay minerals are for ion like PO_4^-, Cl^- and OH^-. Adsorption of acidic herbicides with organic matter *via* ion exchange is not anticipated but it may occur when molecular species dominate at low pH by weak hydrogen or hydrophobic bonding. Thus the adsorption of acidic herbicides would be more for soils with low pH and high organic matter content. However, it is still of much lower magnitude than the adsorption of cationic or basic herbicides with organic matter and it is in competition with water. Adsorption with organic matter does not deactivate acidic herbicides although movement through soil may be impeded (Wed and Weber, 1974).

The weak acid herbicides such as dinoseb (pKa= 4.40) are adsorbed more than the stronger acids like picloram (pKa = 1.90), 2, 4-D (pKa= 2.80) and dicamba (pKa = 1.9). It is because there are proportionately more molecular than anionic species in solution at a given pH for the weaker acids as compared to the stronger acids (Weber 1972, 1980; Weber and Scott 1966).

(d) Non-ionic herbicides

Some herbicide molecules do not ionize significantly in aqueous or soil system, for example, substituted aniline, anilide, phenyl-carbamate, phenylurea, phenylamide, benzonitrile, and ester herbicides. Considerably less work has been reported on mechanism of adsorption on the non-ionic herbicides as compared to the ionic ones. For relatively nonpolar compounds, adsorption on organic matter is through hydrophobic bonding, while for polar, nonionic herbicides, adsorption occurs in greater competition with water.

Normally, adsorption of non ionic herbicides by soils is positively correlated with soil organic matter content but the effect of adsorption on the activity and movement of herbicides vary widely. This variability appears to be related to solubility, polarity and to a restricted extent, molecular size of the herbicide. The hydrophilic-lipophilic balance (HLB) of both adsorbate and adsorbent strongly influence adsorption (Wed and Weber, 1974).

Factors affecting herbicide adsorption

Soil pH

Soil pH can have a significant effect on the adsorption of many herbicides. Some herbicides, principally the triazines such as atrazine and simazine, are particularly affected by soil pH. Lesser amounts of these herbicides are adsorbed or held to soil colloids at higher soil pH, so more amount is present in the soil solution. Herbicides in the soil solution are available for plant uptake. As the pH decreases below 7 the concentration of hydrogen ions found in the solution increases. Many herbicides can incorporate hydrogen ions into their molecular structure, therefore tend to change the charge of the herbicide molecule. At soil pH's below 7, atrazine may pick up hydrogen ions from the soil solution causing the atrazine to take on a positive charge. The positive charge on the atrazine molecule under acid conditions increases the attraction between the herbicide molecule and negatively charged soil colloids. At soil pH's above 7 most of the atrazine maintains a neutral charge and thus the herbicide is less tightly adsorbed and more available to plants.

The greater persistence of atrazine at high pH's is due to the herbicide being more susceptible to degradation when it is bound to soil colloids than when it is in free solution. The adsorption and persistence of several sulfonylurea herbicides is also strongly influenced by soil pH. Certain members of the sulfonylurea group (chlorsulfuron and chlorimuron) can also persist in higher-pH soils because rates of chemical breakdown are decreased. Low pH (less than 7.0) affects the persistence of clomazone and the imidazolinones (imazamox, imazaquin and imazethapyr, imazamethabenz). Imidazolinone herbicides tend to be more adsorbed under acidic or low soil pH which reduce their availability for microbial degradation (Bresnahan *et al.*, 2000; Sondhia, 2008). They are broken down primarily by chemical hydrolysis and this process occurs much quicker under acidic conditions. Therefore, they tend to degrade faster when soil pH is low. Adsorption of herbicides, therefore, is basic to understand the behavior of herbicides in soil. Less imazethapyr was adsorbed initially at high soil pH, but it did not readily desorb. It was concluded that imazethapyr was more desorbable and bioavailable in low pH soil, which resulted in injury to rotational canola and sugar beet crops. Sulfonylurea herbicides react in the opposite manner to the imidazolinones. They are broken down primarily by chemical hydrolysis and this process occurs much quicker under acidic conditions. Therefore, they tend to degrade faster when soil pH is low.

Much informations are not available on the effect of soil pH on the persistence of other herbicides. Research shows that various nutrients and cations in the soil affect both herbicide activity and degradation. The CEC, principally a function of clay type and organic-matter content, is directly involved in herbicide adsorption. Some herbicides are more available in the presence of certain cations, whereas others may not.

Soil moisture

Soil moisture plays two important roles in herbicide performance. The amount of herbicide in solution is directly related to soil moisture content. But in high adsorption soils, irrigation may move some reversibly adsorbed herbicide into the soil solution even within available soil moisture range thus improving weed control on such soils. But in low adsorption soil solution without inducing desorption of herbicides, which leads to poor weed control. The amount of space available for herbicides to go into solution decreases as soils dry out, thus less 'free' herbicide is present in dry soils. Under dry conditions, plants are exposed to less herbicide and therefore be less likely to absorb toxic herbicide concentrations. The non-ionic herbicides such as CDEC, CDAA and EPTC are adsorbed more on dry soil than on wet soil. Moisture competes for sites on colloidal surface with the herbicides and induce desorption. The volatile compounds thus desorbed are lost to atmospheres. When soil moisture is replenished, herbicide will desorb from the colloids and re-enter the soil solution.

Soil organic content

Soil organic content is thought to be the best determinant of herbicide adsorption rates (Farmer and Aochi 1987, Que Hee and Sutherland 1981, Helling *et al.*, 1971). Adsorption of pesticides by soils has frequently been found to be correlated with organic matter and clay contents. It is generally accepted that this effect is due to the high adsorptive capacity of these soil constituents for herbicides. The large surface and the chemical nature of humic substances can explain their high adsorptive properties. Carboxylic and hydroxylic groups are important (Li and Felbeck, 1972) as these can form ionic bonds with ionized molecules (Burns *et al.*, 1973, Khan 1974) and participate in cation–dipole bonds (Khan and Mazurkevitch, 1974). The organic nature of soil organic matter is also important in the formation of charge transfer complexes (Hayes, 1970). Hydrogen bonding is also possible either directly between solute and adsorbent or indirectly through hydration of water molecules. It should be noted that the properties of organic material with respect to adsorption are not constant and vary as result of the action of soil chemical and biological processes.

Solubility of a herbicide

Adsorption is also related to the water solubility of an herbicide, with less soluble herbicides being more strongly adsorbed to soil particles (Helling *et al.*, 1971). Solubility of herbicides in water generally decreases from salt to acid to ester formulations, but there are some exceptions. Glyphosate is highly water-soluble and has a strong adsorption capacity. Weber (1972) reported an inverse relationship between water solubility of acidic herbicides and their adsorption when other soil variables were constant.

Soil temperature

Adsorption of herbicides by ionic exchange mechanism is independent of changes in soil temperature. In variance with it, physical adsorption is generally weakened by rise in soil temperature. This is particularly pronounced for volatile herbicides and herbicide with negative heats of solution i.e. the ones which exhibits less solubility in the soil solution at higher than at low temperature for example, thiocarbamate herbicides (Weber, 1972).

Molecular structure of herbicides

The molecular properties such as electronic structure, molecular volume play an important role on adsorption of herbicides. The presence of charge is an important characteristic as it permits adsorption by ion exchange. As some molecules bear permanent charges (paraquat and diquat) and some may be ionized according to the acidity of the medium (triazines) and other are neutral (ureas). Ionization depends on molecular structures. The possibility of a molecule to be ionized or neutral explains why herbicide-soil interaction may vary greatly depending on the pH of the soil. Theoretically a complete knowledge of electronic structure would enable prediction of the adsorption behaviour of herbicides. But this is not the case since several types of interaction act simultaneously with a given surface.

Volatilization

Volatilization is a process wherein a condensed phase such as liquid or solid is transformed into a vapour by elevation of temperature or reduction of external pressure. The term vapour describes a substance in the gaseous stage below its critical temperature. Herbicides vary widely in volatility and loss to the atmosphere as a gas. Herbicide volatilization increases as the temperature rises. The tendency of a substance to volatilize is expressed by its vapour pressure. The vapour pressure is usually expressed in millimeters of mercury (mmHg) or occasionally in micron of mercury. Herbicides with vapour pressure of more than 1 x 10^{-5} mmHg at 20^{0}C are generally considered volatile.

Usually, very little volatilization occurs once the herbicide is mixed into the soil either by mechanical incorporation or rainfall. Volatility loss is greater when herbicides are applied to a moist soil surface than a dry soil surface. Volatility generally increases with increasing temperature and soil moisture, and with decreasing clay and organic matter content (Helling *et al.*, 1971). The use of a surfactant can change the volatility of a herbicide (Que Hee and Sutherland 1981). In extreme cases, losses due to volatilization can be up to 80 or 90% of the total herbicide applied (Taylor and Glotfelty 1988). 2, 4-D and triclopyr can present significant volatilization problems in the field (Table 3).

Table 3. Volatilization of some herbicides under field conditions

Name of herbicides	Days after application	Volatilization (%)	Reference
Alachlor	35	14	Gish *et al.*, 1995
Atrazine	35	9	Gish *et al.*, 1995
Ethalfluralin	24	10	Smith *et al.*, 1997
EPTC	2	75	Cliath *et al*, 1980
Metolachor	10	22	Pruger *et al.*, 1999
Triallate	24	21	Smith *et al.*, 1997
	14	29	Bor *et al.*, 1995
Trifluralin	24	18	Smith *et al.*, 1997
2,4-D iso octyl ester	5	21	Grover *et al.*, 1985

Volatilization is considered one of the primary pathways for herbicide dissipation from the site of a herbicide application. Under unfavorable conditions, losses that result from volatilization can reach 80 to 90 % within a few days although such rates are dependent upon climate and microclimatic conditions (Taylor and Glotfelty, 1988). The process of volatilization is controlled by two mechanisms; the evaporation of the active ingredient from the soil or plant surface to which it has been applied or has migrated and dispersion of the resulting vapour through the overlaying atmosphere by diffusion and turbulent mixing. Though these two processes are different in character and are controlled by different chemicals and environmental factors, they are not dependent on one another and in general should be considered as integrated one.

The first process of evaporation involves a phase change from its initial state of solids or liquid to its corresponding vapour. Thus one of the main factors controlling the rate of loss by volatilization is the rate at which surface residues evaporate. The rate of evaporation is controlled by the vapour pressure of the exposed residues. It is either the vapour pressure of the active ingredient or lower values which causes reduction in the interaction of the compound with surface. For residues that remain no longer on the surface but drawn away to underlying by capillary action or diffusion after application, the rate of volatilization becomes dependent upon the rate of back diffusion of the material to the upper layers (Taylor and Glotfelty, 1988).

The second process of dispersion into the atmosphere acts through the turbulent flow of air over the evaporating surface. The flow of air continuously replaces the air around the evaporating surface and mixes and dilutes the herbicides vapour in the surrounding air. The movement and dispersion of vapour within the turbulent layer are considered rapid to those at the evaporative surface. The flow of air over plant and soil surfaces is generally turbulent except for a shallow layer of air directly above the evaporating surface where flow is regarded as laminar. The actual rate of volatilization away from the evaporative surface through the stagnant layer is proportional to the molecular diffusion coefficient and the vapour pressure of the herbicides at the evaporating surface. The depth of the stagnant layer in turn is affected by both the surface geometry and the air flow of the system. Surface geometry or roughness alters the turbulent flow of air over the evaporative surface. Thus an increase in either air flow or turbulence increases the rate of vaporization (Spencer *et al.*, 1982) as vapour are removed from the evaporative surface. Volatile herbicides such as the thiocarbamates (EPTC, butylate) must be incorporated immediately to avoid gaseous losses. These herbicides are less likely to persist than herbicides with low vapour pressures. Vaporization depends on number of factors as below:

Soil organic matter

Once a herbicides has been applied to a soil the material either remains on the surface, diffuse into the soil by capillary action or becomes sorbed to the soil matrix. The extent of sorption to the soil matrix is influenced by the carbon content of the soil, which in turn affects the rate of volatilization from the soil. Soil carbon content controls the concentration of pesticide present in the soil solution. Spencer (1970) has demonstrated that vapour density of a herbicide over a soil is proportional to its concentration in the soil solution. In general, the higher the soil carbon content the lower the concentration of herbicide in the soil solution. As a result of its reduced concentration in soil solution, the rate of volatilization for a herbicide from a soil with a high organic matter content is lower than that from one with low organic matter content (Jury *et al.*, 1980; Jury 1986; Getzin 1981; Glotfelty *et al.*, 1989).

Empirically the concentration of a herbicide in the soil solution is equal to the ratio of its vapour pressure and water solubility. This concept has been further developed by Swann and coworkers (1982) as an empirical rate to its vapour pressure (P), water solubility (S) and soil adsorption coefficient (K_{oc})

$$\text{by } K_v = Q(P/K_{oc}.S)$$

Where Q is an empirically determined coefficient.

Soil moisture

The moisture content of soil has been determined to be one of the most significant environmental parameters which influence the rate of herbicide volatilization. In general herbicides volatilize more rapidly from moist than dry soils. The effect of low soil moisture content can be dramatic, effectively halting volatilization of a herbicide from a given soil. Increased binding of a herbicide to a soil has been estimated to occur once the soil moisture content decreases to one to three molecular layers after the application on the surface of a soil particle. However, the point at which increased herbicide binding occurs varies and is also dependent upon soil composition which in turn, influences soil moisture capacity. The binding process is revertible and with increased soil moisture content such as after rainfall or dew fall, an increase in herbicide flux from the soil can be detected (Harper *et al.*, 1976; Taylor *et al.*, 1976; White *et al.*, 1977; Grover *et al.*, 1985; Spencer *et al.*, 1970, 1982). In the field, similar influences of soil moisture on herbicide flux have been observed. The flux of trifluralin from soil (Harper *et al.*, 1976) decreased shortly however, when soil moisture is no longer limiting, volatilization increases proportionally with increased energy input in the form of sunlight (Glotfelty *et al.*, 1989).

Temperature

The vapour pressure of a herbicide varies proportionally with temperature. Thus with an increase in ambient temperature a corresponding increase in volatilization from soil or plant surfaces is anticipated. When sorption sites on the soil are saturated with herbicide the vapour density or vapour pressure for that herbicide above the soil may be equivalent to that of the pure active compound (Spencer and Cliath, 1969) and varies proportionally with temperature. At lower concentrations, the temperature effect on the vapour pressure for a herbicide no longer occurs but is directly influenced by the energy of sorption to the soil matrix, as a result becoming less well defined (Specncer 1970). A higher temperature will cause greater volatilization of volatile herbicides and at the same time hasten microbial action.

Herbicide formulation

The herbicides formulation also affects the rate of volatilization from a site. Use of emulsifiable concentrations (ECs) results in direct application of the active ingredient to the soil, allowing for sorption to the soil matrix. In wettable formulation (WP) active ingredient remains on the soil with limited sorption to the soil matrix. Glotfelty *et al.*, (1989) reported that simazine and atrazine applied as WP formulations were displayed from the soil on dry days by the wind as compared to alachlor applied as EC formulations. A change in formulation can be utilized to reduce volatilizations of a herbicide (Turner *et al.*, 1978).

Leaching

Leaching is the movement of herbicides through soil by water. Leaching may occur in any direction (down, up, sideways), depending on water movement, but usually refers to the downward movement of a herbicide as water percolates down through the soil following precipitation or irrigation. The movement of a herbicide by leaching is important to weed control effectiveness, herbicide carryover, and the potential for environmental problems. Herbicides that readily leach may be carried away or carried to rooting zones of susceptible plants. When a herbicide is leached downward, the concentration of herbicide near the soil surface is reduced, lessening the chances for herbicide carryover problems. Leaching is greater in coarse textured soils (sandy loams, etc.) than in finer-textured soils (clay loams, etc.), and is limited by lack of rainfall and by compaction layers. The tendency of a herbicide to leach is determined by its solubility in water and its soil adsorption tendencies. Highly soluble herbicides are prone to leaching.

The dinitroanilines, for example, have very low water solubility and moderate soil adsorption, and thus have a low leaching potential.

Herbicides that are strongly adsorbed to soil particles, like glyphosate and paraquat, cannot be leached unless the soil particles are moved by the water. Soil pH can play a major role and depends how tightly a herbicide is adsorbed to soil particles. Sandy soil would have a higher leaching potential than a clay soil due to larger pore spaces and lower CEC.

Herbicides may move in soil by the physical process of diffusion or mass flow in the liquid or vapor phase. Transport by diffusion is slow relative to mass transport by soil water movement but it may be excessively important over short distance. Soil and other factors that discourage adsorption aid leaching of herbicides in soils. High adsorption herbicides are relatively immobile. When a surface applied herbicide fails to leach even 2-3 cm depth in soil, it will bring about negligible amount or very poor pre-emergence control of weeds. On the other hand, if herbicide leaching is excessive, crop seedling may be injured. Very deep leaching may be desirable for controlling deep rooted weeds. Usually leached herbicides reside in sub-surface soil for a long period in active forms because at that location the conditions for degradation of herbicides are less favorable. Eventually, from sub-surface either a herbicide may return to soil surface during deep ploughing or it may contribute to ground table. Herbicide leaching in soil water can move herbicides out of the tillage and root zone of subsequent crops. Herbicide leaching is greatest in coarse-textured soils with low levels of organic matter.

Herbicide leaching in soil water can move herbicides out of the tillage and root zone of subsequent crops. Soil water affords the driving agency in leaching of a herbicide and then it down under the impact of gravitational pull. That way, solubility of herbicides in water should be an important factor in their leaching behaviours. However, water is never a limiting factor in irrigated or adequate rainfall areas to dissolve even a very low solubility herbicide like simazine at its usual rates of application. Moreover, herbicides with similar water solubility but with different ionic and adsorption properties have been known to differ widely in their leaching behaviours. For example, benefin, trifluralin and nitralin though almost equally soluble in water, exhibit leaching in advancing order of their listing. On the contrary, phenylcarbamate herbicide are more water soluble than substituted anilines, but are equally immobile in the soil system (Weber, 1972). The relative water solubility of herbicides with similar ionic and adsorption properties may, however relate to a great extent with their movement in soils and aqueous solutions (Table 4). For example fenuron (2900 ppm) and diuron (42 ppm) both are non-ionic and weakly adsorbed by soil but fenuron is much more mobile than diuron.

The amount of water applied to a soil at a given time seems to influence the rate of leaching of herbicides more than their extent (depth) of leaching. Variation in rapidity of application of water to a herbicide treated soil likewise, alter the total amount of herbicide that leaches under a given delta of water but not to depth of leaching. Usually slow application of water will move larger quantity of a herbicide down the soil column than its fast application. This is so because slow moving water remains in contact with adsorbed herbicide for a longer duration to initiate desorption. For the same reason, split application of water at frequent intervals induces more leaching than its one time application. But when the interval between intermittent rainfall or irrigation is wide, upward capillary movement of herbicide, considerably may negate its leaching effect. Heavy rainfall in monsoon will cause greater leaching and runoff.

Table 4. Water solubilities (at 25 °C or within 10 to 25 °C) and soil organic carbon sorption coefficients (Koc) of some soil applied herbicides (source; Herbicide handbook of the Weed Science Society of America, 1994).

Herbicides	Average Koc (ml/g)	Solubility in water (ppm)
Dicamba (amine salt)	2	720000
Clopyralid (amine salt)	6	300,000
Picloram	16	200,000
Hexazinone	54	33,000
Chlorsulfuron	40	31,800
Tebuthiuron	80	2500
Metribuzin	60	1100

Contd...

Triasulfuron	65-191	815
2,4-D (amine salt)	20	796
Metolachlor	200	488
Diuron	42	480
Chlorimuron	110	450
Prometryne	33	400
Sulfometuron	78	300
Alachlor	242	124
Ametryn	300	200
Imazaquin	20	60
Atrazine	100	33
Butachlor	700	23
Simazine	130	6.2
Oryzalin	600	2.6
Pendimethalin	17,200	0.30
Oxyfluorfen	100,000	0.1

Leaching is considered as the main cause of groundwater contamination by herbicides and the physical and chemical properties of herbicides are responsible for the leaching behaviour of herbicides. It is observed that the intrinsic mobility of herbicides in soil is inversely related to its degree of sorption to soil surface (Gustafson, 1995). It was observed that top 10 cm soil layer showed higher retention of herbicides. Further it is reported that herbicide mobility is inversely related to clay and organic matter content and the Freundlich sorption constant. Herbicides have been known to move upwards in the soil. If water evaporates from the soil surface, water may move slowly upward and may carry with it soluble herbicides. As the water evaporates, the herbicide is deposited on the soil surface. In arid areas where irrigation is applied, lateral movement of herbicides on soils also occur.

The extent to which a herbicide is leached is determined principally by:

1. Adsorptive relationship between the herbicide and the soil
2. Solubility of the herbicide in water.
3. Amount of water passing downward through a soil.

Solubility is sometimes cited as the principal factor affecting the leaching of a herbicide. The salts of 2, 4–D are water soluble and readily leach through porous sandy soil. Leaching potential of some herbicides are given in Table 5.

Table 5. Relative leaching behaviour of different herbicides in soil under tropical conditions.

Herbicides class	Water solubility	Soil retention	Leaching potential
Chloroacetamide	Moderate	Low	Low-moderate
Dinitroanilines	Very low	Moderate -high	Low
Dipyridiums	Very high	Very high	
Diphenyl ether	Moderate	high	Low-moderate
Imidazolines	Moderate-high (pH dependent)	Very low-low	Low-moderate
Phenoxys/ benzoics/ picolinates	Low-high	Very low-low	Low-moderate
Substituted ureas	Very low-low	Low-high	Low-moderate
Sulfonylurea ureas	Very low-high	Very low-low (pH dependent)	Moderate-high
Thiocarbamate	Low-moderate	Low-moderate	Low-moderate
Triazines	Very high	Very high	Moderate-high

Factors influencing dissolution and leaching

A number of chemical, physico-chemical and biological factors influence the processes that govern dissolution of contaminants and thus the leachability of contaminants from solid materials. The important factors are given below:

pH

Many solids exhibit a marked increase in solubility at both low and high pH values (amphoteric behaviour). Other solids may exhibit maximum solubility at neutral pH values. The pH of the leaching system before leaching occurs is described as the initial pH of the leachant. The leaching system pH at equilibrium is usually governed by the dissolution of the major soluble phases in the solid material. Frequently, the initial and equilibrium pH values differ widely, particularly if the liquid to solid ratio (L/S) is low and the solid phase dominates the system. When L/S ratios are high, the reverse can hold true; the pH of the initial solution is close to the equilibrium pH because the solution phase dominates. Except for the direct influence of pH on dissolution, pH does also influence redox conditions, complexation and sorption. Uptake of carbon dioxide (carbonation reactions) or oxygen (oxidation reactions) from the atmosphere may influence the pH of a leaching system exposed to ambient conditions.

Redox condition

Redox reactions occur like other chemical reactions with the exception that electrons, like protons or cations or anions, participate in the reaction. Reactions can be described thermodynamically (equilibrium) or kinetically (reaction rate). A reduced solid is considered to be electron-rich. An oxidized solid is considered to be electron-poor. The presence or absence of oxygen or the presence or absence of oxidized or reduced mineral species can have an important influence on the stability of solids. Under oxidizing conditions, metal oxides remain an important phase for immobilizing metals. Under reducing conditions, metal sulphides become an important phase for immobilizing metals. The redox conditions in a system also strongly influence the biological degradation of organic constituents/contaminants. Some organic components are degraded only under oxidizing conditions, others under reducing conditions, and some components may be degraded under both oxidizing and reducing conditions.

Soil complexation

Frequently, constituents in the solution phase of a system can bind to solid phase constituents and form a complex that remains in solution in a dissolved state. Such phenomena, governed by equilibrium reaction constants are susceptible to system pH, and may increase the mass distribution of the solute in the solution phase and thus promote solubilisation. In the complexed state, the bound solute is not available to participate in dissolution equilibrium. The solid will then further dissolve to satisfy dissolution equilibrium. The unbound free and complexed solute, when combined, increases the mass distribution in the solution phase. Complexing constituents can be organic as well as inorganic.

Sorption reactions

Many of the solid phases that occur in soils and wastes are very sorptive and are capable of immobilizing dissolved constituents onto the solid surface *via* a number of sorption reactions. Sorption includes a number of reactions; adsorption, desorption, absorption, ionic exchanges and surface complexation. Sorption reactions can involve the formation of bonds that range from relatively weak to relatively strong. Sorption is a reversible process and the sorption process will normally approach equilibrium relatively fast. Many sorption processes are highly pH dependent as a proton or a hydroxyl ion may exchange many sorbing constituents.

Soil colloids

The word colloid is used for microscopic organic or inorganic particles suspended in a solution phase. Colloids are normally defined on the basis of their particle sizes. Particles larger than 10 μm settle faster and are used as the upper limit for colloids. The lower size limit for colloids is generally 1-10 μm (Stumm, 1992). Colloids may constitute a separate phase in a leaching system, and due to the relatively large surface area per unit mass the sorption capacity/complexation capacity of the colloid fraction is often much higher than for the solid phase itself. The presence of colloids may have following effects on leaching and transport of a certain compound/contaminants:

1. Binding of solutes to colloids can increase the mass distribution of the solute in the solution phase;
2. Colloid-associated solutes may be mobilized, as suspended colloids may be transported with the pore-water;
3. The transport of solutes is not only governed by the dissolution of the components from the solid phase but also on processes that govern the stability and transport properties of the colloids.

Presence of microorganisms

Microorganisms may affect organic constituents and the redox conditions in the lysimeter. Microorganisms will primarily be associated with the solid phases but will to some extent also appear in the leachant (e.g. percolating rainwater). To which extent the presence of the microorganisms will affect the degradation of the constituents studied depends on the degradability of the compounds under the prevalent conditions in the lysimeter. If the leachability of easily degradable constituents is being studied, it may be necessary to take precautions to inhibit or minimize the effect of the microorganisms. However, it is extremely difficult to control the microbial activity in large field.

Sondhia and Yaduraju, (2004, 2005) demonstrated leaching potential of atrazine and metribuzin in sandy clay loam soil under laboratory conditions and reported high mobility of atrazine and metribuzin and concluded that atrazine and metribuzin could leach upto the depth of 52 cm in the soil column. Rajana *et al.*, (1991) evaluated movement of atrazine and simazine in soil columns containing red, black or laterite soil and reported that most of the herbicides remained in the top layer. However, movement of simazine was more restricted in the laterite soil compared to red and black soils. The movement of atrazine was slow in black and laterite soils and maximum amount of the atrazine was retained in the 0-5 cm layer. High clay content restricts the movement of atrazine and simazine in black and laterite soil.

Leaching studies indicated that black soil retained more alachlor (1.5 kg/ha) than red soil in the top 0-10 cm and it moved upto 10 cm soil depth in black soil and 20 cm in red soil. Butachlor moved upto 15 cm downward only in the sandy loam soil of Karnataka. Overall about 31 to 68 percent leaching of the different herbicides was reported in sandy loam soils of Karnataka. Metolachlor leached down to a depth of 15-30 cm soil layer and its residues were detected up to the harvest of sunflower crops at both the 0-15 and 15-30 cm soil layers. Prakash *et al.*, (1999) evaluated the relative mobility of butachlor in three soils viz., Hiriyur clay, Mandya sandy clay loam and Mudigere sandy clay loam soil and found that the mobility was more in Mudigere sandy clay loam soil than in other two soils and the relative mobility was influenced mainly by water solubility, sorption, soil texture and hydraulic conductivity of the soils. Movement of pyrazosulfuron was observed up to 60 and 40 cm in Hebbal and Katangese soil in the soil column (AICRP-WC Annual Report, 2006).

Leaching studies with oxyfluorfen (0.10 and 0.15 kg/ha) showed that black soil retained more oxyfluorfen than red soil in top 0-5 cm soil layer. The results of the mobility study after oxadiargyl application showed most of the herbicide in 0-5 and 5-10 cm depth in clay loam soil (AICRP-WC Annual Report, 2006).

Pendimethalin is more tightly bound to soil particles and is therefore moves little in the soil column. Gowda *et al.*, (1993) reported that mobility of pendimethalin in the 0-5 cm layer was greater than 5-10 cm depth and very little movement occurred at 10-15 cm depth.

Prakash *et al.*, (1999) evaluated the relative mobility of ^{14}C-labelled pendimethalin in three soils *viz*. Hiriyur clay, Mandya sandy clay loam and Mudigere sandy clay loam soil and found that the mobility was more in Mudigere sandy clay loam soil than in other two soils and the relative mobility was influenced mainly by water solubility, sorption, soil texture and hydraulic conductivity of the soils. Kulshrestha *et al.*, (2000) found that residues of pendimethalin applied at 1.5 kg/ha rate continuous for five years in maize-wheat cropping system were recovered from various surface (0-15 cm) and sub surface (15-90 cm) layers.

Sondhia (2005, 2007) evaluated leaching potential of pendimethalin and butachlor under lab conditions in 60 cm soil columns and reported that approximately 80 % of applied pendimethalin was found distributed in 0-12 cm soil depth indicating slow mobility of pendimethalin in clay loam soil. However, approximately 45 % applied butachlor was found distributed in 0-4 cm and 7 % butachlor could leach to the 20-24 cm soil depths which showed greater adsorption of butachlor at the surface soil as compared to subsurface soil in clay loam soil.

Thiobencarb, dicamba and anilofos moved upto 20 cm depth in sandy loam soils of Karnataka. 2, 4-D moved upto 20 cm in the soil column under laboratory conditions, although maximum quantity of 2, 4-D was retained in the top 5 cm soil layer. Napropamide did not move beyond 10 cm in the soil.

Sondhia (2006b, 2008c) evaluated leaching potential of sulfosulfuron in loamy sand soil and clay loam soil under simulated rainfall of 200 mm and found that sulfosulfuron and metsulfuron were mostly absorbed in the surface layer of the soil (0-10) but small amount was leached down upto 80 cm. Between the two soil evaluated, clay loam soil retained maximum amount of sulfosulfuron and metsulfuron in the surface soil (0-10 cm) as compared to loamy sand soil (Fig. 3 and 4).

Surface runoff

Herbicides are most vulnerable for transport when excess rain is received soon after their application (Wauchope, 1978; Gaynor *et al.*, 1995). Intact herbicide molecules may be carried by surface water flow. They may be adsorbed to soil particles that move, dissolved in surface runoff water, or carried intact by surface runoff water. On erodable soils, runoff may be followed by wash off. Wash off is the movement of herbicide ions or molecules adsorbed on soil particulate matter that is suspended in the run off water. Water solubility and adsorption characteristics of herbicides affect runoff and wash off losses of herbicides, respectively. Other factors being constant, wash off of herbicides occurs in direct proportion to their rates of application and percent area of the watershed treated. The wash off and run off losses of soil applied herbicides is usually greatest in early stage of a storm as well as from bare soil. Factors that encourage herbicide leaching reduce their runoff and wash off. Tillage practices also influence runoff on the farm (Isensee and Sedeghi, 1992).

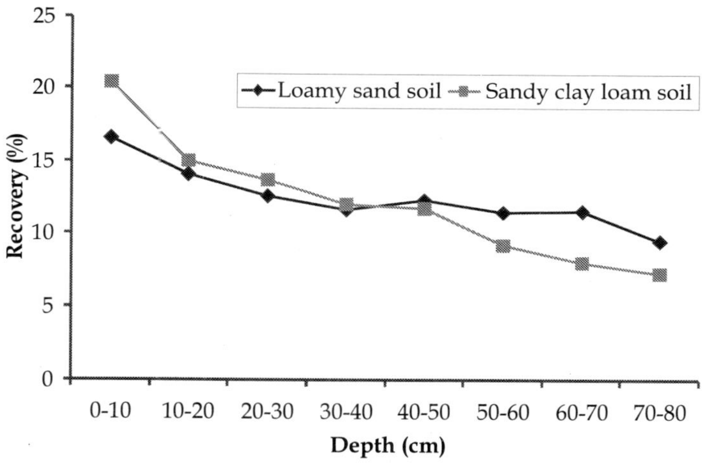

Fig. 3. Distribution of sulfosulfuron in loamy sand and sandy clay loam soil

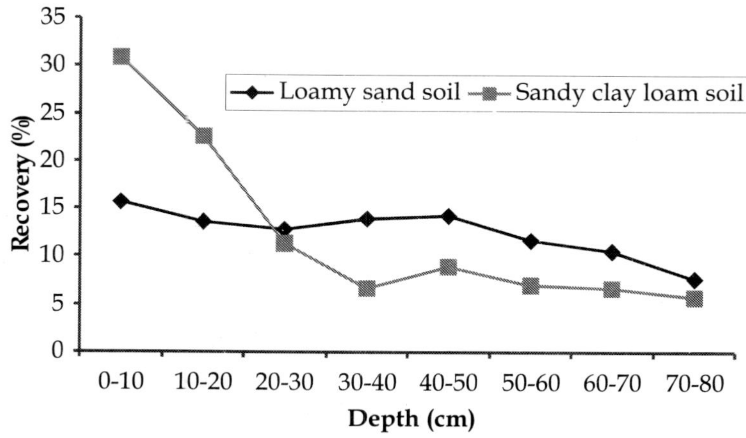

Fig. 4. Distribution of metsulfuron in loamy sand and sandy clay loam soil

Tile can reduce herbicide transport by surface runoff. For example, tiling a silty clay loam reduced losses of metolachlor by 90% and trifluralin by 57% compared with no tile drains (Southwick *et al.*, 1997). In an earlier study, Southwick *et al.* (1997) reported that tile drains reduced metolachlor and atrazine losses by 56% compared with no tile drainage because of a 38% reduction in surface runoff volume. Installation of tile drains reduces herbicide loss by channelling surface runoff to tile drainage, which increases herbicide adsorption to soil, thereby promoting degradation by soil microorganisms or chemical reactions. The effectiveness of tile drains to mitigate herbicide loss in runoff is a function of management practice, soil type, soil hydraulic properties, herbicide solubility, and environmental factors (Jury, 1986).

The primary mechanisms of herbicide dissipation in and on soils are presented in Table 6.

Table 6. Soil factors responsible for slow degradation of some residual herbicides

Group name	Herbicides	Mechanism of degradation	degradation rate
Dinitroanilines	Trifluralin ethafluralin	Microbial	Slow in drought condition
Imidazolinone	Imazethapyr, Imazamox, Imazamethabenz	Microbial	Slow in drought and soil pH < 7.0
Protophyrinogen oxidase inhibitors	Sulfentrazone	Microbial	Slow at soil pH, drought low CEC
Pyridine carboxylic acid	Clopyralid	Microbial	Drought low soil organic matter
Sulfonylurea	Ethametsulfuron, Metsulfuron Sulfosulfuron	Chemical hydrolysis and microbial	Slow at drought low soil organic matter soil pH > 7.0
Sulfonylaminocarbonyl-triazolinone	Flucarbazone	Microbial	Slow at low soil organic matter Soil pH > 7.0 drought
Triazolopyrimidine Sulfonamide	Florasulam	Microbial	Slow at low soil temperature

2. Degradation

Degradation is the process of destruction of the original herbicides molecule and usually loss of herbicidal activity. After degradation parts of the original herbicides structure remain as different molecule, these breakdown products ultimately/ may be decomposed further to simple organic molecules, but more complex breakdown products may be incorporated into organic residues. Degradation processes include biological degradation by soil organisms and abiotic chemical and photochemical transformations.

Microbial decomposition

Microbial decomposition is one of the most important methods by which herbicides are decomposed in soil. Microorganisms in the soil metabolize organic herbicides either aerobically (with oxygen) or anaerobically (without oxygen). Certain groups on the pesticide molecules, like –OH, -COO, $-NH_2$ and $-NO_2$ provide position of attack for the microorganism. In most cases, the microorganisms consume the herbicide molecules and utilize these as a source of energy and nutrients for growth and reproduction. Microbes can also degrade herbicides by a process called co-metabolism, which occurs when the organic herbicide is not used by the microorganism for growth but is metabolized in conjunction with another substrate used for growth. Some herbicides are decomposed easily by microorganisms while others are not. For example, microbial degradation of 2,4-D occurs very quickly in the soil, whereas microbial degradation of atrazine is slow. Fig. 5 depicts the degradation pathway of 2, 4-D by soil bacteria.

Microorganisms in the soil include fungi, actinomycetes, and bacteria. The population levels and activity of these microorganisms depend on food supply, temperature, soil moisture, oxygen, soil pH, and organic matter content. When a herbicide is applied to a soil, microorganisms may immediately attack it. The population of the particular microorganism that uses that herbicide for an energy source will increase (Hutzinger 1981) (Fig. 4). After the herbicide is degraded, the microbial population may return to the original level, or it may stabilize at a level greater than before application (Fig. 5). The increased population could cause more rapid herbicide degradation upon subsequent herbicide applications. Different microbes can degrade different herbicides, and consequently, the rate of microbial degradation depends on the microbial community present in a given situation (Voos and Groffman 1997, McCall *et al.,* 1981). Higher humidity enhances the soil microflora proliferation.

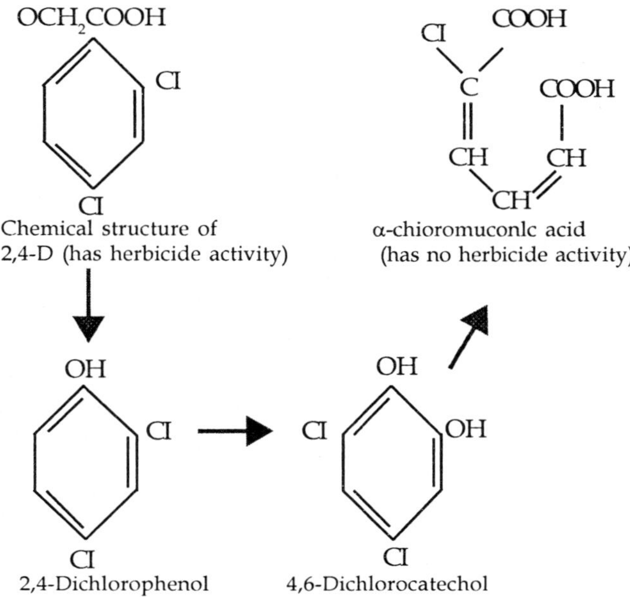

Fig. 5. Degradation pathway of 2,4-D by soil bacteria

There is sometimes a lag time before microbial degradation proceeds. This may be because the population of appropriate microbes or their supplies of necessary enzymes start slightly, and take time to build-up (Farmer and Aochi 1987, Kearney and Karns 1960). If this lag time is long, other degradation processes may play more important roles in dissipation of the herbicide (Farmer and Aochi, 1987). Degradation rates of co-metabolized herbicides tend to remain constant over time.

Factors affecting microbial decomposition

Soil conditions that maximize microbial degradation include warmth, moisture, and high organic content. The optimum temperature for microbial activity generally is 26–32°C. As soil temperatures decrease, soil microbial activity declines, with minimal activity below approximately 4.4°C. Therefore, maximum microbial herbicide breakdown occurs in the summer when soils are warm. The rate of microbial breakdown decreases in the fall as soils cool, and virtually ceases as soil temperatures drop below 4.4°C.

Soil moisture is essential for soil microbial activity. Soil moisture levels between 50-100 percent of field capacity are optimum for microbial activity and, therefore, herbicide breakdown. When soil moisture is limited throughout the sowing season, the rate of microbial degradation of a herbicide is reduced, and the herbicide is more likely to carry over and injure later rotational crops. Similarly, microbes are most active in warm soils (with adequate moisture), when soil temperatures fall below about 40°F, microbial activity becomes negligible. The majority of microbial degradation of herbicides is by aerobic organisms, which are very sensitive to the oxygen supply. Flooded and compacted soils with poor aeration will reduce microbial activity and herbicide breakdown.

Soil pH also affects microorganisms activity. Soil bacteria and primitive fungi called actinomycetes usually are most active in soils with pH above 5.5. Other fungi are less sensitive to soil pH and predominate at pH 5.5 and below. Soil organic matter content is important to soil microbial populations and activity. Organic matter is the primary source of energy and nutrients for soil microorganisms. The highest microbial populations and the majority of microbial herbicide breakdown will be in the surface soil where organic matter content is highest. Small increases in soil organic matter content can increase microbial activity and the rate of herbicide breakdown.

A warm, moist, well-aerated soil with pH between 5.5 and 7.0 generally is most favorable for rapid microbial breakdown of herbicides. Any adverse condition, such as cold temperatures or dry soils, will reduce the rate of herbicide decomposition by microorganisms and lengthen the soil persistence (half-life) of a herbicide.

Certain extenders are used to inhibit microbial decomposition to extend the persistence of a herbicide in soil. Such chemicals prevent the enrichment of enzyme necessary for biodegradation. Boron is a long known extender of 2,4-D. More recently, Vorlex a fungicide is found to act as an extender for linuron. Similarly R-33865 (O, O-diethyl-O-phenolphosphorothioate) is a reported extender for EPTC. Sharton *et al.*, (1993) found phosphate fertilizers acted as potential inhibitors of terbutryne degradation.

Chemical decomposition

Decomposition of herbicides in the soil purely by chemical (non-biological) processes is common for some herbicides. Chemical decomposition is dependent not only on the chemistry of the herbicide but also on soil and climatic factors. Chemical breakdown of a herbicide involves reactions such as hydrolysis, oxidation, reduction etc. The occurrence of these reactions and the rates at which they take place vary with soil type and climatic conditions. These reactions, along with microbial degradation, are important processes in the decomposition of herbicides. Oxidation-reduction reactions involve the transfer of electrons (negatively charged particles) from one substrate to another. A variety of compounds exist within the organic matter matrixes that have the potential to receive electrons from or donate electrons to herbicide molecules. These reactions are not well understood but apparently occur in soil, resulting in reduced herbicidal activity.

Hydrolysis is a process in which the herbicide molecules react with water. Usually, chemical bonds in the herbicide molecule are broken and one or more atoms or groups of atoms in the herbicide molecule are replaced by hydroxyl ions (OH-) from water. This change in molecular structure may inactivate the herbicide. Even very dry field soil has enough moisture for some hydrolysis to occur.

Hydrolysis is the major method by which sulfonylurea herbicides such as metsulfuron-methyl, sulfosulfuron and imazosulfuron degrade in soils. The major hydrolytic pathway in soil for the sulfonylurea herbicides is the cleavage of the sulfonylurea bridge, resulting in sulfonamide and heterocyclic amine molecules (Fig. 6). Cleavage by hydrolysis is pH sensitive. As the soil pH increases, the rate of chemical hydrolysis in the soil decreases. Therefore, sulfonylurea herbicides degrade more rapidly at a lower soil pH than at a higher soil pH. For this reason, there is more potential for chlorimuron or chlorsulfuron herbicide carryover in higher pH soils than in lower pH soils. In addition, chemical hydrolysis of sulfonylurea herbicides is fast during the summer when soil temperatures are warm than in the fall and winter when soils are cool.

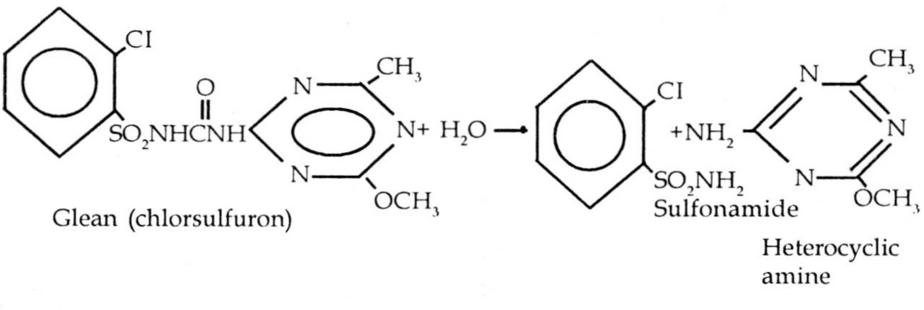

Fig. 6. Breakdown of chlorsulfuron in soil by chemical hydrolysis

Photodecomposition

Photodecomposition is the degradation of herbicides by sunlight. In the photodecomposition process, the herbicide molecule absorbs energy from sunlight, causing chemical reactions that result in herbicide inactivation. Certain dinitroaniline herbicides, for example, are readily degraded by sunlight if left on the soil surface. Soil incorporation will reduce or eliminate photodecomposition.

Photodecomposition or photolysis is decomposition of herbicide under the direct effect of solar radiation. Photodecomposition by sunlight is often the primary degradation route for herbicide present in the exposed environment compartments including soil and plant surface, water and the atmosphere. All photolytic reactions begin with the absorption of light radiation by some chemical species which can then undergo a variety of chemical or physical process. Photodecomposition is often viewed as a detoxifying mechanism, producing products of lower environmental significance than the parent molecule, although some notable exceptions have been observed (Miller and Herbert, 1987).

The phenomenon of photosensitization also must be important in pesticide photodecomposition. In this process a light absorbing compound passes on its added photochemical energy to another substrate which otherwise could not itself become activated. Paraquat solution is not degraded in sunlight because the UV absorption maximum lies sharply at 257 nm (Slade, 1965). When absorbed to a surface as on soil or a leaf the maximum is shifted to about 275 nm and broadened to include the lower end of the sunlight spectrum and photodecomposition occurs. On the other hand, the closely related bipyridium herbicide, diquat already absorbs in the sunlight region (about 310 nm) and so is readily degraded under both circumstances (Salde, 1967).

Photolysis can be grouped as direct or indirect. In direct photolysis, the herbicides itself absorb light energy and then react. The rate of photolytic reaction is directly correlated to the overlap of the absorption spectrum of the herbicides and the spectral distribution of sunlight. Many herbicides absorb light most readily in the near ultraviolet (UV) region; thus the intensity of natural sunlight in this region will greatly enhance photodecomposition rates. Irradiance is dependent upon latitude, molecular scattering and particulate diffusion (Wolfe *et al.*, 1990; Konstantinou *et al.*, 2001). Direct photolysis reactions can be significant in any environment compartment that receive light depending upon the particular chemical and environment conditions.

Indirect photolysis occurs when a chemical species other than the compound of interest absorbs light energy and initiates a series of reactions that eventually degrades the pesticides. These other chemicals species are known as photosensitizers and can include naturally occurring organic and inorganic species, including humic materials, clay minerals, transitions metals, ozone and various free radicals produced by the interaction of sunlight with air and water.

In natural water system, the dissolved humic materials present can serve as strong photosensitizer (Burkhard and Guth, 1976, 1979; Zepp *et al.*, 1981, 1985). In addition, direct photodegradation may be important depending upon the properties of the chemical (Mc Call and Gavit 1986). On soil surface, there has been evidence of both direct and indirect photolysis of pesticides (Burkhand and Guth, 1979; Gohre and Miller, 1983; Choudhary and Webster, 1985), although investigations have not been extensive. Since laboratory water system is homogenous and simple to characterize, photodegradation in water has been extensively studied in the laboratory (Choudhary and Webster, 1985) . In the atmosphere, photodecomposition is thought to be a major degradation pathway, although very little is known about actual reaction rates and their significance in the dissipation of herbicides (Miller and Crossby, 1982) Fig. 7 & 8.

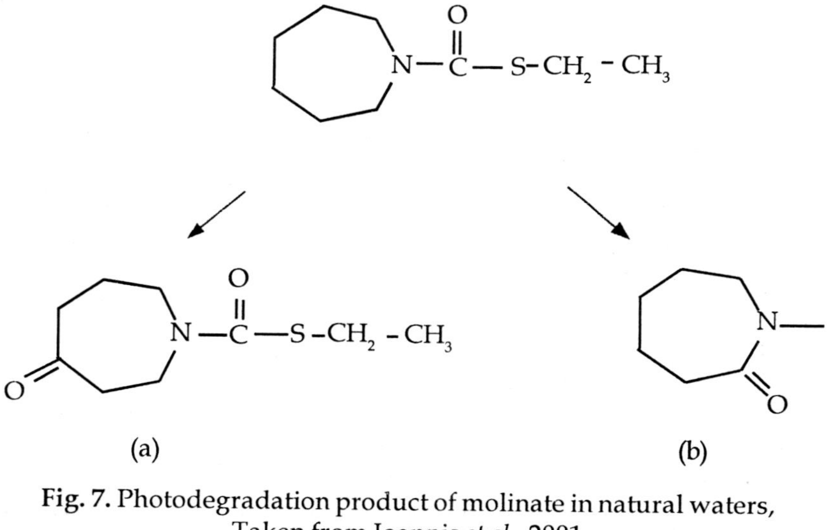

(a) (b)

Fig. 7. Photodegradation product of molinate in natural waters, Taken from Ioannis *et al.*, 2001

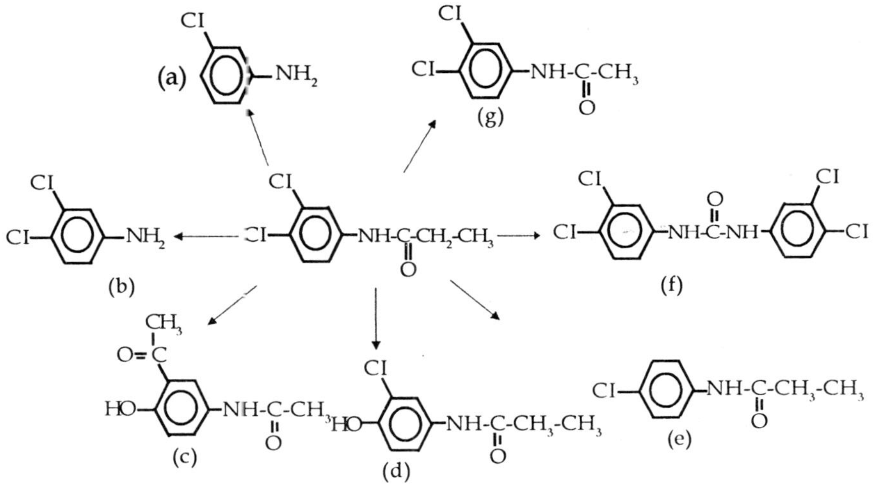

Fig. 8. Photodegradation product of propalin in natural waters, Taken from Ioannis *et al.*, 2001

In soils, photolysis will occur within a shallow surface zone, the depth of which depends on soil characteristics and the mechanism of photodegradation. Light absorption and photolysis of organic contaminants will be influenced by sorption reactions that are related to the soil matter content and by singlet oxygen formation. Both direct and indirect process could be occurring depending on the depth.

Hebert and Miller (1990) concluded that the vertical depth of direct photolysis on the soil surface will be restricted to a region of approximately 0.2 to 0.3 mm. Several studies (Takahashi *et al.*, 1985; Lartiges and Carrigues, 1995; Zepp *et al.*, 1985; 1981; Zafiriou *et al.*, 1984; Jensen-Korte *et al.*, 1987; Vaughan and Blough, 1998) have indicated that humic acids are capable of acting as sensitizers for the production of reactive intermediates such as singlet oxygen (½O_2), hydroxyl radicals (OH ·), superoxide anion (O_2^-), hydrogen peroxide (H_2O_2), and peroxy radicals (ROO ·). Sediment-sorbed DDE photolyzes faster than when it is dissolved, and the product mixture differs in a way that suggests an environment rich in H-donors, probably organic matter (Zafiriou *et al.*, 1984; Miller and Zepp 1979). Such reactive species can potentially diffuse to depths approaching 1mm depending on moisture depth, soil porosity, and thermal gradients on the sunlight exposed soil interface. Finally, metal oxides presented in the soil, such as ZnO, Fe_2O_3, and MnO_2, absorb radiation in the sunlight wavelength range and could accelerate degradation by reaction of OH* and $^{-}O_2$ through the well-known mechanism of semiconductor photometric chemistry.

Moreover, in many cases, the electronic structures, absorption spectra, and excited state and lifetimes of sorbed compounds are much different from their solution properties, making difficult to predict effects that may result from sorption. Nitrogen-containing compounds often show different spectra when sorbed on clay minerals compared with solution spectra. This is probably due to protonation–deprotonation effects (*s*-triazines), charge transfer phenomena, or excited state alterations. Triplet lifetimes for many aromatic compounds are greatly extented when they are adsorbed on silica or alumina, which means they may be more susceptible to photo chemical reactions (Larson and Weber, 1994). Photolysis rates and phototransformation products are actually dependent on the intensity and wavelength distribution of the light used. To accelerate photolysis testing, high-pressure Hg lamps with a wavelength output of approximately 254 nm were typically used in laboratory studies (Durand *et al.*, 1994). These studies pose difficulties for extrapolation to environmental conditions, as only wavelengths of light at 290 nm would be useful for fate measurements (Wan *et al.*, 1994). Although modeling pesticide behavior under laboratory conditions is a very useful tool for environmental studies, there is a need to conduct photodegradation studies in natural conditions taking into consideration the factors such as the daily temperature variation and seasonal variation of natural sunlight intensity (Chiron *et al.*, 1995).

Typically phenoxy herbicides undergo oxidation, reduction and substitution (hydrolysis) in aqueous medium when activated by sunlight in air (Crosby and Wong, 1973). Photodecomposition is the most important environmental reaction. Molecular oxygen is abundant in air, soil and water and generally occurs in diradical (triplet) forms. The photodecomposition is often rapid, NAA, paraquat and diquat can be photolyzed in hours under UV light (Crosby and Tang 1969). In monuron photodecomposition in aqueous solution occurs through oxidation (Crosby and Tang, 1969). By one route the N-methyl group was removed successively as also observed with amides such as diphenamid (Rosen, 1967) and amines such as trifluralin (Crosby and Leits, 1973). Although aromatic halides normally are very stable to nucleophilic displacement reaction, absorption of UV light by the resonating system activates the ring towards nucleophilic attack. Some examples of radical reactions are the stepwise photo-reduction of dichlobenil to 2-chlorobenzonitrile and benzonitrile (Plimmer 1971); amiben to the monochloro derivative in methanol, while the reductive declorination of chlorobenzoic acids (Crosby and Leitis, 1969) and reduction of the aromatic nitro group of nitrofen to the amine and azobenzene show the photo-chemical reducing conditions of irradiated aqueous solutions.

Plant uptake and metabolism

Plant roots may absorb (take up) herbicides from soils and plant foliage may intercept foliar-applied chemicals. Crops and plants tolerant to a herbicide often metabolize the chemical into non-toxic substances. Differential metabolism is often the basis for herbicide selectivity. For example, corn and sorghum absorb and metabolize (detoxify) atrazine, herbicides are incapable of metabolizing and detoxifying the chemical rapidly enough to escape the herbicidal activity. These factors result in herbicide breakdown or deactivation; however, herbicide concentrations in soils are also affected by the movement of intact herbicide molecules *via* leaching and surface runoff.

Degradation experiments of simazine in pine seedlings have shown the presence of three unknown metabolites. In wheat, hydroxy-simazine was found as a metabolite of simazine (Hamilton, 1964). Resistant species such as maize degrade simazine to hydroxysimazine nonenzymatically. A compound named benzoxazinone was isolated from maize extracts, and appears to be responsible for apparent catalytic degradation of simazine. It is possible that red pine seedlings contain very little benzoxazinone. This was evident from the fact that the amount of simazine degraded by red pine seedlings was small and there was no hydroxysimazine (Dhillion *et al.*, 1968).

Absorption of amitrole and 2,4,5-T after bark application was considerably superior to that after foliar application in red maple (*Acer rubrum* L.) and white ash (*Fraxinus americana* L.) (Leonard *et al.*, 1966). When applied to leaves, herbicide may either migrate in the apoplast following penetration of the cuticle or passage through stomata or other openings; such transport proceeds towards the margins of laminas. A portion of the herbicide may be absorbed into living cells and then be translocated in the symplast.

Norris and Freed (1966a, b, and c) studied absorption and translocation of several phenoxyalkyl acid and chlorophenoxyalkyl acid herbicides in bigleaf maple (*Acer macrophyllum.*). In an experiment in leaves were treated with 2,4-D or 2,4,5-T as the acid, triethanolamine salt or 2-ethylhexyl ester formulations. The absorption increased with decreasing polarity of the chemical (Norris and Freed, 1966a). In general, an inverse relationship between the polarity of the compound and its degree of absorption exists. Also, inherent toxicity of the compound to plant transport mechanisms, the degree to which it is accumulated by non-transport tissues in the vascular system of the plant, and the possible influence of water solubility of the translocated compound are factors which might interact to produce differential translocation. However, the relative importance of each of these factors is not yet known. Norris and Freed (1966b) concluded that phenoxy herbicides were poorly translocated by bigleaf maple. They suggested that those herbicides might be toxic to phloem transport tissues, which could be a limiting factor in translocation. In the later study, they tested 2, 4-(2,4-dichlorophenoxy) butyric acid (2,4-DB), which is supposed to have superior translocation characteristics (Norris and Freed, 1966c). Intact maple seedlings were treated with 2,4-DB and after 3 days, the plants were harvested, sectioned, and extracted. The results showed that 2,4-DB was oxidized to 2,4-D in several plant parts. While β-oxidation is not unknown in plants, β-oxidation probably resulted in this case. Roots of maple seedlings were found to have a greater ability to convert 2,4-DB to 2,4-D than stem phloem. This agrees with a previous study by Norris and Freed (1966b) that 2,4-D was not translocated in maple seedlings. They found relatively low rate of transformation from 2,4-DB to 2,4-D in the leaves and stems.

Lund-Hoie (1969) studied uptake, distribution, and fate of simazine in 4-year old seedlings of Norway spruce (*Picea abies* L. Karst). Simazine was found to be readily absorbed and distributed in the spruce seedlings. There was a positive correlation between the amount of simazine absorbed and the growth of the plants. This indicates as one of the final products. Three possible reasons may account for tolerance of spruce to simazine. Firstly, simazine is very insoluble in water and could be strongly adsorbed to soil. Secondly, the root system of spruce may present a barrier to the uptake of simazine due to its lipophilicity. Thirdly, the resistance of spruce may be correlated with metabolic inactivation or detoxification of simazine. In this study, simazine was absorbed by the roots and hydroxysimazine, the degradation product, was translocated to the needles, with a direct relationship between the amount of simazine absorbed and the transpiration intensity. As expected, seedlings at apparent dormancy translocated less simazine compared to seedlings at intense internode elongation (Lund-Hoie, 1969). An important conclusion is that in northern latitudes, spruce may serve as a useful species to assist clean up of simazine. Simazine taken up by spruce was fixed in roots and stems. Fixed simazine was degraded by roots and stems to hydroxysimazine and two other unknown metabolites. Since no simazine was observed in spruce needles, this indicates a very rapid rate of metabolism in the roots and stems. However, there was a very slow degradation of metabolites to CO_2.

Erickson and Lee (1989) have reviewed the pathways for degradation of simazine and atrazine. The pathways appear to be the same for degradation in soil and in plants; however, the rate of the hydrolysis process to form hydroxyatrazine in soil is influenced by nucleophilic compounds that are present in soil. Within plants, other compounds appear to be important as described below for simazine.

A study was conducted to investigate factors such as uptake and degradation of simazine by black walnut (*Juglans nigra* L.) and yellow poplar (*Liriodendron tulipifera* L.) trees (Wichman and Byrnes, 1975). Mono-dealky lated simazines CEAT and other degradation products were found in leaves of both species. Gysin and Knusli (1960) reported that monodealkylated simazine is phytotoxic and an effective inhibitor of the Hill reaction in isolated chloroplasts. Since simazine is a photosynthetic inhibitor and its degradation product, CEAT is also an inhibitor, tolerance of a plant should be inversely related to the concentration of phytotoxic compounds in the leaves of the plant. Consequently, it shows that black walnut was more tolerant than yellow poplar to simazine. Walnut is a high value species that might profitably be used in riparian buffer strips.

The results of Wichman and Byrnes (1975) showed that both black walnut and yellow poplar degrade simazine rapidly and have a considerable degree of simazine tolerance. Four other degradation products were identified as 2-chloro-4, 6-(diamino)-s-triazine (CAAT), as hydroxysimazine (EEOT), and two other unidentified degradation products. It was believed that they represent the products of further hydroxylation or conjugation reactions. From this study a major pathway of simazine degradation was found to be N-dealkylation in both species.

Paterson and Schnoor (1992) found that atrazine and alachlor degraded more rapidly in a riparian zone soil under hybrid poplars (*Populus deltoides* × nigra cv. *"Imperial Carolina"*) than in soils where herbicides were applied to row crops. In a recent study, Burken and Schnoor (1996) applied atrazine, another triazine herbicide, to the same hybrid poplars grown in silica sand. They showed that 91% of atrazine could be assimilated in less than 10 days yielding less toxic metabolites. The evapotranspiration rate of poplar trees has a significant influence on contaminant uptake. The formation of bound residues from atrazine is considered to be a detoxification mechanism. Burken and Schnoor (1997) proposed atrazine fate mechanism in poplar tissues, with major degradation steps involving hydroxylation, dealkylation, and oxidation. Twelve organic contaminants including atrazine were later studied for uptake by hybrid poplar trees. Translocation and subsequent transpiration of volatile organic compounds from the leaves to the atmosphere was a significant pathway for several organic compounds. However, for atrazine, volatilization from leaves was found to be negligible.

In the aquatic plants simazine was primarily accumulated in the leaves of parrot feather (*Myriophyllum aquaticum*) and canna (*Canna hybrida* L.) when those plants were grown in simazine-contaminated water (Knuteson *et al.*, 2002). They concluded that four-week old parrot feather and canna were more tolerant of simazine than two-week old plants. The tissue burden with simazine, normalized for plant biomass, was less for the four-week old plants. The authors attributed simazine biodegradation to dealkylation of simazine and suggested that simazine could be metabolized either to CEAT or to EEOT followed by storage of end products in the vacuoles. However, they also concluded that the main form of detoxification was through conjugation to glutathione using glutathione-S-transferases. In general, the foliage of plants has the highest levels of glutathione and glutathione-S-transferases. Hatton *et al.*, (1996) had demonstrated that the levels of glutathione in maize plants declined after 30 days of growth. However, Knuteson *et al.*, (2002) found that plant tolerance for simazine increased with plant age and size. Both parrot feather and canna were more vulnerable to simazine shortly after emergence.

Summary

As soon as a herbicide is applied a number of processes immediately begin to remove the compound from the original site of application. This is the process of environmental fate. For the herbicide which is intercepted by plants, the chemicals may be taken up by the plant itself, may be washed off by precipitation on to the soil, may undergo decomposition by biological (microbial) and chemical processes, photodegradation on plant surface or may volatile back into the air. Transport of herbicides within the soil compartment can occur downward into the soil profile (leaching), across the soil surface (runoff), or into the air (volatilization) each can be a combination of more fundamental processes including adsorption, convection and diffusion. Soil organic matter has a wide range of physical, chemical, and biological characteristics. It has been demonstrated that humified matter is more chemically reactive with pesticides than nonhumified matter (Kozak *et al.*, 1997). Herbicides persistence in the soil is expressed as half life or time required to degradate fifty percent of the original molecule. However, the half life is not absolute because it depends on the soil type, temperature, and concentration of the herbicide applied. The persistence varies with the nature of a chemical and on soil and climatic conditions.

Plant root may absorb herbicide from soil and plant foliage and may intercept foliar applied herbicides. Crop and plants tolerant to a herbicide often metabolize the chemical into non-toxic substances. Differential metabolism is often the basis for herbicide selectivity.

□□□

Chapter 8

Adjuvants

Adjuvants are materials that, when added to the spray solution facilitate or modify the action of the herbicide. The earliest work using adjuvants in 1889 indicated that addition of soap increased the toxicity of arsenical insecticides. Before the turn of the century, whale oils or fish oils were most commonly used as adjuvants. Thus, adjuvant is an all-inclusive term used to define all spray tank additives, from wetting agents to foaming agents. Since the development of synthetic organic herbicides in the late 1930s, the importance of effective foliar penetration has been recognized as an important factor for effective weed control. Prior to 1950s, household detergents and various oils were used as adjuvants. Soaps form scum in presence of calcium, magnesium, and iron in hard water hence are undesirable. Household liquid detergents are also undesirable due to the lower surfactant content. The average detergent contains only 10% - 20% surfactant and may contain additives that may interfere with the activity of a herbicide. It wasn't until the 1950s that the advantages of specialized adjuvants in spray solutions were recognized. By 1960s, the need for agricultural adjuvants had escalated to the point that household detergents were no longer recommended and investigators began looking into new formulations that would be more applicable to agriculture.

Today the technology of adjuvants includes the use of complex synthetic formulations designed for specific purposes. Small changes in the chemical structure of adjuvants can greatly alter their action. There are more than 3,000 chemicals presently being marketed as adjuvants, and more than 80% of the herbicides marketed in the United States today contain anywhere from 2% -15% emulsifier in their formulation to affect flowability, wettability, dispersibility, suspendibility, foaming, and physical and chemical storage stability.

Adjuvant terminology can be confusing, sometimes because of incorrectly used terms and also, due to the complexity in adjuvant chemistry. Adjuvant chemistry is complicated because most adjuvants are mixtures of many materials and the exact composition is considered proprietary information and thus unknown. This makes adjuvant classification difficult.

A term commonly used interchangeably, but incorrectly, with adjuvant is surfactant. Surfactant, or surface-active agent, is defined by the Weed Science Society of America (WSSA) as any material "that facilitates and accentuates the emulsifying, dispersing, spreading, wetting or other surface-modifying properties of liquids." However, a surfactant is an adjuvant.

In general speaking, the term adjuvant should be used to group all spray tank additives and the term surfactant would be used to describe those additives that modify the distribution of spray on plant foliage.

Often adjuvants are being considered biologically inactive by themselves, and therefore we can assume that it is only the herbicide that acts in the plant. The reasons for using adjutants could be to save active ingredient of an expensive herbicide (save money), to lower the environmental adverse effects of herbicides, avoid spray drift, etc. It is obvious that the selectivity of a herbicide may change when we use adjuvants, and consequently it is important that the end user is aware of this fact, otherwise it may be a costly experience. Manipulating herbicide spray solutions with various adjuvants represent an opportunity to improve performance of foliar-applied herbicides under a variety of environmental conditions; however, there are limits. Adjuvants can change a number of parameters that may improve herbicide performance. The term adjuvant encompasses a wide range of compounds that have been used to improve herbicide performance. Adjuvants can improve herbicide performance by influencing a number of factors that are involved in herbicide absorption.

Properties of Adjuvants

Increase droplet coverage and spray retention

One of the best-known properties of surfactants is their ability to reduce the surface tension of the spray solution. Reducing the surface tension of the spray solution means that spray droplets will spread beyond their initial contact area. Increasing the coverage area increases herbicide absorption. Materials that increase the stickability of the droplets on the leaf surface rather then acting like little ping-pong balls are called stickers. Stickers are often water-soluble polymers, acrylic latex or resins. Wetting or spreading agents are often combined with stickers to improve coverage. The term spreader-sticker was often used to describe these adjuvants, but does not represent a logical term since 'sticking' is the opposite of 'spreading'.

Increase spray droplet drying time

A material that increases the drying time of a spray droplet is called a humectant. Humectants are commonly water-soluble materials that resist drying even after the aqueous solution has evaporated. They accomplish this by drawing water from the atmosphere. Glycerine, propylene glycol, diethylene glycol, polyethylene glycol, urea and ammonium sulfate are common humectants. Oil-based compounds, like crop oil concentrate (COC) or methylated seed oil (MSO) also resist drying. Since absorption can occur only when the herbicide is in solution; slowing the rate at which droplets dry will allow more time for absorption. Not all herbicides are compatible with oil-based humectants. Theoretically, adding a humectant should improve herbicide performance under hot, dry conditions, however, herbicide performance rarely increases with the addition of a spray adjuvant if weeds are under severe moisture stress.

Increase cuticle penetration

Crop oil concentrates (COC) and vegetable oils methylated seed oil, (MSO) fall into the category of penetrants. This type of surfactant can improve cuticular penetration by softening, plasticizing, or dissolving cuticular waxes and allowing herbicide movement to the more hydrophilic regions underneath. Penetrant adjuvants are often complex mixtures of surfactant and oil (paraffinic petroleum or modified vegetable).

Increase cellular accumulation

The cellular absorption of weak acid herbicides can be improved with the addition of UAN (urea ammonium nitrate) or AS (ammonium sulfate) to the spray solution. Weak acid herbicides have a carboxylic acid or other functional group that accepts or loses a hydrogen ion depending on the pH of the surrounding solution. This means that the herbicide's solubility will change with changes in solution pH. Under normal conditions, ammonium ions (NH_4^+) are actively transported into the cell across the cell membrane. The addition of UAN or AS to the spray solution increases ammonium accumulation in the cell. This accumulation alters the cell's pH because once inside the cell the ammonium ion becomes ammonia (NH_3^+) plus a free hydrogen (H^+). The increase of hydrogen inside the cell would decrease cellular pH if allowed to accumulate. To maintain cytoplasmic pH in the range of 7.5-8.0, hydrogen ions are pumped across the cell membrane into the cell wall. The movement of ammonium into the cell and subsequent pumping of hydrogen ions into the cell wall causes the cell wall pH, which is already acidic, to become even more acidic, as low as pH 4.5. For weak acid herbicides, acidic conditions cause more of the herbicide molecules to be present in their non-ionized, lipophilic form. The nonionized form passes through the cell membrane and into the cell in response to a concentration gradient, increasing the herbicide concentration in the cell due to a process called ion trapping.

The bottom line is that adjuvants can help improve herbicide performance under adverse weather conditions; however, it is not possible to substitute adjuvants for good growing conditions. Adjuvants may improve herbicide movement in the cuticle or leaf, but the problem lies in the effects of moisture stress on the plant's physiology. If the herbicide requires translocation to meristemic regions or requires active metabolic process in the plant, moisture stress will significantly reduce herbicide efficacy regardless of the adjuvant system.

Use of adjuvants

Adjuvants are commonly used in agriculture to improve the performance of pesticides. Broadly defined, "an adjuvant is an ingredient that aids or modifies the action of the principal active ingredient." The use of adjuvants with agricultural chemicals generally falls into two categories:

(1) formulation adjuvants are present in the container when purchased by the dealer or grower; and

(2) spray adjuvants are added along with the formulated product to a carrier such as water.

The liquid that is sprayed over the top of a crop, weeds, or insect pest often will contain both formulation and spray adjuvants.

Formulation adjuvants are added to the active ingredient for a number of reasons including better mixing and handling, increased effectiveness and safety, better distribution, and drift reduction. These traits are accomplished by altering the solubility, volatility, specific gravity, corrosiveness, shelflife, compatibility, or spreading and penetration characteristics. With the large number of formulation options available (solutions, emulsions, wettable powders, flowables, granules, and encapsulated materials), adjuvants become even more important in assuring consistent performance.

Spray adjuvants are added to the tank to improve pesticide performance. Literally hundreds of chemical additives are now available that fall into this category. Spray additives can be grouped into two broad categories: *activator adjuvants* include surfactants, wetting agents, stickers-spreaders, and penetrants; *special purpose* or utility modifiers such as emulsifiers, dispersants, stabilizing agents, coupling agents, co-solvents, compatibility agents, buffering agents, antifoam agents, drift control agents, and nutritionals. Descriptions of the more common types of special purpose adjuvants follow.

Classification of adjuvants

Adjuvants can be grouped into three categories, activator adjuvants, spray-modifier adjuvants and utility-modifier adjuvants. The following is a description of the different types of additives grouped into categories according to the type of action;

1. Activator adjuvants

As the name indicates, this group of adjuvants increases the activity of the herbicide. This effect can be attributed to:

(a) particle size of the herbicide spray,

(b) distribution of the spray on the plant,

(c) viscosity of the spray,

(d) evaporation rate,

(e) rate of herbicide uptake by the plant, and

(f) solubility of the herbicide in the spray solution.

Activator adjuvants are by far the most common type of additives used to enhance herbicide performance. Although some products are sold to alter pesticide-soil interactions, the emphasis of this discussion will be on foliar-applied materials. The primary use of activator adjuvants is with post-emergence herbicide applications.

Before any foliar-applied herbicide can perform the desired biological function, it must be transferred from the leaf surface into the plant tissue. The aboveground portions of plants are covered by a continuous noncellular, nonliving membrane called cuticle (Fig. 1). Cuticle is the first barrier that any herbicide must overcome to be effective. The plant cuticle is composed of water-repellent waxes and less water-repellent cutin and pectins that can provide pathways for more water-soluble pesticides. The structure of plant cuticle can be likened to a sponge where the matrix of the sponge corresponds to the cutin and the holes correspond to the embedded wax. The surface of the sponge is also covered with wax (epicuticular wax). Cuticle is extremely diverse and varies greatly between different species of plants.

Waxes are the principal barrier restricting herbicide movement into plant foliage. The chemical or physical properties of the wax appear to be more important than thickness in restricting penetration. Surface wax high in hydrocarbons and other repellent molecules is less permeable to water and most herbicide sprays than cuticle membranes with lower amounts of water-restrictive waxes. For example, lambsquarters cuticle wax is known to be a strong barrier to the penetration of many herbicides. Lambsquarters cuticle is high in chemical substances called aldehydes, which may help to prevent the passage of more water-soluble herbicides. Not only does cuticle composition vary between species, but also the age of the plant has been associated with differences in leaf wax chemistry over time.

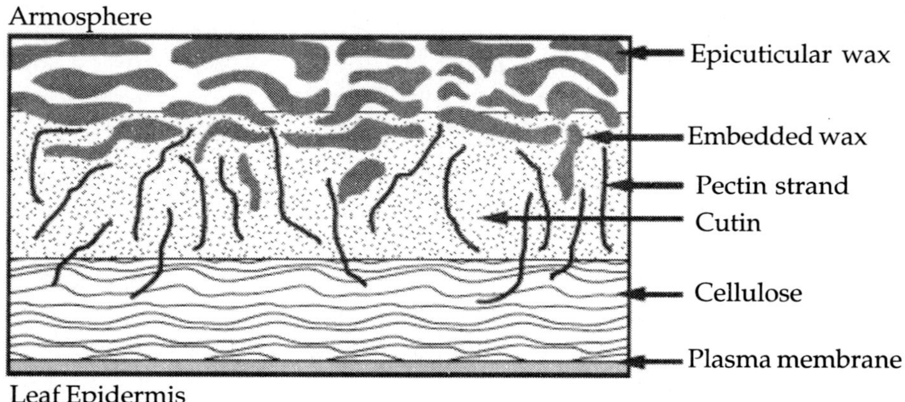

Fig. 1. Simplified plant cuticle (after Hess, 1991, 1999).

Some of the example of activator adjuvants are given below:

(i) Wetting Agents

The term "wetting agent" applies to any spray additive that increases the ability of water to displace air or liquids from a plant surface. This displacement helps spread of the spray solution over the entire plant surface more evenly. Some degree of wetting is also a property of all surfactants, however, the extent of wetting may vary greatly among products.

(ii) Oils

Oil adjuvants are petroleum or vegetable oil in origin. Petroleum oil adjuvants are selected for non-toxicity to plants and can vary in chemical composition, viscosity and emulsifier content. Petroleum oils with 17% or more of a specific emulsifier were called "crop oil concentrate" because they have equal efficacy with atrazine when used at 2.3 L/ha compared to 9.2 L/ha required for petroleum oil with 5% or less emulsifier. Vegetable oils are from several plant sources and are either refined oil or esterified oil, usually the methyl ester.

Petroleum and vegetable oil efficacy with nicosulfuron generally increased as emulsifier percentage in the adjuvant increased from 3 to 25 or 35% (Nalewaja *et al.*, 1995). However, percentage emulsifier in the methylated oil adjuvant did not greatly influence efficacy. Thus, methylated oil was effective independent of the emulsifier and would only need enough emulsifier for good dispersion in the spray carrier water.

Spray retention generally increased as the percentage of oil adjuvants in the spray mixture increased (Bruns and Nalewaja 1998; Hall *et al.*, 1997; Western *et al.*, 1998; Coret *et al.*, 1993; Ditomoso 1999). However, spray retention with an oil adjuvant was less with oil at 4% v/v than with surfactants at 0.25% v/v (Bruns and Nalewaja, 1998). Spray retention for oil adjuvants, all with 15% v/v of the same emulsifier, generally esterified more in vegetable than petroleum oil, (Bruns and Nalewaja 1998). Thus, the function of oil adjuvants is more to aid absorption than spray retention. The above results were for adjuvants in water without herbicide, but addition of a sulfonylurea herbicide did not influence retention. Oils increase in effectiveness with increased concentration in the spray mixture, especially at low spray volumes.

Petroleum oils used with herbicides have a 70 or 110 second viscosity. These oils were used commonly as horticultural spray oils and were adapted for herbicide use because they were not toxic to plant tissue (Nalewaja 2002). Oil viscosity influenced efficacy of sethoxydim (Matysiak and Nalewaja 1999). Sethoxydim applied at 10 °C was more efficacious when mixed with 70 second petroleum oil adjuvants than 110 second oil but the opposite occurred when applied at 25 °C. High temperature appeared to cause the low viscosity oil to spread excessively, thereby reducing the sethoxydim deposit concentration for absorption. Conversely, at low temperature the high viscosity oil did not spread and gave poor droplet deposit contact with the leaf surface. Esterified vegetable oil is equally effective at both temperatures, providing deposits that spread with close contact to the leaf surface. The greater solvency of esterified oils than petroleum oils in cuticle may account for their efficacy, even though the deposit is thin (Manthey and Nalewaja, 1992). Esterified vegetable oil are more efficacious than petroleum oils, regardless of temperature.

Solvency and viscosity characteristics of oil adjuvants do not always relate to efficacy with all herbicides (Nalewaja 1994). If cuticle solvency is primary to herbicide absorption, then esterified vegetable oil adjuvants would be most effective with all herbicides. However, for grass control, methylated sunflower oil are only equal to petroleum oil with diclofop, quizalofop, and fluazifop, but the methylated oil are more effective with sethoxydim, fenoxaprop and haloxyfop. For broadleaf control, methylated oil are more effective than petroleum oil with acifluorfen, fomesafen and imazaquin, equally effective with bentazon, and less effective with lactofen. Sunflower oil are less effective than petroleum oil with all of these grass control herbicides, but equal with sethoxydim. For broadleaf control with lactofen, a soybean oil adjuvant is only less effective than petroleum oil. These oils have the same emulsifier (15% Atplus 300F). In general, methylated vegetable oils were equally or more effective than petroleum oil. Vegetable oils are equal to petroleum oil, except less effective with quizalofop, fluazifop, and lactofen. (Nalewaja 1994).

Water emulsions with many herbicides can increase weed control. The oil-water emulsion may increase retention time of sprayed material and enhance uptake through leaf surfaces. Most marketed oil-surfactant concentrates contain around 80% oil and 20% surfactant in their formulations. The rate of application varies, but a rate of 1 pt per acre to 2 gal per acre is most common. Many herbicides are formulated as emulsions and have a milky appearance when added to water. Adjuvants that are primarily oil based are very popular with pesticide applicators (McWhorter 1982).

Crop oils

Crop oils are probably the oldest group within this category. Crop oil is a misnomer because the material actually is from petroleum (paraffin or naphtha base, not vegetable derivative), a phytobland (non-phytotoxic), non-aromatic oil of 70 to 110 second viscosity (water = 1 and 30 w motor oil = 300). Crop oils are 95 to 98 percent oil with 1 to 2 percent surfactant/ emulsifier. Crop oils are believed to promote the penetration of pesticide spray through waxy cuticle or the tough chitinous shell of insects. Traditional crop oils are more commonly used in insect and disease control than with herbicides. Crop oils are typically used at 1 to 2 gallons per acre.

Crop oil concentrate

Crop oil concentrate contains 80 to 85 percent phytobland emulsifiable crop oil (petroleum based) plus 15 to 20 percent non-ionic surfactant. The purpose of the surfactant in this mixture is to emulsify the oil in the spray solution and lower the surface tension of the overall spray solution. Crop oil concentrates attempt to provide the penetration characteristics of the oil, while capturing the surface tension reduction qualities of a surfactant. Crop oil concentrates are also important in helping to solubilize less water-soluble herbicides such as quizalofop, sethoxydim, fluazifop-p-butyl, clethodim, and atrazine on the leaf surface. Crop oil concentrates are used at 1 to 3 pints per acre or at 1%v/v (1 gal/100 gal) unless otherwise directed.

Vegetable oil concentrates

Vegetable oil concentrates have performed less consistently than their petroleum-based counterparts. However, manufacturers are attempting to improve plant or vegetable-based oils by increasing their non-polar or lipophilic characteristics. The most common method has been through esterification of common seed oils such as methylated sunflower, soybean, cotton, and linseed oils. The methylated forms of these seed oil concentrates are comparable in performance to traditional (petroleum) crop oil concentrates so their importance has increased. In taking it one step further, organosilicone-based methylated vegetable oil concentrates are also available. These adjuvants boast the surface tension-reducing properties of silicone but have the advantages of a methylated vegetable oil concentrate. The more widely available oil based additives are given in Table 1.

Table: 1. Important oil based additives

Additive type	Trade name	Manufactures
Crop oil	Cenex spray oil	Cemex/land O Lakes
	Domant oil	Riverside/Terra
	Knock Down crop oil	Cemex/land O Lakes
Crop oil concentrates	Activate oil adjuvant	Drexel Chemical
	Agri Dex	Halena
	CLASS 17% concentrate	Cemex/land O Lakes
	Herbimax	Loveland industries
	Peptoil	Drexel chemicals
	Prim oil	Riverside/Terra
Vegetable oil concentrates	CLASS destiny (methyl soybean)	Cemex/land O Lakes
	Dyne Amic (silicone methyl vegetable)	Halena
	Meth oil (methyl soybean)	Riverside/Terra
	Sun-it II/Scoil (methyl vegetable)	Halena
	Vegetable oil concentrate (vegetable)	Agsco
	Vegetable oil	Drexel chemicals

(iii) Surfactants

A surfactant (surface active agent) is material that improves the emulsifying, dispersing, spreading, wetting or other properties of a liquid by modifying its surface characteristic. Surfactants are to be considered as inactive ingredients even though they can have a pronounced effect on the performance of the product (Fig. 2). The increase in leaf penetration is associated with a reduction of surface tension of liquids, which improves wetting of the leaf surface. Increased penetration may also be due to the surfactant dissolving leaf tissue components. Surfactants are adjuvants that reduce surface tension within the external surface layers of water.

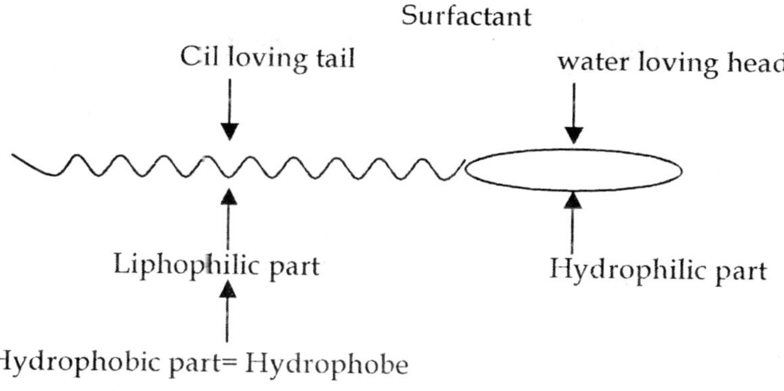

Fig. 2. Component of a surfactant

There are four different types of surfactants:

Anionic surfactants

Anionic surfactants are negatively charged, and enhance foaming and other spreading properties. Normally anionic surfactant are not used with herbicides. For example, shampoo for hair contains sodium or ammonium laureth sulfate, which is the preferred anionic surfactant for hair. Using an anionic surfactant in the greenhouse can cause problems with sprayers that have an agitator, or any system where the foam could disrupt water flow or pump suction.

Cationic surfactants

Cationic surfactants are positively charged, and are often very toxic to plants as they can disrupt membrane ion balance. Cationic surfactants are not widely used for pest control, but they are more commonly used in heavy-duty cleaning compounds. Engine wash surfactant used to clean the tractor should not be used. The results can be devastating to plant material.

Amphoteric surfactants

Amphoteric surfactants, depending upon the pH of the solution, will form either a positive or negative charge in water. Their use in horticulture crop protection is rare. These products are used very specifically to match the properties of specific pesticide formulations to carrier components or other materials, and are generally not available for use in the greenhouse as a stand-alone product.

Non-ionic Surfactants

Nonionic surfactants are the most commonly recommended and used adjuvants. Nonionic surfactants do not have a charge in solution and are the most commonly used surfactants in agriculture. When used properly, do not harm plants, remain stable, and do a good job of breaking water surface tension. However, application rate is critical. When applied at too higher rate plant injury may result. Nonionic surfactants are now being researched in relation to controlling fungal infection, especially in hydroponic systems. When a fungal zoospore is coated with a nonionic surfactant, it does not function well. They are generally biodegradable and are compatible with many fertilizer solutions. Some nonionic surfactants are waxy solids and require addition of acosolvent (such as alcohol or glycol) to solubilize into liquids. Glycol cosolvents are generally preferred over alcohols, as the latter are flammable, evaporate quickly, and may increase the number of fine spray droplets (making the formulation likely to drift when sprayed) Table 2.

Table: 2. Some important non-ionic surfactant

Trade name	Manufactures
Activate plus	Riverside/Terra
Activator 90	Loveland industries
Adspray 80	Halena
Dash (surfactant + fatty acid)	BASF
Induce	Halena
Kinetic (Organosilicone)	Halena
Latron AG-98	Rohm and Haas
Sikin (Organosilicone)	Riverside/Terra
SilwetL-77 (Organosilicone)	Loveland industries
Spray Booster S	Cenex/Land O Lakes
Spret	Halena
Surf-Aid	Riverside/Terra
Surf -Ac 820	Drexel chemicals
Surf -Ac 910	Drexel chemicals
Triton AG-98	Rohm and Hass
X-77	Loveland industries

Most surfactants molecules are composed of a lipophilic long chain hydrocarbons group (alkyl) and a hydrophilic polar group. Surfactants are generally classified according to the nature of the polar segment of the molecule. Among the types, surfactants are cationic (positive charge) and anionic (negative charge), Zwitterion (having both a positive and negative charge, depending on the water pH), and nonionic (no charge). Nonionic surfactants dissociate little in water, whereas the others are charged when dissolved in water. The most common surfactants for use with herbicides are nonionic, and most emulsifiers are blends of ionic and nonionic types. In general a blend with a high proportion of the anionic types will improve performance in cold water and soft water, whereas a bend with a predominance of nonionic type will usually perform better in warm water and hard water.

Surfactants concentrate at the surface of the liquid in which they are dissolved because their molecules have both polar and non-polar segments. The polar segment is attached to water (hydrophilic) and the nonpolar segment is attached to oil like compounds (lipophilic). Most agriculture nonionic surfactants have chains of ethylene oxide (EO, $-CH_2-CH-O$) also called oxyethylene or ethoxylate as the polar groups. The number of EO units in the polar portion are referred as the moles of ethylene oxide; common nonionic surfactants are alcohol, alkylamine and alkylphenol ethoxylates. Propylene oxide [PO, $-(CH_2)_3-O-$] or butylenes oxide [$-(CH_2)-O-$] can be built into the EO chain to reduce its hydrophilic nature. The more EO or PO units on the surfactants, the more polar the surfactants, the constituents at the end of the EO or PO chain (called the end group) can further modify the polarity of the surfactant. For example a methoxy ($O-CH_3$) is a less polar end cap than a hydroxy (-OH) end cap.

At very low concentration surfactants are soluble in water; however, as the surfactant concentration is raised to that commonly used in weed control, the lipophilic group associate with one another to form micelles. The surfactant concentration where micelle formation occurs is called the critical micelle concentration (CMC). These micelles can emulsify lipophilic substances, including herbicides, oils and perhaps cuticular components. In a mixture of oil and water, the lipophilic portion orients itself into the oil droplets and the hydrophilic portion is within the water. This is how an emulsifier a type of surfactant, facilitates the suspension of an oil like herbicide in a water carrier. Another type is a wetting agent, which is oriented with the polar segment in the water droplets and the lipophilic segment producing from the water droplets allow the herbicides spray to spread over a normally repellent leaf surface.

Surfactants are often assigned a hydrophilic lipophilic balance (HLBV) value. HLB is a quantitative measure of the polarity of surfactant molecules. HLB uses a scale of 0 to 20, with higher numbers being more hydrophilic than lower numbers (Table 3). The HLB can be determined experimentally. It is also possible to estimate the value by calculation or by observing disposability in water. Lipophilic surfactants are assigned HLB numbers of 8 and below. A surfactant with optimal HLB for a particular herbicide can be predicted on the basis of the water solubility of the herbicide viz. low HLB surfactant for water insoluble herbicides, and high HLB surfactants for water soluble herbicide. HLB values are very useful in the selection of surfactants for formulating emulsions (Table 3). They also serve to emphasize that all surfactant are not equally suitable for all uses. It is most important to select a surfactant appropriate for the intended use, and the herbicide label can assist in the proper selection.

An important function of surfactant used as adjuvant is to reduce the surface tension of a spray solution. This allows increased wetting of leaves and spreading of the spray to achieve more intimate contact between the spray droplet and plant surface. Surface tension is the tendency of surface molecule of a liquid to be attracted toward the centre of the liquid body. Spray droplet spreading occurs when the surface tension of the droplet is less than the surface tension of the leaf surface. The degree of effectiveness of the surfactant on droplet spreading can be determined by measuring the contact angle between the droplet and the surface. Any substance that will bring the herbicide into intimate contact with the leaf surface and keep it in a soluble form has potential of adding absorption. Surfactants achieve this by;

1. Causing a more uniform spreading of the spray solution and a uniform wetting of the plant,
2. Helping spray droplets to stick to the plant, resulting in less runoff,
3. Ensuring that droplet do not remain suspended on hairs, scale or other surface projections,
4. Partially solubilizing the lipoidal plant cuticle substance,
5. Preventing crystallization of the active ingredient on the leaf surface by acting as a solvent,
6. Slowing the dryness and increasing the water retention in spray droplets once on the leaf surface

Non-ionic surfactants are commonly used with contact herbicides such as paraquat and many post-emergence grass and broadleaf specific herbicides to enhance activity at 0.1 to 0.5 % by volume of the spray mixture. Surfactants are usually lying at 50 to 100 % active ingredient. Surfactant having a dominant lipophilic character (HLB 0 to 8) increase the fluidity of cuticle component and result in easier diffusion of liquid soluble herbicides (ECs, WPs, flowables and dispersion) suspended in the spray water across the cuticle. Surfactants having a dominant hydrophilic character (HLB 11 to 20) increase water retained in the cuticle, resulting in increased presence of hydrophilic routes for herbicide entrance and result in easier diffusion of water soluble herbicides (salts and acid) dissolved in the spray water across the cuticle.

Table 3. HLB value and surfactant properties

HLB	Property	Behaviour in water	Uses
3-6	Invert (water in oil emulsion)	Immiscible at room temperature	Adding water based mixture to oil
7-9	Wetting agent	Stable but opaque dispersion	Adding oil based mixture to water
8-15	Normal (Oil in water emulsion)	Give hazy solution	Adding oil based mixture to water
13-15	Detergent	Give clear solution	Adding oil based mixture to water
15-18	Solubilizer	Give clear solution	Adding oil based mixture to water

Oils used in agriculture are of two types, refined oil (petroleum based) and seed oil/vegetable oil to increase foliar activity of certain herbicides. Originally the oil surfactant mixture contained 2-5 % surfactant. However, crop oil use with herbicides has largely been replaced by crop oil concentrate (COC). Crop oil is used as 1% in water as herbicide spray mixtures. COC can also reduce volatile and photodegradadative loss of some herbicides. Highly purified paraffin based non-phytotoxic oil have been used as carriers for oil soluble herbicides applied post-emergence; however, the application volumes originally used are no longer practical.

There are two main type of seed oil used. The first are triglycerides and second are methylated oils. Seeds oil are primarily triglycerides when isolated and are generally not directly used in agriculture applications, the methylated form is obtained by hydrolysis of the triglyceride to yield free fatty acids which then esterified by reaction with methyl alcohol and this form is combined with a surfactant for use as an adjuvant. The composition of triglycerides oils varies, depending on the seed source and fatty acid composition which can influence efficacy (Nalewaja 1994, Shinoda 1967, Shinoda and Friberg 1986, Rosen 1978) Fig. 3.

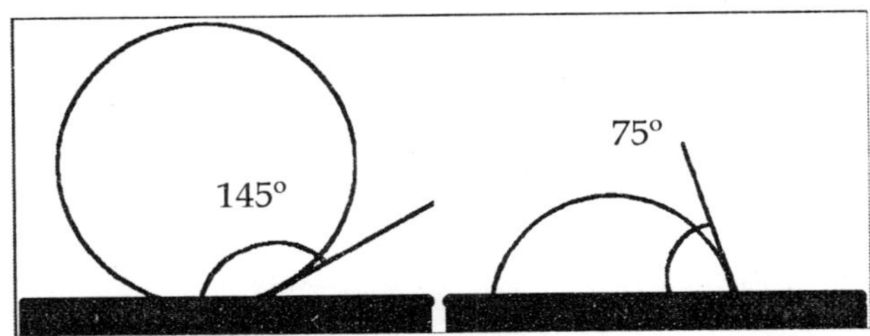

Fig. 3. Contact angle for droplets without (left) and with surfactant (right).

Organosilicone surfactants are often compound of trisiloxane backbone (lipophilic or hydrophobic portion), with an ethylene oxide chain (hydrophilic portion) attached to one of the silicon atoms using a mix of EO and PO or modifying the EC and cap can reduce the lipophilicity. Organosilicone surfactant cause a tremendous reduction in the surface tension of water based spray solution and cause substantial greater spreading of the spray droplet than would be predicted by the reduction in surface tension. This increased spreading is thought to be due to the compact size of the lipophilic portion of the trisiloxane moiety, allowing it to transfer readily from the liquid/air interface to the leaf surface as the drop moves across the leaf (Roggenbuck *et al.*, 1993; Ananthapadmanabhan *et al.*, 1990; Gaskin and Holloway, 1992; Gaskin and Stevens 1993). These surfactants provide improved rain fastness of the spray which, in most cases be thought to increase the rate of droplet evaporation, tend to slow the drying of the droplet. One problem with organosilicone is that these are unstable when the pH of the spray solution is not within the range of 6 to 8. Hydrolysis of silicon oxygen bonds occurs under acidic as well as basic conditions. Buffering of spray solution to the neural pH can reduce this effect. Other disadvantages are that the degree of spreading is lost when such surfactants are mixed with other non organic silicone adjuvants, the extreme surface activity can cause excess spray tank foaming, therefore, the eye and skin contact must be avoided. These surfactants are expensive and are not effective with all herbicides. Examples of some herbides used with surfactants are given below Table 4.

Table 4. Surfactant class, commonly used with herbicides

Surfactant class	Commonly used with
Non-ionic surfactant(NIS)	fluazifop-p-butyl and paraquat
Crop oil concentrates(COC)	atrazine, quizalofop, fluazifop, fluazifop-p-butyl, sethoxydim, and clethodim
Nitrogen-surfactant blends (NSB)	glyphosate, chlorimuron, thifensulfuron and imazethapyr
Esterified seed oils (ESO)	imezethapyr, chlorsulfuron, metsulfuron
Organo-silicates (OS)	bentazon, acifluorfen, lactofen, fomesafen

Ammonium sulfate has been shown to reduce the precipitation of glyphosate on the plant surface (MacIsaac *et al.*, 1991). Ammonium sulfate has been used as a surfactant to overcome decreased herbicide activity due to antagonism caused by cations (Ca^{++}, Na^{+}, K^{+} and Mg^{++}) in the water used as a spray carrier for certain herbicides such as 2,4-D, bentazon, dicamba, acifluorfen, imazethapyr, glyphosate, nicosulfuron and clethodim (Nalewaja and Matysiak, 1993 a, 1993b, 1995, 2000, 2001; McMullan, 1994, 2000; Nalewaja *et al.*, 1995; 2001). Adding ammonium sulfate to the spray solution has often been shown to reduce the antagonism between herbicides such as bentazon plus sethoxydim, primsulfuron plus atrazine and dicamba plus bentazon. Structure of some important surfactant are represented in Fig. 4.

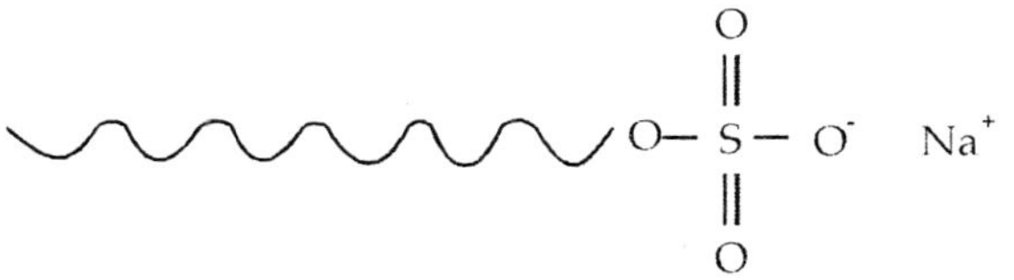

O–S(=O)(=O)–O⁻ Na⁺

Sodium dodecylsulfate (SDS)

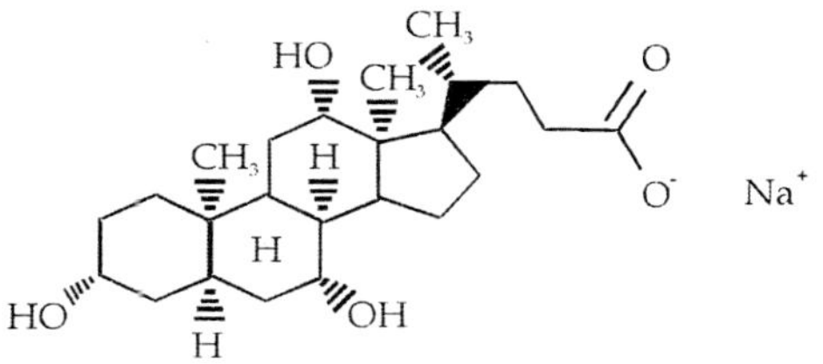

HO, CH3, CH3, CH3, H, H, HO, H, OH, O, O⁻ Na⁺

Sodium cholate

Fig. 4. Contd...

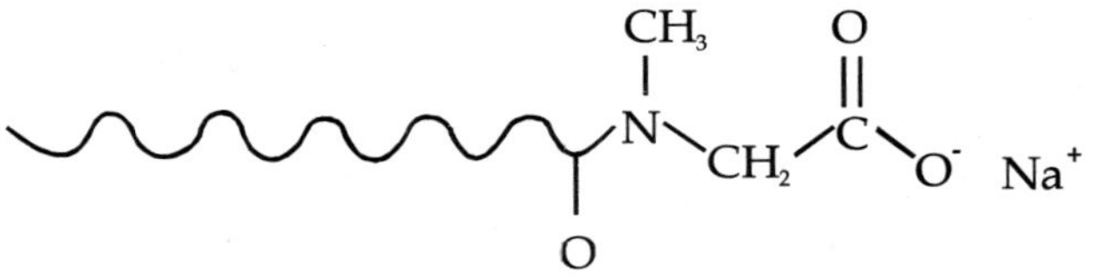

N-Lauroylsarcosine Sodium salt

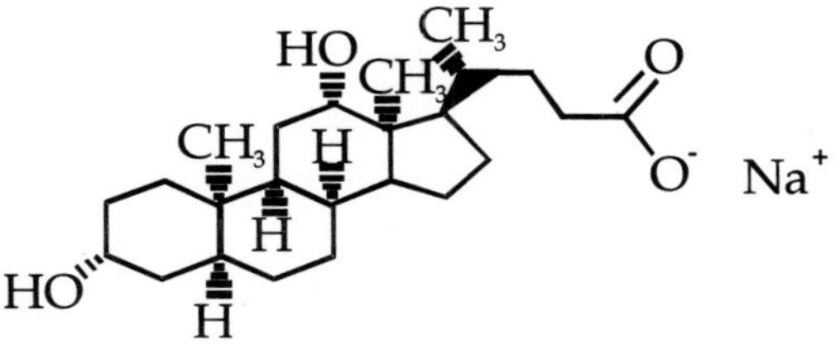

Sodium deoxycholate (DOC)

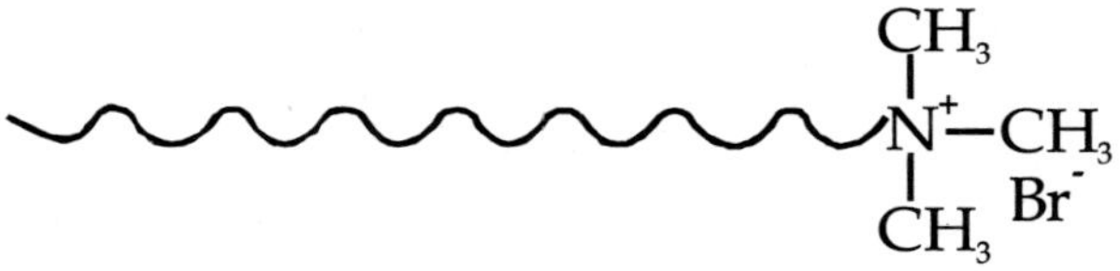

Cetyltrimethylammoniumbromide (CTAB)

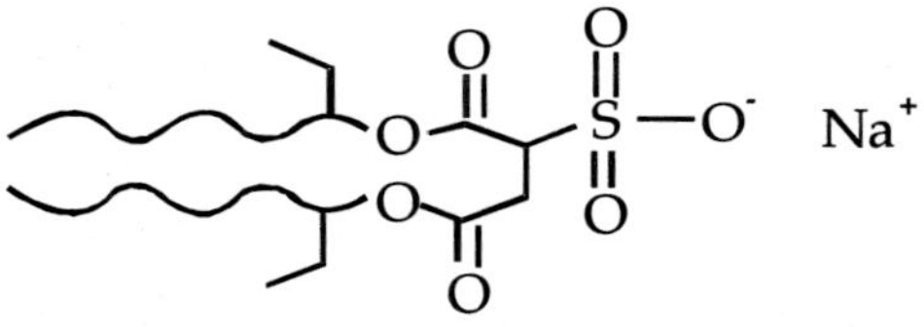

Bis (2-ethylhexyl) sulfosuccinate Sodium salt

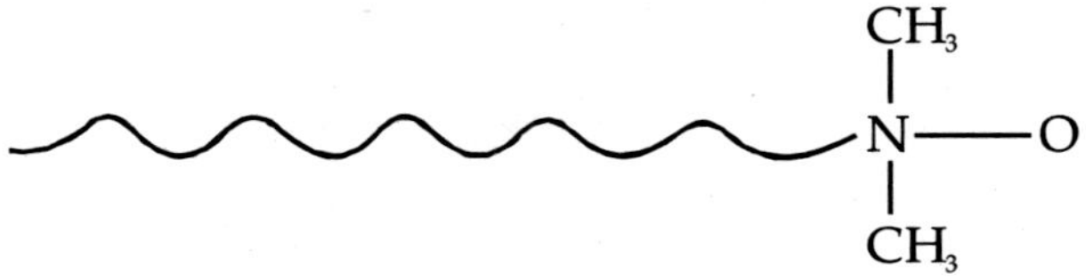

Lauryldimethylamine-oxide (LDAO)

Fig. 4. Structures of common surfactants

Due to the limited solubility of surfactants in water, aggregates are formed in which the hydrophobic or hydrophilic sections of the surfactant are stuck together. The micelle may be represented as a globular cylindrical orellipsoidal cluster of individual surfactant molecules in equilibrium with its monomers. The reverse orientation of the hydrophilic and hydrophobic portion of the surfactant in a hydrocarbon medium leads to reversed micelles. Planar bilayers or small unilamellar vesicles are also formed. The extraordinary variety of the phase behavior of surfactants in solution can further be broadened by the inclusion of additives or co-surfactants Fig. 5&6.

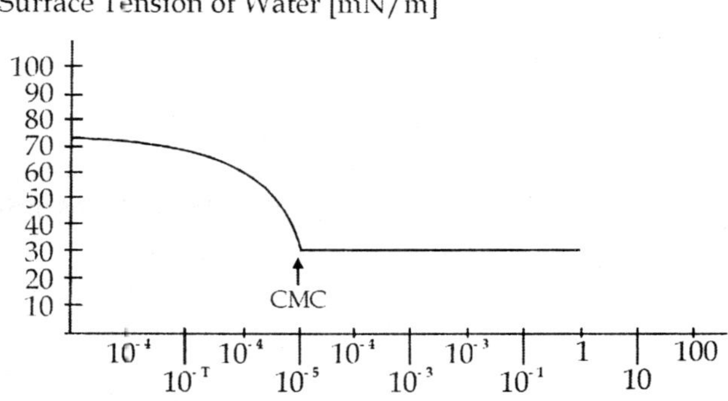

Fig. 5. Typical decrease of the surface tension of water by surfactants (Singh *et al.*, 1984)

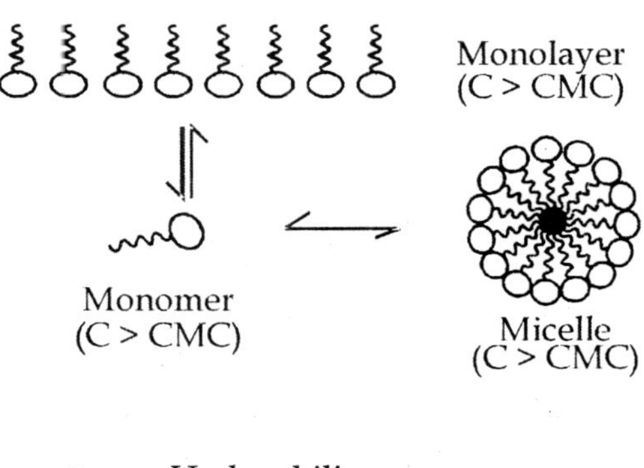

Fig. 6. Equilibrium between the surfactant monomer, surface monolayers and micelle in an aqueous medium

2. Spray-Modifier adjuvants

This group of adjuvants has its greatest influence on the delivery and placement of the spray solution. Additives which alter or modify the spray, such as polymers, inverts, and foams, are included in this group.

Stickers, Spreaders and Spreader-stickers

Spreaders are compounds that cause the surface tension of the pesticide to be reduced in such a way that it easily spreads into a very thin film over a surface. Like surfactants, spreaders and stickers increase the efficiency of the pesticide dramatically. Spreaders are blends of surfactants, primarily non-ionic (uncharged particles), used for spreading and adhering spray droplets to foliage. Spreader-stickers simply combine the two for additional retention during wet conditions.

Stickers, spreaders and spreader-stickers may contain fatty acids, vegetable gels, resins, latex, waxes, aliphatic alcohols, crop oils such as cottonseed, or inorganic oils. Each formulation is very different. Stickers are very much like thickening agents or oils in that they cause the pesticide solution to adhere to the leaf surface, resisting rain, evaporation and runoff. Some products use emulsified polyetheylenes, others use polymerized resins, fatty acids or petroleum distillates. Stickers are commonly used in field crops (like corn and soybeans) where residue on leaves is not a problem. However, in greenhouses, they can cause a mess, especially on leaves with indentations such as Pileas, or hairy plants such as Dusty Miller. Some spreader/sticker cause phytoxicity in tender annuals and herbs. These materials are often more expensive than surfactants and therefore have limited use with herbicides; most are marketed for wettable powder formulations of fungicides and insecticides.

Foaming agents

When used with specially designed nozzles, at a spray mixture of 0.1% - 4.0%, air induced foam is created in a venturi effect. Larger particle sizes are created using air as the conveying medium compared to use of water. The resulting effectiveness of drift control is not well understood.

Polymers

Used in aquatic plant control primarily for drift control, these long-chain carbon molecules (up to 40,000 carbon units per polymer strand), are also used to help break surface tension of the water surface to aid in sinking the herbicide. Depending on the type of polymer formulation and the extent of drift control desired, the rate is usually between 0.1% and 1.0% of the total spray volume. Polymers marketed as sinking agent are designed for that particular purpose. These have higher molecular weight and require good agitation. Polymers formulated for drift control are easier to mix, require less agitation and are generally formulated as solutions. Solution polymers are not formulated as sinking agents and generally dissociate when come in contact of the water surface.

Inverting oils

When properly blended with water, inverting oils form droplets of water surrounded by oil. This viscous blend is referred to as an invert emulsion and resembles mayonnaise (which itself is an invert emulsion). The ratio of oil to water is usually one part oil to 10-30 parts water (1:10-1:30). Inverts are used to reduce drift, increase contact time of the sprayed solution on target foliage and to sink herbicides. To make an invert requires the proper equipment, inverting oil, blending sequence and experience.

Drift reduction agents (Thickeners)

These are used to reduce drift of sprays. Drift control agents are designed to reduce spray drift, which most often results when fine (< 150 μm diameter) spray droplets are carried away from the target area by breezes, including those caused by the aircraft or vehicle carrying the spray equipment (Downer *et al.*, 1998). Drift control agents alter the viscoelastic properties of the spray solution, yielding a coarser spray with greater mean droplet sizes and weights, and minimizing the number of small, easily-windborne droplets (Hewitt 1998, 2001).

Drift reduction agents or thickeners may contain polyacrylamide, polyethylene polymers, polysaccharides (long-chain sugars), or vegetable oils. These products not only help volatile pesticides become less volatile, but also cause the carrier solution to become more viscous and heavier. These agents are beneficial when spraying is done in fields near a housing complex. These agents reduce drift, odor, and waste. However, when thickeners are applied at too high a dosage, can cause surface deposits on foliage, excessive exposure to the pesticide, or clog spray nozzles. These adjuvants need to be used only in outdoor situations and only after reading the label or information about compatibility and phytotoxicity. Examples are sodium alginate, hydroxyethycellulose and dacagin. Dacagin resists both spray and vapour drift. It is used at 2.8 kg/100 gal spary Table 5.

Table 5. Influence of adjuvants on drift reduction

Adjuvants	Spray solution	Experiment	Distance downwind	Reduction of drift potential	Reference
Lectin based	Glufosinate	Wind tunnel	2.3	76	Mass and Krasel (1988)
Polymer guar gum blend	Glyphosate	Wind tunnel	2	80	Hewitt *et al.*, (2001)
Liquid ammonium sulfate/ invert suspension blend	Glyphosate	Wind tunnel	2	70	Hewitt *et al.*, (2001)

3. *Utility-Modifier adjuvants*

Utility modifiers are materials that when added to the spray solution improve the conditions in which the formulated mixture is useful. By themselves, these do not directly enhance herbicidal activity (McMullan 2000). Instead, change the physical or chemical properties of the tank mix in ways that make it easier to apply to the target plant(s), minimize unwanted effects, and broaden the range of conditions under which a given herbicide formulation can be effective.

Types of modifiers include emulsifiers, dispersants, stabilizing agents, coupling agents, co-solvents, compatibility agents, buffering agents, and antifoam agents. Buffering agents and antifoam agents are perhaps the only two of aquatic plant management significance.

Most utility adjuvants are typically not used in wildland situations, since herbicides applied in wildlands are generally not applied aerially, with large booms, or in tank mixtures with several herbicides and other additives. Utility adjuvants may be of different types based on their functional categories but there may be some overlap of these functional categories. Some activator adjuvants are also utility adjuvants and some even have herbicidal effects of their own.

Buffering agents

Buffering agents may increase the dispersion and solubility of herbicides in alkaline or acid water when used in the spray mixture. Buffering agents should be added to the spray tank before adding the herbicide. Buffering agents usually contain a phosphate salt or more recently citric acid, which maintains a slightly acid pH when added to alkaline waters. These are added to higher pH solutions to prevent alkaline hydrolysis (a chemical reaction) of some organophosphate (OP) and carbamate insecticides. Some acidifying agents are also sold to enhance herbicide uptake and performance. However, there is little evidence to support the need for these acidifying agents for this purpose with most herbicides. Some buffering agents are also "water softening" agents that are used to reduce problems with hard water. In particular, calcium and magnesium salts may interfere with the performance of certain pesticides. Ammoniumsulfate (AMS) is sometimes added to reduce hard water problems. For determining the need of a buffering agent in a herbicides solution, the type of herbicide and water has to be ascertained first.

Antifoam agents

These materials are most commonly silicone based and are used for eliminating foam in the spray tank. Generally used at 0.1% or less of the total spray volume, these additives are especially useful when mixing herbicides with soft water, where foaming problems are usually more.

Many herbicides already contain adjuvants in their formulation and may not need any additional material added to the spray tank; however, many of these herbicide labels may suggest that additional adjuvant, such as a surfactant, be added. Most adjuvants are optional and may be added to help modify the final spray solution, but may have little or no effect on the activity of the herbicide.

Compatibility agents

Compatibility agents allow simultaneous application of two or more ingredients and to correct the problems such as suspension fallout, precipitation, viscosity and clabber that arise due to the spray mix incompatibilities resulting from the mixing of herbicides in fertilizer spray solutions (Witt 2001). For instance, if the herbicides bentazon and sethoxydim are mixed, that may react to form precipitates, resulting in reduced rates of sethoxydim penetration (Wanamarta *et al.*, 1993). In most cases, the herbicide label will state which herbicides may or may not be mixed together. Unless the pesticide label states that it can be mixed with liquid fertilizers, a compatibility agent should be included. They are usually phosphoric esters of alkyls, aryl, polyoxy ethanol or ethylene glycoplus and alchohal solubilizer. Extra phosphatic acid may be added for buffering (acidifying effects).

Herbicide labels often specify a jar test to determine the need for a compatibility agent when mixing herbicides with fertilizers. Most herbicides can be applied in nitrogen solutions without any compatibility problems, but compatibility may be poor when the water contains high levels of various salts (hard water), or when the water is unusually cool. When 2,4-D is applied with liquid-nitrogen fertilizers the solution may separate even if mixed vigorously unless a compatability agent is added to the mix.

Dyes

Dyes are commonly used for spot or boom spraying and are recommended for most herbicide treatments in wildlands even if applied with small handheld sprayers or wicks because the presence of a dye makes it far easier to see where the herbicide has been applied and where it has dripped, spilled or leaked. Dyes make it easier to detect missed spots and to avoid spraying a plant or area twice.

Penetrants

Penetrants dissolve or penetrate waxy layers on leaves and allow other chemicals to interface with plant cells or insect epidermal tissue and enter the spaces between the epidermal cells, sometimes called"cell free space." Penetrants can contain petroleum by-products, crop oils, complex alcohols, and other hydrocarbon-based materials. Penetrants work well under very specific conditions and with very specific crops grown outdoors. In the greenhouse, stand-alone products are rare.

Anti-condensates

Anti-condensates greenhouse walls made of various films, especially in poly houses with poor ventilation, have condensate problems and drip in the most inopportune places. Anti-condensates act like surfactants, making it easier for water to spread throughout the surface of the film instead of collecting as droplets on the plastic and dripping on plants. By creating a film, the water can be carried off as a thin sheet along the pipe or plastic film. They work great, but should never, ever be used on plants or on soil.

Emulsifiers

Emulsifier agents work by coating tiny particles or groups of the liquid molecules and preventing them from coagulating with other like molecules. The emulsifiers allow oil and water solutions to mix. These products are usually added by the manufacturer and are commonly mixed with petroleum-based pesticides to help them mix more readily with water.

Herbicide safener

Herbicide safeners selectively protect crop plants from herbicide damage without reducing activity in target weed species. They are used commercially to improve herbicide selectivity between crop and weed species and can be applied either as a mixture with the herbicide or as a seed-treatment to the crop seed prior to sowing. The phenomenon of herbicide safening was discovered by Otto Hoffman in 1947, following the accidental exposure of tomato plants to vapour of the herbicide 2, 4-D. Although previously treated with another herbicide 2, 4, 6-T, plants curiously did not develop any herbicide symptoms. Further experiments showed that foliar treatment with 2, 4-D, also protected wheat from injury caused by the herbicide, barban. However, these interactions could not be exploited commercially, as foliar sprays also protected target weed species while treatment of wheat seed with 2,4-D damaged the crop.

The first commercial herbicide safener is 1,8-naphthalic anhydride (NA), which was patented by the Gulf Oil Company in 1971, for use as a seed-treatment to protect maize from injury by thiocarbamate herbicides. However, Stauffer Chemical Company, who marketed the thiocarbamates, subsequently patented their own safener, dichlormid. Unlike NA, dichlormid, with its superior selectivity, could be formulated with the herbicide thus reducing application costs. This undermined the commercial value of NA, which, despite being active in a number of species and against a diverse selection of herbicides, was withdrawn from the market (Hatzios, 1983). Nevertheless, the discovery of NA initiated intensive industrial research, which, over the last 30 years, has led to the discovery of the oxime ether safeners, namely cyometrinil, oxabetrinil and fluxofenim by Ciba-Geigy (now Syngenta), and the 2,4-D substituted thiazole carboxylates, such as flurazole, by Monsanto.

The majority of these early compounds were active in cereal crops, namely maize, sorghum and rice, while protection was only provided against pre-emergence applications of herbicides representing a narrow spectrum of chemistries and modes of action. However, trends towards post-emergence herbicide treatments and the use of high activity herbicide molecules, has led to the development of safeners to protect winter cereals against post-emergence herbicides. Safener activity is now widely reported against many herbicide classes including aryloxyphenoxypropionate, sulfonylurea, imidazolinone, cyclohexanedione and isoxazolidinone herbicides. Those safeners that have been used commercially are listed in Table 6.

Table 6. Herbicide safeners available commercially

Safener	Crop	Herbicde	Application method
Benoxacor (CGA 154281)	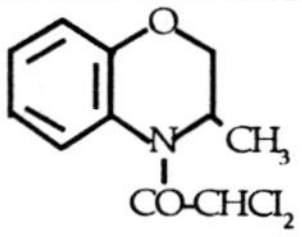 Maize	Metolachlor	Spray as mixture with herbicide
Cloquintocet-mexyl (CGA 184927)	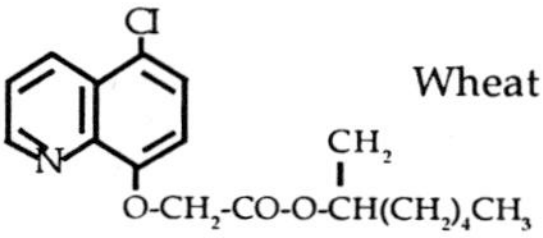 Wheat	Clodinarop-progaryl	Spray as mixture with herbicide
Cyometrinil (CGA 43089)	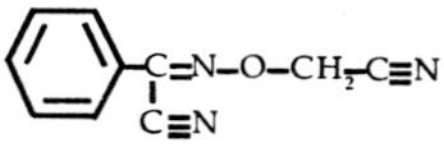 Sorghum	Metolachor	Seed-treatment
Dichlormid (DDCA,R25788)	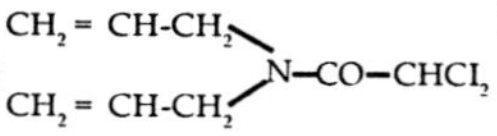 Maize	EPTC, butylate, vernolate	Pre-plant incorporate with herbicide
Fenchlorazole-ethyl (HOE 70542)	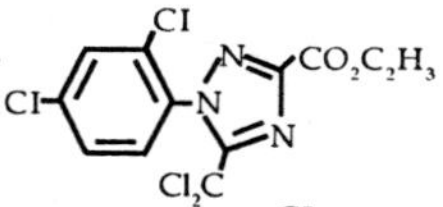 Wheat	Fenoxaprop-ethyl	Spray as mixture with herbicide
Fenclorim (CGA 123407)	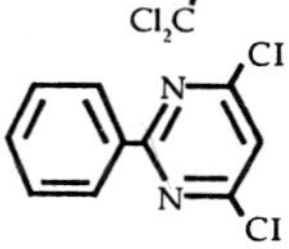 Rice	Pretiachlor	Spray as mixture with herbicide
Fluraxole (MON 4606)	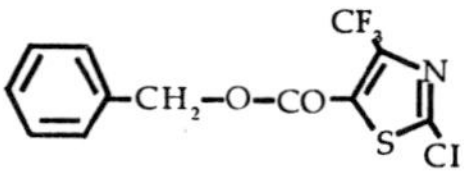 Sorghum	Alachlor	Seed-treatment
Fluxofenim (CGA 133205)	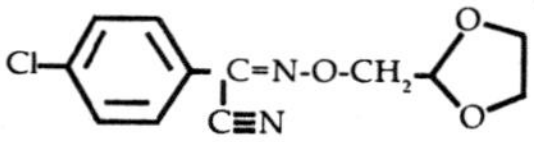 Sorghum	Metolachlor	Seed-treatment
Furilazole (MON 13900)	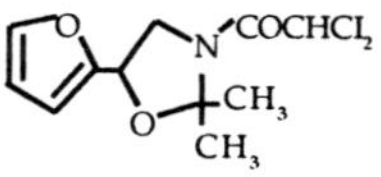 Cereals	Halosulfuron methyl	Sprya as mixture with herbicide
Mefenpyr-diethyl	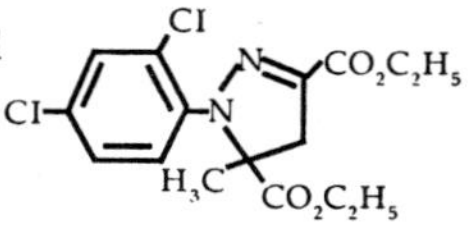 Wheat, rye, triticale, barley	Fenoxaprop-ethyl	Spray as mixture with herbicide
MG 191	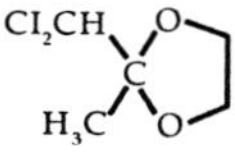 Maize	Thiocarba-mates	Spray as mixture with herbicide
Naphthalic anhydride (NA)	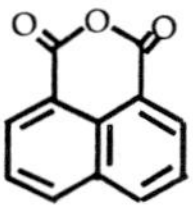 Maize	EPTC, butylate, vernolate	Seed-treatment
Oxabetrinil (CGA 92194)	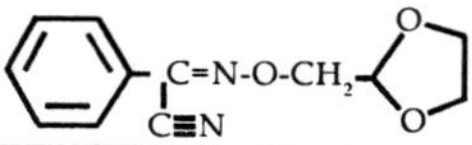 Sorghum	Metolachlor	Seed-treatment

Safeners have many potential uses. Their major role enables the development of molecules with favourable environmental toxicology, whose use would otherwise be limited by poor selectivity. To a lesser extent, they are used to extend the use patterns of currently available herbicides. Other potential uses include protection of crops from damaging levels of pesticide residues to allow greater flexibility in the choice of crops grown in rotation, and the use of herbicides under adverse environmental conditions, where crop damage is likely to occur. In addition, safeners can be used to address difficult weed problems, which, for technical and economic reasons, are unlikely to be solved by the development of conventional selective herbicides. For example, herbicide safeners could be used to facilitate the control of weeds in crops or provide weed control options for minor crops, which, due to their small market value, are not generally targeted for the development of new products. The commercial success of safeners is clearly reflected in the number of herbicide-safener products now on the market. However, possibly their most important application lies in their use as powerful research tools with which to identify and manipulate the biochemical and physiological mechanisms controlling herbicide selectivity. It is for this reason, that considerable research has been devoted to the study of safener mode of action.

Safeners work by reducing the ability of herbicides to reach and inhibit their target sites, by interacting directly with the biochemical targets or receptor proteins of herbicides in crop plants. Alternatively, they could reduce the amount of herbicide reaching this target in an active form either by reducing herbicide uptake or translocation, or increasing the degradation of herbicides to less active or immobile metabolites. In reality, safeners have been found to affect all of these processes and the elucidation of their primary mode of action has been a long process. The ability of safeners to prevent the detrimental effects of herbicides originally led to suggestions that safeners interact with those biochemical processes or target proteins whose activity would normally be inhibited by the herbicide.

The structural similarity of several herbicide-safener combinations led to development of the competitive antagonist hypothesis whereby safeners are believed to compete with herbicide molecules for binding sites on target proteins. Recent comparisons of the 3D similarity between 28 safener and 20 herbicide molecules suggest that structural similarity between herbicides and safeners could provide a useful guide in the design of new safeners, (Bordas *et al.*, 2000). However, where isolation of known herbicide receptors has been possible, safeners have not directly influenced herbicide interactions with these sites. Instead, safeners have been found to elevate levels of herbicide target enzymes and, in some cases, reduce the susceptibility of these enzymes to herbicide inhibition. For example, levels of acetolactate synthase, an enzyme involved in branched chain amino acid biosynthesis and the target of sulfonylurea and imidazolinone herbicides, were not only elevated in maize plants reacted with NA but were also less sensitive to inhibition by the sulfonylurea herbicide, metsulfuron-methyl. Generally these studies have produced inconsistent results and have failed to explain how a single safener, such as NA, can prevent injury from several herbicides which act at different target sites.

Safeners could act by reducing herbicide uptake or translocation to sensitive tissue within the crop plant. However, studies into the effects of safeners on these processes have produced a series of contradictory results, with the majority suggesting that uptake is unaffected or actually enhanced by safener treatments. Where reductions in uptake have been seen, they are usually considered a consequence of safener interactions with other processes. For example, reductions in the uptake of the herbicide metolachlor, seen following treatment with the safener, cyometrinil, can be attributed to decrease in transpiration rate. This reflects the ability of cyometrinil to prevent the inhibition of epicuticular wax formation associated with metolachlor treatment. Further evidence dismissing the relevance of effects on uptake and translocation comes from the fact that, in some cases, safeners work even when applied after the herbicide. Under these circumstances, safeners cannot interfere with herbicide uptake processes (Davies and Caseley 1999, Davies *et al.,* 1967).

Several herbicide safeners increase crop tolerance to herbicides known to undergo oxidation, suggesting that safeners could work by increasing the activity of the cytochrome P_{450} enzymes. At least 10 safeners have been shown to increase the oxidative metabolism of several, mainly sulfonylurea and imidazolinone herbicides in several crop species, including maize, sorghum and rice. In many cases, this increased degradation has been attributed to increases in the activity of enzymes with the characteristics of cytochrome P_{450} monooxygenases. When incubated with herbicide in the presence of NADPH and molecular oxygen, membrane fractions extracted from safener-treated plants generate oxidative metabolites faster than membranes extracted from untreated plants. In some cases, safeners cause up to 30-fold increases in the amounts of oxidative metabolites produced *in vitro*. Furthermore, metabolite production can be inhibited by the addition of carbon monoxide (CO) or compounds such as 1-aminobenzotriazole (ABT), piperonyl butoxide and tetcyclacis, which are known to inhibit P_{450} activities. These observations strongly suggest that such degradation is catalysed by cytochrome P_{450}-dependent systems and that safeners work by promoting the activity of these systems, (Davies and Caseley 1999, Ohkawa *et al.*, 1999).

During Phase II metabolism, conjugation of herbicides, or their Phase I metabolites, with glutathione is catalyzed by glutathione *S*-transferases (GST). In 1997, 35 genes encoding GST enzymes were reported to have been identified from 13 plant species and seven of these were known to be involved in herbicide metabolism in maize (Droog, 1997). Alternatively, some herbicides undergo conjugation with glucose during Phase II. These reactions are usually catalyzed by glucosyl transferase enzymes (Cole, 1994). Herbicide safeners also enhance crop tolerance to several chloroacetanilide and thiocarbamate herbicides which undergo glutathione conjugation during Phase II. Safeners known to promote glutathione content include, benoxacor, dichlormid, flurazcle, and fenclorim. In some cases, increases can be attributed to the increased activity of enzymes involved in glutathione biosynthesis. However, safener activity is not always accompanied by elevated glutathione levels and efficacy is often better correlated with increases in the activity of GST enzymes. At least 12 safeners have been demonstrated to enhance GST activities in maize, rice, sorghum or wheat (Davies and Caseley 1999, Farago *et al.*, 1994, Collier *et al.*, 1986, Delvelder *et al.*, 1997). In addition to enhancing glutathione conjugation, safeners are also reported to enhance the conjugation of oxidative metabolites with glucoside and promote the activity of vacuolar membrane pumps responsible for the sequestration of metabolite-conjugates in the vacuole. In particular, the safener cloquintocet-mexyl is reported to enhance the glucosylation of the herbicide, clodinafop, in wheat. This effect may result from promotion of UDP-glucosyl transferase activities (Kreuz *et al.*, 1991, 1996). Furthermore, treatment of barley with the safener, cloquitocet-mexyl, not only accelerates GST activity but has also been reported to enhance the activity of a vacuolar transporter for glutathione- and glucoside-conjugates (Gaillard *et al.*, 1994).

Summary

Adjuvants are materials that, when added to the spray solution facilitate or modify the action of the herbicide. Thus, adjuvant is an all-inclusive term used to define all spray tank additives, from wetting agents to foaming agents. The technology of adjuvants includes the use of complex synthetic formulations designed for specific purposes. Small changes in the chemical structure of adjuvants can greatly alter their action. Adjuvants can improve herbicide performance by increase droplet coverage and spray retention, increase spray droplet drying time, increase cuticle penetration and increase cellular accumulation. Adjuvants can be grouped into three categories; activator adjuvants, spray-modifier adjuvants and utility-modifier adjuvants. Activator adjuvants have two class; wetting agents, oils and surfactants.

Oil adjuvants are petroleum or vegetable oil in origin. Oil adjuvants vary greatly. Petroleum oil adjuvants are selected for non-toxicity to plants and can vary in chemical composition, viscosity and emulsifier content. Oil adjuvants are comprised of crop oils, crop oil concentrate and vegetable oil concentrates. Surfactants are adjuvants that reduce surface tension within the external surface layers of water. There are four different types of surfactants; anionic surfactants, cationic surfactants, amphoteric surfactants and nonionic surfactants.

Spray-modifier adjuvants have greatest influence on the delivery and placement of the spray solution. Additives which alter or modify the spray, such as polymers, inverts, and foams, are included in this group. Utility modifiers are materials that when added to the spray solution improve the conditions in which the formulated mixture is useful. Buffering agents, antifoam agents, compatibility agents and dyes fall under this category.

Chapter 9

Management of Weeds in Major Field Crops

Weed control is essential for sucessful crop production. All crops grown in the field are subject to weed competition. Weeds intensity may vary from one crop to another, from one region to another and from one farm to another. In general weed growth is more intensive in warm, humid and high rainfall areas than in hot dry and low rainfall conditions. Since weeds can adopt even to extreme climatic condition, they are always compete with crop plants in any situation. Achievement of the desired level of weed suppression requires the use of specific weed management pratices. Weeds management in major field crops is discussed below:

(1) Cereals and millets

Cereal crops are members of the grass family grown for their edible starchy seeds. The important cereal crops are wheat, rice, barley, oats, rye, corn and grain sorghum etc. Cereal crops may be divided into spring-sown types; which complete their life cycle in one season (summer annuals) ex. rice, corn, soybean and fall-sown types, which require an over wintering period (winter annuals) such as wheat, barley, oat. Millet is a collective term for the grain of a large number of small-seeded grasses that are grown as cereal crops. Millets grow well in arid and semi-arid environments, requiring less water than any other grain.

Rice *(Oryza sativa)*

Rice is the most important staple food crop in India. Losses caused by weeds are estimated at 10% of India's rice crop. Reduction in yield to the tune of 34 percent in transplanted rice, 45 percent in direct seeded low land rice and 67 percent in upland rice are reported. Weed competition is more severe in the direct-seeded than in transplanted crop. The major weeds that normally infest a rice crop are *Echinochloa crus-galli, E. colona, Panicum spp., Setaria glauca, Cyperus rotundus, C. deformis, Cynodon dactylon, Panicum spp., Sagittaria* spp., *Chara spp., etc.* Generally, the weed spectrum in rice is dominated by grasses and sedges, such as *Echinochloa crus-galli* and *Cyperus rotundus*. Chemical weed control through herbicides in the crop is highly efficient and economical. The various chemical weed control options in rice are as follows:

Nursery

In nursery, application of any one of the pre-emergence herbicide such as butachlor, thiobencarb, pendimethalin or anilofos at 1-2 kg/ ha upto 8th day after sowing to control weeds in the lowland nursery is recommended. Nursery beds should be sufficiently moist for effectivity of the herbicide. Drainage of water from the field should be avoided.

Transplanted rice

The herbicide which has been more safe and effective for rice is 2, 4-D (0.75-1.0 kg /ha), at 20-25 days after transplant for the control of broad leaf weeds. 2,4-D may be applied as a follow- up application with pre-emergence herbicides. The fields may be drained before application and reflooded within a few days. It gives good control against water hyacinth and *Monochoria*. Application of anilofos (400g /ha) at 3-5 days after transplanting, on saturated soil is recommended but the field should not be flooded for next 2-3 days for the control of many annual grasses and some broadleaved weeds. 2, 4-D supplement with almix, pyrazosulfuron, fenoxaprop-ethyl as post-emergence is also recommended for broad spectrum weed control.

Butachlor (1.0-1.5 kg/ha) application as early post emergence after sowing can effectively control some annual grasses and broadleaved weeds but it requires sufficient moisture at the time of application. Butachlor can be supplemented with 2, 4-D, almix, pyrazosulfuron, cyhalofop-butyl or fenoxaprop-ethyl as post-emergence for broad spectrum weed control. Butachlor+ propanil (1.12 +1.12 kg/ha) application at 10-12 days after sowing, provide excellent control of annual grass weeds and some dicot weeds. Almix (4 g/ha) a mixture of chlorimuron+ metsulfuron ethyl is applicable as post-emergence, for control of broad spectrum of weeds including annual broad leaved weeds. Cyhalofop-butyl at 60-75 g/ha or fenoxaprop-ethyl application at 80-120 g/ha as post-emergence effectively control annual grasses particularly the barnyard grass. Fenoxaprop may be applied as a follow up application with all pre-emergence herbicides. Pendimethalin at 1.0-1.5 kg/ha as pre-emergence may also be applied. Pretilachlor+ safener which is available as Sofit at 750 kg/ha, or pretilachlor dose at 3-7days after sowing or thiobencarb at 1.0 to 1.5 kg/ha after 8-10 days after sowing are good for control of annual grasses and some broad leaved weeds. These herbicides can also be useful for weed control in dry seeded nursery as well. Pyrazosulfuron at 0.70 kg/ha at 20-25 days after sowing gives effective control of *Echinochloa spp.*

Wet and direct seeded rice

2, 4-D (0.75-1.0 kg/ha) at 20-25 days after transplanting is beneficial for the control of sedges and broad leaf weeds. 2,4-D may be applied as a follow- up application with pre-emergence herbicides. It gives good control against water hyacinth and *Monochoria*. Application of anilofos (400g/ha) at 3-5 days after transplanting, on saturated soil is recommended but fields should not be flooded for next 2-3 days for the better control of many annual grasses and some broadleaved weeds. 2, 4-D supplement with almix, pyrazosulfuron, fenoxaprop-ethyl as post-emergence can also be used for broad spectrum weed control.

Butachlor (1.0-2.5 kg/ha) application at 6-7 days after sowing can effectively control some annual grasses and some broadleaved weeds. It requires sufficient moisture at the time of application. The herbicides may be applied after light rains before germination of weeds. Butachlor can be supplemented with 2,4-D, almix, pyrazosulfuron, cyhalofop-butyl or fenoxaprop-ethyl as post-emergence for broad-spectrum weed control. Butachlor+ propanil (1.12 +1.12 kg/ha) application at 10-12 days after sowing, is excellent for control of annual grass weeds and some dicot weeds. Almix (4 g/ha) is useful as post-emergence, for controlling a broad spectrum of weeds including annual broad leaved weeds. Cyhalofop-butyl at 60-75 g/ha or fenoxaprop-ethyl at 80-120 g/ha application after 25-30 days control annual grasses particularly the barnyard grass. Fenoxaprop may be applied as a follow up application with all pre-emergence herbicides. Pendimethalin may be applied at 1.0-1.5 kg/ha as pre-emergence, or pretilachlor at 1.0-1.5 kg/ha /ha as post-emergence for control of annual weeds., Pretilachlor + safener at 750 g/ha, or pretilachlor as pre-emergence after sowing or thiobencarb at 1.0 to 1.5 kg/ha as pre-emergence are good for control of annual grasses and some broad leave weeds and can also be useful for weed control in dry seeded nursery as well. Pyrazosulfuron at 700 g/ha as post-emrgence gives effective control of *Echinochloa spp.* Oxadiargyl at 90 g/ha can be used as pre-emergence for control of grasses and some sedges but less effective against broad leaved weeds. Oxadiazon at 500-700 g/ha, can be used as pre-emergence or early post emergence for control of broad spectrum of weeds. Oxyfluorfen at 150-250 g/ha may also be used as pre-emergence.

Oxadiargyl, pendimethalin, thiobencarb, pyrazosulfuron and oxyfluorfen can be supplemented with 2, 4-D, almix, cyhalofop-butyl or fenoxaprop-ethyl as post-emergence for broad spectrum weed control. Sufficient moisture should be ensured at the time of application hence may be applied after light rains before germination of weeds.

Wheat *(Triticum aestivum)*

It is the second important food crop next to rice grown mostly all over India. Weeds are major problem in wheat, causing reduction in yield by 30-50%. The most critical time for crop weed competition in wheat is first 30-40 days after sowing of the crop. Cultivation of short structured, high fertilizer and irrigation responsive semi dwarf Mexican wheat varieties as compared to tall growing indigenous varieties have change the microclimate in favour of weeds particularly grassy weeds like *Phalaris minor* and *Avena ludoviciana*. In the indigenous tall wheat, broad leaf weeds are the main problem. The major weed species that generally infest wheat fields are *Chenopodium album, Avena fatua, A. ludoviciana, Phalaris minor, Fumaria spp., Melilotus alba, Cirsium arvensis, Convolvulus arvensis, Rumex spp.,* and *Carthamus oxycantha,* etc.

A large number of herbicides are available for controlling weeds in wheat and their choice depends upon the nature of the weed infestation. Broadleaf weeds in wheat are successfully controlled by post-emergence application of 2, 4-D at 0.40-0.60 kg/ha after 25-30 days of sowing. The herbicides may be applied after first irrigation. It should not be used in sensitive cultivars such as HD 2009. Application may be avoided on very young seedlings and crop at shoot stage. Application during morning hours when foliage is wet due to dew fall should also be avoided (Singh and Sharma, 1984; Balyan *et al.*, 1987; Johri and Singh, 1991; Balyan and Malik, 1992, Panwar *et al.*, 1994).

Isoproturon at 0.75 to 1.0 kg/ha after 25-30 days of sowing is good for control of many annual broad leaved and grassy weeds. It is suitable for application in wheat when mix/intercropped with mustard. metsulfuron-methyl at 4-6 g/ha also controls several annual broad leaved weeds and sedges but does not control grasses. Pendimethalin (1.0 kg/ha) can be used upto 0-3 days after sowing (DAS) for control of several annual grasses and some broad leaved weeds. Pendimethalin may be applied on finely prepared seed bed. Before application of herbicides ensure adequate soil moisture in the surface soil. It is suitable fo application in wheat when mixed/ intercropped with mustard. Clodinofop, fenoxaprop and diclofop-methyl also give excellent control of wild oats but does not control broad leaved weeds. These herbicides are suitable for application when mix/intercropped with mustard.

Clondinofop and fenoxaprop (100 to 120 g/ha) applications at 60 g/ha, after 4-6 weeks and diclofop-methyl at 1.0 kg/ha after 25-30 days of sowing is ideal for controlling isoproturon resistant (HR) *P.minor*. Sulfosulfuron at 25 g/ha can be used after 25-30 days after sowing, for control of HR-*P.minor* and provides control of other broad lreaved weeds to some extent. Sulfosulfuron is not suited to wheat inter or mix cropped with mustard. Metribuzin at 175-210 g/ha after 25-30 day of sowing is quite effective on herbicide resistance *P.minor* and broad leaved weeds. During morning hours when foliage is wet due to dew fall application of herbicides may be avoided. Enough care is required in application as it may cause crop phytotoxicity in some cases.

Maize *(Zea mays)*

Maize is an important cereal crop of India. It is most sensitive to weed competition during its early growth period and can cause 50-60 % reduction in yield. Traditionally it is a rainy season crop but its cultivation during winter season is also being done in Bihar, Karnataka and Andhra Pradesh and very recently in Punjab and Haryana. Maize crop is infested with grassy weeds *viz, Echinochloa colona, Dactylocteium aegyptiacum, Eleusine indica, Setaria glauca, Cynodon dactylon, Phragmites karka, Cyperus rotundus, Sorghum halepense* etc. and broad leaved annual weeds such as *Celosia argentea, Commelina benghalensis, Phyllanthus niruri, Solanum nigrum, Amaranthus viridis* and *Portulaca oleracea* etc. Maximum weed competition in maize occurs during the period of 2-6 weeks after sowing.

Herbicides, which prevent weed establishment at least during the first 6-week period, are very useful in maize. A large number of herbicides belonging to triazine group (simazine, atrazine, cyanazine, terbutryne each 1.25 kg/ha), substituted urea (linuron 1-2 kg/ha), carbamates (butylate and EPTC 2-3 kg/ha) and quarternary compounds (paraquat 0.5-0.75 kg/ha as directed spray) have been found effective. 2, 4-D (0.75 kg/ha) or atrazine (0.75 to 1.0 kg/ha) either prior to emergence of crop or within 2 weeks after emergence can be used for the control of broad leaved weeds only and not suited to maize intercropped with grain legumes. Alachlor at 2.0-2.5 kg/ha and metolachlor at 1.0 to 1.5 kg/ha as pre-emergence, control many annual grasses and some broad leaved weeds and also suited to maize intercropped with grain legumes. Alachlor and atrazine may be tank mixed with, pendimethalin or metolachlor at half of their recommended dose for broad spectrum weed control (Gill and Brar 1974; Mukhopadhyaya and Halder 1979; Singh *et al.* 1985; Saraswat 1988).

Sorghum *(Sorghum vulgare)*

Grain sorghum is usually grown as a rain-fed crop. Weeds are most competitive when moisture is limiting, placing crop plants at disadvantage. Sorghum is a poor competitor when young and hence weeds always have a competitive edge. Weeds reduce the grain yield to an extent of 75 percent under uncontrolled conditions. However, if sorghum is infested with *Striga* spp. even cent percent losses can be expected (Hosmani 1982, Bandhe and Nalamwas, 1984). Losses due to weeds are higher in *kharif* crops than in *Rabi* crops because environment for growth of weeds is more congenial during *kharif* season. The important weeds that occur predominantly in sorghum fields are the grass weeds viz., *Cynodon dactylon, Echinochloa colona, Dactylocteium aegyptium, Digitaria sanguinalis* etc., the sedges include *Cyperus rotundus, Scirpus sp.* etc., and the broadleaved weeds *Trianthema portulacastrum, Portulaca oleracea, Boerhavia erecta, Amaranthus viridis, Portulaca quadrifida* etc. under irrigated conditions. In dry lands *Celosia argentea, Cynotis cucullata, Aerva lanata, Digera arvensis, Tephrosia sp., Phyllanthus maderaspatensis, Leucas sp.*, etc. are dominant.

The pre-emergence herbicides found safe in sorghum are propazine, atrazine, simazine, noruron, fluometuron, terbutryne, alachlor, propachlor and isoproturon and post-emergence herbicide such as 2, 4-D. These herbicides are applied as pre-emergence and their doses vary between 0.75 to 2.0 kg/ha depending upon soil and agroclimatic conditions. Dicamba and 2,4-D sodium or ammonium salt at 0.75 kg/ha after 5-6 weeks of sowing are recommended for the control of striga and the other broadleaved weeds only. Herbicides may be applied when sorghum is between 12 and 35 cm tall. Early or late application may cause crop injury. Atrazine (0.5-0.75 kg/ha) or pendimethalin (0.75-1.0 kg/ha) application as pre-emergence control many annual broad leaved and some grass weeds. However, atrazine is not suited to sorghum intercropped with grain legumes. Rain immediately after application may damage the crop. Therefore care should be taken to apply the atrazine in light soils particularly when their is a likelihood of occurance of rains after application.

Striga can be controlled effectively by stimulating germination of its seeds in the field and then destroying by tillage or by spraying paraquat or 2,4-D or MCPA (Hosmani, 1982). Growing crops like soybean, cotton, cowpea, sunflower, groundnut, chickpea, linseed and pigeonpea which stimulate germination of Striga seeds but are not parasites by the weed in rotation with sorghum help managing the weed.

Pearlmillet *(Pennisetum typhoides)*

Pearlmillet is cultivated for grain as well as for fodder in arid and semi-arid parts of the country as a rainfed crop. It is an important food crop in Rajasthan, Maharashtra, Gujarat and Uttar Pradesh. Poor weed management is one of the factors limiting the productivity of the crop. The weeds normally found in these millet crops are *Cyperus rotundus, Cenchrus biflorus (sandbur weed), Tribulus terrestris* (puncturevine), *Tephrosia purpurea, Pulicaria wightiana, Euphorbia hirta, Phyllanthus niruri, Crotolaria burhia, Avena tomentosa, Heliotropium sublatum, Digera arvensis, Celosia argentea* etc

The safety margin for herbicides is low in millets and as such very few herbicides are available for controlling weeds in this crop. Atrazine (0.50-0.75 kg/ha) or pendimethalin (0.75-1.0 kg/ha) application as pre-emergence, control many annual broad leaved and some grass weeds. Atrazine is not suited when intercropped with grain legumes and thus application of herbicides in very light soils should be avoided. Other herbicides such as simazine, cyanazine, terbutryne and linuron (0.25-0.50 kg/ha) as pre-emergence application also hold good promise for controlling weeds in this crop. Rain immediately after application may damage the crop, therefre care should be taken to apply these herbicides. 2,4-D sodium or ammonium salt at 0.75 kg/ha after 5-6 weeks of sowing for the control of striga and other broadleaved and few grasses is suited to pearlmillet intercropped with grain legumes (Gautam 1970; Rathod *et al.*, 1983; Sandhu and Gill 1973; Sandhu *et al.*, 1987; Umrani and Bio, 1982).

(2) Grain legumes

Blackgram *(Vigna mungo)*, Pigeonpea *(Cajanus cajan)*, Chickpea *(Cicer arietinum)*, Lentil *(Lens esculentus)* and Field peas *(Pisum sativum)*

Grain legumes normally grow very short and therefore, dense growth of weeds can smother the crop plants severely. Grains legumes such as greengram, blackgram, pigeon pea, chick pea, lentil and field peas etc. have identical weed problems. These crops are more heavily infested by weeds in *kharif* than in *rabi*. The important annual grass weeds and sedges that occur predominantly in pulses field in *kharif* are *Echinochloa colona, E. crus-galli, Elucine indica, E. aegyptium, Dactyloctenium aegyptium, Digitaria samguinalis, Setaria glauca, C. iria* and Panicum maxicum. The predominant broadleaf weeds in kharif are, Trianthema portula castrum, Celosia argentea, Digera arvensis, Physallis minima Amaranthus viridis, Cleome viscose, Commelina benghalensis etc. Perennial weed consists of Cyperus rotundus cynodon dactylon, Sorghum halepense, Saccharum spontaneum etc. The major weeds in rabi consists of P. minor, Avena ludoviciana, Poa annua, Chenopodium album melilotus spp., Vicia sps., Lathyrus aphaca, Convolvulus arvensis, Anagallis arvensis, Asphodelus tenuifolius, Medicago denticulata, Rumex dentatus, Fumaria pariflora, Cuscuta spp., Circium arvense, Cyperus rotundus, Cyanodon dactylon etc. The critical period of crop weed competition in these grain legume crops is generally 4-6 weeks from sowing. Thus weed control practices during this period are essential for getting good yield. The techniques of chemical weed control options in some important pulse crops can be summarized as below.

Urd bean and Mung bean are important pulse crops grown throughout in India. These are grown in *kharif* and summer season in North India. Several annual and broad leaved weeds are associated with urd bean and mung bean criops which reduce the yield drastically. Pigeonpea (*Cajanus cajan*) is comparatively longer duration grain legume crop than other grain legume crops. After gram, (chick pea), arhar (pigeon pea) is the second important crop in India. It is a rich source of protein. Herbicides are now available which can replace 1-2 manual weedings in this crop. During rainy season weeds pose a serious problem. The period of first 60 days is very critical in the life cycle of the pigeonpea. Alachlor at 2.0-2.5 kg/ha as pre-emergence controls many annual grasses and some broadleaved weeds. Alachlor may be supplemented with quizalofop as post-emergence for grass weed control. Jaiprakash and Pahwa (1979) observed that alacholor, fluchloralin and nitrofen at 1.0 and 1.5 kg/ha in blackgram reduced all types of weeds.

Lentil is one of the important crop of *rabi* pulses and grown in northern plains, central and eastern parts of India. The major lentil producing areas are situated in Madhya Pradesh, Uttar Pradesh, Bihar and West Bengal. Lentil being slow in growth in early stages suffers adversely from competition with weeds. The period of 4-6 weeks after sowing is most crucial for competition with weeds. Pea is one of the most important crops of the world. The period of first 4-6 weeks is very critical in the life cycle of the plants of pea. The crops are badly infested with weeds because of their short growing nature.

There are many good herbicides which can be used to control weeds in pulses. Alachlor (2.0-2.5 kg/ha) application as pre-emergence controls many annual grasses and some broadleaved weeds and can be supplemented with quizalofop as post-emergence for grass weed control. Fluchloralin (1.0-1.5 kg/ha), metolachlor (0.75-1.0 kg/ha), pendimethalin (0.75-1.0 kg/ha) and oxadiazon (250 g/ha) can efficiently control many annual grasses and some broad leaved weeds. These herbicides can be supplemented with quizalofop as post-emergence for grass weed control. Fluchloralin may be incorporated at pre-planting into the surface soil immediately after application. Oxyfluorfen at 100-125 g/ha controls wide spectrum of weeds and may be supplemented with quizalofop as post-emergence for grass weed control. Quizalofop-ethyl at 40-50 g/ha provides excellent control of annual grasses but less effective on broad leaved weeds and sedges.

(3) Oilseed Crops

Oilseed crops are grown primarily for the oil contained in the seeds. The oil content of oilseeds ranges from about 20% for soybeans to over 40% for sunflowers and rapeseed (canola). The major world sources of edible oils seed are soybeans, sunflowers, ground nuts, rapeseed and mustard. The weeds account for heavy loss in oilseed crops. The chemical weed management in major oilseed crops is given below;

Groundnut *(Arachis hypogea)*

Groundnut, normally a rain-fed crop, is sown without sufficient tillage due to frequent rains. As a result the crop is infested with broadleaf weeds and grasses. It has been estimated that yield of groundnut may be reduced upto 77 percent due to weed infestation depending upon the intensity, nature and weed competition (Drennan and Jennings 1977). Weed competition is critical upto 45 days after sowing of ground nut, and weed-free environment during this period gives highest pod yield.

The major weeds are *Echinochloa colona, Dactyloctenium aegyptium, Cynodon dactylon, Phyllanthus madraspantensis, Digera arvensis, Cyperus rotundus* and *Amaranthus viridis.*

There are many herbicides which are found safer in groundnut. Pre-emergence herbicides such as alachlor (1.5-2.0 kg/ha) or imazethapyr (100 g/ha) or metolachlor (1.0 to 1.5 kg/ha) or oxadiazon (750 g/ha), and oxyfluorfen (250g/ha) control many annual grasses and some broad leaved weeds (Sivannarayan *et al.*, 1990; Brar and Mehra, (1989); Selvamani and Sankaran (1989). Butachlor (1.0-1.25 kg/ha) as pre-emergence, fluchloralin (0.75-1.0 kg/ha) or trifluralin (1.0 kg/ha), as pre plant incorporation within 3-4 days of planting of groundnut control many annual grasses and some broad leaved weeds (Vairavan *et al.*, 1992; Ahuja and Yaduraju, 1992). Pre-emergence application of pendimethalin at 0.75-1.0 kg/ha controls many annual grasses and some broad leaved weeds. Post-emergence application of quizalofop-ethyl or fluazifop-p-butyl (40-50 g/ha) provides excellent control of annual grasses but less effective on broad leaved weeds and sedges.

Soybean *(Glycine max)*

It is very important oilseeds crop of India. The critical period of crop-weed competition is generally during the first 30 days after sowing. Weed infestation in soybean fields may reduce yield by 40-45 % depending upon the intensity, nature, and the duration of weed competition. The common weeds like *Echinochloa colona, Dactyloctenium aegyptium, Eleusine indica, Digitaria spp., Celosia aregentea, Cyperus rotundus, Cynodon dactylon, Sorghum halepense* etc. are generally associated along with soybean.

Pre-emergence application of alachlor (1.5-2.0 kg/ha) or imazethapyr (100 g/ha) or metolachlor (1.5-2.0 kg/ha) or metribuzin (350-525 g/ha) or oxadiazon (750 g/ha), are effective to control many annual grasses and some broad leaved weeds. Chlorimuron application at 8-12 g/ha as post emergence controls many annual grasses and broad leaved weeds and sedges. Fenoxaprop application at 80-100 g/ha as post-emergence provides excellent control of annual grasses but poor control of other weeds. It can be used as a sequential application with all pre-plant and pre-emergence herbicides.

Preplant incorporation of fluchloralin (0.75–1.0 kg/ha) or trifluralin (1.0 kg/ha) or pre-emergence application of pendimethalin (0.75-1.0 kg/ha) provides effective control of many annual grasses and some broad leaved weeds. Pre-emergence application of oxadiazon (250 g/ha) or oxyfluorfen (100-125 g/ha) provides effective control of wide spectrum of weeds and may be supplemented with quizalofop as post-emergence to provide control of grass weeds.

Sunflower *(Helianthus annus)*

Sunflower is cultivated throughout in India. Weed causes on an average 33-63% loss in seed yield of sunflower. The major weeds during *kharif* include *Echinochloa crus- galli, Brachiaria reptans, Cynodon dactylon, Sacharum spontaneum, Dactyloctenium aegyptium, Dinebra spp., Cyperus rotundus, Celosia aregentea, Commelina benghalensis, Alysicarpus longifolius, Psoralia corylifolia, Caesulia axillaries* and *Corchorus oiltorius.* In *Rabi* season, the major weeds are *Eragrostis diarrhena, Chenopodium album, Melilotus spp., Vicia sativa, Convolvulus arvensis, Amaranthus viridis* etc.

Pre-emergence application of alachlor at 1.5-2.0 kg/ha controls many annual grasses, broad leaved weeds and sedges. Granular formulations can also be applied at the time of last cultivation. Pre-emergence application of butachlor (1.0-1.5 kg/ha) or, metolachlor (1.0-1.5 kg/ha) or oxadiazon (0.75 kg/ha) or pendimethalin (0.75-1.0 kg /ha) or oxyfluorfen (0.25 kg/ha) provides effective control of annual grasses and broad leaved weeds (Gill *et al.*, 1983; Reddy *et al.*, 1998).

Pre plant incorporation of fluchloralin or trifluralin at 1.0 kg/ha is carried out into the surface soil immediately after application for control of many annual grasses and some broad leaved weeds (Nalayini *et al.*, 1992). Suresh *et al.*, (1992) controlled *Cynodon dactylon* with fluazifop p-buytl (0.25 kg/ha) as post-emergence application.

Rapseed *(Brassica compestris)* and Mustard *(Brassica lcuhcea)*

These are major types of oil-yielding cruciferous crops. The reduction in yield varies from 30 to 70 % depending upon the intensity and nature of weeds. *Phalaris minor* weed ecosystem reduces yield significantly under irrigated mustard and rapeseed, grown either in rotation as sole crop or inter cropped with wheat. Generally, farmers delay hand weeding to obtain some green fodder from weeds, with the consequent there is an adverse effect on crop productivity.

The major weeds which cause substantial losses in mustard-rapeseed include *Melilotus spp, Vicia sativa, Lathyrus aphaca, Chenopodium album, Cichorium intybus, Medicago denticulata, Digitaria adscendens, Cynodon dactylon, Cyperus rotundus* etc.

There are several herbicides which are safer to the crop. Preplant incorporation of fluchloralin at 1.0 kg/ha into the surface soil before sowing of the crop is recommended for control of many annual grasses and some broad leaved weeds. Isoproturon (0.75-1.0 kg/ha) as pre-emergence controls many annual grasses and broad leaved weeds. Untimely rain immediately after herbicide application may damage the crop. For post-emergence application isoproturon can be mixed with 20-25 kg dry soil and broadcast uniformly. However, care should be taken while applying on wet foliage. Fluchloralin (0.68 kg/ha), or nitrofen (1.5 kg/ha) or oxadiazon (0.50 g/ha) or pendimethalin (0.75-1.0 kg/ha) application as pre-emergence control many annual grasses and some broad leaved weeds (Tiwari *et al.*, 1988; Ghosh and Bera, 1983, Pandey and Kumar, 1984). Post-emergence application of quizalofop-ethyl (40-50 g/ha) provides effective control of annual grasses, including *Phalaris minor* and wild oats but is less effective on other weeds.

Linseed *(Linum usitatssimum L.)*

It is an important oilseed and fiber crop. It is grown both for its seed as well as fiber which is used for the manufacture of linen. Morphologically, linseed plant is a poor weed competitor due to its slender stem, small leaves, short stature and slow growth at early stage.

The weed infestation in linseed causes 30-75 % loss in seed yield. The drop in yield up to 74 % was reported due to *Avena fatua* by Bell and Nalewaja (1967). The loss in seed yield due to *Phalaris minor* and its codominant associates vary from 33 to 74 % under irrigated conditions in rice-wheat-rice-linseed sequence (Tiwari & Bisen, 1985). Weeds cause greater loss during 20-60 day stage. Hence the weeding in third week is essential. The major weeds infesting the linseed crops are *Convolvulus arvensis, Chenopodium album, Medicago denticulata, Asphodelus tenuifolius, Carthamus oxycantha, Sphaeranthus indicus, Saccharum spontaneum, Cyperus rotundus, Phalaris minor, Avena fatua,* etc.

Preplant incorporation of fluchloralin at 1.0 kg/ha can be done to control many annual grasses and some broad leaved weeds. Isoproturon at 0.75-1.0 kg/ha as post-emergence controls many annual grasses and broad leaved weeds. It also gives satisfactory control of parasitic weed *Cuscuta*. Oxadiazon application at 500 g/ha or pendimethalin at 0.75 to 1.0 kg/ha as pre-emergence can control annual grasses and broad leaved weeds. Linuron (0.3 kg/ha) and monouron (0.5 to 1.0 kg/ha) can be applied as pre-emergence (Anonymous 1986, Tosh and Jena, 1983).

Sesame *(Sesamum indicum)*

It is an important oilseed crop in India and mainly grown in Uttar Pradesh, Rajasthan, Madhya Pradesh, Andhra Pradesh, Gujarat, Tamilnadu, and Orissa. Sesame is grown in almost all seasons in different parts of the country. The weed problem is more associated with *kharif* crop as compared to *Rabi* or *Zaid* crop. If the weeds are not managed properly, the losses in yields are caused from 10-60%. The critical period for weed competition is first 20-30 days after sowing. Hence, the weeds must be controlled within 20-30 days after sowing. The major weeds which are associated with sesame crop are: *Alysicarpus longifolius, Celosia cregentea, Digera alternifolia, Psoralia corylifolia, Acanthosperumum hispidum, Commelina benghalensis, Echinochloa crus-galli, Dactyloctenium aegyptium, Brachiaria reptans etc.*

The selective herbicides for effective control of annual and broad leaved weeds include fluchloralin (1.0 kg/ha) or trifluralin (0.50 kg/ha) at pre- plantation stage, alachlor (1.0 kg/ha), butachlor (2.0 kg/ha), asulam (1.0 kg/ha) and diuron (1.0 kg/ha). Isoproturon (0.75-1.0 kg/ha) as post-emergence controls many annual grasses and broad leaved weeds. It also gives satisfactory control of parasitic weed *Cuscuta* but untimely rain immediately after herbicide application may damage the crop. Oxadiazon at 500 g/ha or pendimethalin at 0.75-1.0 kg/ha as pre-emergence can control many annual grasses and some broad leaved weeds.

Niger *(Guizotia abyssmica L.)*

It is grown mainly as a sole or inter crop with Paspalum, Kutki or finger millets on light soil. The major weeds of niger crop are: *Alysicarpus longifolius, Celosia aregentea, Digera alternifolia, Psoralia corylifolia, Acanthosperumum hispidum, Commelina benghalensis, Echinochloa crus-galli, Dactyloctenium aegyptium, Brachiaria reptanas etc.* A parasitic weed dodder (*Cuscuta chinensis*) grows as shoot parasite and destroys the whole crop. The seedling of cuscuta makes contact with the niger shoots at early stage by penetrating the haustoria in phloem of niger and extract the photosynthate from the host for its growth resulting in the poor growth and ultimately death of niger. The niger has greater competitive ability with weeds as compared to other crops. However, depending on weed species and density 10-35% loss in yield was recorded. Major herbicides that can be used for control of annual grasses and broad leaved weeds are preplant incorporation of fluchloralin at 1.0 kg/ha or isoproturon application at 0.75-1.0 kg/ha. These herbicides also give satisfactory control of parasitic weed *Cuscuta*. Care should be taken as untimely rain immediately after herbicide application may damage the crop. Oxadiazon at 500 g/ha or pendimethalin at 0.75-1.0 kg/ha as pre-emergence can control many annual grasses and broad leaved weeds. For control of cuscuta, clorprophom (granules) 4 g/ha 6 days after sowing on moist soil or porpnamide (1.5 - 2.0 kg/ha) 15 days or fluchloralin (1.0 kg/ha) can be applied (Tosh *et al.*, 1977).

(4) Fibre Crops

Fibre crops that produce filamentous matter from the bark tissue or other parts of plants, are used for industrial purposes. Fibre crops include a wide variety of plants of fibrous biomass that are grown annually or seasonally at specific weather and climatic conditions. These crops are considered as important agricultural commodities. The herbicides used for the control of weeds in major fibre crop are given below:

Jute (*Corchorus olitorius L.*)

Jute is an important fibre crop next to the cotton. Weeds pose the greatest problem in jute cultivation. Weed competition may reduce fibre yield to the extent of 40-60%. The weeds observed in jute fields under different soil conditions are as follows:

In upland common grassy weeds are *Eleusine indica, Dactylocteium aegyptiacum, Echinochloa colona, Brachiaria repens, B. ramose, Imperata cylindrica, Leptochloa chinensis, Eragrostis tenella,* etc. and common sedges are *Cyperus rotundus, C. iria, C. compressus etc.* Broad leaved weeds are *Euphorbia hirta, Croton bractiferus, Phyllanthus niruri, Portulaca oleracea, Physalis minima, etc.*

In Midland common grassy weeds *are Paspalum commersonii, Eragrostis unioloides, Echinochloa colona, Sporobolus diander, Setaria glauca, Cynodon dactylon, Panicum repens, Chloris gayana, Commelina benghalensis etc.* common sedges are *Cyperus alulatus, Fimbristylis aestivalis, Bulbostylis barbata,* Broad leaved weeds are *Triumfetta rhomboidea, Euphorbia linifolia, E. microphella, Launea pinnatifida etc.*

In lowland, common weeds are: *Axonopus compressus, Eragostis uniloides, Paspalum scorbiculatum, P. commersonii, Panicum repens* and *common sedges* are *Fimbristylis aestivalis, Cyperus alulatus, Eclipta alba, Triumfetta rhomboidea, Desmodium trifolium, Scoparis dulcis, Launea pinnatifida etc.*

Pre plant incorporation of fluchloralin (1.0 kg/ha) or pre-emergence application of pendimethalin (0.75-1.0) kg/ha provide effective control of many annual grasses and some broad leaved weeds.

Cotton (*Gossypium spp.*)

Cotton occupies prominent position in Indian agriculture having approximately 77.8 million hectare in the country except north-eastern region. Studies revealed 45-85% reduction in yields due to weed competition in cotton crop. Cotton is a wider spaced, long duration and heavily manured crop. Luxuriant weed growth is a major problem in addition to other pests. Major weeds present in cotton field are : *Dactylocteium aegyptium, Digitaria sanguinalis, Ischaemum polosum, Alloteropesis cimicina, Eragrostis tanella, Eragrostis japonica, Eragrostis major, Paspalum disticum, Sorghum halepense, Brachiaria reptans, Eleusine indica, Cynodon dactylon, Cyperus rotundus, Vernonia cinerea, Spharamthus indicus, Caesulia axillaries, Abutilon indica, Phyllanthus madraspantensis, P. niruri, Vicia indica, Convolvulus arvensis, Oscimum basilicum, Hibisucs malvastrum, Lencus aspera, Conyza stricta, Digera arvensis, Corchorus tridens, Echinochloa colona etc.*

Butachlor at 1.0-1.25 kg/ha applied as pre-emergence within 3-4 days of planting of cotton controls many annual grasses and broad leaved weeds. Glufosinate at 450 g/ha or glyphosate at 1.0 kg/ha as post-emergence provide effective control of most of the weeds as directed spray when cotton is over 15 cm high. Herbicide may be sprayed when it is not windy so as to cause least drift on the crop. Soil incorporation of fluchloralin at 1.0 kg/ha at pre- plant stage into the surface soil before sowing of the crop controls many annual grasses and some broad leaved weeds.

Pre-emergence application of oxadiazon at 0.50 g/ha or pendimethalin at 0.5-1.0 kg/ha can control many annual grasses and some broad leaved weeds. Preplant incorporation of trifluralin at 1.0 kg/ha can also be used before sowing to control annual grasses and broad leaved weeds.

(5) Sugar Crops

Sugarcane (*Saccharum barbari*)

Sugarcane is the most important crop producing sugar globally. Approximately 60 % sugar come from sugarcane. It is cultivated over an area of 3.64 million ha with a production of 232.32 million tones. (Agricultural Research Data Book 2006). In sugarcane losses due to weed competition are particularly high because of certain factors which specially favour the growth of weeds in this crop. It is because of the reason that sugarcane crop takes about 25-35 days to come up to the soil surface, this time the whole field gets covered with all kinds of weeds. Further, the wide spacing between the rows provides good opportunity for the weeds to cover more surface area than the crop. Moreover, the critical period of growth of sugarcane coincides with the monsoon period which also favours the growth of weeds. Also, the heavy manuring and irrigation of sugarcane aid the weeds in their growth.

Losses in the yield of sugarcane in India have been recorded from 20-63% (Mukhopadhyay, 1959; Thakur and Negi, 1954; Verma and Bhardawaj, 1958) under different climatic conditions. Common weeds are *Dactylocteium aegyptium, Cynodon dactylon, Digitaria sanguinalis, Echinochloa colona, Eleusine indica, Eragrostis sp., Sorghum halepense., Cyperus rotundus, Cyperus deformis, Ammannia baccifera, Marsilea quadrifolia, Acalypha indica, Amaranthus viridis, Phyllanthus niruri, Ageratum conyzoides, Argemone maxicana, Cardiospermum helicacabum, Commelina benghalensis, Commelina mediflora, Eclipta alba, Euphorbia hirta, Heliotropium indicum, Leucas aspera, Ludwigia parviflora, Physalis minima, Portulaca oleracea, Panium repente, Sida acuta, Sonehus arvensis, Corchorus accutangulus, Convolvulus arvensis, Digitaria marginata, Spharanthus indicum, Solanum nigrm, Tridase procumbens etc.*

Pre-emergence herbicides which are very effective in sugarcane are atrazine, simazine, alachlor, metoxuron and diuron. In a long duration crop like sugarcane, best results are obtained by sequential application of pre-emergence and post-emergence herbicides or repeat application of post-emergence herbicides. 2, 4-D at 1.0 kg/ha as post-emergence after 5 weeks of planting provides good control of broad leaved weeds and sedges but not grasses. Atrazine at 2 kg/ha as pre-emergence controls many broad leaved weeds and some grasses. Soil moisture enhances the herbicide activity. Tank mix atrazine with alachlor at 2.0 kg/ha may be recommended for broad specutrum control of weeds. Diuron at 1.5 kg/ha as pre-emergence controls wide range of weeds which includes grasses and broad leaved weeds. Metribuzin application at 1.0-1.5 kg/ha after 3-5 days of planting provides effective control of broad spectrum of weed. Target spray of glyphosate at 1.0 kg/ha as post-emergence provides good control of all weeds including perennial weeds.

Glyphosate (0.8-1.6 kg/ha), dalapon (3-4 kg/ha), paraquat (0.4-0.8 kg/ha), 2,4-D (1-1.5 kg/ha), MCPA (1-3.5 kg/ha) and asulam (3-4 kg/ha) are effective post-emergence herbicides. Paraquat at 500 g/ha or 0.30% spray to emerged weeds as directed spray when the crop is over 30 cm high controls all weeds. Paraquat can be applied as tank- mix with 2, 4-D at 1.0 kg/ha to increase spectrum of weed control. Effective weed control and maximum sugarcane yield may be obtained with the combination of asulam and isooctyl ester of 2, 4-D (3.4+ 0.75 kg/ha).

(6) Vegetable Crops

Vegetables are high value crops and exert less competition to weeds. Dense weed growth at any stage of crop growth affects yield, however, the first four weeks of crop growth are the most critical in affecting crop yields. Hand weeding and hoeing are the usual practice but too expensive. The main weed problem in vegetables is the growth of annual weeds which are removed usually by hands. Perrenial weeds, if any are removed by digging at the time of harvst or field preparation. The different chemical weed control options in some important vegetables crops are given below.

Potato (*Solanum tuberosum* L.)

Potato is the most important vegetable supplement food crop. Weeds pose a serious problem due to good soil moisture and high fertility of the soil. Potato requires nine weeks of weed free period to prevent yield loss (Zimdhal, 1980). The unchecked weeds can bring about yield losses from 46-82% (Singh *et al.*, 1984; Mani *et al.*, 1967; Nankar, 1980). Manual weeding is the most common method practiced in potato crop in India. Straw mulching resulted in maximum mortality of weeds and highest yield (Chakraborty, 1971). Common weeds are *Poa annua, Lepidium spp., Spergula arvensis, Chenopodium album, Artemisia* etc. Fluchloralin at 1.0 kg/ha as pre- plantation stage controls many annual grasses and some broad leaved weeds.

Herbicides have been used extensively for managing weeds in potato, the most commonly used are MCPA, 2, 4-D, propanil, nitrofen, linuron, prometryne, as pre plant pre emergent sprays (Gill *et al.*, 1976; Randhawa *et al.*, 1979; Bhan 1972, Mani *et al.*, 1967). These herbicides give effective weed control and also increase yield. Butachlor at 0.75-1.0 kg/ha as pre-emergence after 3-4 days of planting of potato tubers controls many annual grasses and some broad leaved weeds. Isoproturon at 0.75-1.0 kg/ha as pre-emergence controls many annual grasses and broad leaved weeds. Post-emergence application spray should be carried out after earthing-up operations. Incorporation of pendimethalin at 1.0 kg/ha before sowing in to surface soil, or application 2-3 days after planting provide effective control of several annual grasses and broad leaved weeds. Metribuzin at 1.0-1.5 kg/ha and metolachlor at 0.75-1.0 kg/ha, 3-5 days after planting of potato provide good control of broad spectrum weed control.

Cabbage / Cauliflower *(Brassica sps.)*

The cabbage needs at least two weeks of weed free period after emergence to prevent yield reduction. *Cyperus spp.,* and *Cynodon dactylon* are common weeds of cabbage and cauliflower. The yield of cauliflower could be reduced by 36-59 % if weeds are not controlled properly.

Cauliflower needs four weeks of weed free period after emergence to prevent yield reduction. Alachlor at 1.5 kg/ha, nitrofen 3.4 kg/ha or propachlor 4.5 kg/ha as pre-emergence followed by one hand weeding are recommended in cabbage and cauliflower. Incorporation of fluchloralin at 1.0 kg/ha into the surface soil immediately after application or just before planting or application of pendimethalin at

1.0 kg/ha as pre or post transplanting stage followed by irrigation control many annual grasses and some broad-leaved weeds. Leela (1982) found that oxyfluorfen at 0.25 kg/ha controlled all dicot and monocot weeds including *Cyperus spp.* and *Cynodon dactylon* upto a period of 60 days in cabbage and cauliflower.

Tomato (*Lycopersicom esculentum*) / Bringal (*Solanum melongena*)/ Chilli (*Capsicum annuum*)

Unchecked weeds can cause a yield loss of 40-60% in the above crops (Bhan *et al.*, 1980; Singh *et al.*, 1984; Zimdhal, 1980; Nair *et al.*, 1982). For good production of tomatoes, it is necessary that organic manure and fertilizers be applied in adequate amounts besides giving irrigation at regular intervals. This situation also encourages growth of a broad spectrum of monocots and dicot weeds to a great extent. Herbicide application in tomato is found cheaper than mechanical or manual weeding. Chilli crop can be grown throughout the year. The irrigated chili crop is weeded and hand hoed 3 or 4 times.

There are several safe herbicides which can be used in tomato, brinjal and chillli. Alachlor at 1.5 -2.0 kg/ha at pre-transplanting stage or before transplanting followed by irrigation controls many annual grasses and some broad leaved weeds and sedges. Incorporation of fluchloralin or trifluralin at 1.0 kg/ha at pre or post transplanting stage into the surface soil immediately after application or just before planting followed by irrigation controls many annual grasses and some broad leaved weeds. Pendimethalin at 1.0 kg /ha before or immediately after transplanting followed by irrigation controls several annual grasses and some broad-leaved weeds. Alachlor or butachlor or nitrofen at 2 kg/ha in tomato and chillies as pre-emergence after transplanting are very effective and showed no residual toxicity in the following crop.

Metribuzin at 0.525 kg/ha at pre-transplanting stage or early post emergence one week before transplanting followed by irrigation or post-emergence two weeks after transplanting provides good control of many annual grasses and some broad leaved weeds in tomato. It can also be applied as early post-emergence by avoiding direct contact with tomato plants. Flood jet nozzle may be used for directed spray in case of post-emergence application. Oxadiazon at 0.5 kg/ha at pre-transplanting stage or early post-emergence controls many annual grasses and some broad leaved weeds (Ali, 1985).

Onion (*Allium cepa*) and Garlic (*Allium sativum*)

Onions exhibit greater susceptibility to weed competition than many other crops. Without weed control, onion yield sometimes, shrinks to zero. Because of onions slow germination at early growth and the absence of dense foliages, initial competition tends to be severe. Unchecked weeds can reduce the production of onion by 42 to 55 % (Nair *et al.*, 1980). Very close spacing and shallow root system of onion crop make the mechanical methods difficult proposition. Usage of herbicides has become popular in onion crop as it grown on large scale both for internal consumption and export especially to middle East countries.

Garlic can tolerate weeds only up to 3 weeks after seedling. Uncontrolled weeds can bring decline yields up to 60% (Sharma *et al.*, 1980). Nitrofen, oxadiazon, and fluchloralin have been widely used in onion crop by several workers (Bhan and Singh, 1970; Randhawa and Bhalla, 1976; Nair *et al.*, 1980; Sandhu and Randhawa, 1980; Patel *et al.*, 1983, Gupta and Singh, 1958).

Incorporation of fluchloralin at 1.0 kg/ha as pre or post transplanting stage into the surface soil immediately after application or just before planting followed by irrigation controls many annual grasses and some broad leaved weeds in onion crop. Butachlor application at 1.0-1.25 kg/ha as post-transplanting either before or after irrigation controls many annual grasses and some broad leaved weeds. Pendimethalin at 1.0 kg/ha before or after transplanting, followed by irrigation controls several annual grasses and some broad leaved weeds. If direct sown, spray may be done within 2-3 days of sowing. It can be applied as pre-emergence in direct seeded crop. Haloxyfop-butyl at 100 g/ha at 2 weeks after transplantation gives excellent control of grasses and some sedges, but does not control broadleaved weeds. Oxadiazon at 0.50-0.75 kg/ha or oxyfluorfen at 250 g/ha or oxadiargyl at 90 g/ha before or after transplanting followed by irrigation provide good control of wide spectrum of weeds. Oxyfluorfen can also be applied up to 25-30 days after transplantation.

Beet root (*Beta vulgaris*)/ Carrot (*Daucus carota*)/ Radish (*Raphanus sativus*)

Beet root require 2-4 weeks weed free period to prevent yield losses (Zimdhal 1980). Weed account for 30-90 percent yield loss in onion crop (Singh *et al.*, 1974; Bhan *et al.*, 1977). Carrot possesses the capability to recover from competition when weeds are removed early. However, yield can shrink to 30-60 % under severe competition (Gautam *et al.*, 1965). Carrots can tolerate weeds only upto five weeks after emergence (Zimdhal, 1980). In radish the prominent weeds are *Cyperus spp., Cyanodon spp., Poa spp., Chenopodium spp., Amaranthus spp., Anagalis spp., Melilotus spp., Convolvulus spp., etc.*

Incorporation of alachlor or butachlor at 1.0 kg/ha or fluchloralin at 1.0 kg/ha as pre or post transplanting stage into the surface soil immediately after application or just before planting followed by irrigation control many annual grasses and some broad leaved weeds. Pendimethalin at 1.0 kg/ha before or after transplanting, followed by irrigation controls several annual grasses and broad leaved weeds. If direct sown, application can be done within 2-3 days of sowing. Oxadiazon, metribuzin, pendimethalin and oxyfluorfen also efficiently control weeds (Singh and Malik, 1993). Linuorn at 1.0 kg/ha followed by chloroxuron at 1.0 kg/ha as post-emergence can give good control of weeds in radish crop.

(7) Spices

Spices constitute an important group of agricultural commodities which, since antiquity, have been used for flavouring foods. Some species are used in the pharmaceutical, perfumery, cosmetics and related industries, and others possess colourant, preservative, antioxidant, antiseptic and antibiotic properties. India is one of the leading spice producing and exporting countries in the world. Weeds account for potential yield losses of spices. Chemical weed control in some important spices are described hereafter.

Cumin (*Cuminum cyminum*) /Coriander (*Coriandrum sativum*)/ Fenugreek (*Trigonelle Foenum-graceum*)

Cumin is largely sown by broadcast. The emerging crop seedlings are badly infested with weeds. *Polygonum pumila* (Jeeri) is a very notorious weed in cumin. Its seeds are resembled with cumin and difficult to separate from the crop seeds. In a slow growing and frequently irrigated crop like fenugreek several winter weeds grow profusely and cause upto 50 % yield reduction. Weed seeds pose a serious problem in the cultivation of coriander. Incorporation of fluchloralin at 1.0 kg/ha at pre or post transplanting stage into the surface soil immediately after application or just before planting followed by irrigation controls many annual grasses and some broad leaved weeds. Pendimethalin at 1.0 kg/ha before or after transplanting and followed by irrigation controls several annual grasses and some broad leaved weeds. In direct sown crop, herbicide may be sprayed within 2-3 days of sowing. Triallate 1.7 kg/ha as pre sowing incorporation in the soil or atrazine at 1.5 kg/ha as pre-emergence on moist soil or prometryne at 1.7 kg/ha as post-emergence are very effective to manage weeds in coriander.

❑❑❑

Chapter 10

Residues in Soils Food Chain, Water and its Management

Herbicides are increasingly used for control of weeds in crop and non crop situations. With increased awareness on environmental pollution and health hazards the data on residual toxicity of recommended herbicides become indispensable for approving a herbicide for large scale commercial use. The knowledge of herbicide residues is of great concern to all research and field workers to make meaningful recommendations to the farmers on the application of herbicides.

After application, herbicides move to the soil along with uptake by weeds. Decomposition takes place and part of it is taken up by the plants accumulating in the economic produce. Accumulation of herbicide residues should not exceed the maximum residue limits prescribed by the national and international standards. There are several evidences of hazards associated with the residues of herbicides, therefore there is need of intensive studies of herbicide residues in soil, food chain and water. Herbicides are relatively non-toxic to soil organism at concentration ordinarily used for weed control. Herbicides from soil are removed by leaching, retention by soil colloids, volatilization, microbial and chemical decomposition. Aqueous solubility affect leaching but possible differential fixation of formulations and the carrier used may also have an influence.

Herbicides commonly used have practically no toxicity problem. Extensive toxicological tests are conducted before a herbicide is registered ensuring that no toxicity problem arise with its commercial use. Problem may be associated with the use of herbicides due to contamination of soil, surface or ground water, phytotoxicity to succeeding crop, bioaccumulation in food chain etc. if not used judiciously. There has been a growing public concern about adverse effect of herbicides on the environment and on safety and quality of the food. Question has also been raised that whether an agricultural production system is sustainable over a long term. Thus in this chapter herbicide residues in soil, food chain and water is discussed in details as follows.

1. Aryloxy phenoxy propionate

Aryloxy phenoxy propionate herbicides are used for selective grass weed control in cereals. These herbicides are applied as post-emergence to grass foliage, absorbed and move through symplast to meristems and stop meristematic activity. Acetyl-CoA carboxylase (ACCase) is bound by the herbicide and stops lipid synthesis. Differential metabolism and/or insensitivity of ACCase (i.e. genetic resistance) appears to be the major mechanism of selectivity for the aryloxy phenoxy propionates. In weeds the ACCase is altered so that herbicides do not bind.

These herbicides inhibit fatty acid biosynthesis as their primary mode of action. These are highly selective and are toxic to many annual and perennial grasses when applied at recommended dosages. In general, broadleaf crops and weeds are highly tolerant or resistant to these herbicides and some cereal and turf grass species also exhibit varying degrees of tolerance. Fatty acid synthesis in susceptible plants is blocked when aryloxy phenoxy propionates inhibit acetyl-coenzyme A (acetyl-CoA) carboxylase (ACCase), an enzyme responsible for the first committed step in fatty acid biosynthesis. Most herbicides in this category are broken down by microbial action. Anaerobic conditions may delay breakdown. These herbicides show low mobility in soil and have little or no soil activity. Fate of some of the prominent herbicides of aryloxy phenoxy propionates in soil, food grain and water is discussed hereafter.

Clodinafop-methyl

Clodinafop-methyl is a post-emergence herbicide and is used for control of grassy weeds in cereals. Clodinafop-propargyl inhibits the acetyl co-enzyme A carboxylase (ACCase), which is essential for the production of lipids (fatty acids) needed for plant growth. Clodinafop-propargyl selectivity is based on the difference in the rate of herbicide breakdown in the crop versus weeds. Clodinafop-propargyl converts from the ester form to the active acid and then to biologically inactive compounds. A safener, cloquintocet-mexyl, is added to the formulation to accelerate rate of clodinofop break down in wheat, thus preventing the accumulation of a lethal dose.

Clodinafop degraded to safe level by 100 days in the soil of wheat field at doses of 60, 120 and 240 g/ha (Sondhia and Mishra, 2005). Persistence of clodinofop-methyl evaluated at Ludhiana showed that it degraded to safe level by 60 days at 0.03 to 4.0 g/ha application rates and at higher doses *viz.*, 11 and 22 g/ha, residues persisted for more than 80 days. Whereas, Singh *et al.*, (2004) reported that clodinofop at 60 and 120 g/ha rates in wheat crop degraded completely by harvest and hence residues were not detected in wheat grains and soil at harvest. Clodinafop residues were not detected in the wheat grain and straw at doses 60 and 120 g/ha however, 0.0089 ppm residues were detected in wheat grains at 240 g/ha treatment (Sondhia and Mishra, 2005).

Fluazifop-p-butyl

Fluazifop-p-butyl is a selective herbicide used for post-emergence control of annual and perennial grassy weeds. It is used on soybeans and other broad-leaved crops such as carrots, spinach and potatoes etc. Fluazifop-p-butyl is actively taken up by plants and translocated throughout the plant. Fluazifop-p-butyl accumulates in the actively growing regions of the plant (shoots, root rhizomes, stolons of grass) where it interferes with the plant cell's ability to produce energy. In plants, fluazifop-p-butyl is rapidly dissipiated in the presence of water to fluazifop -p.

Fluazifop-p-butyl, applied to soybeans, at 0.25 and 0.50 kg/ha, dissipated to 0.1 ppm in 30 days from both the dosages and was below detectable limits (0.08 ppm) after 60 days (Singh *et al.*, 1999). Fluazifop-p-butyl can move upto 15 cm soil (Kulshrestha *et al.*, 1995). At harvest 0.012 and 0.036 ppm residues were found in the soil of soybean field with 0.250 kg/ha and 0.500 kg/ha rates, respectively (Kulshrestha *et al.*, 1995). Persistence of fluazifop-p-butyl in sandy loam soil at 10 and 20 ppm rates was found inversely related to soil temperature. Higher dissipation of fluazifop-p-acid occurrs at 40°C as compared to 35 or 28°C (Raut and Kulshrestha, 1991). Fluazifop-p-butyl degradates to the corresponding acid *viz.*, fluazifop-p in soil. Leela (1993) reported that fluazifop-p- butyl persisted upto 30 days in sandy loam soil at Karnataka. Sondhia (2007) found 0.051, 0.064 and 0.079 μg/g residue of fluazifop in soil of soybean field at application rates of 0.125, 0.250 and 0.500 kg/ha respectively. Sondhia (2007) found 0.472, 0.554 and 0.702 μg/g residues in soybean straw and 0.297, 0.300 and 0.312 μg/g residues in soybean grains at 0.125, 0.250 and 0.500 kg/ha rates respectively at harvest, however, application of fluazifop-p-butyl at 0.5 kg/ha rate in soybean crop resulted in the translocation of 0.005 and 0.001 ppm residues to soybean grains and cake, respectively (Kulshrestha *et al.*, 1995).

Fenoxaprop-p-ethyl

Fenoxaprop -p-ethyl is a selective post-emergence herbicide with contact and systemic action. It absorbed principally by the leaves, and translocates both acropetally and basipetally to the roots or rhizomes. It is used for post-emergence control of annual and perennial grass weeds in potatoes, beans, soybeans, beets, vegetables, peanuts, flax, oilseed rape, and cotton. It is also used for control of annual and perennial grassy weeds and wild oats in wheat, rye and barley when applied with the herbicide safener mefenpyr-diethyl. Fenoxaprop -p-ethyl is non-phytotoxic to broad-leaved crops.

Laboratory studies showed that fenoxaprop-p-ethyl degrade in the loamy sand soil to safe level by 40 days at 0.03 to 0.04 kg/ha application rates and can persist upto 63 days at higher application rates. Fenoxaprop-p residue in the soil of wheat field was found below the detection level (0.001 ppm) at 70-100 g/ha rates, however 0.0011 and 0.0011 μg/g residues were detected at 200 and 400 g/ha application rates, respectively (Sondhia, 2006).

Haloxyfop

Haloxyfop is a selective pre and post-emergence herbicide and has proved effective for the control of broadleaf and grass weeds in various crops. Because of the distinctive chemical structural linkage of two phenol groups, haloxyfop and its analogs are often called phenoxy-

phenoxy compounds. Because this group of herbicides and various derivatized cyclohexane compounds display almost identical herbicidal selectivity and injury symptoms, these are considered as being similar compounds on biological basis. Necrosis in meristematic tissue and necrosis and chlorosis in developing leaf tissue are the most common injury symptoms in the field following application of these herbicides (Asare *et al.*, 1983; Chow *et al.*, 1978). Haloxyfop show moderate persistence under tropical conditions. Haloxyfop has been shown to inhibit lipid synthesis and respiration, and has no affect on cellular sugar or ATP levels. Sondhia and Dixit (2007) found 0.091 µg/g residues in the soil of onion crop after 129 days and 0.005 and 0.003 µg/g, residues in the green and mature onion bulbs collected at 50 days and at harvest (129 days), respectively.

2. Chloroacetamide

Chloroacetamides are commonly known as growth inhibitors. These growth inhibitors are applied as pre-emergence or occasionally as pre-plant incorporated. Affected plants show stunted root development, swollen root tips, chlorosis and whitening of lower leaves, leaf tip burn, and stunted growth. Chloroacetamide kill germinating seedlings instead of established plants. Chloroacetamides reduce cell division and enlargement and thus inhibit growth. Inhibition increases with increasing concentrations and varies with different species. Chloroacetamides are readily absorbed by plants and appear to be translocated primarily in the apoplast but some symplastic movement may also occur.

This family of herbicides exhibit moderate water solubility, low soil retention, and low to moderate volatility. Soil texture and organic matter are important in determining chloroacetamide application rates. Volatilization rates are generally quite low. However, if no rainfall occurs within seven days following application, mechanical incorporation may be needed to retain adequate pre-emergence activity. Soil persistence ranges from less than 10 to 90 days. The major means of dissipation of chloroacetanides is microbial decomposition which showed longer persistence under anaerobic and cooler conditions. Fate of some of the prominent herbicides of chloro-acetamide class in soil, food grain and water is given below.

Alachlor

Alachlor is an acetanilide herbicide of low acute toxicity, but repeated exposure has been reported to cause hepatotoxicity, irreversible uveal degeneration and tumour formation in rats. Alachlor is a selective systemic herbicide, which absorbed primarily by shoots but also by the roots with translocation through out the plant and accumulation mainly in vegetative parts rather than in reproductive parts. It acts by inhibition of protein synthesis and root elongation. It has low to moderate leaching potential in soil and good translocation in plants demand careful adherence to good agricultural practices to avoid potential contamination of food and water resources. Alachlor is being recommended to crops like cotton, maize, chickpea, soybean, sugarcane and onion.

Because alachlor is adsorbed by soil and organic colloidal particles it is not subject to excessive leaching in most soils. It is primarily degraded by soil micro-organisms. Persistence is shorter in sandy soil, which is low in organic matter than in heavy soils.

Adsorption of alachlor increased with increase in concentration, time of incubation, rise in activation temperature, lowering of pH and increase in the organic matter content. (Sethi and Chopra, 1975). Alachlor residues were found below the detection limit in soil, seeds and straw at harvest at 4.0 kg/ha rate in soybean crop (Annual Report AICRP-WC 1999). Dissipation of alachlor at 1.5 kg/ha rate in cotton and its movement in alfisol under laboratory conditions showed that, rate of dissipation of alachlor increased progressively with time and the semilogarthimic plot for alfisol followed the first order reaction kinetics. Alachlor dissipated in soil within 40 days and the residues were not found in the soil at harvest. Residues of alachlor applied at 1.0 or 1.5 kg/ha were not found in post-harvest soil in cotton at Hyderabad. However, alachlor persisted upto 40 days in soil of maize field. (AICRP-WC, 1999)

Alachlor applied to fingermillet (0.6 kg/ha) persisted in the soil upto the harvest in fingermillet-cotton-sorghum cropping system. In groundnut field alachlor persisted upto 70 days at Faizabad. Sondhia, (2002a) reported that alachlor residues were not detected at harvest at 1.0 kg/ha rate in the soil of soybean field but at 1.2 and 1.5 kg/ha rates, showed 0.01 and 0.02 ppm residues at harvest. Whereas in sandy loam soil, alachlor persisted for 60 days at 1.5 kg/ha application rate applied as pre-emergence in vegetable crops (Leela, 1993, 1987). Alachlor residues were found at trace level in cotton plant, cotton lint and oil, water and fish at 2.5 and 5.0 kg/ha application rates under field conditions (Ramesh and Maheshwari, 2004).

Butachlor

Butachlor is a pre-emergence, selective, systemic herbicide, absorbed primarily through germinating shoots and secondarily by roots. Butachlor translocates throughout the plant, with higher concentration in the vegetative parts than the reproductive parts. Butachlor is absorbed by young sprout and young second shoot of weed to control the synthesis of protein in weed, which can make weed swollen and deformity until weed is dead. It is used in rice, wheat, rape, cotton, flax, peanut and vegetables to control annual and broad leaf weeds. Butachlor dissipate at faster rate and show moderate persistence under tropical conditions.

Studies conducted in different textured soil showed that butachlor accumulation is more in heavy textured soil as compared to medium and light textured soil. Butachlor residues did not persist in red and black soil at various locations of Coimbatore at harvest. Butachlor residues were not found at 0.75 and 1.5 kg/ha application rates in the post harvested soil of rice-rice and finger millet-groundnut cropping system in Bangalore. Butachlor dissipate at faster rates and residues were not found beyond 20-30 days in 0-15 cm depth. Residues of butachlor at 0.75 kg/ha application rate did not persist beyond 21 days at recommended level with FYM and 28 days without FYM in Bangalore. Butachlor residues were evaluated in transplanted onion field soil at harvest. Soil samples contained 0.0152 μg/g butachlor residues when applied at 1.0 kg/ha rate (Sondhia, 2006b).

Devi *et al.*, (1997) reported that butachlor residues applied at 2.0 kg/ha in rice crop were found below the maximum permissible residue limit (0.25 ppm) in soil. Sondhia *et al.* (2006) reported that butachlor residues were not detected after 120 days in clay loam soil applied at 1.0 kg/ha in transplanted rice crop. Continuous use of butachlor at 1.5 kg/ha in rice crop did not leave any residues in grains and straw at harvest. Residues of butachlor at 0.75 and 1.5 kg/ha rates were not detected in groundnut, rice, fingermillet, potato tubers and sorghum grains (Annual Report, AICRP WC 2000). However, all residues were below the maximum residue limit of 0.25 mg/kg (Reddy *et al.*, 1998). Deka and Gogoi (1993) found 0.012 and 0.007 ppm residues in rice grains and straw after treatment with butachlor at 2.0 kg/ha rate. Sondhia and Dubey (2006) found 0.004 and 0.0007 μg/g residues in green and mature onion respectively at 1.0 kg/ha application rate. Butachlor residues ranged between 0.001 to 0.093 ppm in the water of rice field.

Metolachlor

Metolachlor is a pre-emergence selective herbicide, which is mostly absorbed by the hypocotyls shoots and used to control certain broadleaf and annual grassy weeds in corn, cotton, sunflower, groundnut, soybeans, peanuts, grain sorghum, potatoes, cotton, safflower, stone fruits, nut trees, and woody ornamentals. Metolachlor acts by inhibiting germination and protein synthesis. High-protein crops can be adversely affected by metolachlor applications (Wiegand *et al.*, 1981).

Metolachlor appears to behave similar to the other cholroacetamide herbicides in soil. Leaching is restricted by adsorption onto clay and organic colloidal particles, with organic matter being the more important restraint. Dissipation of metolachlor in soil is relatively rapid. The half life of metolachlor is generally one to four weeks in Indian conditions.

Metolachlor showed higher adsorption in black soil followed by alluvial and red soil and followed the adsorption isotherm. Metolachlor at 3.0 kg/ha application rate showed 0.0018 ppm residues in soybean seeds and residues were below the detectable limit in soybean straw and soil (Annual Report AICRP WC 2001). Metolachlor applied as pre-emergence at 1-2 kg/ha application rates was dissipated almost 100 % in the soil at harvest under field conditions (Singh *et al.*, 1997). Metolachlor did not persist for more than 40 to 60 days, in the red sandy loam soil of soybean field at 1.0 and 1.5 kg/ha, application rates at Hyderabad with half-life of 8.00 and 9.52 days respectively (Devi *et al.*, 1997).

Sanyal *et al.*, (2003) evaluated dissipation of metolachlor in soil under both field and laboratory conditions and found that metolachlor showed moderate persistence with a half-life of 27 days in the field conditions and leached to a depth of 15-30 cm in soil. It was found that fungi *Aspergillus flavus* and *Aspergillus terricola* rapidly degraded metolachlor applied at 10 kg/ha up to 92 % and 87 % after 20 days in sterile and non-sterile soils, respectively (Sanyal and Kulshrestha, 2003).

Pretilachlor

Pretilachlor is a selective pre-emergence broad-spectrum herbicide with excellent action against annual grasses, sedges and broadleaved weeds. Pretilachlor is taken up by the shoots and to a lesser extent by the roots of germinating weeds, which die shortly after emergence. It is recommended in rice crop for controlling annual weeds. Dissipation of pretilachlor was evaluated in the soil of rice field at Coimbatore. Pretilachlor at 0.75 kg/ha application rate showed 0.081 ppm residues after 7 days in the treatment that received 100 % inorganic nitrogen in the soils of rice crop, which gradually reduced to 0.013 ppm at 45 day and residues were not found in the post-harvest soil. Half-life of pretilachlor in soil was found 10.3 to 11.4 days. Application of pretilachlor at recommended levels in rice-rice system and aquatic system did not result in build up of any residues in the soil. Balasubramanian *et al.*, (1999) reported that pre-emergence application of pretilachlor in rice crop showed residues below the MRL value and did not influence the germination and yield of black gram (*Vigna mungo*) raised subsequently. Deepa and Jayakumar (2006) evaluated persistence of pretilachlor under lab conditions and reported that residues were not found after 30 days with half life of 3.3 days in different water having pH range of 5.5 to 8.5 at 0.5 μg/ml fortification level.

3. *Cyclohexanediones*

Cyclohexanedione herbicides are used for grass weed control in broadleaf crops and generally applied as post-emergence to grass foliage and translocate by absorption by foliage and moved through symplast to meristems. These herbicides inhibit fatty acid biosynthesis as their primary mode of action. These are highly selective and are toxic to many annual and perennial grasses when applied at recommended dosages. In general, broadleaf crops and weeds are highly tolerant of or resistant to these herbicide, and some cereal and turf grass species also exhibit varying degrees of tolerance. Differential metabolism and/or insensitivity of ACCase (i.e. genetic resistance) appear to be the major mechanism of selectivity for the cyclohexenediones. The fate of some cyclohexanediones herbicides in soil is discussed hereafter.

Tralkoxydim

Tralkoxydim is a foliar-applied, translocated post-emergence herbicide and used for the selective control of wild-oats, black-grass, rye-grasses and other grassy weeds in winter and spring cereals. The activity of tralkoxydim is by foliar uptake only and independent of soil type, soil Kd levels or organic matter. However, weeds germinating after application will not be controlled.

Tralkoxydim is rapidly absorbed through the leaves and moves to the growing point. Uptake of tralkoxydim takes place through the leaves then translocated to growing points of roots, shoots and leaves. The growth of susceptible weeds ceases within a few days of application, preventing further weed competition. Decay of growing points is seen after 2-3 weeks. The persistence of tralkoxydim in soil of wheat crop was evaluated by Srivastava *et al.*, (1994, 2006) at two application rates (400 g/ha and 800 g/ha). Tralkoxydim persisted up to 28 and 45 days in soil at 400 and 800 g/ha, but became non-detectable only after 45 days in the crop. Dissipation of the tralkoxydim from soil occurs in two phases each phase followed first-order kinetics.

4. Diphenylether

These herbicides act by free radicle formation *via* inhibition of protoporphyrinogen oxidase (Protox) which is the enzyme that converts Protoporphyrinogen IX to Protoporphyrin IX. Protox is a membrane bound protein located in the chloroplast envelope where it is involved in chlorophyll biosynthesis and heme synthesis. Heme is an iron containing porphyrin which regulates chlorophyll formation. Diphenylethers herbicides compete for the Protoporphyrinogen IX binding site on Protox.

Diphenylethers are foliar applied annual broadleaf herbicides. Diphenyl ether activity is expressed as foliar necrosis after 4-6 hours of sunlight following post-emergence application. The first symptoms are a water soaked appearance (dark green spots on the foliage) followed by the water soaked areas. Water soaking is indicative of membrane damage and cellular substances leaking into intercellular spaces, which changes the refractive index of the tissue. With pre-emergence applications, tissue is damaged by contact with the herbicide as the plant emerges above the soil surface. As with post-emergence applications, the damage is tissue necrosis. Membrane damage induced by diphenylether herbicides is the result of lipid peroxidation.

Diphenylether degradates by photo and microbial degradation mechanism. After application of diphenylether plant leaves turn yellow then brown and die. Reddish-colored spotting on the leaf surface may appear shortly after application. Plants that do not die may be stunted for a week or so. Crop oils and other additives may increase plant injury. These herbicides are generally not prone to resistance.

Oxyfluorfen

Oxyfluorfen is a selective pre- and post-emergent herbicide, absorbed by the foliage, shoots and roots. It is used to control certain annual broadleaf and grassy weeds in rice, chickpea, lentil, peas, groundnut, sunflower, onion, garlic, vegetables, fruits, cotton, ornamentals, and on non-crop areas. It inhibits protoporphyrinogen oxidase. It is a contact herbicide and light is required for it to affect target plants. Like the other diphenyl herbicides, oxyfluorfen is strongly adsorbed on soil colloids and is not subject to leaching. It undergoes detoxification in soils, but microbial degradation apparently is not a major factor. Diphenylether shows short to moderate persistence in soil. Oxyfluorfen at 0.25 kg/ha application rate persisted for 60 days in sandy loam soil. Oxyfluorfen persisted upto 30-100 days in different textured soil (AICRP-WC, 2006).

Sondhia and Dixit (2006, 2007) and Sondhia, (2008d) reported 0.0015 μg/g residues of oxyfluorfen at harvest in the soil of transplanted onion crop at 300 g/ha, however residues were below the detection limit at 150-250 g/ha rates. Oxyfluorfen residues applied to cabbages at 0.1 to 0.4 kg/ha application rates were not found in soil at harvest (Sundararajan *et al.*, 1993). Leela (1993) reported that oxyfluorfen at 0.5 kg/ha application rate applied as pre-emergence in vegetable crops persisted for 30 days in the sandy loam soil. Whereas Devi *et al.*, (1998) reported that oxyfluorfen activity at 0.2 kg/ha application rate was lost by 60 days in red soil and by 80 days in black soil. Sondhia and Dixit (2006) found 0.041, 0.041, 0.040 and 0.063 and 0.0034, 0.0035, 0.0050, 0.0460 μg/g residues at 150-300 g/ha application rates, respectively in the green onion and mature onion bulbs.

5. Dinitroanilines

These herbicides are very low in water solubility, moderate to high in soil retention, and moderate to high in volatility. For adequate weed control, incorporation of these herbicides is recommended within 24 hours to 7 days after application. Soil composition is important in determining the rate of application, because soil texture and organic matter determines the degree of soil retention. Dinitroanilines are breakdown by photodegradation and soil microbes. Dinitroaniline activity is largely determined by clay and organic matter content. Greater clay and organic matter content requires a greater quantity of dinitroaniline. Sunlight can cause herbicide loss. If the active ingredient is unprotected on the soil surface, a large portion may be lost from ultraviolet degradation.

Microbial degradation is more rapid under anaerobic conditions. Persistence is longer under cool and dry conditions. Enhanced degradation may occur where microbes can rapidly breakdown herbicide to reduce efficacy. Dinitroanilines herbicides inhibit growth of the entire plant when absorbed by the roots. However, this is apparently largely brought about by initially limiting root growth, especially the development of lateral or secondary roots. The roots which develop are often only the primary roots which are somewhat thickened, stubby and are devoid or have only a limited number of secondary roots. Dinitroanilines cause characteristic root tip swelling. This swelling is due to abnormal cell enlargement in the zone of elongation. In this zone, cells of treated roots enlarge to a spherical shape rather than a normal cylindrical shape. This is caused by the loss of microtubules. Microtubules are responsible for the orientation of the cellulose microfibrils that provide support to the cell wall. Without microtubules, the microfibril orientation is random and the cell wall has no preferential support in any direction. Fate of some of the prominent dinitroaniline herbicides in soil, food grain and water is discussed hereafter.

Fluchloralin

Fluchloralin is used as a pre-plant or pre-emergence herbicide. Fluchloralin degrade rapidly in the vapor-phase by reaction with photochemically produced hydroxyl radicals (half-life of about 5 hours). If applied to soil surfaces, fluchloralin dissipates rapidly through photodecomposition and volatilization, however, immediate soil incorporation greatly increases soil persistence. Within soil systems, biodegradation is the probable route of disappearance, but it appears to be much slower than surface photolysis or volatilization. Fluchloralin is tightly absorbed to soil colloids. Under field conditions, fluchloralin has not been found to leach. In water, fluchloralin can degrade through photodecomposition and biodegradation. Volatilization may contribute to its loss from water. It is recommended in soybean, rice, groundnut, sunflower, jute, cotton, onion, cumin crop and vegetables.

Fluchloralin is tightly adsorbed to soil colloids and is not subject to leaching. It is degraded mainly by soil microorganism but may persist more than one season after application. Persistence of fluchloralin is high in black soil as compared to red and alluvial soils due to high adsorption. Fluchloralin residues persisted upto 125 days in the soil of onion crop at Anand. In irrigated cropping system of fingermillet-cotton-sorghum, fluchloralin at 1.0 kg/ha application rate persisted in soil upto the harvest of fingermillet. Fluchloralin persisted upto 20 days in sandy loam soil and 80 days in clay and organic soils and 50-56 days in the soil of maize crop at Faizabad. Sondhia (2002a) did not found fluchloralin residues when applied at 1.0 and 1.2 kg/a rates in soil of soybean crop at harvest, however 0.020 ppm of fluchloralin residues were detected at 1.5 kg/ha application dose.

Singh and Kulshrestha (1995) reported fast degradation of fluchloralin in flooded anaerobic soil than in aerobic soil and found that amendment of fluchloralin with organic matter enhanced degradation of flooded anaerobic soil. The major degradation products of fluchloralin were partially dealkylated fluchloralin, partially reduced fluchloralin and its cyclic product. Patel *et al.*, (1996) evaluated fluchloralin in sandy loam and loamy soils during *rabi* season and found that persistence of the pre-plant incorporated fluchloralin at 0.67, 0.90 and 1.35 kg/ha application rates was longer in the loamy soil as compared to sandy loam soil. Diwan *et al.*, (1999) found that fluchloralin applied at 1.0 kg/ha in a sandy loam soil in onion crop showed first-order rate kinetics in a biphasic mode and residues were not found at harvest.

Fluchloralin residues at 0.5 to 2.0 kg/ha were not detected at harvest in green forage lucerne plant, groundnut, rice, fingermillet and potato tubers, However, 0.040 ppm residues of fluchloralin at 1.125 kg/ha application rate were detected in cumin seed (AICRP-WC 2002). At harvest, 0.009 and 0.006 ppm terminal residues of fluchloralin applied at 0.75 and 1.50 kg/ha, respectively were found in stover and grains (Saikia and Pandey, 1999).

Pendimethalin

Pendimethalin is mostly used as a pre-emergent herbicide for low land rice. Though the main use of pendimethalin is for rice, it has proved useful for other crops as well *viz.*, wheat, maize, sorghum, pearlmillet, chickpea, peas, groundnut, soybean, sunflower, mustard, linseed, jute, cotton and vegetables. Like other dinitroaniline herbicides, pendimethalin binds to tubulin and inhibits the production of microtubules. This disrupts cell division and cell wall formation. Pendimethalin is absorbed by roots and coleoptiles.

Pendimethalin is strongly absorbed on soil clay and organic matter and is not subject to leaching. In contrast to the case with most other dinitroaniline herbicides, soil microorganisms do not appear to play significant role in degradation of pendimethalin.

Persistence of pendimethalin is high in black soil as compared to red and alluvial soil due to high adsorption of pendimethalin in black soil. Kulshrestha *et al.*, (2000) evaluated persistence of pendimethalin at 1.5 kg/ha in a five year long term field trial of maize-wheat rotation and found that repeated application of pendimethalin on the same soil led to rapid degradation of pendimethalin in each successive year with each successive crop (Fig. 1). Pahwa and Bajaj (1997) found that persistence of pendimethalin is directly correlated with temperature and application rate. Lower temperature and high dose results in high persistence of pendimethalin in soil. Residues of pendimethalin at 1.0 kg/ha and 2.0 kg/ha rates were not found in sunflower and cumin seed. Pendimethalin applied at 0.6-0.9 % to tobacco recorded 0.198 to 0.376 ppm residues in tobacco leaves and 0.72 ppm residues in leaves treated with 0.5 % pendimethalin and 0.04-0.079 ppm residues treated with 0.25% pendimethalin (Parmar *et al.*, 1998). Sondhia and Dubey (2006) did not found pendimethalin residues in soil and mature onion, however 0.007 μg/g pendimethalin residues were detected in green onion at 1.0 kg/ha application rate.

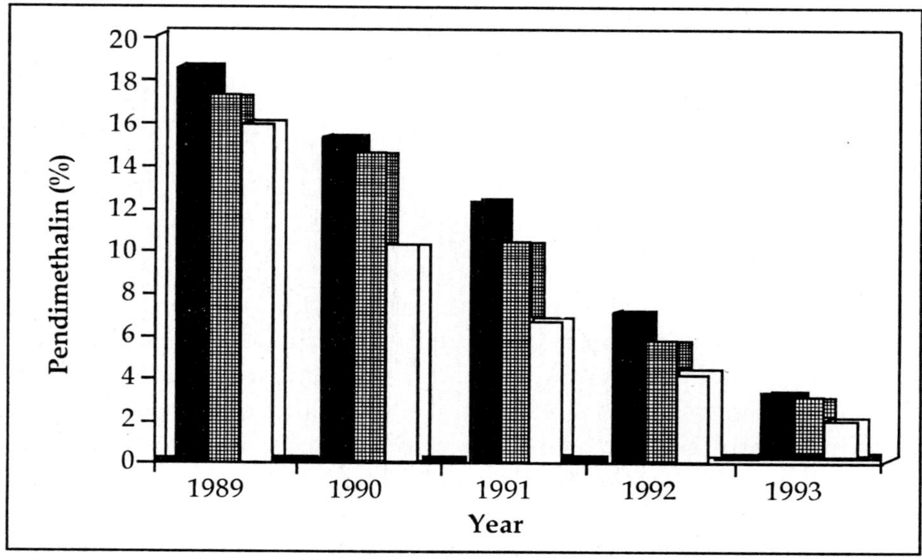

Fig. 1. Harvest time residues of pendimethalin in soil (0-15 cm) from 1989 to 1993 in maize-wheat cropping system.

At Bangalore, degradation of pendimethalin in sunflower-soybean system was monitored for two years. Pendimethalin residues were not found beyond 30 days in a sandy loam soil. Pendimethalin applied at 1, 2 and 4 kg/ha rates in a sandy loam soil in wheat crop showed persistence up to 200 days and caused phytotoxicity to the succeeding sensitive sorghum crop at higher dose (Yadav *et al.* 1995). The half-life of pendimethalin in onion plants and soil varied from 11.8 to 15.5 days and 14.9-15.1 days respectively (Sinha *et al.*, 1996). Fate and persistence of pendimethalin in different cropping system and soil is presented in Table 1.

Diwan *et al.* (1999) found first-order rate kinetics of pendimethalin in sandy loam soil. Pendimethalin dissipated by 85-90 % upto harvest with a half-life varying from 28 to 32 days, at 0.75 to 2.00 kg/ha application rates. Rai *et al.*, (2000) evaluated persistence of pendimethalin in six tropical soils of varying physico-chemical characteristics under flooded and non-flooded conditions and reported that degradation is rapid under flooded than non-flooded conditions in all the soils. Non-flooded conditions showed 36-78 % degradation of pendimethalin after 30 days of incubation.

Sondhia and Dubey (2006) did not found residues of pendimethalin in soil of onion field at harvest at 1.0 kg/ha. It was found that black soil absorbed higher amount of pendimethalin than alluvial and red soil and desorption increased with increase in concentration of pendimethalin in all the soil. Pendimethalin showed high adsorption in the topsoil layer of black soil and the adsorption in sandy loam and clay soils increased with increase organic matter content (Raj *et al.*, 1999).

Trifluralin

Trifluralin was first registered in the US in 1963 (EPA, 1996). Trifluralin is a widely used herbicide, which is a suspected carcinogen. In some countries its use is increasing, while in other countries it is banned for its persistence and its threat to ecosystems. Trifluralin is a selective, pre-sowing or pre-emergence herbicide used to control many annual grasses and broadleaf weeds in a large variety of arable and horticultural crops. Trifluralin prevents weed growth by inhibiting root development through the interruption of mitosis (Grover *et al.*, 1997) *viz.* early developmental cell division. Trifluralin is applied as a soil-incorporated herbicide. Trifluralin has been shown to inhibit the development of the activity of several enzymes in germinating seedlings.

Table 1. Fate of pendimethalin in surface soil (0-20 cm) in different crops and cropping system in different agroclimatic conditions.

Pondimethalin dose (kg/ha)	Crop/cropping system	Soil type	Location	Persistence (days)	Reference
0.5	Lucerne	Sanday loam soil	Anand	upto harvest	AICRP - WC (2000)
1.0	Rice	Red sandy loam soil	Bangalore	60	AICRP-WC, 2000
1.0-2.0	Rice	-	Hyderabad	40-60	AICRP-WC, 2001
1.0-2.0	Onion	Sand loam soil	Anand	125	AICRP-WC, 2000
1.0	Onion	Clay loam	Jabalpur	100	Sondhia and Dubey (2006)
1.0	Onion	-	-	45-60	Sinha (1996)

Four factors are responsible for the dissipation of trifluralin in soil. These include volatilization, photodecomposition, microbial decomposition and chemical decomposition. The relative importance of each factor is influenced by soil type, moisture content, temperature, and microflora, rate of application and method of incorporation. Trifluralin is strongly adsorbed onto soil and is extremely resistant to movement by water. Adsorption on organic matter appears to strongly restrict availability of trifluralin to germinating weeds, thus limiting its efficacy as herbicide.

Pahwa and Bajaj (1997) noticed a direct correlation in the persistence of trifluralin, with dose and temperature *viz.*, lower temperature and higher dose resulted in greater persistence of trifluralin in soil. Goyal *et al.*, (2003) reported that intermittent wetting and drying results in high persistence (90-99%) of trifluralin, whereas with continuous ponding, decrease persistence of trifluralin by 22-40%. Prasad *et al.*, (2006) reported decomposition of trifluralin and metribuzin in water as pH and dose dependent. They reported decompositions of trifluralin faster under alkaline conditions and at low dose of application.

6. Glycine

These herbicides exhibit high acidity, very high water solubility and very low volatility. Glyeine herbicides readily react with clay and metal oxides in soils to form insoluble iron, aluminum and calcium precipitates. As a result, have very high soil retention. Soil mobility of these herbicides is very low. Persistence of these herbicides is very short due to the formation of insoluble precipitates, making these biologically inactive upon contact with mineral soils.

These herbicides inhibit aromatic amino acid biosynthesis. Aromatic amino acid biosynthesis involves a photosynthetic carbon reduction cycle. Within this cycle, PEP (phosphoenolpyruvate) and Erythrose-4-P combine to eventually make Shikimate. Through another reaction, Shikimate is used to make Chorismate, which eventually makes the aromatic amino acids (tryptophan, phenylalanine, and tyrosine). The glyphosate molecule finds its site of action between shikimate and chorismate. After Shikimate is produced, a phosphate group is added to it (now called S3P) and it combines with PEP to make EPSP (Enolpyruvylshikimate-3-P). EPSPS (5- enolpyruvyl-shikimate-3-phosphate synthase) is responsible for catalyzing the reaction; this is the enzyme that glyphosate inhibits, thus causing a buildup of shikimate. Glyphosate is able to bind to an allosteric site near PEP on the EPSPS enzyme. Therefore, PEP can not bind with S3P to produce EPSP. Without EPSP, chorismate concentrations are reduced so as of the aromatic amino acids mentioned above. A reduction in amino acid production will reduce protein synthesis and subsequently cause an inhibition in growth. Fate of some of the prominent herbicides of this class in soil, food grain and water is given below.

Glyphosate

Glyphosate is a non-selective, systemic, non-residual post-emergence herbicide. It is very effective on deep-rooted perennial species. Glyphosate inhibits EPSP synthase (5-enolpyruvyl-shikimate-3 phosphate synthase), an enzyme involved in the production of several amino acids. Both class 2 and class 9 herbicides inhibit amino acid synthesis, but their target sites are different enzymes and cause disruption of the synthesis of different amino acids. Glyphosate is used for control of a great variety of annual, biennial, and perennial grasses, sedges, broad-leaved weeds, and woody shrubs in fruit orchards, vineyards, conifer plantations, many plantation crops (examples coffee, tea, bananas, rubber, coconut, palms, cocoa) and non-crop area.

Glyphosate also interferes with normal carbohydrate translocation in plants. It has been demonstrated to inhibit the import of sucrose to sink leaves, the export of sucrose from source leaves, and the net starch accumulation in source leaves. This disruption in carbon metabolism would eventually starve the plant of compounds required for growth. The appearance of a red coloration on leaves and stem of treated plants is due to the accumulation of the pigment anthocyanin. Anthocyanin accumulation is a typicaly symptom of an interference in carbon metabolism.

Glyphosate is strongly adsorbed to soil, remains in the upper soil layers, and has a low propensity for leaching. Glyphosate readily bio-degrades in soil. Biodegradation in foliage and litter is somewhat faster. It dissipates rapidly from the water as a result of adsorption and possibly biodegradation. Sediment is the primary sink for glyphosate. After spraying, glyphosate levels in sediment rise and then decline to low levels in a few months. Glyphosate is active through the leaves, but has little effect when applied to the soil.

These herbicide are potentially mobile, but leaching is minimized by rapid degradation. Fate of some of the prominent herbicides of this class in soil, food grain and water is discussed hereafter.

2, 4-dichlorophenoxy acetic acid (2, 4-D)

One of the hormone weed killers, 2, 4-D (2, 4-dichlorophenoxy acetic acid) is an aryloxyalkanoic acid, also known as a 'phenoxy herbicide. It is a selective herbicide and toxic to broad-leaf plants but less harmful to grasses. 2, 4-D has complex mechanisms of action against weeds, resembling those of auxins (growth hormones). After absorption, 2, 4-D is translocated within the plant and accumulates at the growing points of roots and shoots where it inhibits growth. The principal use is for the control of broad leaf weeds in cereal crops including wheat, maize, rice, sorghum and grassland and turf areas. It is also widely used in mixtures with other herbicides to provide weed control in forestry, orchards, non-crop areas, and for the control of aquatic weeds.

Soil type and formulation of 2,4-D influences its leaching in soil. Water soluble forms leach more readily than those that are slightly soluble. It is adsorbed by soil colloids and less leaching occurs in clay and organic soils than in sandy soils. It is mainly decomposed by microorganism.

2, 4-D residues did not persist in red and black (post harvest) soils collected from various locations in TamilNadu. Continuous use of 2, 4-D at 0.5 kg/ha application rate in rice –wheat cropping system did not leave residues above the maximum residue level (MRL) of 0.25 mg/kg) in soil (AICRP-WC, 2001). An application of 2, 4 -D sodium salt at doses ranging from 0.25 to 1.25 kg/ha as post-emergence did not leave any residues in post harvest soil of wheat field. 2, 4-D persisted for less than a month in the rice and maize field. At Bangalore 2, 4-D persisted in red soil for 3 months. However, residues were not detected or found below maximum residue limits in the soil of rice crop at Coimbatore.

Sushilkumar *et al.*, (2003) reported that 2, 4-D residues dissipated fully after 45 days with 1.5 kg/ha application rate and 60 days with higher application rates *viz.* 2.0 and 2.5 kg/ha in the black soil of Jabalpur. Whereas Kumari *et al.*, (2004) found that 2,4 -D residues dissipated by 92% after 90 days of its application in wheat crop at Hisar. 2,4-D at 0.4 kg/ha alone and in combination with anilofos persisted upto harvest with half-life of 18-22 days (Jayakumar and SreeRamulu, 1993). In rice-rice system and aquatic system application of 2,4-D at recommended levels did not result in build up of any residues in soil.

Balajee and Mahadevan, (1993) reported that *Azotobacter chroococcum* efficiently cleaved 2, 4-D and the cleavage was profoundly influenced by bacterial cell concentration, moisture content of the soil and temperature at which the soil was incubated and increase in cell concentration enhanced degradation of 2, 4-D. Adsorption of 2, 4-D at 15°C and 25°C temperature increased with increase in 2, 4-D concentration and organic matter of the soil and adsorption decreased with increases in temperature and decrease in pH (Bhardwaj and Singh, 1991). Continuous use of 2, 4-D at 0.5 kg/ha application rates in wheat crop did not leave any residues in grains and straw (Annual Report, AICRP-WC, 2000). Devi (2004) evaluated degradation of 2, 4-D in water. She reported that even though 2, 4-D dissipated initially at a faster rate but at higher application rates of 2, 4-D (2.0 and 4.0 kg/ha) a waiting period of more than 4 months should be insisted for safe consumption of fish. 2, 4-D residues were detected at lower level than the acceptable daily intake (0.01 mg per kg body weight) in fish samples at all the sampling interval in Thrissur where 2, 4-D was applied at recommended rate (1.0 kg/ha). Hence 2, 4-D at 1.0 kg/ha application rate could be considered as safe as far as fish consumption is concerned.

11. Pyridine carboxylic acid

Pyridine carboxylic acid is degraded by photodegradation and microbial breakdown. Pyridine carboxylic acids are degraded more slowly than the phenoxys or benzoics. These herbicides are very important agrochemicals for selective weed management in grass and cereal crops and are generally phytotoxic to broadleaf plants, causing little or no damage to monocots. Monocot crops are tolerant, in general, because of restricted translocation, enhanced metabolism, the lack of a vascular cambium in their phloem tissue. Broadleaf species can be tolerant because they can metabolize herbicides to a less toxic form. These herbicides are considered mimics of the natural plant auxin and are thought to induce changes in gene expression, which lead to plant death.

After exposure to a high dose, symptoms include cupping and stunting of leaves, brittleness, stunting and twisting of stems, and general abnormal growth of susceptible species. Terminal leaf growth by these herbicides also ceases. Tissue proliferation along the stem takes place first at the stem tip, then the nodes and finally along the length of the stem resulting in disintegration or blockage of phloem and xylem tissue. Meanwhile, epinasty, bending and splitting of the stem occur. The roots become thickened and stunted. Adventitious roots or callus tissues develop which crush the phloem and cortex, eventually resulting in rupturing of the epidermis of stem tissues. Symptoms normally appear within a few hours or days although death may not occur for several weeks or months. Fate of some of the prominent herbicides of this class in soil and food grain is discussed hereafter.

Dithiopyr

Dithiopyr (*S*, *S'*-dimethyl 2-difluoromethyl-4-isobutyl-6-trifluoromethylpyridine-3,5-dicarbothioate) is safe to apply to most turfgrasses including creeping bentgrass (spring, pre-emergence use only), Kentucky bluegrass, tall fescue, creeping red fescue, perennial ryegrass, bermudagrass and zoysiagrass. Granular forms of dithiopyr effectively control crabgrass, goosegrass and foxtails. The emulsifable formulation of dithiopyr provides more consistent control when applied twice per season. Dithiopyr also provides effective post-emergence control of crabgrass. It inhibits cell division and cell growth of meristematic regions (growing points of roots and shoots). Dithiopyr is lost from soil by chemical and microbial degradation. Saikia and Kulsherestha (2003) reported that dissipation of dithiopyr applied as pre- or post- emergence applications at two rates, *viz* 100 and 200 g/ha in wheat crop followed first-order decay curve irrespective of rate or method of application with half-life of 17.3 and 25.0 days and residues at harvest (150 days after application) ranged between 4.0 and 8.8% of amount applied.

Fluroxypyr

Fluroxypyr is a post-emergence herbicide used for the control of annual and perennial broadleaf weeds and woody brush. Fluroxypyr induces auxin-type responses in susceptible annual and perennial broadleaf weeds. After predominantly foliar uptake, the ester is hydrolyzed to the parent acid, which is the active form, and translocated rapidly to other parts of the plants. Fluroxypyr acts by inducing characteristic auxin-type responses e.g. leaf curling. Fluroxypyr persisted upto 21 days at 0.1 kg/ha application rate whereas it persisted upto 42 days at higher application rate of 1.0 and 1.5 kg/

ha at Ludhiana under laboratory conditions (AICRP-WC, Annual Report, 1999).

12. Sulfonylureas

Sulfonylurea herbicides control weeds through inhibition of the acetolactase synthase enzyme (a vital enzyme in the synthesis of amino acids). Affected plants exhibit chlorosis, necrosis, stunting, epinasty and shortening of internodes. Sulfonylureas kill germinating seedlings. These herbicides also exhibit foliar activity. Herbicide symptoms develop slowly. Some deformation of new growth can occur.

These herbicides are very important because of their low cost, effectiveness against a large number of weeds (mostly broad leaved plants), good crop and operator safety, and activity at low concentrations. These herbicides have moderate to high acidity and very low to low volatility. Herbicide retention ranges from very low to low, and water solubility from very low to high. Both water solubility and soil retention are pH dependent. Retention is directly related to organic matter, and mobility is inversely related to it.

Sulfonylureas are primarily broken down by hydrolysis and microbes. Sulfonylurea herbicides are more tightly adsorbed to soil particles and soil organic matter at low pH. Persistence ranges from short to long and is greatly influenced by environmental conditions. Water and soil pH is very important. Sulfonylurea herbicides carryover is more in higher pH soils since acid hydrolysis ceases at high pH levels. The rate of hydrolysis is greater at pH below 6.8. However, these herbicides appear to be more persistent under basic than acidic conditions. Sunlight plays a minor role in sulfonylurea losses. If the active ingredient stays unprotected on the soil surface, a relatively small portion of the herbicide may be lost due to ultraviolet degradation. Fate of some of the prominent herbicides of sulfonylurea class in soil, food grain and water is described hereafter.

Chlorsulfuron

Chlorsulfuron is one of the sulfonylurea herbicides, which affect amino acid synthesis and inhibit cell division in plants by inhibiting the enzyme acetolactate synthase (ALS). Chlorsulfuron is used to control broadleaf (dicot) weeds and some grassy weeds. It is used as pre- or post-emergence to crop for some annual grasses and broadleaf weeds in wheat, barley and oats. Chlorsulfuron degrade at faster rate in low pH soil rather than in high pH soil and showed high GR_{50} value in low pH soil as compared to high pH soil (Amarjeet *et al.*, 2003).

Imazosulfuron

Imazosulfuron is a new post-emergence sulfonylurea. It is highly active at low application rates to control annual and perennial broad-leaf weeds and sedges in rice. Imazosulfuron is having moderate persistence in soil. Sondhia *et al.*, (2006, 2007) evaluated imazosulfuron residues in transplanted rice field and reported 0.002, 0.006, 0.0075 and 0.010 µg/g residue of imazosulfuron in soil after 60 days at 30, 40, 50 and 60 g/ha application rates, respectively, however, no residues were found after 90 and 120 days. In rice grains 0.006, 0.009 µg/g of imazosulfuron residues were found at 50 and 60 g/ha application rates. Whereas 0.009 and 0.039 g/µg residues of imazosulfuron were detected at 50 and 60 g/ha application rates in rice straw. Imazosulfuron residues were not detected at 30 and 40 g/ha application rates in rice grains and straw, respectively (Sondhia *et al.*, 2007, 2008).

Metsulfuron-methyl

Metsulfuron-methyl is a residual sulfonylurea compound used as a selective pre- and post-emergence herbicide for control of broadleaf weeds and some annual grasses. It is a systemic herbicide with foliar and soil activity, and it works rapidly after it is taken up by the plant. It is used in wheat, barley, rye, and pastures. Metsulfuron-methyl is absorbed through the roots and foliage and moves rapidly through the plants. It inhibits cell division in the roots and shoots. Metsulfuron-methyl dissipated more rapidly in acidic silty loam soil as compared to high pH soil and light did not play any role in altering the persistence (Yadav *et al.*, 1995). Sushilkumar *et al.*, (2003) reported that metsulfuron- methyl residues were not detected after 60 days at 16 g/ha application rate, but at 20 - 24 g/ha, 0.002 and 0.011 ppm residues were found in back soils. No residues of metsulfuron-methyl were detected in soil at harvest applied at 2-8 g/ha in wheat and rice crop as post-emergence (Sondhia and Singhai, 2006, 2008).

Sulfosulfuron

Sulfosulfuron is used for the control of various annual and broad leave weeds in cereal crops. Sulfosulfuron provides very good control of *P. minor* and partial control of broad leaved weeds. Sulfosulfuron persisted upto 15 days in wheat soil at 25 g/ha rate and upto 30 days at 40-50 g/ha application rates (Sondhia, 2002b). Sulfosulfuron follows first order dissipation kinetics in soil and residues were not detected at harvest under wheat cropping system (Ramesh and Maheshwari, 2003). Sondhia *et al.,* (2006, 2007) reported dissipation of sulfosulfuron in the wheat with half life of 8.4 days and no residues of sulfosulfuron detected at harvest in the soil at 25 g/ha rate. Sulfosulfuron residues were not found in wheat grains, straw and subsequent vegetables in natural ecosystem as well as in model ecosystem at recommended rates (25 and 50 g/ha) in wheat crop (Ramesh and Maheshwari, 2003, Sondhia *et al.,* 2008).

Sondhia, (2007b) and Sondhia and Singhai, (2007, 2008) evaluated persistence of sulfosulfuron in water at 25, 50 and 100 g/ha rates and found 8.940, 9.00 and 10.150 ppm sulfosulfuron residues after one day which got down to 0.498, 0.588 and 1.140 ppm after 5th day in 25, 50 and 100 g/ha treatments respectively and showed initially fast dissipation of sulfosulfuron residues. After 5th day sulfosulfuron dissipate slowly and by 90, 120 days residues got down to 0.094, 0.108, 0.121 and 0.049, 0.062, 0.098 ppm, respectively in 25, 50 and 100 g/ha treatments. Sondhia (2007) reported half life of 17-25 days in surface and sub-surface soil at 25-100 g/ha doses.

Triasulfuron

Triasulfuron is a new post-emergence sulfonylurea. It is highly active at low application rates generally 3-40 g/ha to control annual and perennial broad-leaf and grassy weeds in cereal crops. Triasulfuron dissipate slowly in alkaline soil primarily by microbial degradation (Sarmah *et al.,* 1999). Singh and Kulshrestha (2003) reported persistence of triasulfuron at two rates of application *viz.,* 15 and 20 g/ha in soil under wheat field and laboratory conditions and found that dissipation of triasulfuron followed a first-order-rate kinetics indicated biphasic degradation with faster rate initially ($t_{1/2}$ = 3.7 days), followed by a slower dissipation rate at the end ($t_{1/2}$ = 9.4 days). Triasulfuron residues dissipated rapidly from field soil with half-life of 5.8 and 5.9 days at two rates of application.

13. Substituted urea

These herbicides are photosynthetic inhibitors, selective, soil-applied and absorbed through roots and translocate in the xylem. These are applied as pre or post-emergence and control many annual grasses and a few annual broadleaf species. Plants show chlorosis, drying of green tissue, yellowing of leaf tips and margins. Lower leaves may be affected first, and newer leaves may be unaffected. These herbicides exhibit very low to low water solubility, very low to moderate volatility, and low to high soil retention and display a wide range of behaviors in soils, depending on texture, moisture and organic matter. These herbicides have short to moderate persistence and are degraded in soil by microorganisms. Fate of some of the prominent herbicides of this class in soil, food grain and water is discussed hereafter.

Isoproturon

Isoproturon is a selective herbicide and acts principally after root absorption. It is applied either pre-emergence or post-emergence. It inhibits photosynthetic electron transport. It is effective against several broad-leaved weeds and numerous grassy weeds. It is safe to all varieties of winter wheat, winter barley and rye. Isoproturon is strongly absorbed by soil and shows negligible leaching. It appears to be degraded in soil mainly by microorganism and drops to non-phytotoxic level within a year under warm, moist soil conditions.

Continuous use of isoproturon at 0.94 kg/ha application rate in wheat crop did not leave any residues in soil, wheat grains and straw (Annual Report AICRP WC, 2000). Isoproturon persisted upto 50-56 days in the soil of maize crop. Whereas it degraded to non-detectable level within 60 days at 0.94 kg/ha rate (Walia *et al.*, 2000). Isoproturon applied at 1.0 kg/ha rate in wheat crop degraded completely at harvest in black soils of Jabalpur (Sondhia, 2002). Similarly, no residues of isoproturon in sandy loam soils of Ludhiana were detected at harvest at 0.5 to 1.0 kg/ha application rates (Randhawa and Sandhu, 1992, 1989). Isoproturon at 0.5 and 1.0 kg/ha application rates were evaluated at 1, 5 and 11 days after sowing of potatoes in sandy-loam soils. Isoproturon residue level was found 0.0213 ppm after 70 days and 0.0201 ppm after 120 days (Yaduraju *et al.*, 1993).

Sondhia and Singh (2005) found rapid dissipation of isoproturon in the soil of wheat crop and recorded 1.480 µg/g residues after 2 hrs which got down to 0.912, 0.113, 0.083 and 0.0005 µg/g at 30, 60 and 100 days respectively. Isoproturon dissipated almost 100 % after 120 days with half-life of 11.31 days. Isoproturon dissipates completely by 120 days in the soil of wheat crop applied at 1.0 kg/ha and residues were not detected in wheat grains and straw at harvest.

Metoxuron

Metoxuron is a selective and systemic post-emergence herbicide absorbed both by the leaves and roots. It acts by inhibiting photosynthesis. It is effective for the control of broad leaf weeds and grassy weeds in cereals and vegetable crops and also used for weed control in non-crop areas. Metoxuron residues applied at 1.0-1.5 kg/ha application rates as post-emergence in wheat were not found after 80 days in loamy sand soil (Randhawa and Sandhu, 1997).

14. Triazines

The inhibition of photosynthesis by the triazines herbicides is a well established fact. The blockage of light reactions of photosynthesis by the triazine herbicides results in a lack of reducing power, which is required for carbon dioxide fixation. Although the triazine herbicide inhibit CO_2 fixation in the light but that do not inhibit non-photosysnthetic CO_2 fixation (Zweig and Ashton, 1962, Sikka and Davis, 1969, Gogoi *et al.*, 2006). These herbicides move only in the xylem (upwards only). Injury symptoms occur after the cotyledons and first true leaves emerge. Injury symptoms include yellowing of the leaf margins or tips and yellowing between the leaf veins in broadleaf plants. Older and larger leaves are affected first. Injured leaf tissue eventually turn brown and die. There is greater injury on higher pH soils (greater than pH 7.2). These herbicides are prone to resistance when used injudiciously.

All chlorine-substituted triazines are more crop selective than the sulfur substituted triazines. The base strength of these herbicides ranges from very low to low. As a consequence, water solubility is dependent on soil pH. These herbicides range from low to moderate in solubility, very low to low in volatility, and low to high in soil retention. Soil retention is generally highest at low soil pH. Higher pH soils exhibit low herbicide binding and increase the potential for crop injury and leaching. Both soil organic matter and clay mineral content are important in binding these herbicides to soil. Soil persistence ranges from very short to moderate. Chemical decay is a major avenue of herbicide degradation, which is dependent on soil pH. Degradation generally occurs faster under acidic than under neutral or basic conditions. Sunlight plays a minor role in triazine losses. The active ingredient stay unprotected on the soil surface, a relatively small portion may be lost due to ultraviolet degradation. Persistence is also greater in soils with low organic matter, low clay content, and higher pH.

Several workers have shown that the rate of absorption and translocation of the triazine herbicides from roots to shoots is proportional to amount of water absorbed and/or the transpiration rate (Sheets, 1961, Vostral *et al.*, 1970). Fate of some of the prominent herbicides of triazine class in soil and food grain is discussed hereafter.

Atrazine

Atrazine is a selective systemic herbicide used for pre- and post-emergence control of annual broad-leaved weeds and grasses. It inhibits photosynthesis and interferes with other enzymatic processes. It is mainly absorbed through the plant roots, but can enter through the foliage, and accumulates in the apical meristems and leaves. Globally, atrazine is used in maize, sorghum, sugarcane, pineapples, grassland, macadamia nuts, conifers, and for industrial weed control, with its biggest market in maize production.

Adsorption of atrazine was more in medium black soil as compared to deep black and red soil. Atrazine adsorption is high at high clay and organic matter contents. At Bangalore atrazine was not found beyond 45 -60 days in 0-60 cm soil depth, and in sugarcane cropping system where atrazine was applied at 1.0 kg/ha continuously for over 10 years. Atrazine residues were not found in juice, jaggery and sugarcane plant on all sites of Coimbatore. However, atrazine residues ranged from 0.0007 to 0.0012 ppm in the soil samples at harvest. Atrazine residues were not found in post harvest soil of maize crop at 1.0 kg/ha rate but at 2.0 kg/ha rate, 0.032 ppm residues were found in the soil. Similarly no atrazine residues at 0.125 and 0.187 kg/ha application rates were detected at the time of harvest, which was applied to control *Phalaris minor* in wheat crop (Annual Report AICRP WC, 1999).

Atrazine degraded at faster rate in red soil followed by alluvial and black soils and persisted up to 63 days in the soil of maize field. Kulshrestha *et al.*, (1977) reported that more than 95 percent of atrazine dissipated from the field at the time of crop harvest. Pre-emergence application of atrazine at 1.5 kg/ha and simazine at 1.5 kg/ha application rates persisted up to 47 and 83 days respectively (Sharma and Angiras, 1997). Kausik and Moolani (1974) reported that atrazine (1.0 kg/ha) dissipation occurred as a first-order reaction in the greenhouse experiment and within 4 months, about 97 % of the herbicide dissipated from the soil in which maize plants were growing whereas about 83 % dissipated from un-cropped soil. Atrazine residues were not found in maize field at 0-15 cm depth when it was applied at 0.5 and 1.0 kg/ha but at higher doses *viz.* 1.5 and 2.0 kg/ha, 0.012 and 0.020 ppm of residues were found at harvest (Sondhia, 2001, 2002a). Atrazine residues in the 0-10 cm soil depth decreased with passage of time and after 90 days, no residues were detected at 1.0 and 1.25 kg/ha rates as pre-emergence and post-emergence to the sugarcane (Sandhu *et al.*, 1994).

Atrazine applied at 1.0 kg/ha rate in maize crop degraded completely by harvest and hence residues were not found in maize grains but at 2.0 kg/ha rate, 0.088 ppm of residues were detected (Sondhia and Saraswat, 2000a,b). 0.050 ppm of atrazine residues were found in sorghum grains, at 1.0 kg/ha rate where as atrazine residues were below the MRL value (0.25 mg/kg) at 0.5 and 1.0 kg/ha application rates in sorghum grains (Annual Report AICRP WC, 2000).

Metribuzin

Metribuzin is a selective triazinone herbicide that inhibits photosynthesis. It is absorbed by the roots and leaves with acropetal translocation in the xylem. It is used for the control of annual grasses and numerous broadleaf weeds in field and vegetable crops, in turfgrass, and on fallow lands. Metribuzin residues were not found in the soil, grains and straw following application at 210 and 420 g/ha in wheat crop (Dubey *et al.*, 1998). In a study carried out in soybean wheat cropping sequence in *kharif* it was observed that dissipation of metribuzin (at 0.50 or 0.75 kg/ha) was more pronounced during the first 15 days and after 45 days, rate of dissipation ranged from 49 to 52 %, suggesting the high level of persistence during first 30 to 45 days of crop growth (Kewat and Pandey, 2002). Sondhia (2002a) reported that metribuzin persisted upto harvest in black soil at 0.85 and 1.20 kg/ha. *Pseudomonas straita* can degrade metribuzin up to 86 % within 20 day of herbicides application (Gopal *et al.*, 2005).

15. Thiocarbamates

These growth inhibitors are pre-plant incorporated in the soil. The thiocarbamate herbicides inhibit lipid synthesis, but the exact site and mode of action is unclear. These herbicides may have multiple sites of action.

Thiocarbamates affect shoot development. Plants have twisted shoots and a compact appearance. Thiocarbamates primarily kill germinating seedlings, not established plants. These herbicides exhibit low to moderate water solubility, low to moderate soil retention, and high to very high volatility. Major atmospheric losses could occur if thiocarbamates are not immediately incorporated. As with the dinitroanilines, the rate of application of these herbicides is highly dependent on soil texture and organic matter content. Higher clay and organic matter content means a higher rate of application, because more herbicide is retained on soil surfaces. In addition, soil texture, temperature and moisture influence volatilization. Higher temperature and soil moisture speed volatilization, as does coarser (sandier) soil. Major losses can occur from volatilization. The prominant means of dissipation is microbial decomposition. Fate of some of the thiocarbamate herbicides in soil and food grain is discussed hereafter.

Thiobencarb

Thiobencarb was developed by Kumiai Chem. Ind. Co. Ltd. and introduced in the market in 1970 (Ishikawa, *et al.*, 1976). Thiobencarb is a systemic, pre-emergent herbicide that acts by inhibiting shoots of emerging seedlings. The expected site of activity for thiocarbamate is in the aboveground meristematic tissue (Ashton *et al.*, 1981). Thiobencarb absorbs by roots and shoots of grass seedlings, and then translocates upward in the apoplast to locations where it inhibits cell division and enlargement (Jordan and Codney, 1987). It is being recommended in rice crop. Thiobencarb is rapidly adsorbed by soil and not readily leached with water. Adsorption onto soil minimizes its loss from volatilization and photodecomposition and mostly decomposed by microorganism. Thiobencarb persistence in the soil varies 20-80 days under tropical conditions.

Kumar *et al.*, (1993) reported that thiobencarb persistence (at 1.5, 3.0 and 4.5 kg/ha application rates) in soil decreased with time and showed direct correlation between temperature and concentration. Lower temperature and higher concentration resulted in greater persistence. Thiobencarb persist upto harvest and follow 1.5 order kinetics with half-life of 24 days at 1.25 kg/ha application rate to transplanted rice (Jayakumar and Ramulu, 1993). Residues of thiobencarb were found below the MRL value in soil and plant parts at recommended dose (Balasubramanian *et al.*, 1999).

16. *Unclassified*

There are a number of herbicides which do not fit into any of the class of herbicides because their mode of action is not known or that is uncertain and are categorized as unclassified such as anilofos an organophosphorous herbicide. Fate of some unclassified herbicides in soil and food grain is discussed hereafter.

Anilofos

Anilofos is a pre-emergence and early post-emergence selective herbicide used for the control of annual grasses, sedges and some broad-leaved weeds in transplanted and direct seeded rice. Anilofos is absorbed by the weeds primarily through the roots and to some extent through newly emerging shoots and young leaves. Anilofos acts on meristematic tissues of weeds and severely affecting cell division and cell elongation.

Anilofos residues did not persist in red and black soil and accumulation of anilofos residues was more in heavy textured soil as compared to medium and light textured soil. Anilofos can persist upto 20 days in sandy and clay soils. Anilofos dissipate rapidly in organic and sandy loam soil. Anilofos applied in puddle broadcast rice did not leave any toxic residues in post harvest soil, at Faizabad. Anilofos residues were not detected or found far below maximum residue limits at Coimbatore in rice crop. Anilofos degraded within 56 days at 0.2 to 0.4 kg/ha application rates in clay loam soil during the winter, summer and *kharif* [monsoon] seasons in rice crop. Anilofos residues were detected below the prescribed maximum residue limits in rice grains and straw at harvest at 0.8 kg/ha (Jayakumar and Sankaran, 1995). Anilofos at 0.4 kg/ha application rate with 2, 4-D and without 2, 4-D in *rabi* crop (herbicides rotation) showed a slight build up of residue in soil (0.0025 ppm) after 8th crop as compared to that of 1st crop at Coimbatore. The pre-emergence application of anilofos applied at recommended doses continuously for four seasons in rice crop showed residues in soil, rice grains and plant parts below the maximum allowable level (Balasubramanian *et al.*, 1999). Anilofos at 0.4 kg/ha application rate persisted upto 56 days in direct seeded rice field (Radhamani *et al.*, 1997). Rai *et al.*, (2000) evaluated persistence of anilofos in six tropical soils of varying physico-chemical characteristics under flooded and non-flooded conditions and found rapid degradation (40-61 %) after 30 days of incubation under flooded than non-flooded conditions.

Herbicides residues in water

Herbicide residues can reach water by several routes such as direct application for weed control, spray drift, runoff from treated fields and leaching into ground water. Only a small number of herbicides are registered for controlling aquatic weeds. Initial concentration of herbicide applied to floating and submerged weeds are typically of the order of 1 ml/litre. Most of the herbicides are applied as spray but a few are applied as granules such as dichlobencil. After application the herbicides partition between the water, sediment and weeds, following the same principles as for the soil, the major difference is the much higher ratio of water to solid phase. Consequently a much higher proportion of the residues is located in the water. For example if a soil contains 50% water a herbicide has an adsorption coefficient of 1, two-third of the herbicide will be adsorbed. In a pond 1 metre deep, assuming the herbicide is equilibrate with a layer of sediment 1 mm deep, only 0.1 % would be adsorbed. Some herbicides such as diquat are rapidly adsorbed by weeds and suspended sediments and the concentration in water falls to <0.02 mg/l in <10 days.

Degradation processes in the aquatic environment are the same as in soil except photochemical degradation of residues on plant surface and the water is more important for herbicides which are decomposed by sunlight. The label on aquatic herbicides indicates which are decomposed by sunlight. The label on aquatic herbicides indicates the safety interval before treated water can be used for irrigation crops. The range may be from nil up to 4 to 5 weeks for residual herbicides such as dichlobenil.

Traces of herbicides can reach surface waters in runoff, from spray drift and sub surface land drainage. In UK, for most herbicides < 0.5 % of the amount applied is carried off the field in run-off when severe runoff event occurs within 1 or 2 weeks of application, this can be upto 5% particularly for wettable powder formulations. The maximum concentration of residues in runoff is highly variable but is typically in the range of 0.001 to 1 mg/litre. Most herbicides are lost mainly in the water phase, even when concentrations are 2-3 orders of higher magnitude in the sediment than the water, because of the large ratio of water to sediment. The traces of residues which reach surface waters are subjected to the same adsorption and degradation processes as aquatic herbicides, although the relative importance of the different mechanism and the rates are not fully understood.

Dilutions also play a major part in reducing the concentration. The recent development of analytical methods sensitive to residues down to 0.0001 mg/litre or lower has made it possible to monitor residues in water. For example, monitoring of surface water in East Anglia, where herbicides are widely used, has shown residues of most herbicides to be <0.001 ml/litre and in most cases to be non detectable (<0.0001 mg/litre). A small number of water samples contained traces of herbicides such as mecoprop, atrazine, and simazine but always <0.001 mg/litre. It is suspected that the trace of atrazine and simazine are due to use on industrial sites, railways, and road edges rather than from agricultural use. The toxicological properties of herbicides are investigated in great detail to establish their safety to users and consumers of treated crops.

A few studies conducted at AICRP weed control centres to evaluate herbicide residues in ground water, drainage water and canal water showed that butachlor residues ranged between 0.001 to 0.093 ppm in the water of rice field. Residues of paraquat were not detected after 20 days at 0.80 kg/ha application rate to control *Eichornia* but application of 1.8 kg/ha showed 0.069 and 0.028 ppm residues in pond and canal water. Increase in pH, electrical conductivity, carbonates and free CO_2 were noticed after treatment with 1.0 and 2.0 ppm of 2, 4-D but the dissolved oxygen decreased and the 2, 4 -D residues become non-detectable after 42 days in glass tank under laboratory conditions at Pantnagar.

Butachlor residues were not detected in underground water and drainage water near the paddy grown area at Coimbatore. Isoproturon residues were not found in the ground water in all the water samples collected from different districts of Hissar where isoproturon in rice-wheat cropping system was applied at recommended dose of 0.94 kg/ha for many years. Sushilkumar *et al.*, (2003) evaluated 2, 4-D at 1.5, 2.0, 2.5 kg/ha, glyphosate at 2.0, 2.5, 3.0 kg/ha and metsulfuron- methyl at 0.016, 0.020, 0.024 kg/ha rates against alligator weed (*Alternanthera philoxeroides*) in non-crop area and reported that 2, 4-D and glyphosate were not detected in water after 45 and 60 days, however, 0.024 ppm metsulfuron-methyl residues were detected after 60 days.

Devi (2004) evaluated degradation of paraquat and 2, 4-D in the water and reported that even though paraquat and 2, 4-D dissipated initially at a faster rate but at higher application of 2, 4-D

(2.0 and 4.0 kg/ha) a waiting period of more than 4 months should be insisted for safe consumption of fish. Similarly paraquat residues in the fish samples were also detected below the acceptable daily intake of 0.002 mg per kg body weight. It is further reported that only 0.80 to 1.11 % of the applied paraquat remained in the sediment fraction. Paraquat at 0.8, 1.6 and 3.2 kg/ha application rates increased the pH and electrical conductivity of water.

Sondhia (2008a,b) evaluated persistence of sulfosulfuron in water applied at 25-100 g/ha under laboratory conditions and reported faster dissipiation of sulfosulfuron upto 5 days only and later on dissipiation was slow, and persisted upto 120 days in water.

Herbicide leaching into soil is subjected to both adsorption and degradation which minimizes the potential for leaching into the groundwater. Extensives field studies have shown that maximum concentration of residues normally remain in the top soil. Nevertheless, traces of herbicides might be leached and persist when a compound is applied in the spring. Small amount of residues might also be washed down into the channel and fissures in the subsoil and underlying rocks. Monitoring of groundwater has shown, that in most cases residues are lower in surface water as compared to ground water.

Herbicide residues management

There has been lot of concern on the effect of herbicides. Herbicide enters the environment in the form of vapours and drifts from the application to the soil and plants. Herbicides in the soil are subjected to various processes on the environment such as leaching to lower layers of the soil or surface runoff with irrigation or rain water to stream, lakes etc. causing water pollution. Herbicides with vapour pressure greater than 1x 10^{-5} mm Hg at room temperature are volatile. These move in soil in vapour form and dissipate slowly into atmosphere. Certain herbicides may be subjected to photochemical decomposition and/or volatile loss. All herbicides from soil surface can be eroded by wind or sudden erosive storm.

On agriculture land a major problem arising from herbicide persistence is injury to following crops. With triazine herbicides, rotation crop may be injured even 6 to 24 months after their application at normal rates. These herbicides are therefore called persistent herbicides. Several substituted ureas, dichlobencil and diphenamid also often pose phytotoxic residues problems on cropland. Rotation corps like cereals, soybeans, sugarbeet and tobacco have been reported injured (Crafts and Drever, 1960; Smith, 1969; Burnside, 1965). Sometimes even non-phytotoxic residues of a previously applied herbicide may damage the rotation crop by interacting with the herbicide applied to the present crop such as injury to lucerne from EPTC is noted in fields treated with atrazine in previous season. Likewise triazine will injure wheat sown after maize that was treated with simazine or atrazine.

High level of chemicals remaining at the site into the following growing season generally refered as carryover are called herbicide residues. Such residues can result in injury to subsequent crops planted on the site that are not tolerant of the herbicides. Carryover can occur over an entire field or more likely in spots throughout the field and can result in stunted or malformed plants or lack of seedling emergence. Growers should be aware of herbicide soil residual properties, crop tolerance to particular herbicide and the proper application procedure in designing their herbicide and crop rotation program to avoid carryover problems.

Most residue problems can be associated with the use of long lived soil-residual herbicides. Factors contributing to carryover involve misapplication, adverse environmental conditions that reduce herbicide breakdown in the soil (cold temperature, moisture variations) and ignoring replant restriction for sensitive crops.

Herbicides drift may be eliminated largely by using proper nozzle, spraying techniques, proper time of application and correct formulation. Volatility is the characteristic of few herbicide classes such as anilines, carbamates etc. and this can be minimized to certain extent by proper soil incorporation and management techniques: The easiest way to minimize the residue problem is lower the application rate of herbicides but this may not always be possible because of inadequate weed control. Although many herbicides do not pose residues problem in the soil, it is important to know the corrective measures that need to be taken to meet any eventuality. This means applying the lowest rate of the chemical consistent with obtaining the desired effect. To accomplish this goal, accurate acreage determination, accurate chemical measurement, proper sprayer calibration, and uniform application are essential. Information available as the packing should invariably be read and followed before applications. The method and time of application can be important in avoiding herbicide carryover. Herbicides that have the potential to persist longer than desired will more likely remain longer if incorporated than if surface-applied without incorporation. Incorporating the herbicide makes it less susceptible to loss by volatilization and photodecomposition. In addition, an incorporated herbicide is immediately exposed to soil particles and may be tied up temporarily through adsorption and later released. Decreased environmental losses (volatilization and photodecomposition) and increased adsorption favor herbicide carryover. Band herbicide applications can reduce carryover potential because less total herbicide is applied than in a broadcast application. Post-emergence and late soil applications have greater potential than earlier applications for being present the following season. Herbicide persistence in soil may injure succeeding crop. For example, injury to pea from sulfosulfuron is noted in field treated with sulfosulfuron in the previous year (Sondhia, 2006; Sondhia and Singhai, 2008).

Herbicides may interact with one another or with other pesticides and may enhance crop injury when applied in the same year or in consecutive years. For example, a soybean crop may tolerate a certain level of atrazine carryover. However, if another photosynthetic inhibitor, such as metribuzin, is applied to soybeans after atrazine-treated corn, injury is more likely. The amount of tillage affects herbicide persistence. At the recommended dose of herbicide application, the problem may not arise and will kill selectively the weeds. But when the dose is more than the recommend rates that happens due to indiscriminative use and improper calibration and method of application, there is possibility of residual hazards in soil particularly in persistent herbicides such as triazines and uracils. Repeated application of the same herbicides in a mono crop sequence may cause accumulation of residues in soil, which in turn will affect the sensitive crops. Leaching of herbicides can cause crop injury due to transport of herbicide into the absorption zone of susceptible crop plants and accumulation of herbicides in toxic level in tolerant crop plants.

Management of herbicide residues in soil

Soil applied herbicides should have sufficiently longer persistence to give better weed control but its persistence in the soil should not affect the growth of succeeding crops. Various management techniques have been developed, which can help to minimize the herbicides residue in soil. Basically there are two approaches to reduce herbicide residues in soils. The first one is indirect one. It aims at reducing the dose of the herbicide without losing on its weed control effects. This can be achieved by:

(i) the use of activators and surfactants to enhance the herbicidal activity against weeds,

(ii) selection of the most appropriate time and method of application of the herbicide

(iii) selective stimulation of crops to smother weeds and

(iv) supplementing herbicidal control with tillage and other methods of weed control.

The second approach to reduce herbicide residues in soil is by direct action is as below:

(i) Structural approach

At equal rates, volatiles, photo-decomposable and biodegradable herbicide structures are less persistent than other types of structures. In case of volatile herbicides, their dilution in the air is sought as solution to their long persistence in soil but these may contribute to air pollution. Photo-decomposable herbicides, however, do not contribute to air. Further biodegradation of herbicides is a more reliable process of dissipation of herbicides from soil than their chemical degradation. Thus developing volatile, photo-decomposable, and biodegradable soil active herbicide prove helpful in lowering herbicide residues in soils.

(ii) Enhanced degradation approach

Provision of adequate soil moisture and, addition of easily decomposable organic matter enhances microbial activity in soils, leading to rapid biodegradation of residual herbicides. The amount of tillage affects herbicide persistence. Tillage encourages herbicide decomposition indirectly through increased microbial and chemical breakdown. Minimum-till and no-till, which leave crop residue on the soil surface, also tend to leave a greater concentration of herbicide near the surface zone. Persistent herbicides present in this concentrated zone may affect susceptible crops. In addition, higher rates of herbicides are often used in reduced-tillage systems to maximize weed control and adjust for greater amounts of crop residues. If a herbicide-carryover problem already exists, some tillage to dilute the chemical may help. Repeated deep tillage of the treated soil will hasten both volatile loss and degradation of herbicides and it may also dilute the herbicide by mixing it with a larger volume of soil to innocuous levels.

Triazines can be degraded faster when the soil pH is lowered by the application of sulphuric acid but this may not be permissible under all conditions. Further band application of herbicides rather than broadcast will reduce the rate of herbicides application. This can be performed in line-sown crops raised along ridges *viz.*, maize, cotton, sugarcane, sorghum etc. Addition of farmyard manure is also an effective way to minimize the residues in soil. Applications of FYM not only increase the microbial population but also enhance the herbicides degradation at faster rate. A long-term study on the effect of soil organic matter on herbicide persistence in rice-wheat cropping system at Hisar, revealed that isoproturon, 2, 4-D and butachlor disappeared more rapidly in no manuring soil than green manuring soil. Sodium and calcium polysulphides have been found effective in deactivating simazine and atrazine in laboratory by replacing chlorine radicals of herbicides ions. A 50 t/ha of muck and peat added to soil can also eliminate triazines residues from soil.

Plants absorb herbicides from the soil in which they are growing. Persistence may be reduced if the herbicide is metabolized (broken down) by the plant or if the plant containing the absorbed herbicide is harvested and removed from the field. Plant extraction of the herbicide from the soil may not be an important factor under most situations, but it has been used in some cases to help remove persistent herbicides from treated soils.

Planting cover crops increase soil organic matter and biological activity. These two factors promote herbicide degradation. High nitrogen levels and soil moisture increases soil biological activity and further accelerate herbicide degradation.

Presence of some herbicides (pendimethalin, fluchloralin etc.) affects germination of susceptible crops. If cultivation of such crops is inevitable, then transplanting seedlings raised on healthy soil would be a good alternative. Ploughing with disc plough or inter-cultivations also reduce the herbicidal toxicity (Fig. 2). Dinakar *et al.*, (1986) reported that ploughing could reduce damage caused by herbicides. Crop rotation is also effective to minimize the residues in soil. Vasmani and Vijyakumar (1978) also reported no injury to oats and wheat grown in rotation with cotton, which received diuron at 1.0 kg/ha.

If herbicide carryover is suspected, a soil chemical test or biological assay can help determine if harmful levels of herbicide residue are present. Chemical analysis can be expensive, so a bioassay may be more feasible. Either can help determine if herbicide residues exist and if a tolerant crop or variety should be planted into a problem area.

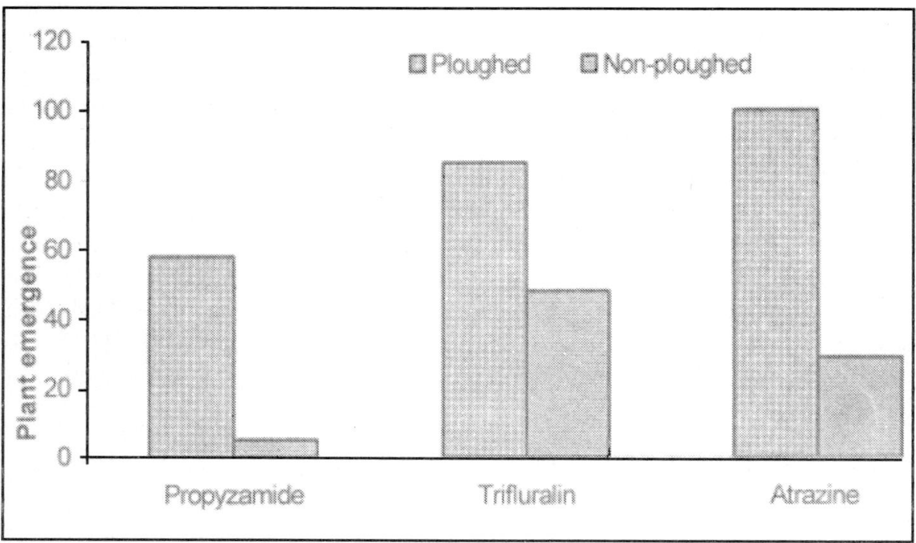

Fig. 2: Effect of ploughing on plant emergence.

(iii) Toxicity prevention approach

Adsorbents and antidotes that are used for preventing injury from direct application of pre-plant and pre-emergence herbicides can also be used to protect crops from injury of many herbicides. Organic matter is a good source by which toxicity could be prevented of any herbicides in soils. Adding organic matter to sandy soil has been reported to reduce herbicidal activity of fluometuron and fenuron in growth chamber studies as well as in field experiments (Weber *et al.,* 1972). Enhancement in metabolism of herbicides absorbed in the crop plants can also be an approach to reduce the herbicidal toxicity. Benzyladenine has been found to enhance the metabolism of amitrole in plants. Addition of phosphorous to soil reduced the phytotoxicity of diuron to cotton.

Activated charcoal has long been used as a strong adsorbent of herbicides like 2, 4-D and EPTC. Activated carbon is very effective in inactivating residues of non-ionic herbicides with low water solubility. Most herbicides fall into this category (Weber, 1980). Activated carbon used is of high value and replacing another crop is not an option (example garden, landscapes, golf courses and other high value sites). Activated carbon is thoroughly mixed into soil by disking or rototilling to deactivate the herbicides residues. In some cases activated carbon slurry is used to protect transplant roots or seed from a toxic herbicide such as root dips of strawberry transplant roots to protect against simazine. Activated slurry has been used to coat seed or have been sprayed as a band over seeded row for protection against diuron. It can be applied as a band over crop row or by coating it on the crop seeds with the help of thick gum. In case of transplanted vegetables like tomato and brinjal, the seedling roots could be dipped in slurry made of activated charcoal. Singh and Sarah (1979) reported that atrazine residual toxicity can be reduced by mixing it with non-toxic oil, which also enhances the weed control efficiency. Use of activated carbon can also reduce the herbicidal toxicity because activated carbon has a high adsorptive capacity because of its tremendous surface area, which vary from 600-1200 m^2/g. Jayakumar *et al.*, (1990) found that an application of charcoal at 5.0 kg/ha along the seed line reduced the residual toxicity of atrazine in soybean crop.

Eleftherohorinos (1987) reported that the incorporation of 50 kg/ha of activated charcoal inactivated completely chlorsulfuron applied at 1.25 and 2.50 kg/ha and did not affect the yield of maize compared to untreated control. The use of safeners and antidotes along with herbicides protect crop from possible damage caused by the herbicides. Incorporation of FYM at 10 t/ha or neemcake at 3 t/ha can mitigate pendimethalin (1.0 kg/ha), fluchloralin (1.0 kg/h), trifluralin (1.0 kg/ha) and oxadiazon (0.50 kg/ha) residues applied to preceding crop of cumin at Anand.

(iv) Tolerant crops cultivars

Prior knowledge about the susceptibility of crops (to be grown in rotation) to a recommended herbicide applied to the preceding crop will be handy in choosing the right crop, depending on the residue level of the herbicide in the soil at the time of harvest. Quite often, economics dictates crop rotation; however, there are varietal differences that might affect the likelihood of serious crop injury. For example, some soybean varieties are more sensitive to the triazine herbicides than others and should not be used if the potential for triazine injury exists. For instance growing chickpea or pea would be a better proposition than wheat, barley or mustard in a field contaminated with triazines herbicides. Varshney (2005) however observed high level of toxicity of atrazine in following chickpea crop. Certain herbicide tolerant crops can reduce herbicide residues in soil by absorbing these and deactivating them in their tissues. Sulfosulfuron applied at 25, 50 and 100 g/ha in wheat crop did not show any adverse effect on succeeding maize and sorghum crop however significantly affected the growth of lentil and pea (Sondhia and Singhai, 2007). Maize, for instances is a good candidate for decontaminating a field containing triazine herbicides, similarly pea and lentil can be used to detect sulfosulfuron residues in soil (Sondhia, 2007). Also, as a general rule, smaller seeded crops and varieties have a greater potential for injury from persistent herbicides than do larger seeded species.

(v) Controlled mobility approach

Mobility of herbicides in soils can be manipulated to dissipate the herbicides by leaching and volatilization. Heavy irrigation may leach a mobile herbicide beyond the root zone of rotation crops. In the case of EPTC and carbamates, irrigation can induce dissipation by volatilization. Repeated deep tillage of the treated soil will hasten both volatile loss and degradation of herbicides and it may also dilute the herbicide by mixing it with a larger volume of soil to innocuous levels.

In India intensive cropping is practiced. The herbicide residues in the soil from one crop may adversely affect the succeeding crops in a cropping sequence. Very little research has been made in this regard, hence intensive studies on herbicides residues management in different cropping system have to be developed. The effective amendments such as FYM, compost, activated charcoal and antidotes may be well utilized for residue management and the dose, time and method of application of such amendments and antidotes have to be standardized. Potentialities on the use of bio-agent which act on residues and help in enhanced degradation of herbicides may be explored.

Summary

Herbicide enters the environment in the form of vapours and drifts from the application to the soil and plants. Herbicides in the soil are subjected to various processes on the environment such as leaching to lower layers of the soil or surface runoff with irrigation or rain water to stream, lakes etc. causing water pollution. Herbicides drift may be eliminated largely by using proper nozzle, spraying techniques, proper time of application and correct formulation. Volatility can be minimized to certain extent by proper soil incorporation and management techniques. On agriculture land a major probloem arising from herbicide persistence is injury to following crops. Sometimes even non-phytotoxic residues of a previously applied herbicides may damage the rotation crop by interacting with the herbicides applied to the present crops such as injury to lucerne from EPTC is noted in fields treated with atrazine in previous season. The easiest way to minimize the residue problem is to lower the application rate of herbicides but this may not always be possible because of inadequate weed control. This means applying the lowest rate of the chemical consistent with obtaining the desired effect. To accomplish this goal, accurate acreage determination, accurate chemical measurement, proper sprayer calibration, and uniform application are essential.

□□□

Chapter 11

Effect on Soil Health Succeeding Crops and non-target organisms

Herbicides are designed to control unwanted vegetation but considered to have the greatest potential for causing phylotoxicity if not properly used. Inert ingredients in herbicide formulations may also cause phytotoxicity. Herbicides vary in their potential to persist in soil. Herbicides that can persist to the next season may injure subsequent crops and require close monitoring. It is important to distinguish between the terms persistence and bioavailability. A herbicide may persist in the soil and may be detectable by chemical extraction; however, this does not necessarily mean that it is available to cause injury to sensitive crops. The herbicide paraquat has a relatively long half-life in the soil but is strongly adsorbed to soil colloids making it unavailable. A herbicide must be present in the soil solution in order for it to cause injury to sensitive rotational crops.

The potential risk to a succeeding crop from herbicide residues depends on the persistence of the herbicide, the susceptibility of the following crop to the residues and the position of the residues in the soil in relation to the seed and root system. Residues tend to be near the surface thus the risk of damage is greater when seeds are sown without ploughing into the herbicide residues. Ploughing greatly reduces the risk of damage because it removes most of the herbicides from the vicinity of the seeds and young roots and also dilutes the residues. The persistence of a herbicide in soil depends on its inherent stability, the physical and chemical properties of the soil, moisture content and the temperature.

Environment has a major effect on the persistence of herbicides. Soil moisture and temperature are very important factors in determining the rate of herbicide breakdown. There are a number of mechanisms which determine the environmental fate of a herbicide such as volatilization, photolysis (breakdown by sunlight), leaching and others. However, the two primary mechanisms of herbicide degradation are microbial and chemical hydrolysis. Both these processes are dependent on soil water and temperature; however, soil moisture is more critical with herbicides that require microbes to degrade (Streck, 2005). Soil microbes thrive in warm, moist soils, which result in faster degradation. It was estimated that there is two to three-fold increase in chemical half-life with a 10°C decrease in temperature and a one and one-half to two-fold increase in chemical half-life if soil moisture content is reduced by a factor of two (Walker, 1987).

Herbicides differ in their persistence under different environmental conditions. Soil less in organic matter content has more herbicides available and cause more persistence. Herbicides are more strongly adsorbed in acidic soil than alkaline soil. Dinitroanilines herbicides applied with irrigation water or as pre-plant in sandy loam soil having low organic carbon showed longer persistence. An increase in organic carbon content is reported due to application of FYM over a period of time in rice-rice cropping system. Different application rates of atrazine at 0.5, 1.0, 2.0, and 4.0 kg/ha did not affect the yield of succeeding crop of wheat, oat, barley and gram by using soil amendments such as FYM at 2.5 t/ha, oil cakes at 25 kg/ha and seed coating by activated charcoal. Similarly, study conducted at Ludhiana showed that there was no build up of isoproturon residues in the soil when applied continuously for four years in rice-wheat crop rotation. The persistence of isoproturon was more in low pH soils as compared to high pH soils (Annual Report AICRP WC, 2002).

Palaniappan and Balasubramanian (1985) evaluated invertase [beta-fructofuranosidase], dehydrogenase, amylase and urease activity in the laboratory and reported that fluchloralin inhibited the activity of all enzymes at 5 and 10 ppm, but at the recommended rate fluchloralin had no significant effect. Kumaran and Devi *et al.*, (1996) studied the behaviour of atrazine in Devanur (fine montmorillonitic isohyperthermic Typic Haplustert), Vijayapura (fine kaolinitic isohyperthermic Typic Kandiustalf) and Arikunte (loamy skeletal isohyperthermic Typic Haplustalf) soil series of Karnataka and reported that the addition of organic amendments increased atrazine adsorption but decreased its desorption and concluded that the behaviour of atrazine was influenced not only by the addition of organic matter but also the nature of organic colloidal fractions. Study conducted at Bangalore clearly indicated that the weed management practices or sources of fertility did not change the soil physico-chemical properties significantly in fingermillet-groundnut cropping system (Annual Report AICRP WC, 2000) (Table 1).

Table 1. Influence of herbicides on physico-chemical properties of soil under fingermillet-ground nut cropping system

Treatment	Soil properties			
	pH	EC (dsm^{-1})	OC (%)	BD (g/cc)
Metolachlor without FYM	6.20	0.02	0.40	1.46
Metolachlor with FYM@ 10 t /ha	6.00	0.03	0.38	1.43
Pendimethalin without FYM	6.00	0.04	0.30	1.46
Pendimethalin with FYM @10 t /ha	6.20	0.03	0.40	1.49
Hand weeding without FYM	6.10	0.04	0.30	1.50
Hand weeding with FYM @ 10 t /ha	6.10	0.04	0.45	1.40
Initial soil sample	6.10	0.03	0.30	1.60

(*Source: Annual Report, 2001-2002, AICRP Weed Control*)

Soil properties such as organic matter content, soil texture, and soil pH also play an important role in the carryover potential of residual herbicides. Generally, there are fewer problems with herbicide carryover when soil organic matter is high. It is the most important variable in controlling adsorption and the distribution of the herbicide between solid and solution phases (Walker, 1987). Herbicide adsorption to organic matter may reduce its bioavailability and the moisture holding capacity of high organic matter soils makes them conducive for increased microbial activity. The effect of clay content on herbicide residues is similar to organic matter in that it tends to adsorb the herbicide as well as improve water-holding capacity. Soil pH is another factor affecting the residual characteristics of some herbicides. A low soil pH (less than 7.0) tends to increase the persistence of imidazolinones herbicides such as imazethapyr, imazamethabenz, and imazamox. Imidazolinone herbicides tend to be more adsorbed under acidic or low soil pH causing reduction in their availability for microbial degradation (Loux and Reese, 1992). Extended carryover of imidazolinones herbicides in acidic soils may also be related to their sorption-desorption characteristics. A higher level of imazethapyr was sorbed at low soil pH than at high pH, but it was also more readily desorbed at a low soil pH. Less imazethapyr was sorbed initially at high soil pH, but it did not readily desorb. Sulfonylurea herbicides react in the opposite manner to the imidazolinones. Sultonylurea herbicides are broken down primarily by chemical hydrolysis and this process occurs much quicker under acidic conditions. Therefore, tend to degrade faster when soil pH is low.

The use of residual herbicides prior to a period of drought may influence sowing decisions for the next crop due to reduced herbicide breakdown. Herbicides of concern include the group B and C herbicides including sulfonylureas (SU), and triazines. Herbicide carryover is of particular concern in alkaline soils. Whenever herbicides are used in an intensive and diversified crop rotation, the herbicides applied to one crop should not persist in soil at concentration high enough to affect the succeeding sensitive crops. Studies showed that in most of cases no residues or very minute quantity of herbicides were left in the soil and did not affect the subsequent sensitive crops.

It is possible to assess the relative persistence of herbicides for a particular soil by determining their half-lives under controlled laboratory conditions of temperature and moisture. However, even under such controlled condition, persistence can vary widely between soils. For example, in a study of the persistence of simazine in 15 soils from eight countries, the half-life under similar conditions was found to vary by a factor of six. The variability was reduced by excluding soils from countries with a very different climate from UK but was still a factor of nearly three. In the field conditions rate of degradation vary greatly through out the season which depends on moisture and temperature conditions. Summer rates of degradation are generally around three to four times faster than the winter but are much less than this when the surface soil dries out during extended dry periods so there is also variability between seasons. Because of all these variables it is impossible to give a precise forecast for the persistence of active residues of a herbicide.

The susceptibility to residues of simazine does not vary greatly between crops. Recently introduced herbicides have tended to be increasingly selective. Residues of propyzamide, which is widely used on oilseed rape are safe to many crops but can seriously damage winter cereals unless sown after ploughing, residues of isoxaben applied to winter cereals are safe to most crops but are damaging to brassicaas and can cause complete crop failure of oilseed rape sown without ploughing. The sulfonylureas are unlike other herbicides previously in use. Although only moderate persistence in acid soils, some can be highly persistent in alkaline soils (Sondhia, 2007b). They have both foliar and soil activity and unlike most other soil acting herbicides they are only slightly adsorbed. They are mobile in the soil and do not remain so near the surface as most other compounds so the risk of damage may not be minimized by ploughling. Some crop notably onions and beet are extremely sensitive to residues of sulphonylureas and can be serioully damaged by <1 % of the recommended dose (Rillay and Eagle, 1990).

Residual effect of atrazine, butachlor, pendimethalin, alachlor and oxyfluorfen applied at recommended level were observed on succeeding crops like chickpea, green gram, sunflower and cucumber at Hyderabad. Application of atrazine at 1.0 kg/ha application rates as pre-emergence to maize followed by 0.25 kg/ha application rate as simazine pre-emergence to potato (in rotation with maize) did not affect the germination and seedling growth of oats and cowpea at Ludhiana (Annual Report AICRP WC, 2000). Singh and Mani (1981) found that atrazine and alachlor in maize–pulse rotation did not affect the grain production of rabi pulses by the herbicides applied to maize crop continuously for three years.

Application of fluchloralin (1.0 kg/ha), oxyfluorfen (1.0 kg/ha) and butachlor (2.0 kg/ha) in soybean and isoproturon and oxadiazon at both 1.0 kg/ha in wheat did not adversily influence the productivity of the system following continuous use for five years at Jabalpur. Effect of various herbicides on succeeding crop is summarized in Table 2.

Table 2. Effect of herbicides residues on the succeeding crop

Crop	Herbicide	Dose (Kg/ha)	Toxic	Nontoxic	Reference
Wheat	Pendimethalin	1-1.5	-	Sorghum	Kulshrestha and Yaduraju, (1987)
Wheat	Sulfosulfuron	0.25-100	Pea, lentil	Sorghum, maize	Sondhia (2007b, 2008)
Sorghum	Atrazine	0.25-1.0	-	Fingermillet Cotton	Jayakumar *et al.*, (1988) Latchanna (1987)
Cotton	Fluchloralin Oxadiazon Oxyfluorfen Pendimethalin	1.0 0.5-1.5 0.1-0.2 1.25	Cucu mber	Fingermillet Foxtail millet Mungbean	Jayakumar *et al.* (1988)
Groundnut	Imazethapyr	0.75	-	-	Selvamani and Sankaran (1989)
		>0.75	Cotto n Sorgh um	-	
	Pendimethalin Oxyfluorfen Fluchloralin	1.0 0.15 1.0	-	Cotton Sorghum	
Onion	Oxyfluorfen Oxadiazon Alachlor Fluchloralin	0.2-0.3 0.5-1.0 1.5-2.0 0.9	-	Field bean Cowpea	Patel and Pacet (1989)
Rice (Transplanted)	Thiobencarb Butachlor Fluchloralin Pretilachlor	2.0 1.5 1.0 0.75	-	Wheat	Singh and Vaishya, (1994) Singh *et al.*, (1996)
Maize 1) Sown before Aug 15	Atrazine Simazine	1-3 1-3	-	Wheat	Sharma and Sandhu, (1985)
Maize 1) Sown after Aug 15	Atrazine Simazine	1-3 1-3	Wheat	-	
Soybean	Fluchloralin Oxadiazon Oxyfluorfen Butachlor	1.0 1.0 0.2 2.0	-	Wheat	Tiwari *et al.*, (1996)
Lentil	Pendimethalin Metolachlor	1.0 1.0	-	Cucumber	Singh *et al.*, (1996)
Potato	Oxyfluorfen Metribuzin Methabenzthia-zuron	0.1-0.2 0.5-1.0 1-1.5	-	Urdbean	Lal (1990)
Sunflower	Fluchloralin Butachlor Alachlor Pendimethalin	1.0 1.0 1.0 1.0	Mung bean	Groundnut Cowpea Cucumber	Basavarajappa and Nanjappa (1994)

Residual effect of 2, 4-D applied to wheat on succeeding sensitive cotton crop was observed at Anand. No residual effect of sulfosulfuron applied in winter wheat crop applied at 25, 50, 75 and 100 g/ha rates on succeeding rice crop was observed at Pantnagar. Similarly, fluchloralin (0.90 kg/ha), pendimethalin (0.75 kg/ha), metolachlor (1.0 kg/ha) and oxyfluorfen (0.10 kg/ha) applied in bhindi, cotton and blackgram did not exert any adverse effect on sensitive cucumber crop (Annual Report, AICRP WC, 2000).

Germination and grain yield of green gram and sesamum remained unaffected due to absence of herbicide residues of EPTC, nitrofen, 2, 4-D and metoxuron applied to previous crop of potato. However, there was some adverse effect of fluchloralin on the germination of both crops (Hoda *et al.*, 1988). At Pantnagar, fluazifop-p-butyl (0.125, 0.250 and 0.500 kg/ha), metolachlor and butachlor (1.0, 1.5 and 2.0 kg/ha) and oxadiazon (0.5, 1.0 and 1.5 kg/ha) applied in soybean did not cause toxicity on the succeeding wheat crop but atrazine (1.0 kg/ha) applied to maize crop affected the following soybean crop.

Pre-sowing fluchloralin (0.75 kg/ha), pre-emergence oxadiazon (0.75 kg/ha) and diuron (0.6 kg/ha) applied in cotton crop and pre-emergence atrazine (0.5 kg/ha) and simazine (0.5 kg/ha) applied to sorghum crop affected the germination of following groundnut crop at Parbani. In sugarcane, atrazine (1.0 and 2.0 kg/ha) left 0.1 to 0.32 ppm terminal residue that affected the subsequent groundnut germination (Annual Rprot , AICRP WC, 2000). Yadav *et al.*, (1995) evaluated the effect of metsulfuron-methyl at 2- 32 g/ha at 2 moisture levels (25 and 75 % available soil moisture) on transplanted onions and found that residual phytotoxicity of metsulfuron-methyl increase with increasing herbicide concentration.

The pre-emergence herbicides *viz.* thiobencarb, butachlor, pretilachlor and anilofos applied at recommended doses continuously for four seasons in rice crop did not influence the germination and yield of urdbean raised subsequently (Balasubramanian *et al.*, 1999). Pretilachlor (0.50 to 1.00 kg /ha), 2, 4-D (1.50 to 2.50 kg/ha), anilofos (0.40, to 0.60 kg/ha) and pendimethalin (1.50 to 2.00 kg/ha) applied as pre-emergence to transplanted rice did not leave any residual affect on succeeding wheat and peas crops but cucumber germination was reduced by 28 % in 2.5 kg/ha 2, 4-D treatment indicating a greater level of persistence in soil (Gupta *et al.*, 2000). At Parbhani pendimethalin (0.50 to 1.50 kg/ha), butachlor (1.5 kg/ha), anilofos (0.40 kg/ha), benthiocarb (0.5 kg/ha) and pretilachlor (1.00, 1.5 kg/ha) applied as pre-emergence in upland rice did not leave any residual effect on succeeding crop of chickpea. Whereas atrazine at 0.50 and 0.75 kg/ha showed phytotoxic effect on sensitive cucumber crop. Fenoxaprop-p-ethyl (45 to 90 g/ha) applied in transplanted rice during rainy season showed no residual effect on succeeding winter wheat crop. Similarly, there was no residual effect of fenoxaprop-p-ethyl applied at 45 to 90 g/ha in rice and soybean in *kharif* on succeeding wheat crop at Pantnagar. The application of oxadiazon (1.0 kg/ha), thiobencarb (3.0 kg/ha), butachlor (2.0 kg/ha) each superimposed with 2, 4-D (1.0 kg/ha) in rice crop and subsequently isoproturon (1.0 kg/ha) and oxadiazon (1.0 kg/ha) in wheat crop had no adverse effect on germination of wheat even after continuous use of the herbicides for five years at Jabalpur. At Faizabad, butachlor and thiobencarb applied to rice crop did not show any phytotoxic effect on the crop that followed the rice.

Application of glufosinate ammonium at 337 to 412 g/ha applied in mustard crop did not significantly affect the pearlmillet and green gram growth whereas pendimethalin and fluchloralin at 500 g/ha application rate to mustard crop were found toxic to succeeding pearlmillet crop (Annual Report AICRP WC, 2002).

Sulfosulfuron applied at 25, 50 and 100 g/ha in wheat crop did not show any adverse effect on succeeding maize and sorghum crop however, significantly affected the growth of lentil and pea (Sondhia, 2007). Pendimethalin, fluchloralin and metolachlor applied at 1.0 kg/ha to groundnut showed that herbicides did not have any adverse effect on the succeeding pearlmillet and fingermillet crop (Subrahmaniyan *et al.*, 1998). Devi *et al.*, (1997) reported that green gram and maize can safely be grown following a tomato crop treated with pendimethalin, metolachlor, butachlor and oxyfluorfen, as no residual effects of these herbicides could be detected. Evaluation of oxyfluorfen under field condition revealed that maize, groundnut, tomatoes, chickpea and peas were tolerant to oxyfluorfen at 0.2 kg/ha. However, green gram, fingermillet and Indian mustatard were susceptible to oxyfluorfen residues.

Because of the risk of serious damage from residues of persisting herbicides, it is very important to follow label recommendations on safety periods before sowing a sensitive crop and to plough when advised. These precautions are particularly important in seasons when residues are higher than normal owing to slow degradation during an extended summer drought or an exceptionally cold winter. It is also possible to predict using weather data and a computer program when the risk from residues is different from the norm.

Analysis of soil for herbicide residues can also helpful in assessing the risk to a crop. However, damage is generally worst in spray overlap areas so the maximum residues is likely to be at least twice the average level for a field. Bioassay of the soils to detect residues can also useful in some circumstances. Growing a susceptible crop such as tomato will detect the presence of active residues of hormone weed killers such as 2,4-D, clopyralid and picloram. The characteristic distortions caused by these herbicides are readily recognized. Bioassays are also useful for testing the residues of sulfonylureas by growing for example lentil and pea (Sondhia, 2007, 2008) which are extremely sensitive. However, the assay must be carefully done under controlled condition with careful watering. In lentils sulfonylureas caused stunting and root abnormalities, which are carefully recognized. Compaction of soil which can readily develop under glasshouse condition may interfere with the test by restricting root growth. As sulfonylurea are mobile in soil, it is necessary to test sub surface soil as well as topsoil. The activity of residues of herbicides such as the triazine, uracils and substituted ureas which act by inhibition of photosynthesis is greatly affected by growing conditions, phytotoxic effects in the glasshouse may be greater than in the field because growth is more rapid, moisture conditions nearer to optimum and all the roots are in treated soil. A negative response would be clear evidence of a lack of damaging residues but interpretation of positive response is less straightforward, particularly as cultivations and especially ploughing can greatly reduce or eliminate risks from residues. Because of these difficulties and the time taken to do bioassays, chemical analyses, which are quick and reliable, have generally been found to be more useful for herbicides in these groups.

Effect of non-target organism

Herbicides are essential inputs of modern agriculture. A side effect of usage of some herbicides may result in unfortunate consequences to non-target organisms if not applied judiciously. Before making a herbicide application, it is important to become familiar with the area to be treated and its surroundings. Some herbicides are more "environmentally friendly" than others and may be selected for sites where there are special concerns. The effect of herbicides on non-target organism is summarized hereafter.

Soil microflora and fauna

Soil organisms are responsible for contributing to the decomposition of dead animal and plant material into organic matter, an important component of soil fraction. Some microbes are involved in the natural control of soil pests. Aside from their direct effects on pest organisms, soil microbes are a major agent in degrading herbicides. The breakdown of pesticides is beneficial for crop rotation and food residue concerns, and provides herbicide selectivity in some instances. The value of certain soil bacteria that have a symbiotic relationship with leguminous plants in fixing nitrogen translates into reduced synthetic nitrogen fertilizer inputs and increased crop yields. Fortunately, the effect of soil-applied herbicides is short-lived; in fact, in some instances, may enhance the population of certain soil microorganisms.

The microflora and fauna in soil are large and diverse. The principal micro-organisms in the soil are algae, fungi, actinomycetes and bacteria. Organic compounds of the soil provide this food supply, the exception being a very small group of organism that feed on inorganic sources. If a herbicide is applied to the soil, microorganism immediately attack it. Those can utilize the new food supply will likely to flourish in number. In fact, this hastens decomposition of organic substance. Herbicides may affect certain species of microbes and soil borne organisms. Generally negative effect of herbicides is reversible, which indicates that population level or composition of species is decreased for a while but subsequently improves. Microorganisms are quite adept in decomposing herbicides of diverse chemical structure. Soils with a higher organic matter and clay content tend to have a higher moisture holding capacity. Generally, the population of soil microbial flora and fauna is larger and more active in soils with a higher soil moisture level, organic content and temperature. The breakdown of herbicide residues are accelerated in situations where microbial populations flourish.

It is therefore not surprising that some herbicides should affect certain species of microbes and soil borne fauna. Generally, negative effects of herbicides are reversible, meaning that the population level or composition of species is decreased for a while but subsequently improves. Beneficial organism known to be affected negatively by specific herbicides includes nitrogen fixing bacteria (Rhizobium) and some mycorhizal fungi [(rhizophagus), Barka *et al.*, 2000)]. In addition the incidence of some plant pathogen has increased as a result of herbicide use.

Actinomycetes have been investigated to a relatively small extent, considering their outstanding contribution to species equilibrium, antagonistic relationship and the degradation of organic, after in soil. Actinomycetes are relatively resistant to herbicides and affected at high concentration only. Suppression of growth of actinomycetes was reported at doses recommended for field use with methurin at 3 kg/ha (Tulabaev, 1972) however, stimulation of actinomycetes growth also occurred with 8 kg/ha of simazine (Akopyan and Agaronyan, 1968).

Adverse effect of herbicides on soil bacteria are shown in relatively less reports. Beck (1970) and Spiridonov and Spiridonova (1971) observed a stimulation in growth of bacteria by simazine at 100 ppm and at 5 to 10 kg/ha, respectively, whereas Balicka and Sobieszczanski (1969) using excessively high dose i.e. 150 and 1000 ppm of simazine and chlorpropham in soil, did not detect any inhibitory effects. The later herbicides was reported to increase counts considerable at 1000 ppm (Matsuguchi and Ischizawa, 1969); however, Chandra (1964) reported inhibition by several herbicides at concentration approximately field doses i.e. aminotriazole at 9 kg/ha, 2,3,6-TBA at 2.5 kg/ha and diallte at 1.1 kg/ha with effects persisting for 20 weeks. EPTC also reduced total population count at 4-6 kg/ha. Spore forming bacteria were often as susceptible as the non-sporting form or even so (Audus, 1970, Marillesy and Aragno 1999, Martin 1964). Metabromuron inhibited the growth of spore forming bacteria i.e. *Bacillus megaterium* at 0.3 ppm level (Maèek and Milharèiè, 1972). However, spore forming bacteria are normally more resistant at these levels, in general tolerate adverse conditions better than non-spore forming species.

Fungi have been investigated more extensively than any other microbes. Fungi are probably more sensitive to the majority of herbicides than are bacteria. Some of substituted ureas, carbamates, and quaternary ammonium compounds reduce the total fungi population in soil and inhibit vegetative growth and spore formation in pure cultures (Ishizawa *et al.*, 1961, 1976; Bartha *et al.*, 1967, Chauhan *et al.*, 1997). Gorssbard and Marsh (1974) observed that though metoxuron was antifungal at 50 ppm this effect was transient only. However, at 500 ppm number of propagules was drastically curtailed for 37 weeks. Similarly, linuron reduced fungal counts in soil only at high concentrations. Compared with bacteria and actinomycetes increase in fungal propagules are less frequent. Effect of herbicide on mycorrhiza has been neglected in spite of its great importance. Kiss (1967) examined atrazine, cycluron, chlorbufam, pyrazon and chlorpropham on mycorhizal fungi. Atrazine was the least active herbicides both *in vivo* and *in vitro*. Sobotka (1970) found most persistent inhibition of the vegetative growth of mycelium at both concentrations with Gramoxone.

Uptake by fungi of ^{14}C derived from labelled herbicides has been observed both in pure culture and in soil (Grossbard, 1975) though it is not clear whether labeling of fungal hypae was due to adsorption or genuine uptake. The composition of growth medium greatly affects the degree of toxicity. An important factor is pH. The toxicity is greater at lower pH in many instances (Balicka, 1969). Allen *et al.*, (1993) observed effects of the terbuthiuron (0.36, 0.6, and 1.01 kg/ha) on vesicular-arbuscular mycorrhizal fungi. Only the highest level of terbuthiuron application showed significant effects on mycorrhizal fungi compared to the untreated control. The herbicide did not significantly affect mycorrhizal root infection of *Sitanion hystrix*, a short-lived perennial grass, at any level of application. There was no significant effect of any level of terbuthiuron on germination of mycorrhizal spores collected 6 months after herbicide application. Glyphosate (0.54 and 3.23 kg/ha) and hexazinone (1.2 and 8 kg/ha) did not reduce soil microbial population or carbon dioxide evolution upto 6 months. However, there was a significant short-term (2 months) effect of glyphosate on both fungal and bacterial counts at the 0.54 kg/ha treatment. In *in-vitro* tests, *Cenococcum graniforme, Hebeloma crustuliniforme* and *Laccaria laccata* were more susceptible than *Suillus tomentosus* and *Paxillus involutus*. The growth of ectomycorrhizal fungi *viz. Cenococcum graniforme, Hebeloma crustuliniforme* and *Laccaria laccata, Suillus tomentosus* and *Paxillus involutus* was significantly reduced when subjected above the recommended level of glyphosate and hexazinone formulation (Chakravarty and Chatarpaul, 1989). Positive and negative response of some of the herbicides on microorganism is presented in Table 3.

Table 3. Effect of herbicides on microorganism

Herbicide	Dose (Kg/ha)	Microbial Response		Reference
Pendimethalin	0.75, 1.0	+ve		Tripathi *et al.*, 1993
Methurin	3.0		-ve	Tulabaev, 1972
Simazine	8.0	+ve		Akopyan and Agaronyan, 1968
Simazine	5-10	+ve		Spiridonov and Spiridonova, 1971
EPTC	4-6		-ve	Balkova and Semikhatova, 1969
Metobromuron	0.3 ppm		-ve	Maèek and Milharèiè, 1972
Gramoxone	-		-ve	Sobotka, 1970
Glyphosate	0.54	+ve		Araújo *et al.*, 2003 Sannino and Gianfreda, 2001
Paraquat	-	+ve	-ve	Langkramer, 1970
Terbutryne, methabenzthiazuron and pendimethalin	-	+ve		Kumar *et al.*, 1985

+ve- stimulation; -ve- inhibition

The atmosphere is the principal source of nitrogen for plant growth. The microorganism involved in fixing of nitrogen are either free living such as blue green algae, *Azotobactor* spp. and *Clostridium pasteurianum* or like *Rhizobium spp*. exist in symbiosis with a higher plant (Labeda *et al.*, 1976). Among the algae the blue green algae make the greatest contribution to soil fertility-especially in lowland rice cultivation-by fixing atmospheric nitrogen. Propanil and EPTC stimulated algal growth at 0.01 ppm and nitrogen fixation (Ibrahim, 1972). Azotobactors are more resistant for herbicides as compared to algae. Very high concentrations are required to affect the growth and nitrogen fixation of the organism (Vindard, 1952; Mickovski, 1966; Zharasov, 1969; Chaulakov and Zharasov, 1971). Paraquat caused marked inhibitory effects in pure cultures at low rates of application (Langkramer, 1970). The effect of ioxynil, mecoprop and dichlorprop were short lived. Symbiotic nitrogen fixation by rhizobia is the most efficient natural nitrogen fixation process. In assessing the influence of herbicides on this activity a distinction must be made between the response of the bacterium and that of its host. *Rhizobium spp* have been studied extensively in pure culture with special emphasis on selectivity of single herbicides (Gillberg, 1971). Selectivity of herbicides with respect to different species of *Rhizobium* is very pronounced. DNOC, dinoseb and linuron showed the most inhibitory activity on *Rhizobium* (Grossbard, 1975).

In spite of the extensive studies made on the effect of herbicide on biological transformation in soil, little is known of the response of the enzyme involved. Herbicides inhibit or stimulate soil enzymes though some may exert only a negligible effect. Simazine rarely has adverse effects on enzymatic activities in soil, inhibited urease activity at 10 kg/ha but not at 2 kg/ha. Dehydrogenase activity has been reduced even in field conditions by chlorthiamid and dichlobencil (Walter, 1970). Protease is also involved in the transformation of nitrogen activity, which is decreased slightly by dalapon and paraquat. Proteolytic enzymes were also inhibited by simazine at 10 and 20 kg/ha in field trials reducing the activity to 60 % or even 30 % of control (Spiridonov and Spiridonova, 1971).

Nowadays soil health and microbial diversity have become vital issues for the sustainable agriculture. Loss of microbial biodiversity can affect the functional stability of the soil microbial community.

Natural ecosystems are characterized by high biological diversity so that mass multiplication of any one organism is impossible due to competition, predators and parasites. This enables disturbances and balance shifts to be self-regulated internally. Continuous plant cover also results in increases in the soil organic matter, leading to improved soil structure, nutrient storage capacity, water infiltration, water holding capacity and microbial density (Gulick *et al.*, 1994, Bugg and Van Horn 1997, Pinamonti *et al.*, 1996). Soil health has been described as implying "ecosystem sustainability, diversity, functional connectedness, and resilience in response to a disturbance or stress" (Van Bruggen and Semenov, 1999) with soils with low microbial biodiversity being less sustainable and resilient.

The effect of herbicides on soil microbial populations is likely to involve not only the direct effect of the herbicide but also the indirect effect of reduction in rhizosphere exudates and organic material due to the lack of vegetation after repeated herbicide applications. Glyphosate has been shown to reduce soil bacterial populations in peach orchard, soybean field soils (Araújo *et al.*, 2003) and in wheat soils (Mekwatanakarn and Sivasithamparam, 1987). It reduces bacterial, actinomycete and fungal populations in forest soil (Gorlach-Lira *et al.*, 1997) and vineyard soil (Encheva and Rankov, 1990). On the other hand, some studies report increased populations of soil actinomycetes and fungi with glyphosate (Araújo *et al.*, 2003), increased soil microbial biomass (Haney *et al.*, 2002) or no long-term change in microbial populations (Busse *et al.*, 2001).

Glyphosate and paraquat were reported to cause activation in soil urease and invertase soil enzymes but suppression of phosphatase enzymes (Sannino and Gianfreda, 2001) whilst diquat and paraquat increased fungal populations, decreased the population of some fungal antagonists to take-all and increased the take-all incidence in wheat soil (Mekwatanakarn and Sivasithamparam, 1987). In a cropping field study, paraquat increased populations of bacteria, fungi and actinomycetes at normal application rates but at higher application rates were toxic to fungi (Camper *et al.*, 1973). Few studies have demonstrated long-term impacts of herbicide applications on soil microbial populations. Encheva and Rankov (1990) have studied the long-term effect of glyphosate in viticulture.

Griffiths *et al.*, (2001) reported that soils of reduced biodiversity due to low plant species diversity were less 'resistant' (i.e. able to withstand the immediate effects of) and 'resilient' (i.e. able to recover from) to perturbations such as copper sulfate application and heat stress.

Soil contains significant populations of microorganisms with the ability to attack or suppress plant pathogenic fungi. Fungistasis has been shown to be strongest in resistant soils with high organic matter content and microbial activity, with soil bacteria such as cellulolytic bacteria and pseudomonads, but also fungi such as *Trichoderma* spp. implicated as major causes of such fungistasis (Sturz *et al.*, 1997, Whipps, 2001).

Apart from soil microflora, herbicide may have adverse or stimulatory effects on soil fauna. The animal organism living in the soil in many respects form a lightly heterogeneous group. Their size varies from hundredths of a millimeter to more than 20 cm. The smallest Protozoa, Rotatoria, Nematoda and Tardigrada live in water filled spaces between soil particle and in organic matter. The micro arthropods (mites and springtails) live in the air filled spaces between soil particles and organic matter. The group of macro fauna includes Gastropoda (snails and slugs), Enchytraidae and Lumbricidae (earthworms). Earthworms are among the most important soil inhabitants. With the exception of the social insects little research has been done on the behaviours of soil fauna (Christiansen, 1970). Edwards (1970) found that sub lethal doses of pesticides can cause hyperactivity in ground beetle. Müller (1971) observed that the primary fraction of the ground beetle *Bembidion femoratum* to all treatments with herbicides was a violent attempts to escape, but Adams and Drew (1965) observed sluggish behaviours in Coccinellidae after a treatment with 2, 4-D, 2, 4, 5-T at 1 ppm and bromacil, chlorpropham and diuron led to significant numerical reduction of rotifers at doses of 5 ppm (Casely and Eno, 1966). 2 ml/m^2 had a paralyzing effect on the springtail, total paralysis being reached within 3 hours. Sometimes effect on micro fauna is rather radical as found by Rapoport and Cangioli (1963), who saw not only an appreciable change in the abundance of a number of large groups but also marked changes in the species compositions. The complex influence of herbicides on soil micro fauna is not caused exclusively by the complex relationship between soil organisms. Indirect effect of herbicides *via* its influence on plants may also contribute to this complexity. Various authors mention this as the cause of their findings respecting a decline of the soil fauna (Fox, 1964; Curry, 1970). According to Apt *et al.*, (1960), amitrole, dalapon, monuron and MH affected the heading of bent grass. This interrupted the life cycle of the nematode *Anguina agrostis*.

Thus herbicides such as carbamates, phenylureas and triazines can be degraded to the point of cleaving the carbon ring and the carbon fragment being utilized as food by microorganisms in the soil. At the normal rate of herbicides application in the field the effect on microorganism is of minor importance, it assumes major importance when the microorganisms affected are useful to crops. Action of soil microbes determines to a large extent persistence of herbicides in the soil. At the usual rate of herbicide application on farmlands, the total number of microorganisms is seldom changed to any great extent because the herbicide may benefit one group of organisms and injure another group. When the herbicides are decomposed the microorganism population returns to normal. The effect of herbicides on soil microbes is transient, the population eventually reaching the original level after a few days (Sindhu *et al.*, 1985, Randhawa and Bhalla, 1976, Smith *et al.*, 2001). Gopal *et al.*, (2005) found that *Bacillus subtiles* could dissipate atrazine by 53 % by 20 day of herbicides application.

Fluchloralin, pendimethalin, metolachlor, oxyfluorfen and linuron applied in okra, cotton and blackgram at recommended doses did not affect bacteria, fungi and actinomycetes at Prabhani. Application of butachlor and isoproturon did not show any adverse effect on bacterial, fungal and azotobactor population in rice and wheat cropping system at Pantnagar. Pendimethalin at 0.75, 1.0 and 1.25 kg/ha stimulated the growth of *Trichoderma harzanium* (a potential bio-control agent). The rhizosphere microflora showed an increase in the fungal population from the seedling to the harvesting stage (Tripathi *et al.*, 1993). Ayansina and Oso (2006) investigated the effect of two commonly used herbicides (atrazine and atrazine + metolachlor) on non-target soil microflora over a period of 8 weeks. Herbicide treatments at both recommended and $X_{1.5}$ recommended rates resulted in decreases in microbial counts. Higher concentrations of herbicides treatments resulted in much lower microbial counts compared to soils treated with recommended herbicide does. Herbicide treatments also resulted in the elimination of some microbial species. *Pseudomonas* sp. and *Bacillus* sp. were the most frequently isolated bacteria from herbicide treated soils. While *A. niger,* A. *Flavus, Penicillium* spp. and *Trichoderma* sp. were the most frequently isolated fungi from herbicide treated soils.

Early studies on the environmental fate of atrazine have shown that it is transformed slowly by fungi (Kaufman and Kearney, 1970). Ayansina and Oso (2006) found that at recommended rates, maximum bacterial counts were observed after 3 week in the two herbicide treated soils; Treatments at X1.5 recommended rates resulted in much lower bacterial counts compared to soils treated at recommended rates. The similar results were also reported by Taiwo and Oso (1997). They reported that, mixed pesticide combinations result in reduced microbial compare to soils treated with single herbicide i.e. atrazine (Table 4 and 5).

Weaver *et al.*, (2007) reported that glyphosate caused no harmful effetcs on microbial community, even when applied at greater than recommneded rates. Gigliotti and Allievi (2001) reported that cinosulfuron and bensulfuron, at the normal field application rate and

100 times higher rate, had no effect on the total number of bacteria and nitrifiers, or on the respiration activity in the soil, but they did decrease the nitrification activity. *In vitro* toxicity tests carried out on representative soil microbial strains using bensulfuron at 50 mg/L showed some inhibition of three of the 17 bacterial strains and strong inhibition of almost all the 12 fungal strains. Cinosulfuron had no effect on any of these strains in the previous study.

Table 4. Mean viable bacterial counts in herbicides treated soil

Week	Control	At (X 10^7 cfu/ml)		Atm (X 10^7 cfu/ml)	
	(X 10^7 cfu/ml)	Recommended rate	X1.5 recommeded rate	Recommended rate	X1.5 recommended rate
1	2.8^{ab}	2.8^{ab}	2.7^{b}	2.6^{ab}	2.6^{b}
2	5.2^{a}	3.3^{bc}	3.1^{d}	3.2^{bc}	3.0^{d}
3	5.1^{a}	3.5^{c}	3.3^{e}	3.4^{c}	3.2^{e}
4	5.4^{a}	3.4^{c}	3.1^{d}	3.2^{cd}	2.6^{d}
5	4.3^{a}	3.0^{b}	2.8^{d}	2.8^{b}	2.3^{d}
6	4.0^{a}	2.8^{b}	2.6^{c}	2.7^{b}	2.4^{c}
7	3.5^{a}	2.5^{b}	2.4^{c}	2.5^{b}	2.3^{d}
8	3.1^{a}	2.3^{b}	2.4^{c}	2.3^{b}	2.3^{d}

Means with the same letters are not sighificantly different.
(Source: Ayansina and Oso 2006)

Table 5. Mean viable fungal counts in herbicides treated soil

Week	Control	At (X 10^3 cfu/ml)		Atm (X 10^3 cfu/ml)	
	(X 10^3 cfu/ml)	Recommended rate	X1.5 recommeded rate	Recommended rate	X1.5 recommended rate
1	11.0^{ab}	8.9^{b}	2.7^{b}	2.6^{ab}	2.6^{b}
2	15.5^{a}	9.1^{b}	3.1^{d}	3.2^{bc}	3.0^{d}
3	17.75^{a}	9.5^{b}	3.3^{e}	3.4^{c}	3.2^{e}
4	18.75^{a}	3.4^{c}	3.1^{d}	3.2^{cd}	2.6^{d}
5	18.0^{a}	3.0^{b}	2.8^{d}	2.8^{b}	2.3^{d}
6	17.75^{a}	2.8^{b}	2.6^{c}	2.7^{b}	2.4^{c}
7	17.0^{a}	2.5^{b}	2.4^{c}	2.5^{b}	2.3^{d}
8	16.5^{a}	2.3^{b}	2.4^{c}	2.3^{b}	2.3^{d}

Means with the same letters are not sighificantly different.
(Source: Ayansina and Oso, 2006).

The risk that the presence of atrazine imposes to the photosynthetic microorganisms of the surface waters has been extensively studied (Abou-Waly *et al.*, 1991; Cox *et al.*, 1958; Mayasich *et al.*, 1986; Peterson *et al.*, 1994; Stratton, 1984). In comparative studies, atrazine appears to be more toxic than metolachlor towards different species of algae and cyanobacteria (Kotrikla *et al.*, 1997; Peterson *et al.*, 1994). Significant decrease in microbial population was found upto 15 days following treatment with sulfonylurea herbicides such as chlorsulfuron, sulfosulfuron, metsulfuron, meso+iodosulfuron. However, at harvest the adverse effect on microbial population of all the herbicides became negligible and microbial population reached to equal to that of untreated check in loamy sand soil at Hisar.

Kumar *et al.*, (1985) found that terbutryne, methabenzthiazuron and pendimethalin did not have any adverse effect on soil fungi, bacteria and actinomycetes in loamy sand and slightly alkaline (pH 8.4) soil with chickpea cropping. Ponnuswany *et al.*, (1997) found that pendimethalin at 1.0 kg/ha application rate initially affected population of microbes; however, it was subsequently built up with time. Low application rate of pendimethalin and butachlor did not alter the activity of the microorganisms, but higher application rate affected bacterial and fungal population. Fluchloralin, nitrofen and alachlor at 2.0 kg/ha applications rates found to stimulate the rhizobial growth under laboratory condition. Application of isoproturon at 0.25 and 0.75 kg/ha rates in winter wheat crop in sandy loam soil did not affect soil microbes (Jaryal *et al.*, 1989).

Repeated application of herbicides may lead to change in soil micro flora that increases their ability to degrade it. This is also called adaptation or enrichment at least in part, be associated with increased numbers of microorganism, especially bacteria and also with enzyme induction. The phenomenon was first observed in the laboratory and later in the field with the phenoxyalkanoic acids for example, MCPA, repeated application have been reported to increase degradation rate some threefold. As these compounds do not act primarily through the soil the effect is of little practical concern. However, recently several carbamates, thiocarbamates and amide pesticides have been shown to induce enriched microflora in the field. These compounds are soil acting so the consequences can be serious.

The phenylalkanoic herbicides have only small long-term effects on soils with 2,4-D application. The population of rhizobium and of actinomycetes are reduced but there is little or no effect on nitrification, amonification, denitrifers and overall populations of bacteria and fungi. Atrazine for example has greater effects, especially on population of algae, actinomycetes and of cellulose degraders;, however, ammonifiers, denitrifiers and total nitrogen fixers are little changed. Paraquat, which is rapidly adsorbed into the clay lattice, can be decomposed by several microorganisms and affect soil fungi differently. Fungi such as *Mucor* and *Zygorhynchus* are more susceptible to bipyridium compounds than *Aspergillus* and *Penicillium* and some pathogen e.g. *Fusarium culmorum* are tolerant, consequently a different balance of microorganism may arise.

The persistence of herbicide activity in the soil depends upon intrinsic phytotoxicity and speed of dissipation. Particular chemicals may be persistent because of features in their structure such as the presence of chlorine, quite small chemical changes sometimes have considerable effects on biodegradability. In some instances one microorganism may perform one step in breakdown/degradation and a series of other microorganism carryout successive stages. Herbicides of long persistence and having a major effect on microorganism are most environmentally damaging because of the disruption of the ecological equilibrium. Although each herbicide may lead to particular changes in the soil microflora, recovery of populations at usual field rate of herbicides is normally quite rapid. Indeed the effects of herbicides are not as drastic as those of even the most favoured soil sterilent, which may be used routinely in horticultural production such as methyl bromide. The sterilant results in both chemical and physical changes, with very severe reduction of many microorganism including Rhizobium, nitrifers and actinomycetes. On the other hand, ammonification is less affected than nitrification and nutrients are made available from the killed organisms.

Agricultural practices are found to affect the soil microflora substantially. Ploughing causes large changes in the spatial structure of microhabitats and leads to a considerable reduction of microbial biomass as compared with no tillage. Minimum cultivation affects actinomycetes and nitrifers less than conventional tillage; nitrification is also adversely affected by soil compaction. In many respects, therefore common agricultural practices such as ploughing and liming, affect nutrient availability and also natural perturbations such as drying and freezing have a pronounced effect on microbial activity than many herbicides at field rates. It is also significant that ploughing leads to a considerable reduction in diversity and numbers of soil fauna such as earthworms and also results in a more even distributions of species vertically than in uncultivated soil.

Earthworms

Soil macrofauna populations play an important role in the degradation of organic matter, nutrient recycling, carbon sequestration, soil aggregation and macropore formation through their behavioral activities (Edwards and Bohlen, 1996; Sollins *et al.*, 1996; Groffman and Bohlen, 1999; De Goede and Brussaard, 2002). Earthworms may have a positive or negative impact of variable magnitude on agricultural production through a variety of biological, chemical, and physical mechanisms (Brown *et al.*, 1999; Senapati *et al.*, 1999; Lavelle *et al.*, 2001). As discussed by Brown *et al.*, (1999), biological mechanisms include competition between earthworms and plants for water, increase in microbial biomass and microbial activity, alteration of carbon and nitrogen status, reduced damage from parasitic nematodes, and increased plant nutrient uptake.

Earthworms are important in enhancing the soil structure and fertility. Earthworms play a major role in soil fertility as well as acting as a transferring route from soil to the terrestrial ecosystem. Earthworms are perhaps the most important soil organisms in terms of their influence on organic matter breakdown, soil structural development, and nutrient cycling, especially in productive ecosystems. Earthworm activity accelerates the decomposition of plant litter, increases rates of nutrient transformation and plant nutrient uptake, improves soil aggregation and porosity, and enhances water infiltration and solute transport. Earthworms can consume and incorporate large amounts of organic matter into soil. Earthworms generally improve the aeration and porosity of soil. By increasing rates of water infiltration, earthworms can reduce the amount of surface runoff.

Moslesh *et al.*, (2002) observed no lethal effect of isoproturon on mature earthworm (*Lumbricus terrestris* L.) under lab conditions at the highest concentration tested (1.4 g/kg soil) after 60 days of treatment. However, significant reduction in the growth was observed during the growth period. Atrazine (1.0 kg/ha), pendimethalin (1.0 kg/ha) and oxyfluorfen (0.125 kg/ha) did not affect the population, mortality and migration pattern of the earthworm *Pontoscolex corethrurus* under laboratory conditions in red sandy loam soil at UAS, Bangalore (Annual Report, AICRP on Weed Control 1998-99) (Fig. 1). However, Sarkar (2003) reported that most herbicides were having adverse effect on earthworm (*Eisenia fetida*). Atrazine (2.5 kg/ha) caused 73 %, 2, 4-D (1.5 and 2.5 kg/ha) 43 to 50 %, metribuzin (0.3

and 0.6 kg/ha) and oxyfluorfen (0.2 and 0.4 kg/ha) 33 to 37 % and glyphosate (2.0 kg/ha) 40 % mortality of earthworms. Shrivastava and Palta (2002) evaluated the effect of atrazine and oxyfluorfen against earthworm (*Eisenia fetida*) and reported no mortality after 48 hours but noticed physiological and behavioral changes in earthworms at normal exposure (Fig. 1).

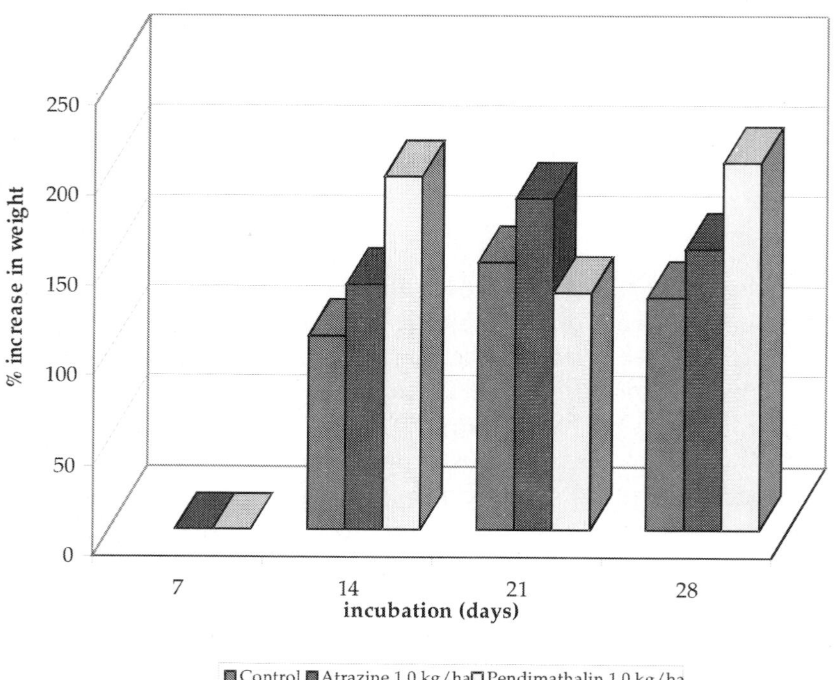

Fig. 1. Effect of herbicide on growth of earthworm (fresh weight/worm).

Muthukaruppan *et al.*, (2005) evaluated toxicity of butachlor against earthworm variety *Perionyx sansibaricus* at doses of 0.065, 0.131 and 0.197 mg/kg and reported that earthworm growth was observed at various exposures over 60 days. While exposing the earthworm to the herbicide, no mortality was observed, not even at a higher dose of butachlor. The mean earthworm biomass was found to be decreased with increasing herbicide concentration. Cocoon production was also reduced by the increasing herbicide concentration. All earthworms in the exposed group were found to have glandular cell enlargement and opined that this may massively affect food intake and indirectly inhibited the earthworm reproductive capacity.

Fishes and aquatic life

The most obvious effects of herbicides on fish and other wildlife are direct effects of acute poisoning. At times, herbicides are solely blamed for fish kills; however, in many cases, indirect effects of herbicides that cause dissolved oxygen depletion are the reason for the kill. Herbicides can enter water sources through drift, runoff, soil erosion, leaching, and occasionally, accidental or deliberate release. Some herbicides which are classified as very highly- or highly toxic to fish are listed in Table 6. These herbicides, ranging in concentrations of less than 0.1 to 1.0 ppm, can kill fish.

The persistence of aquatic herbicides in the aquatic body is important as well as their ultimate fate. The length of time which the herbicide remains in the water determines the length of exposure. The fate of an aquatic herbicide determines whether aquatic life could be exposed at a later time. Continual use of herbicides near the aquatic area has prompted some concern of the effects of these chemicals on the life stages of fish (Skea *et al.*, 1987). If a herbicide which is toxic to the early life stages of fish is used annually, the herbicide could contribute to a decline in the fishery of a lake. Most aquatic herbicides have undergone some toxicity testing to evaluate effects on non-target aquatic organisms (Urban and Cook, 1986).

Table 6. Herbicides that are classified as highly toxic to fishes

Herbicide	LD_{50} value (mg/kg)	Herbicide	LD_{50} value (mg/kg)
Alachlor	930	Pendimethalin	4075
Bromoxynil	779	Endothall	38-51
Oxadiazon	>5000	Butylate	5366
Oxyfluorfen	>2000	Prometryne	4550
Quizalofop-ethyl	6600	Diclofop-methyl	2020-2176

When used according to the recommended specifications, currently available aquatic herbicides are not toxic to fish, birds, or other aquatic organisms. These herbicides are also short-lived in the environment and do not accumulate in organisms. Environmental conditions are not always predictable, however, and under certain circumstances, fish kills can occur usually as an indirect result of aquatic herbicide applications.

Fish kills are only likely to occur as a direct effect of herbicide application if a herbicide formulation known to be toxic to fish, such as the amine salt of endothall, is applied in an enclosed water body. The concentration of copper that is used for most herbicide applications is below toxic concentrations. However, rates recommended for difficult-to-control filamentous algae can be toxic to fish in enclosed ponds and care should be taken when making this type of application. The greatest concern for copper toxicity is in soft water because the toxicity of copper to fish increases as water hardness decreases. This is especially true for most trout species. Most aquatic herbicides have very low toxicity to fish and the concentration that occurs after application of recommended rates is less than concentrations that are toxic to fish.

The most common reason for fish kills due to aquatic herbicide application is the indirect effect of lowered dissolved oxygen (DO) in water. DO in lakes and ponds commonly range between 0.0 and 12 ppm. The lower concentrations occur during early morning hours because aquatic plants consume oxygen during darkness or reduced light. Oxygen is also consumed by other aquatic organisms, especially those associated with decaying organic matter in the water. Fish populations can usually withstand the everyday fluctuations of DO, but many types of fish cannot tolerate prolonged periods of low DO. Natural fishkills can occur in highly productive waters when phytoplankton populations die and cease producing oxygen after prolonged cloudy, still, warm weather.

When large amounts of aquatic plants are killed by a herbicide application the decaying vegetation and lack of oxygen production may cause DO to become so low that fish cannot survive in the water and a fishkill occurs. If a herbicide that is effective on higher plants and not phytoplankton is used, the potential for a fishkill can be minimized because phytoplankton will continue to produce oxygen. Also, the danger of fishkills is less in cooler water because it can hold more oxygen than warm water. Herbicide applications to large weed populations in warm water during periods of prolonged still and cloudy weather, and where fish movement is restricted should be avoided to minimize the potential for fishkills. Large weed populations should be brought under control by a series of applications to portions of the waterbody and treated during the spring when water temperatures are lower. Once under control, weeds should be maintained at low densities.

Herbicide-related fishkills, either direct or indirect, are not likely to occur as a result of partial area applications in large water bodies because fish have avoidance mechanisms and are mobile. If possible, fish will move to other parts of a lake to avoid adverse conditions. When making partial applications of herbicides, such as the diethylamine salt of endothall, that can be toxic to fish at recommended use rates, applications should be started near shore and proceed towards open water. This allows fish to escape to untreated water. All precautions should be taken to avoid conditions that can lead to potential fishkills when applying aquatic herbicides.

Sondhia (2008b) evaluated persistence and bioaccumulation of sulfosulfuron in fishes at 25, 50 and 100 g/ha rates under laboratory conditions in glass aquariums. Result are presented in Table 7. It was observed that sulfsulfuron disipiated at faster rate initially and persisted upto 90 days in fishes.

Table 7. Bioaccumulation of Sulfosulfuron residues in fishes under laboratory conditions (Sondhia 2007)

Dose (g/ha)	Sulfosulfuron residues (µg/g) Days after application				
	10	20	30	60	90
25	1.09	0.248	0.071	0.05	0.044
50	1.67	0.536	0.141	0.079	0.045
100	3.52	0.780	0.195	0.17	0.054

There was generally a poor correlation between pesticide and sediment run-off and even trifluralin, which is well adsorbed, was mainly transported in the water phase. Spalding *et al.*, (1979) monitored ground waters in Nebraska, where 94% of irrigated com was treated with herbicides, mainly alachlor and atrazine and found atrazine in 13 out of 14 samples at concentrations of 0.06-3.12 ppm, Ground waters in Nebraska are used for drinking water supply.

The degradation of acrolein in water was investigated by Bowmer and Higgins (1976). They found that acrolein degraded by a pH-dependent first-order reaction to a non-phytotoxic, non-volatile product. The product gave a positive reaction with dinitrophenylhydrazine consistent with its identity as either, β-hydroxypropionaldehyde or a pentamer polycondensate. The reaction product was itself transitory and dissipated rapidly, probably by microbiological processes, once acrolein concentrations fell below 2-3 mg/l. The rate of reaction of acrolein increased with increasing pH, giving first-order half-lives at pH 6.6 and 8.6 of 50 and 38 h. respectively. In flowing water, the rate of loss was much faster, reflecting the influence of turbulence in increasing volatilization loss (Bowmer *et al.*, 1974). Results for six experiments in flowing water in irrigation channels all indicated a rapid loss of acrolein and a first-order half-life of about 4-5 h (Bowmer and Sainty, 1977). These results were obtained for pulses of acrolein injected at high concentration (2-22 mg 1-1) for a short time (1-8 h), which is currently the standard practice in Australia. The equation derived by O'Loughlin and Bowmer (1975) and O'Loughlin (1975) was used to take into account the effects of dispersion. A disadvantage of acrolein is that it is toxic to fish at very low concentrations (Burdick *et al.*, 1964). Also, it is potentially hazardous in bulk when overheated or contaminated with water, and so requires special facilities for transport, storage and application. Acrolein is not translocated to the roots of plants and is rapidly volatilized, so it gives only temporary control.

Diquat has a very rapid effect on most submerged plants. As an aquatic herbicide diquat has advantages of safety to wildlife and fish, a very rapid effect on weeds and safety to irrigated crops through rapid inactivation in sediments and soil. The algicidal effects of terbutryne are advantageous in retarding development of algal blooms after macrophytes are controlled in fresh waters, but this has led to concern that almost total habitat destruction could occur where the persistence of phytotoxic concentrations is prolonged (Newbold, 1975). However. Murphy (1982) commented that although major changes in the aquatic flora must be expected in the short term, when there is an increase in opportunistic filamentous algae, the changes are not sustained in the long term. Submerged vascular plants regrew within one season of a single treatment of terbutryne with only minor changes in species composition (Murphy *et al.*, 1981). although repeated treatments over several seasons may prevent regrowth and maintain the flora at an early stage of succession (Wade, 1981). Acute toxicity data show that some invertebrates may be rather sensitive to terbutryn. The LD_{50} 48 h for Daphnia magna is 1.4 mg/l. Values for various species of fish, obtained by several authors (Murphy, 1982) range from about 2 to 9 mg/l. Safety margins at normal field dosage rates are considered acceptable in the UK (Tooby, 1978). In a screening test using harlequin fish, the LD_{50} values ranged from 11 to 15 mg 1-1.

Assuming a maximum concentration of 0.1 mg/l in water, this corresponds to a safety factor of about 110 for fish, but only 14 for Crustacea (*Daphnia magna*).

There are few detailed studies of the degradation of glyphosate in the aquatic environment. Reuppel *et al.*, (1977) studied the metabolism of ^{14}C-labelled glyphosate in shake flasks containing a water-soil mixture (20:1), and measured the evolution of $^{14}CO_2$ and the production of extractable metabolites, separated by thin-layer chromatography. Their results suggested that glyphosate was degraded moderately slowly in the soil by microbial co-metabolism, giving 55% transformation of the [^{14}C]glyphosate applied to $^{14}CO_2$ in 28 days. Degradation occurred under both aerobic and anaerobic - conditions. The main product was the non-phytotoxic derivative aminomethylphosphonic acid, which in turn is degraded, but more slowly. Similar results were obtained for soils in many experiments (Torstensson, 1985), although the rates of degradation varied widely, with half-lives ranging from a few days to several months or years. The formation of N-nitrosamines has been investigated because some related compounds are acutely toxic and carcinogenic, but no nitrosation was found at realistic concentration of glyphosate and nitrite (Khan and Young. 1977). The dissipation of glyphosate from water in natural and artificial ponds was reviewed by Brpnstad and Friestad (1985). Sacher (1978), for example, found that glyphosate dissipated from water with a half-life of about 12 days. The other experiments differ vastly in scale and method, so it is unproductive to make comparisons. The rate of dissipation probably depends on turbulence scale and intensity, which will control adsorption by benthic sediments, and suspension of particulates and microorganisms.

Tooby (1985) in reviewing the effects of glyphosate on aquatic organisms concluded that there is very little evidence of any harmful effects, with the possible exception of one species of shrimp. The toxicity to fish of the commercial formulation could mainly be attributed to the surfactant, rather than to glyphosate *per se* (Folmar *et al.*, 1979), and taking this into account the safety factor was 14, based on the ratio of 96 h LC so for rainbow trout and a predicted maximum environmental concentration of 0.6 mg/l. However, a wide variation in the sensitivity of *Daphnia* species to the commercial formulation has been reported, with values ranging from 192 mg/l (Sacher, 1978) to 3.0 mg/l. (Folmar *et al.*, 1979), and this was attributed by Tooby (1985) to variations in experimental conditions such as pH or temperature during the experiment. Hartman and Martin (1984) studied the effects of suspended particulate matter on the sensitivity of Daphnia pulex to glyphosate with surfactant, reporting a population reduction at concentrations of about 1 mg/l and doubling of toxicity in the presence of bentonite. They suggest that the toxicity of glyphosate to this filter-feeding organism is increased when it is ingested along with particulate matter. Glyphosate has been also used for treatment of vegetation in ponds in Poland prior to their flooding (Krueger *et al.*, 1996). This has resulted in the intensive development of food organisms on which carp (*Cyprinus carpio*) feeds. Control of water hyacinth in field trials in the Niger Delta, Nigeria, using glyphosate at 2.88 kg ha^{-1}, resulted in all plants being killed 7 days after treatment. The dead plants sunk or were removed. Subsequent fish catches have shown that the number of some species increased after the treatment, and 28 days after the treatment the catches contained high numbers of juvenile and immature fish (Olaleye and Akinyemiju, 1996). The increase in fish is considered to be the result of improved water quality after the removal of herbicide-treated water hyacinth.

Gunkel and Streit (1980) reviewed the literature on the accumulation of atrazine by aquatic organisms and investigated uptake in a freshwater mollusc and a fish. Accumulation was small, the equilibrium concentration being reached within a few hours. Also, accumulated atrazine was very rapidly lost from the organisms when they were transferred to clean water. Hartman and Martin (1984, 1985) used suspensions of bentonite to simulate the effects of suspended sediments on the acute toxicities of herbicides to *Daphnia pulex* and *Lemna* sp. No influence was observed for alachlor or atrazine, but the toxicity of glyphosate to *Daphnia* was increased by suspended clay, while the toxicity to *Lemna* was reduced.

In bioaccumulation studies fish muscle tissue accumulated only 2-5 times the concentration of terbutryne found in the water. There was rapid accumulation, but also rapid elimination. In ecosystem studies populations of Crustacea recovered rapidly a few months after treating experimental ponds. Although not directly toxic to fish, the dissolved oxygen concentration can be reduced when a high plant biomass is treated, so that terbutryne must be used with caution in fisheries. The problems of deoxygenation after weed decay have been studied by Jewell (1971). among others, and specifically for terbutryne by Wingfield and Johnson (1981). Water containing terbutryne at 0.1 mg/l is generally safe for irrigation (Aston and Reynaert, 1974), although in the UK water cannot be used for irrigation within 7 days of treatment. Concentrations of 0.02 mg/l are safe for other uses (Croll, 1976; Tooby, 1978). Terbutryne is not suitable for pulse addition to flowing water because of the long contact time required for effect, but for weed control in static water it is a most economic effective and safe treatment.

Herbicide treatment, resulting in the removal of aquatic plants, also removes associated epiphyton on which some fish feed, and this will lead to a decline in certain fish species. Where aquatic macrophytes are essential for some species for some part of their life history, such as spawning, as a nursery site, the species will also disappear after a treatment with herbicides. It has been observed that in Irish canals application of dichlorobenil reduced the flora and significantly reduced fish habitats, consequently the stocks (Caffrey, 1993).

Control of water hyacinth on small areas using 2,4-D is effective, and while fish are not being poisoned by the herbicide, the elimination of the weed in some waters may have an effect on fish which feed among its hanging roots and take shelter there. The herbicide 2,4-D is widely used in fish ponds for controlling aquatic weeds and algal blooms. While some studies suggested a negative impact of this herbicide on survival and growth offish (Smith, 1952), others have noted a beneficial influence. such as for bottom-feeding fish (e.g. Sarkar, 1991). The rate of application was found to be of importance, with application of 2.4-D at 4.3 kg ha^{-1} being non-toxic for tilapias (Khogali and Moghraley. 1979). As a response to the treatment of aquatic weeds in fish ponds in India at concentrations of 4.5 to 10.5 kg ha^{-1} there was an increase in the bottom invertebrate fauna. Consequently, the bottom feeding common carp and mrigal (*Cirrhinus mrigala*) showed an increase in the growth rate as compared to the column-feeding fish catla (*Catla catla*) and rohu (*Labeo rohita*). The reduction in the yield of the column-feeding fish which feed on plankton was considered to be due to the destruction of the plankton following herbicide treatment. Also in India. Patnaik (1976) tested the impact of 2,4-D on water hyacinth, *Spirodela polyrhiza, Azolla pinnata* and Salvinia cucullata at concentrations up to 3.0 kg ha^{-1}. He found no harmful effects on fish, i.e. *Notoptrerus notopterus. Barbus ticto, Channa striatus, Cirrhinus mrigala, Labeo rohita, L. bata, Catla catla, Cyprinus carpio, Hypophthalmichthys molitrix.*

Surface application of 2,4-D amine (4.3 kg/ha) directly affects the movement of largemouth bass (*Micropterus salmoides*) (Boyer and Cichra. 1994). Fluridone is a safe, slow-acting herbicide that provides control of rooted aquatic plants such as *Hydrilla, Myriophyllum, Ceratophyllum, Egeria, Potamogeton, Cabomba* without affecting zooplankton. benthic organisms, crayfish, turtles, frogs. water snakes, fish and waterfowl. It becomes toxic to fish only if the normal rate of concentration is exceeded by 76 to 220 times. Fluridone does not affect water quality. It does not accumulate in fish and aquatic birds continue to feed from the treated areas (Arnold. 1979). Control of susceptible plant species normally lasts from 12-18 months. Being a slow-acting herbicide, fluridone requires 11 months to achieve a 100% control of hydrilla tubers. The fluridone-tolerant plant species are water hyacinth, *Pistia, Chara, Nitella,* a number of emergent plants and planktonic algae (Tarver, 1986). At recommended field application rates, herbicides likely to be used in conjunction with grass carp are not acutely toxic to the fish (Tooby *et al.,* 1980, Kracko, 1989). However, grass carp may stop feeding due to changes in the vegetation which makes it unpalatable. Interference with photosynthesis or destruction of chlorophyll in a plant will affect carbohydrate production, and may lower the food quality of the plants (McCowen *et al.,* 1979).

Aquatic herbicides can disappear from treated water by adsorption to bottom sediments, volatilization, absorption by plants and animals or by dissipation. Dissipation refers to the breaking down of a herbicide into simpler chemical compounds. Herbicides can dissipate by photolysis (broken down by light), microbial degradation, or metabolism by plants and animals. Both dissipation and disappearance are important considerations to the fate of herbicides in the environment because even if dissipation is slow, disappearance due to processes such as adsorption to bottom sediments makes a herbicide biologically unavailable.

Aquatic herbicides are non-persistent in treated water, because of their disappearence rapidly. Disappearance is more when spot treatments are made in large bodies of water because the dominant effect is dilution. Aquatic herbicides are water soluble and quickly dilute to non-detectable concentrations. These herbicides disappear at different rates and by different methods. For example, copper is a naturally occurring element and essential at low concentrations for plant growth. It does not break down in the environment, but it forms insoluble compounds with other elements and is bound to charged particles in the water. It rapidly disappears from water after application as a herbicide. Because it is not broken down, it can accumulate in bottom sediments after repeated high application rates. Accumulation rarely reaches levels that are toxic to organisms or significantly above background concentrations in the sediment.

Similarly, 2, 4-D photodegrades on leaf surfaces after foliar applications and is broken down by microbial degradation in water and sediments. Complete decomposition usually takes about 3 weeks in water and can be as short as 1 week. 2,4-D breaks down into naturally occurring compounds. Two pounds of 2,4-D amine will break down into 1 pound carbon dioxide, 1/4 pound water, 1/4 pound ammonia, and 1/2 pound chlorine. Glyphosate is not applied directly to water for weed control, but when it does enter the water it is bound tightly to dissolved and suspended particles and to bottom sediments and becomes inactive. Glyphosate is broken down into carbon dioxide, water, nitrogen, and phosphorus over a period of several months (Folmar *et al.*, 1979).

Dissipation of fluridone from water occurs mainly by photodegradation. Metabolism by tolerant organisms and microbial breakdown also occurs, and microbial breakdown is probably the most important method of breakdown in bottom sediments. The rate of breakdown of fluridone is variable and may be related to time of application. Applications made in the fall or winter when the sun's rays are less direct and days are shorter result in longer half-lives. Fluridone usually disappears from pondwater after about 3 months but can remain up to 9 months. It may remain in bottom sediment between 4 months to 1 year.

Paul *et al.*, (1994) evaluated toxicity of diquat, endothall, and fluridone on fishes and found fish mortality throughout 96 hours of toxicity tests. In almost every species-age combination diquat had a lower LC_{50} than endothall or fluridone. They found diquat more toxic to the fish tested than fluridone which is more toxic than endothall. The NOAECs/LOAECs indicate similar patterns except that very young walleyes appear to be equally sensitive to diquat and fluridone. They reported that walleye tests the very early life stages are the most sensitive. The 8 to 10-days old walleye (*Stizostedion vitreum*), were more sensitive than the 41 to 43-days old walleyes for both diquat and endothall. In addition the 41 to 43-day old walleyes were more sensitive to diquat than the 84 to 86-day old walleyes. Largemouth bass

(*Micropterus salmoides*) and smallmouth bass (*Micropterus dolomieu*), are less sensitive than walleyes to all three herbicides. Smallmouth bass were less sensitive than largemouth bass to diquat for up to 48 hours, but thereafter were comparable in sensitivity. Largemouth bass are less sensitive than smallmouth bass to endothall and fluridone.

A review of the toxicity literature for diquat indicates that diquat is highly toxic to some aquatic animals (Kea *et al.*, 1987, Jones, 1985). *Hyalella azteca*, an amphipod, is one of the most sensitive aquatic organisms tested with a 96-h LC_{50} of 0.048 mg/l (Wilson and Bond 1969, Sunber and Picker, 1962). Gilderhus (1967) found a 96-h LC_{50} of 35 mg/l to bluegills (*Lepomis macrochirus*), and Leung *et al.*, (1983) found a 96-h LC_{50} of 289 mg/l to mosquitofish (*Gambusia affinis*). Many other aquatic organisms have been tested and fall between the extremes of 0.048 mg/l and 289 mg/l (Hiltibran 1967; Hughes 1975; Johnson and Finley 1980; Simonin and Skea 1977; Williams *et al.*, 1984). Young walleyes are the most sensitive fish species tested. Gilderhus (1967) also found walleyes to be the most sensitive of the fish species he tested. He reported a 96-h LC_{50} of 1.2 mg/l for two months old walleyes.

Mammals

A large number of herbicides now used and these herbicides not only kill target weeds but also pollute water, air soil, food stuff, human system and wild life. There are no evidences of acute toxicity resulting in death of mammals due to application of herbicides at recommended doses. Some deaths may occur due to misuse but this is also true with other chemicals. Most herbicides have very less toxicity to mammals (Flecher and Kirkwood, 1982). The use of 2, 4-D and 2, 4, 5-T poses hazards to livestock which eat the weed exposed to sub lethal dose of herbicides. However when sprayed with 2,4-D the sugar content of these weed increase and they become attractive to cattle. Thus the livestock consumes both the toxic component of the weed as well as toxic residues of herbicide. When applied to sudan grass 2,4-D and 2,4,5-T decompose to highly toxic hydrogen cyanide which is then consumed by the cattle grazing on these grass. The death of about 250 reindeers in the spring of 1970 which had fed on coniferous vegetation sprayed with a mixture of 2,4-D and 2,4,5-T the year before has been attributed to nitrate poisoning. Military defoliation of mangrove swamp in Vietnam with a 50:50 mixture of 2,4-D and 2,4,5-T caused scarcity of mallards, quails and partridges.

Other wild life

Herbicides can kill birds in several ways. Direct ingestion of granules, baits, or treated seeds and direct exposure from sprays are obvious ways. Indirect bird kills may result from consumption of treated crops, contaminated water, or feeding on contaminated prey. Birds and other wildlife can be poisoned when baits, such as those targeting rodents, are improperly placed or not recovered in a timely fashion. Pellet and granular-formulated herbicides may be mistaken for food and consumed by birds and other wildlife. Herbicides have bird acute oral LD_{50} values ranging from less than 10 to 50 mg/kg of body weight. Some herbicides have been found negatively affecting the reproductive potential of certain birds and wildlife.

Summary

A herbicide must not be present in the soil solution in order for it to cause injury to sensitive rotational crops. Environment has a major effect on the persistence of herbicides. Soil moisture and temperature are very important factors in determining the rate of herbicide breakdown. There are a number of mechanisms which determine the environmental fate of a herbicide such as volatilization, photolysis (breakdown from sunlight), leaching etc. However, the two primary mechanisms of herbicide degradation are microbial and chemical hydrolysis. Both these processes are dependent on soil, water and temperature; however, soil moisture is more critical with herbicides that require microbes to degrade (Streck, 2005). Herbicides differ in their persistence under different environment conditions. Soil less in organic matter content has more herbicides available and cause more persistence. Herbicides are more strongly adsorbed in acidic soil than alkaline soil. Soil properties such as organic matter content, soil texture, and soil pH also play an important role in the carryover potential of residual herbicides. Whenever herbicides are used in an intensive and diversified crop rotation, the herbicides applied to one crop should not persist in soil at concentration high enough to affect the succeeding sensitive crops. Generally, there are fewer problems with herbicide carryover when soil organic matter is high. Studies showed that in most cases no residues or very minute quantity of herbicides were left in the soil and did not affect the subsequent sensitive crops.

Herbicides are an essential part of agriculture production systems. A side effect of usage of some herbicides may result in adverse consequences to non-target organisms. Before making a herbicide application, it's important to become familiar with the area to be treated and the surroundings. Some herbicides are more "environmentally friendly" than others and may be selected for sites where there are special concerns. Herbicides may affect certain species of microbes and soil borne animals. Fungi have been investigated more extensively than any other microbes. Fungi are probably more sensitive to the majority of herbicides than bacteria. Some of substituted ureas, carbamates, and quaternary ammonium compounds reduce the total fungi population in soil and inhibit vegetative growth and spore formation in pure cultures (Ishizawa *et al.,* 1961). Generally negative effect of herbicides is reversible, meaning that the population level or composition of species is decreased for a while but subsequently improves. The persistence of aquatic herbicides in the aquatic body is important as well as their ultimate fate. The most obvious effects of herbicides on fish and other wildlife are direct effects of acute poisoning. When used according to the recommended specifications, currently available herbicides are not toxic to fish, birds, or other aquatic organisms. Generally these herbicides are also short-lived in the environment and do not accumulate in organisms.

□□□

Chapter 12

Toxicology

The mammalian toxicity of any particular herbicide can be readily be determined and compared with that of other herbicides by appropriate tests on laboratory animals. However, a proper evaluation of any toxic hazard that a compound might present to men, requires additional factors to be taken into account.

Not only the extent and manner of use of a herbicide govern the extent and degree of human exposure that may arise, but also the nature of the toxic effects produced by over exposure. Those indicates the urgency with which steps should be taken to reduce risk of over exposure. For example, it may be uncomfortable or perhaps temporarily dangerous to suffer a disturbance of a physiological process arising from the toxic effects of a compound but the hazards from over-exposure will be much greater where sign of poisoning appears when a serious degree of irreversible issue damage has been produced.

It is necessary to stress these general points at the outset of a discussion of toxic hazards because some people will scan a table of acute LD_{50} data of compound and assume that those with the smallest lethal dose are necessarily the most dangerous compound to use. Attempts are also made to classify pesticide into categories of danger solely on the basis of LD_{50} values for laboratory rats.

The EPA (Environmental Protection Agency) of US has classified pesticides according to toxicity into four categories (I, II, III and IV) (Table 1). The categories are based on oral LD_{50}, inhalation LC_{50}, dermal

LD_{50}, eye effects and skin irritation. EPA has also designated a signal word for each category, which should appear on the label of the pesticide. Furthermore, if pesticides are classified under category I on the basis of its oral, inhalation or dermal toxicity (as distinct from skin and eye effects) the word poison is required to appear in red on a background of contrasting colour together with the symbol of skull and crossbones.

Table 1. Toxicity categories for pesticides as classified by EPA of USA

Toxicity indicator	Toxicity category			
	I	II	III	IV
Toxicity class	Extremely hazardous	Highly hazardous	Moderately hazardous	Slightly hazardous
Oral LD_{50} (mg/kg)	50 or lower	50-500	500-5000	5000 +
Inhalation LC_{50}. 1. Dust or mist (mg/l)	20 or lower	20-200	20-200	200 +
2. Gas or vapours (ppm)	200 or lower	200-2000	2000-20000	20000+
Dermal LD_{50} (mg/kg)	200 or lower	200-2000	2000-20000	20000+
Eye effect	Irreversible corneal opacity at 7 days	Corneal opacity or reversible within 7 days, irritation for 7 days	No corneal opacity, irritation gone within 7 days	No irritation
Skin irritation	Severe irritation or damage at 72 hours	Moderate irritation at 72 hours	Mild irritation at 72 hours	No irritation at 72 hours
Signal word	Danger (Poison)	Warning	Caution	Caution
Colour code	Red label	Yellow label	Blue label	Blue label

Toxicity and exposure

Exposure of any chemical to the organism or biological system followed typical dose response curve. Generally no response is observed until some minimal dosage is reached. Above this level, the organism shows a response proportional to the amount of chemical present so that responses increase with increasing dosage until the maximum possible response is attained. Increasing dosages above this maximum response level should yield no additional effect. The fact that there is a concentration below which no response will be elicited should always be kept in mind when interpreting statements relateu to the effects of herbicides on non-target organisms and human health. In USA, the Environmental Protection Agency (EPA) is assigned the task of overseeing pesticide testing procedures for human and environmental effects. Before a pesticide is sold in the market, it must be registered by the EPA, which may require over 100 safety and environmental test for each potential pesticide. Data submitted to the EPA include physical and chemical properties of the pesticide, its fate in the environment, a determination method of herbicides and metabolites in feed and food crops, toxicological information and ecological studies to determine the effect of non-target plants and animals. These data usually are required not only for the herbicide itself but also for its major breakdown products (Rban and Cook, 1986).

Human health

A large portion of the development of a herbicide is devoted to determining its effect on human health. These effects usually are expressed in terms of toxicity and exposure. Toxicity is a measure of the amount of compound that is harmful or lethal. Exposure is the probability of encountering a harmful dose of a compound. A compound that is extremely toxic but is used in such a way that exposure is low may be less hazardous than a compound that is less toxic but whose application requires greater exposure. Potential adverse effects on human are derived from toxicology studies on experimental animals such as rats, rabbits and dogs. Compounds are also evaluated for their potential to cause reproductive problems and for possible carcinogenic (cancer producing), teratogenic (birth defect producing) properties. These studies are conducted by the company or by independent laboratories of a specific herbicide and all data are submitted to the EPA for review. A product which meets the evaluation criteria is given an EPA registration number and only this product can be offered for sale to the public.

Data submitted to EPA must show the amount of herbicide and breakdown products present in the food, feed crop in order to establish tolerances. A tolerance level is designated as the maximum amount of herbicide allowed on the crop. The tolerance level is multiplied by the amount of the crop or food product an individual might consume over a period of time. This value is compared with data from toxicology tests to determine if there is any human health hazard.

From human health point of view the most dredged herbicide formulation is called 'agent orange'. It is an equal mixture of both 2,4-D and 2,4,5-T which got is name from the bright oranges strip. Most of the literature on the monitoring of humans exposed to herbicides concerns the phenoxy acid group (2,4-D, 2,4,5-T, MCPA, dichlorprop, mecoprop), which are the most extensively used herbicides in USA. Canada and Europe, since mid-1950s. However, field studies from controlled experiments have been reported only relatively recently. There are contradictory reports on cancer risks in connection with phenoxy acid herbicides. A relative risk of 5.3-6.8 for an increased frequency of soft tissue sarcoma in spray operators was reported by Hardell and coworkers (Hardell and Sandstrom, 1979; Eriksson *et al.*, 1979). The first study covered forest sprayers, where 2,4-D and 2,4,5-T were used. Dioxins as impurities in 2,4,5-T may play an important role. The second study was of agricultural workers who had used much less 2,4,5-T, still showed the same magnitue of risk of around 5, Smith *et al.*, (1981) reported a relative risk for soft tissue sarcoma of 1.3 in sprayers exposed to 2,4-D and 2,4,5-T in New Zealand. This study used other cancer types as controls. Wiklund (1986) studied a group of agricultural Swedish workers with regard to cancer in comparison with figures for Swedish cancer incidents during 1961-79 in the Cancer Environmental Registrar. No increase in the relative risk for soft tissue sarcoma was found. The only increase was in skin cancer (malignant melanoma). Of course, these summarized epidemiological studies concerning cancer risks cannot be used for biological monitoring purposes.

Even short periods of spraying, i.e. 1-2 h for 2 days intermittently, may result in measureable amounts of phenoxy acid herbicides in the urine of exposed workers. A single dermal exposure would result in excretion of 2,4-D, 2,4,5-T and MCPA. Five workers were studied by Chester and Ward (1984) during the aerial application of paraquat. Paraquat was collected on filter-paper pads for estimation of both dermal and respiratory exposure. Particle size was studied. Dermal exposure varied between 0.05 and 4 mg/1 per person. Some paraquat 26 Jl gm-4, could be recorded in the breathing zone of the flagman, but no respirable paraquat was detected. No symptoms were recorded.

Literature concerning the monitoring of human exposure to herbicides reveal that biological monitoring will give the most relevant information on exposure and uptake. Estimation of the herbicide in urine or a major metabolite seems to be particularly suitable. The excretion should be followed for several days as it may peak after a few days. Ideally, 24-h samples of urine should be used, but in practice it is difficult in field situations, and morning spot samples will give a good estimate of the excretion pattern.

The determination of potential deposits on skin and inhalable amounts of herbicide reveal the main routes of exposure. Most investigations have shown that dermal absorption is the dominant route. Most information is needed concerning the uptake, metabolism and excretion in humans of many of the herbicides used in the field. Mostly, the dodses ingested or applied to the skin of the volunteers should be kept low to mirror occupational field exposure. Such controlled experiments are very valuable in designing a proper scheme for biological monitoring, as has been shown for the phenoxy acid group. There is little inhalation and dermal uptake of paraquat. However, there are several reports on fatal intoxications among spraymen, some of which were quoted by Wester *et al.*, (1984) and Tungsanga *et al.*, (1983). In some cases the knapsack sprayers had leaked paraquat solution, and the paraquat had been readily absorbed by skin and wounds. Since paraquat seems to be readily absorbed through damaged skin, dermal exposure should be avoided as far as possible, despite the very low penetration rate through intact skin.

Human toxicology

Potential toxicology to the user is generally expressed in terms of the acute toxicity and chronic toxicity. Acute toxicity is the potential of a substance to cause injury or illness shortly after exposure. Chronic toxicity is the potential of a substance to cause injury or illness after repeated exposure over an extended period of time. Acute toxicity is generally expressed as an $LD_{50.}$ subchronic (90 days exposures) and chronic toxicities (12-, 18- and or 24-no exposures) reproductive effects, terataogenicity and several other categories of effects are expressed as NOAELs. In general the safety level for human exposure to a pesticide is set 100 to 1000 times below the NOAEL in order to provide a significant safety margin. This new value is called a Reference Dose (RFD). The RFD serves as the lowest toxicological value such that exposure below the RFD is not expected to cause harm to human.

For example, the NOAEL for chronic toxicity from a herbicide say 'A' when fed to rats over a 2 month period might be 400 mg/kg/d, with the hundred fold safety factor the RFD for the herbicide 'A' is 4 mg/kg/d, therefore, if the tolerance level for that herbicide on a food crop is 6 mg/kg, and an individual with a body weight of 70 kg has the potential to eat 0.5 kg of the crop per day, then the calculation (6 mg/kg times 0.5 kg/day divided by 70 kg) results in value of 0.43 mg/kg/d, well below the RFD (Ross and Lembi, 2000).

Carcinogenicity

Carcinogenicity is the process of induction of malignant neoplasms by chemical, physical or biological agents.

Carcinogenicity classification according to IARC

Classification based on the weight of the evidence and not on potency as follows.

1. Sufficient evidence: Causal relationship has been established between exposure to the agent and human cancer. A positive relationship has been observed between exposure to the agent and cancer in studies in which chance, bias and confounding could be ruled out with reasonable confidence.
2. Limited evidence: Positive association has been observed between exposure to the agent and cancer for which a causal interpretation is considered to be credible, but chance, bias or confounding could not be ruled out with reasonable confidence.
3. Inadequate evidence: Available studies are of insufficient quality, consistency or statistical power to permit a conclusion regarding the presence or absence of a causal association.
4. Evidence suggesting lack of carcinogenicity: There are several adequate studies covering the full range of doses to which human beings are known to be exposed, which are mutually consistent in not showing a positive association between exposure to the agent and any studied cancer at any observed level of exposure. A conclusion of "evidence suggesting lack of carcinogenicity" is inevitably limited to the cancer sites, circumstances and doses of exposure and length of observation covered by the available studies. In addition, the possibility of a very small risk at the levels of exposure studied can never be excluded.
5. Overall evaluation: Total body of evidence is taken into account. The agent is described according to one of the following categories. The categorization of an agent is a matter of scientific judgement, reflecting the strength of the evidence derived from studies in humans and in experimental animals and from other relevant data.

Based on carcinogenic data chemcials are grouped as follows;

Chemical Grop	Toxicological Remark
Group 1	The agent is carcinogenic to humans. This category is used only when there is sufficient evidence of carcinogenicity in humans
Group 2	This category includes agents for which, at one extreme, the degree of evidence of carcinogenicity in humans is almost sufficient, as well as agents for which, at the other extreme, there are no human data but for which there is experimental evidence of carcinogenicity. Agents are assigned to either 2A (probably carcinogenic) or 2B (possibly carcinogenic) on the basis of epidemiological, experimental and other relevant data
Group 2A	The agent is probably carcinogenic to humans. This category is used when there is limited evidence of carcinogenicity in humans and sufficient evidence of carcinogenicity in experimental animals. Exceptionally, an agent may be classified into this category solely on the basis of limited evidence of carcinogenicity in humans or of sufficient evidence of carcinogenicity in experimental animals strengthened by supporting evidence from other relevant data
Group 2B	The agent is possibly carcinogenic to humans. This category is generally used for agents for which there is limited evidence in humans in the absence of sufficient evidence in experimental animals. It may also be used when there is inadequate evidence of carcinogenicity in humans or when human data are nonexistent but there is sufficient evidence of carcinogenicity in experimental animals. In some instances, an agent for which there is inadequate evidence or no data in humans but limited evidence of carcinogenicity in experimental animals together with supporting evidence from other relevant data may be placed in this group
Group 3	The agent is not classifiable as to its carcinogenicity to humans. Agents are placed in this category when they do not fall into any other group.
Group 4	The agent is probably not carcinogenic to humans. This category is used for agents which there is evidence suggesting lack of carcinogenicity in humans together with evidence suggesting lack of carcinogenicity in experimental animals. In some circumstances, agents for which there is inadequate evidence of or no data on carcinogenicity in humans but evidence suggesting lack of carcinogenicity in experimental animals, consistently and strongly supported by a broad range of other relevant data.

Lethal concentration (LC)

The term lethal concentration (LC) refers to the concentration of a chemical required to produce a certain percentage to kill the test organism usually within 24 hours in air (inhalation studies) or water medium. LC_{50} (50 percent of the population is killed) and LC_{95} (95 percent population is killed) are usually used. LC_{50} is expressed either in milligram per liter of air or in part per million (ppm).

Chronic oral toxicity

Chronic oral toxicity is used to refer to the chemical administrated over an extended period of time without causing any change in the test organism. It is usually expressed in ppm of the chemicals in the diet.

Lethal dose (LD)

The term lethal dose (LD) is used to express the quantity of a herbicide (or any other chemical) required to produce a given mortality within a given period of time. The term LD is applied to chemicals administrated directly to the test organism orally or topically. The dosages are expressed in milligram per kilogram body weight. The lower the LD value, the more toxic the material. The LD_{50} is usually used (lethal to 50 percent of the population of test animals) and expressed either as acute oral LD_{50} or acute dermal LD_{50}. In acute oral LD_{50}, the chemical is taken orally as a single dose. In acute dermal LD_{50}, the chemical is applied to the skin, also as a single dose. Whilst the LD_{50} gives an indication of relative toxicity, it must be remembered that it is measured on statistical death rates and therefore will not indicate potential harmful effects of sub-lethal doses (i.e. how much active ingredient will make you sick) or the range of an effect on a population (whether everyone has similar reactions, or there is a wide spread in susceptibility). It is therefore important to treat all chemicals with caution, even if the LD_{50} indicates the moderate or slight toxicity. The categories of acute toxicity for herbicides lebelling are shown in Table 2.

Maximum residue limits (MRLs)

Maximum residue limits (MRLs) are defined as the maximum concentration of pesticide residues (expressed as milligram of residues per kilogram of commodity) likely to occur in or on food after the use of pesticides according to good agriculture practice (GAP). MRLs are established by national regulatory authorities and by the Codex Alimentarius Commission, which maintain the standards that are intended to apply to international trade. The Codex MRLs are based on assessment made by the Joint FAO/WHO Meeting on Pesticide Residues (GMPR).

Acceptable daily intake (ADI)

The acceptable daily intake (ADI) is generally expressed in terms of milligram of a pesticide consumed per kilogram of body weight (mg/kg) per day. It is the amount of residues that if ingested daily over a 70 years lifetime, a human could consume without expecting any health related problems.

Table 2. Probable lethal doses for a 70 kg adult male human for important herbicides

LD_{50} of active ingredient (mg/kg)	Estimated lethal amount of pure active ingredient	Example
<5	A few drops	None
5-50	1 teaspoon	Dinoseb
50-500	30 g or 2 tablespoons	Diquat
500-5000	500 g or 500 ml	Atrazine, alachlor
>5,000	>1000 g or 1000 ml	DCPA, chlorotoluron

Table 3. Toxicity and half-life of various herbicides

Herbicide	Toxicity			Half-life (days)	Rate of Degradation	
	Bluegill (LC_{50} ppm)	Rainbow trout (LC_{50} ppm)	Mallard duck (LC_{50} ppm)	Soil	Air	Water
1	2	3	4	5	6	7
2,4-D	>110	64	>1,000 mg/kg	46	Fast	Very fast
2,4-DB	4	4	>1000	-	-	-
Acetochlor	.3	0.45	>5,620	55	Slow	Slow
Acitofuron	62	17	5,620	14-59	4	2
Alachlor	6.4	4.2	5,000	21	Slow	-

Contd...

Contd...

Ametryn	4.1	8.8	23,000	60	Slow	-
Atrazine	>24	4.5	19,650	60	Slow	-
Benefin	-	-	5000	40	Moderate	-
Bensulide	1.4	0.72	-	120	Slow	-
Bentazon	616	190	>5,000	14	Very Slow	<24hr
Bromacil	71	75	>10,000	150	Very Slow	-
Bromoxynil	0.10	0.17	200	7	--	-
Butylate	6.9	7.2	>40,000	10-20	-	-
Cacodylic acid	133	152	>5,620	50	Very Slow	-
Carfentrazone	2	1.6	>5,620	1	-	8
Chlorimuron	>100	>1,000	>2,510	21-42	Slow	-
Chlorsulfuron	>250	>250	>5,000	28-42	Slow	30
Clethodium	>100	67	>3,978	2	2	28
Clomazone	34	19	>5,620	28-84	Very Slow	-
Clopyralid	125	103	>4,640	12-70	VerySlow	-
Cloransulam	>295	>252	>5,620	13-28	Slow	-
Copper chelate	1.2	0.2	>1,000	-	-	-
Cyanazine	23	9	>2,000	12-25	Slow	Slow
Cycloate	-	5.6	>56,000	28-42	Very Slow	219
Dicamba	>1,000	>1,000	>4,600	14-42	Slow	Slow
Dichlobenil	7	7	>5,200	60-180	Very Slow	10
Dichlorprop	1.1	100-200	-	10	-	-
Diclofop	-	1.98	>20,000	10-30	Fast	-
Diflufenzopyr	135	106	>5,620	30	Slow	Slow
Dmithenamide	6.4	2.6	>5,620	18	-	5
Dithiopyr	0.5	0.5	>5,620	-	-	-
Diquat	-	21	564	-	Fast	<2
Diuron	7.4	4.3	>5,000	90	Moderate	Slow
Endothall	-	-	-	7	Very Slow	-
EPTC	27	21	20,000	7-14	-	-
Ethalfluralin	0.03	0.04	>5000	42-56	Moderate	-
Fenoxaprop	3.3	3.4	Not Toxic	5-14	Slow	-
Fluazifop	0.5	14	>3,528	28-56	Very Slow	-
Fllufenacet	213	3.5	>4,970	29-62	Very Slow	Very Slow
Flumetsulum	>300	>293	>5620	90	Slow	Very Slow
Fluridone	14.3	11.7	>5,000	Moderate	-	21-42
Fluthiacet	0.14	0.043	>2,250	1	-	-
Fomasafen	6,030	680	>5,000	120-240	Fast	Fast

Contd...

Contd...

Glyphosphate Roundup	6-14	8-26	>4,640	3	-	-
Imazapyr	>100	>100	>5,000	90-180	Moderate	2
Imazaquin	>100	>100	>5,000	90-180	2	1
Imazethapyr	420	340	>5,000	60-180	1	1
Isoxaflutole	-	-	-	35	-	-
Lactofen	0.5	0.8	>5,620	14-21	23	23
Linuron	16	16	3,083	60-120	Very Slow	Moderate
MCPA	>10	117	-	10-180	Very Slow	Very-Slow
Mecoprop	-	-	-	7-21	Very Slow	-
Metachlor	10	4	>10,000	25-50	8	Very Slow
Metribuzin	>100	>100	>4000	30-60	Very Slow	Very-Slow
Metsulfuron	>150	>150	>2,510	30	Slow	-
Napropamide	30	17	-	56-100	4	-
Naptalam	354	76	>10,000	21-42	Very Slow	-
Nicosulfuron	>1,000	>1,000	>5,620	21	-	-
Norflurazon	16	8	>10,000	48-180	Moderate	Fast
Oryzalin	2.9	3.6	>5000	20-180	Moderate	-
Oxadiazon	-	>9	>1,000	60	-	-
Oxyfluoren	0.2	0.4	>5,000	30-40	Fast	Fast
Paraquat	Very Toxic	Very Toxic		-	Moderate	Moderate
Pebulate	7.4	7.4	>2000	14	-	-
Pendimethlin	1.0	0.5	10,388	30-60	Moderate	Moderate
Picloram	14	4	>5,000	70-100	3	3
Primisulfuron	>180	210	>2,160	13	Moderate	Moderate
Prodiamine	>0.5	>0.8	>10,000	120	2	
Promenton	>32	20	>4,572	100-200	Moderate	-
Prometryn	10	2.5	39000	30-90	Moderate	Slow
Pronamide	100	72	>20,000	60-270	Slow	-
Propachlor	1.6	0.42	>5,000	30-42	Very Slow	-
Prosulfuron	155	160	>5,000	10	118	-
Pyridate	2	9	>5,000	7-21	Slow	Slow
Quizalofop	0.5-3.0	14	>2,000	7-21	Moderate	-

Contd...

Contd...

Rimsulfuron	>1,000	10	>2,250	-	-	-
Sethoxydim	100	>1,000	>2,510	4-10	Moderate	Fast
Simazin	2.8	32	51,000	30-90	Moderate	-
Sodiumborate	Low	16	Low	>365	Very Slow	-
Sodiumchlorate	Very Toxic	Low	-	-	-	-
Sulfentrazone	94	Very Toxic	>5,220	110-280	Slow	-
Sulfometuron	>12.5	>130	>5,000	20-28	-	-
Tebuthrium	112	>12.5	>2000	360-500	Slow	-
Tebacil	86	144	>56,000	180	-	-
Thifensulfuron	>100	-	>5,600	7-21	-	-
Triallate	1.3	>100	>5,000	48-56	Very Slow	Moderate
Tribenuron	>1,000	>1,000	>5,620	10	Slow	Very Fast
Triclopyr						-
Garlon3A	891	552	>10,000	46	Fast	
Garlon4L	0.9	0.7	>10,000	-	-	
Trifluralin	Toxic	Toxic	>5,000	60-90	Moderate	-

Non observed adverse effect level (NOAEL)

The non-observed adverse effect level (NOAEL) is the highest dose of a substance in experimental animal studies that does not cause any detectable toxic effects. The NOAEL is expressed in milligrams of the substance per kilogram of body weight per day.

Maximum permissible intake (MPI)

Maximum permissible intake of a herbicide can be obtained by multiplying available daily intake with average body weight.

MPI= Acceptable daily intake x average body weight

Theoretical maximum residue contribution (TMRC)

TMRC = Daily consumption of the food commodity **x** residue in ppm

Fate and impact on the environemnt

Toxicology studies are required to be carried out on certain wildlife species of birds, fish and invertrbrats. In addition, the exposure of wildlife to pesticide residues in the environments must be extimated. The estimated enviromnetal concetartion (EEC) is detrmined by measurements of pesticide residues in the foods that birds and ammals might consume and the concentrations of the pesticide in water caused by runoff or spray drift that an aquatic organism might come in contact with. These studies may be conducted in the field and require periodic sampling and analysis, however many of these fate studies also can be done with computer models. The EEC and toxicity data are used in calculation to detemine the potential for harm to non-target species.

Toxicity data for aquatic species are expressed as LC_{50} (concetration of the compound in water in mg/l that causes mortality to 50% of the population after exposure). Compound with an LC_{50} of less than 1 mg/l are highly toxic to fish and should not be permitted to enter directly into water (Table 3). The majority of herbicide pose little risk to wildlife and fish near or in water however, the dinitroaniline herbicides such as trifluralin, benifin, ethalfluralin, oryzalin, pendimthalin, prodiamine are quite toxic to fish. These herbicides are not a problem when used correctly. Level instructions alert the applicator to potential problems with these herbicides and generally fish toxicty occur when these herbicides are misused. For example, when these are sprayed directly over water or are allowed to contaminate water when filling the spray tank (Ross and Lembi, 2000).

In addition to impacts on wildlife, the general question of environmental fate must be provided for all pesticides that are approved by EPA. Data required include the rate of degradation under various environmental conditions the identity of the breakdown products and the likelihood that a pesticide or its metabolites will migrate and or acumulate in the environment. Specific test or computer simulations are conducted to determine hydrolysis (breakdown in water), photodecomposition in water and soil, aerobic soil and anaerobic aquautic metabolism, mobility *via* runoff and leaching and terristeal field dissipation. These data are also used in variours ways to calculate the EEC (Ross and Lembi, 2000).

Summary

Toxicity is a measure of the amount of compound that is harmful or lethal. Exposure is the probability of encountering a harmful dose of a compound. A compound that is extremely toxic but is used in such a way that exposures is low may be less hazardous than a compound that is less toxic but whose application requires greater exposure of the applicator. The EPA (Environmental Pollution Agency) of the US has classified pesticides according to toxicity into four categories (I, II, III and IV). The categories are based on oral LD_{50}, inhalation LC_{50}, dermal LD_{50}, eye effects and skin irritation. EPA has also designated a signal word for each category, which should appear on the label of the pesticide. Before a pesticide is sold in the market, different toxicological and ecological studies are supposed to be conducted by the company to ensure no toxic/ill effect of herbicide on environment and human health.

□□□

Chapter 13

Formulations

Typically, pure herbicide molecules are of limited value to the end user. To have more practical value and use, most of the herbicides are combined with appropriate solvents or surfactants to form a product called a *formulation*. Herbicides are available as formulations and rarely as the pure chemical. The primary reason for formulating an herbicide is to allow the user to dispense it in a convenient carrier, such as water. The primary purpose of the carrier is to enable the uniform distribution of a relatively small amount of herbicide over a comparatively large area. In addition to providing the consumer with a form of herbicide that is easy to handle, formulating a herbicide can enhance the phytotoxicity of the herbicide, improve the shelf-life (storage) of the herbicide, and protect the herbicide from adverse environmental conditions while in storage or transit.

Formulations vary according to the solubility of the herbicide, active ingredient in water, oil and organic solvents, and the manner the formulation is applied (i.e., dispersed in a carrier such as water or applied as a dry formulation itself). The formulations are grouped as follows:

Sprayable formulations

(a) Water-soluble formulations

Soluble liquid (SL)

Water soluble formulations are designed for those active ingredients that dissolve readily in water. The formulation is a liquid and consists of the active ingredient and additives. When herbicides formulated as solutions are mixed with water, the active ingredient will not settle out of solution or separate.

Soluble powder (SP)

Soluble powder formulations are similar to solutions (SL) in that, when mixed with water, these dry formulations dissolve readily and form a true solution. The formulation is dry and consists of the active ingredient and additives. When thoroughly mixed, no further agitation is necessary to keep the active ingredient dissolved in solution. Few formulations of this type are available because few active ingredients are highly soluble in water.

Soluble granules (SG)

Soluble granules are dry and have larger particle size than soluble powder. These are soluble salts of various compounds. Considerable stirring or agitation may be needed to dissolve these herbicides, but once in solution these remain in that state indefinitely and form clear solutions in the sprayer tank and require a surfactant for maximum foliar activity. Typical formulation contain 40 to 95 % active ingredient.

(b) Emulsifiable formulations

Emulsifiable concentrate (E or EC)

Formulations of this type are liquids that contain the active ingredient, one or more solvents, and an emulsifier that allows mixing with water. Formulations of this type are highly concentrated and relatively inexpensive per pound of active ingredient, easy to handle, transport, and store and require little agitation (will not settle out or separate). These are not abrasive to machinery or spraying equipment. Formulations of this type may, however,

(i) have potentially greater phytotoxicity than other formulations.

(ii) exhibit a potential for over- or under dosing through mixing or calibration errors and are more easily absorbed through skin of humans or animals. These formulations contain solvents that may cause deterioration of rubber or plastic hoses and pump parts.

Gels (GL)

Gels are relatively new products that are thickened emulsifiable concentrate packed in water soluble bags. Gel can be formulated so these don't leak from pinhole size tears in the bags. The bags are pre-measure so that user knows exactly how much herbicide is being added to the spray tank.

(c) Liquid suspensions (L or F) to be dispersed in water

Suspension concentrate (SC) and Aqueous suspensions (AS)

Suspension concentrate (SC) are finely divided solids suspended in a liquid such as water or oil. If suspended in water, can be called aqueous suspension. These are added to the water in the spray tank for dispersal. Suspension concentrates contain surfactant to keep the particles dispersed and tend to have smaller amounts of organic solvents than emulsifiable concentrates. These consist of ingredients having low water solubility and can settle out during storage.

Emulsion of a water dissolved herbicide in oil (EO) and emulsion of an oil dissolved herbicide in water (EW)

Emulsion of a water dissolved herbicide in oil consists of fine globules of the herbicide in water dispersed within a continuous liquid oil phase. An emulsion of an oil dissolved herbicide in water is just the opposite, in other words, it consists of a herbicide in fine globules of oil dispersed within a continuous water phase. Although these formulations have many of the same characteristics as other liquid suspensions, and are packaged as liquid droplets in a liquid phase. In contrast of a single oil based phase, these emulsions are packed in two phases, water and oil. The advantage is the ability to mix two unlike or incompatible formulations, in the same package. For example squadron, which is a combination of pendimethalin and imazaquin is water soluble, is an EW formulation.

Microencapsulated formulations (ME) or capsule suspension (CS)

Microencapsulated formulations are small particles consisting of a herbicide core surrounded by a barrier layer, usually made up of a polymer shell. These are also refereed to as capsule suspensions because the capsules are suspended in a liquid medium. In contrast to emulsifiable concentrates, which require a large amount of solvent to ensure suspension, microencapsulation greatly reduces the amount of solvent needed. Thus microencapsulation reduces the cost of formulation and the amount of solvent (sometimes toxic) that will be dispersed into the environment. The polymer shell protect the user from contact with the herbicides, shield the herbicide from degradative agent such as light and reduces the rate and amount of herbicide loss due to volatilization and leaching.

(d) Dry solid to be suspended in water

Wettable powder (W or WP)

Wettable powders are dry, finely ground formulations in which the active ingredient is combined with a finely ground carrier (usually mineral clay), along with other ingredients to enhance the ability of the active ingredient plus carrier to suspend in water. The powder is mixed with water for application.

Wettable powders are one of the most widely used herbicide formulations and offer low cost and ease of storage, transport, and handling; lower phytotoxicity potential than ECs and other liquid formulations; and less skin and eye absorption hazard than ECs and other liquid formulations.

Some disadvantages are that these require constant and thorough agitation in the spray tank, are abrasive to pumps and nozzles (causing premature wear), may produce visible residues on plant and soil surfaces, and can create an inhalation hazard to the applicator while handling (pouring and mixing) the concentrated powder.

Water dispersible granules (WDG, WG, DG) or dry flowables (DF)

Dry flowable and water-dispersible granule formulations are much like wettable powders except that the active ingredient is formulated on a large particle (granule) instead of onto a ground powder. This type of formulation offers essentially the same advantages and disadvantages as wettable powder formulations. However, these formulations generally are more easily mixed and measured than wettable powders. Because these create less dust when handling, cause less inhalation hazard to the applicator during pouring and mixing.

Dry applications

(a) Granules and Matrix granules

Granules are dry formulation of herbicides and other components in discrete particles generally less than 10 mm^3 size. The herbicide is coated on the exterior of the particle. A variety of dry substances have been utilized as granule components including clay minerals, dry fertilizers and ground plant tissue. Herbicide concentrations are typically 2 to 20 %. Generally granular herbicides require slightly more rainfall to leach into the soil than sprayable formulation do. The rate of herbicide loss through volatilization can be reduced if formulated as a granule.

Matrix granules made of starch in which the herbicides is trapped within the particles, have been used to slow the release of certain herbicides into the soil, however slow release technology has not yet been widely adopted by herbicide manufactures.

(b) Pellets (P) or tablets (TB)

Pellets are dry formulation of herbicide and other components in discrete particles usually larger than 100 mm^3. Tablets are in the form of small flat pellets. Pellets and tablets are frequently used for spot applications. Herbicide concentrations typically are 5 to 20%. The primary advantages of this type of formulation are that the formulation is ready to use with simple application equipment (seeders or spreaders), and the drift potential is low because the particles are large and settle quickly. The disadvantages of these formulations are that do not adhere to foliage (not intended for foliar applications), and may require mixing into the soil in order to achieve adequate herbicidal activity.

Summary

Typically pure herbicide molecule are of limited value to the end user. To make herbicides more practical and usable, most herbicides are combined with appropriate solvents or surfactants to form a product called a *formulation*. Herbicides are available as formulations and rarely as the pure chemical. The primary reason for formulating a herbicide is to allow the user to distribute it in a convenient carrier, such as water. Formulating a herbicide can enhance the phytotoxicity of the herbicide, improve the shelf-life of the herbicide, and protect the herbicide from adverse environmental conditions while in storage or transit. Formulations vary according to the solubility of the herbicide active ingredient in water, oil organic solvents, and the manner the formulation is applied (i.e., dispersed in a carrier such as water or applied as a dry formulation itself).

Formulation such as soluble liquid (SL), soluble powder (SP) soluble granules (SG) are water soluble while emulsifiable concentrate (E or EC) gels (GL) are emulsifiable formulations. However, emulsion of a water disoluble herbicide in oil (EO) and emulsion of an oil soluble herbicide in water and microencapsulated formulations (ME) or capsule suspension (CS) are liquid suspensions (L or F) to be dispersed in water. Wettable powder (W or WP), water dispersible granules (WDG, WG, DG) or dry flowables (DF) come under the category-dry solid to be suspended in water. Whereas, granules and matrix granules and pellets (P) or tablets (TB) are applied as dry applications.

□□□

Chapter 14

Application Techniques and Equipments

Herbicides can be applied in a variety of different ways to crop. They can be applied as single component products or as combination products, using the different properties and weed spectrum of the components of the mixtures. Application of herbicide is largely decided by their mode of action and selectivity. An improper method of application of a herbicide results in poor weed control and /or severe crop injury. Important methods of application of herbicides to crop and non-crop areas are given as below;

Application Techniques

(1) Soil application of herbicides

Surface application

Soil active herbicides are commonly applied to the surface of the soil where they may be either left undisturbed or incorporated into the soil. Even when left on the soil surface, the applied herbicide should be able to move into upper 3-4 cm of soil under the influence of rain or irrigation to kill the germinating weeds. Examples, many triazines, ureas and anilides herbicides. Sometimes a surface applied herbicide may have to be physically incorporated into the soil for varied reasons. For example, the triazines perform well in dry weather when these are mixed lightly with the treated soil. Other soil applied herbicides such as carbamates and toludines must be incorporated into the soil to prevent their rapid volatilization or photodecomposition. The physical incorporation of herbicides can be achieved with pick axes in small areas and with animal or power driven implements in larger fields. The optimum depth of incorporation of different herbicides for achieving selective control of weeds varies considerably. For example, EPTC is incorporated 5 to 7.5 cm deep in soil whereas trifluralin and nitralin are incorporated not beyond 5 cm depth. On the other hand, diallate and triallate herbicides are mixed with soil only upto 2.5 cm depth. While most herbicides are incorporated into the soil physically, EPTC and fluchloralin can be incorporated also with irrigation water (Fig. 1).

Fig. 1. Application of herbicides

Subsurface layering

It is the application of a herbicide in a concentrated band about 7-10 cm below the soil surface. This technique has proved effective in controlling certain perennial weeds with conventionally, soil incorporated herbicides, usually employed to control only annual weeds. The concentrated layer of the herbicide inhibits the growth of new shoots of specific perennial weeds such as *Cyperus rotundous* and *Convolvulus arvensis*. The subsurface layering of herbicides is achieved with special nozzles introduced below the soil under the cover of sweep hoods.

Broadcast and band application

Broadcast application of herbicides is characterized with the application over the full surface area, without leaving any intentional gaps. Whereas, band application usually means treating a narrow strip directly over or on the crop row. Band treatment saves on the chemical.

The area not treated by herbicides is normally put under manual and cultural methods of weed control. The band application of herbicide is primarily a cost saving device since it reduces the quantity of herbicide in the ratio of the treated band width to the crop width such as a 30 cm wide band of a herbicide applied over crop rows that were spaced 90 cm apart, will require one third of the quantity of herbicides needed for its application by the broadcast method. The herbicide band width usually 30-35 cm in the band application method, the inter rows must be cultivated later to remove weeds from the unsprayed areas. This will also help in the control of perennial and some other weeds, which may be resistant to the herbicide.

Soil fumigation

Depending upon the nature of soil fumigant it can be applied either by soil injection e.g. chloropicrin, or by releasing it under sealed plastic covers example methyl bromide or by direct soil surface application example metham.

(2) Foliage application of herbicides

Blanket application

Blanket application or over to top application of herbicides is uniform application to standing crops disregard to the location of the crop plants. Only highly selective herbicides are applied by this method example 2, 4-D in wheat, MCPB in pea, 2,4-DB in lucerne, picloram in sugarcane etc.

Directed spraying

In tall growing crops grown in widely spaced row, it is possible to apply herbicides to the weeds or soil without treating the foliage or shoot of the crop, e.g., application of paraquat in sugarcane crop. Directed application is done by covering the spray nozzle under a hood or shield or by carefully directing the nozzles to the inter-row space. This saves the crop plants from herbicide injury and improves weed kill. In potato and soybean, vine lifters are used to improve the accuracy of directed spraying. One important prerequisite of directed spraying is that farmer must get his crop well up over the weeds at the time of treatment. This calls for a good seedbed preparation and planting so that crops will emerge rapidly and gain a height advantage over the weeds. Orchards and plantation are good venues of directed herbicide sprays. Usually the directed spraying is adopted with herbicides that are only partially selective to the treated crops. But when feasible one should apply even selective herbicides by this method to place most of the herbicide on the weed. Herbicide falling on the crop plants is obviously a waste (Fig. 2).

Fig. 2. Directed application of herbicides

Protected spraying

Non-selective herbicides can be employed to obtain selective weed control in distantly planted vegetables and ornamentals by covering the non-target plants before application of the herbicide with plastic or metallic covers. This can be done in 3-4 rows at a time and shifting the covers to the untreated rows each time. Protected spraying is somewhat laborious but it pays itself when weeding of high value vegetables crops and ornamentals are involved (Fig. 3).

Fig. 3. Protected spraying of herbicides

Spot treatment

Spot treatment is usually done on small areas of serious weed infestation to kill it and to prevent its spread. Herbicides are applied either to soil or foliage. When applied to the soil, they are known as soil applied herbicides and when applied to the foliage of the weed, referred as foliage applied herbicides (Fig. 4).

Fig. 4. Spot treatment of herbicides

(3) Methods of treating brush and trees

Brush weeds and unwanted trees are treated with herbicides by different methods, depending upon the situation. Foliage treatment is the most common method of treating brush. The treatment is done when the brush leaves are fully expanded, though still growing actively (Fig. 5). Ground sprayers can coverup to 2.5 m high brush while aerial spraying is needed for more tall brushes and trees. A better method of dealing with tall brushes is to treat their basal 30 cm of stem, preferably after peeling off their bark, to the point of liberal runoff. This is called basal bark treatment. Sometimes the harder to kill brush are first subjected to foliar treatment and then these are retreated by the basal bark method.

The third method of application of herbicides to brush is the cut stump treatment (Fig. 6). It comprises sawing of the tree above the ground followed by liberal application of the herbicide on the cut surface. Other ways by which the concentrated herbicides are applied to unwanted trees are frill, notch and injection methods (ethylene, carbon bisulfide and vernolate). The frills and notches are made with sharp tools into the sap wood at convenient stem height and filled with herbicides. The herbicides injections are made into holes made in the tree trunk. Usually one herbicide injection per 2.5 cm stem thickness is adequate. The frill, notch and injection methods are adopted on thick stem trees which are 8 cm or more in diameter. The treatment of stems by any one of the above methods is practiced when either selective brush control is important or when application of herbicides is not feasible.

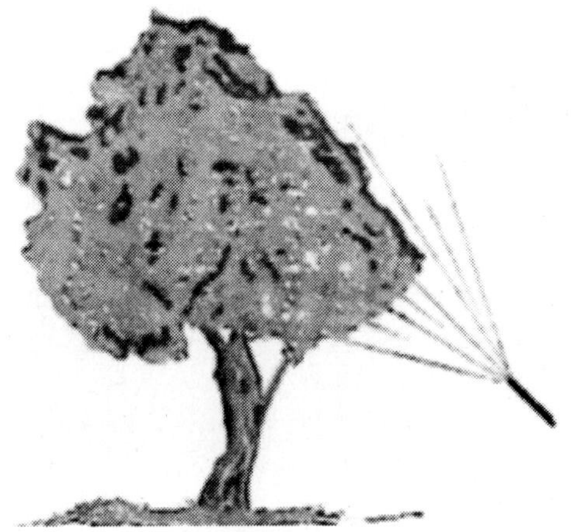

Fig. 5. Foliage Treatment

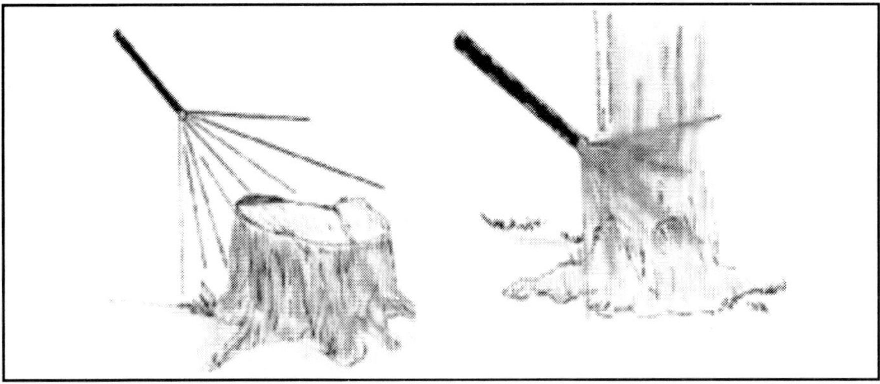

Fig. 6. Stumps and basal bark treatment

Types of herbicide treatments

Pre-plant incorporated

Pre-plant incorporated herbicides are applied before the crop is sown and are incorporated into the soil (Fig. 7). Hence, they are also applied before weeds emerge. The herbicides are incorporated in soil because of being their volatile nature. They would be lost if they are not incorporated. They would be degraded if they remained on the soil surface. Volatility is a useful characteristic as it allows the redistribution of the compound throughout the soil following incorporation. EPTC may be incorporated into soil prior to planting the crop. Pre-plant incorporation of fluchloralin at 1.0 kg/ha is recommended for groundnut. Other pre-emergence herbicides include bromacil, diuron, oryzalin, and tebuthiuron.

Fig. 7. Incorporation of herbicides in the soil

Pre-emergence

Pre-emergence herbicides are applied pre-weed emergence and this will usually mean pre-crop emergence as well (Fig. 6). Herbicides which have greater toxicity on the emerging crop seedlings are applied before the crop is planted. Such compounds are taken up underground by the roots or hypocotyls of the weed. It is important for such compounds to possess some water solubility, in order that they become available to the germinating weed, but not so much that they are leached away from the weed germination zone. They must also be relatively persistent in the soil so that weeds that germinate over a period of time are all controlled (Fig. 8).

Post-emergence

Post-emergence herbicides are applied after the emergence of the weed (and usually, but not necessarily, the crop as well). Compounds such as sulfuric acid or paraquat must be applied in such manner so that they cover all the foliage of the target weed as they are contact herbicides. Others, such as the auxin- herbicides, are taken up by the weed and translocated throughout the target plant. They are systemic and, consequently, it is not so important to ensure that the whole of the weed is covered. Some post-applied compounds are only active through the foliage e.g. bentazon and the auxin-herbicides, while others are taken up through the roots following application, e.g. isoproturon . More recent compounds, such as the sulfonylureas, can be taken up through the foliage and the roots. Hence, the fact that a herbicide is applied post-weed emergence does not indicate that the compound is taken up by the foliage. It is merely a convenient description of the use of the compound. Post-emergence compounds can be applied to the entire crop/weed canopy. This is often described as an over-the-top application. Alternatively, the compounds can be directed away from the crop at the weeds post- directed application. Examples of foliage-absorbed herbicides include 2, 4-D, diquat, fosamine, glyphosate, and triclopyr.

Lay by application

It is application of herbicides after the last cultivation in crops, such as after ridging in sugarcane and cotton. Usually soil-acting herbicides are used, which will prevent further emergence of weeds (Fig. 9).

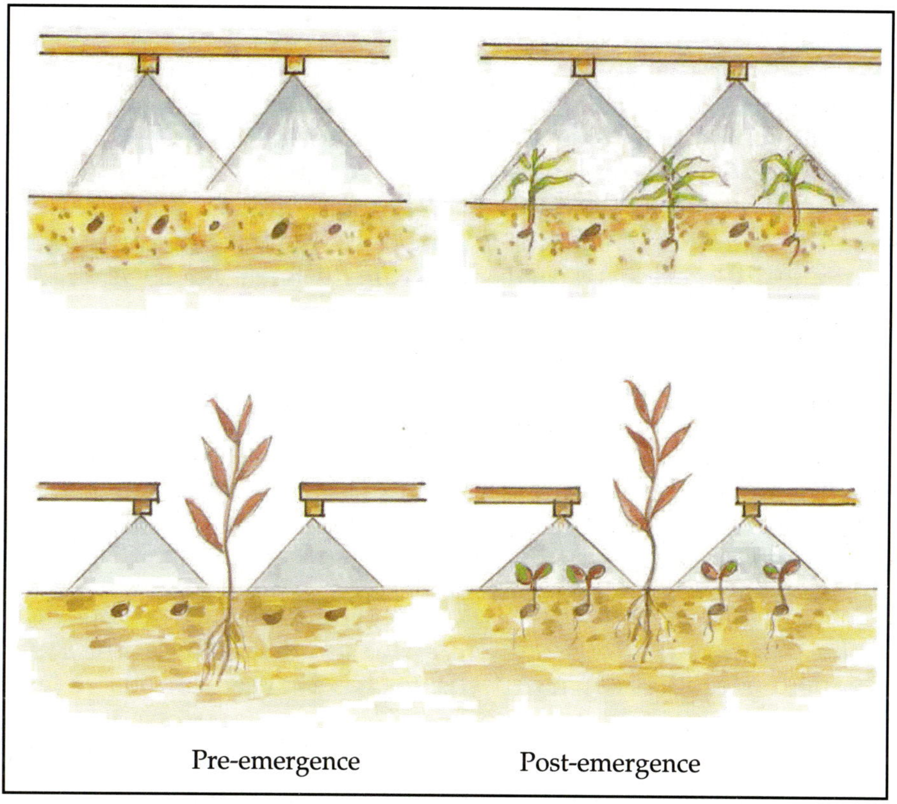

Pre-emergence Post-emergence

Fig. 8. Pre and post-emergence application of herbicides

Fig. 9. Lay by application of herbicides

Herbicides vs. other pesticides

Extensive toxicological tests are conducted before a herbicide is registered ensuring that no toxicity problem arise from its commercial use. In India, out of 181 registered pesticides (as on 2007-08), of the total 49 herbicides, none is classified as extremely hazardous, only 3 have been classified as highly hazardous and 24 as moderately hazardous and 22 as slightly hazardous. Against this, of total 52 pesticides, 9 were classified as extremely hazardous, 32 highly hazardous, 11 moderately hazardous and 2 slightly hazardous indicating the relatively less toxicity of herbicides (Table 1). The list of herbicides fallen under different categories is given in Ann.1 Further the LD_{50} values of other pesticides especially insecticides is very low if compared with the herbicides LD_{50} values which, clearly indicates the less toxic behaviour of herbicides suggesting less harmful effect to the environment and public health. The low toxicity of herbicides for rats is also seen in the results of test on birds and fish.

Many herbicides interfer with biochemical processes important to the plant and of no significance to animals. There are some important exceptions. A few pesticides have caused serious or fatal poisoning of people. While the likelihood that some types of toxic effects seen in animals may be seen in laboratory animals submitted to even more complex toxicity tests whose significance is very difficult to assess. The purpose for which these animals tests are carried out is to provide a basis of assessing risks of toxic effects appearing in people exposed to herbicide under investigation. It is important, therefore to begin by looking at the way in which people may be exposed to herbicides so that if ill effect are produced, the group in which they are most likely to be first seen can be identified.

Table 1: Toxicity of herbicides vs. other pesticides

Toxicity class	Toxicity category	Herbicides	Insecticides
Extremely hazardous	I	-	9
Highly hazardous	II	3	30
Moderately hazardous	III	24	11
Slightly hazardous	IV	22	2

Herbicides are different from other pesticides as far as their application, is concerned. A particular insecticide may be used to control the same insects in different crops; however, different herbicides have to be used in different crops often to control the same weed. This aspect of crop specificity is to be understood properly, otherwise there will be problems of crop injury. Herbicides are invariably used either at planting or immediately after planting (Fig. 10). They are never used directly on the crop at harvest. This longer interaction between application and crop harvest will result in the dissipation of herbicides in the plant system and in soil. The research data suggest that half-life of most herbicides under field condition in India comes from less than 10-60 days. This is due to low rates of application and prevalence of high rainfall and high temperature. Moreover, the residues of herbicides in soil and crop produce are not detected at the time of harvesting, they are detected in food grains only when they are applied in higher amount than recommended rates but of course well below the maximum residue limits. Similarly most of the herbicides are tightly bound to the soil colloids and are therefore not likely to leach in the soil profile to pollute ground water.

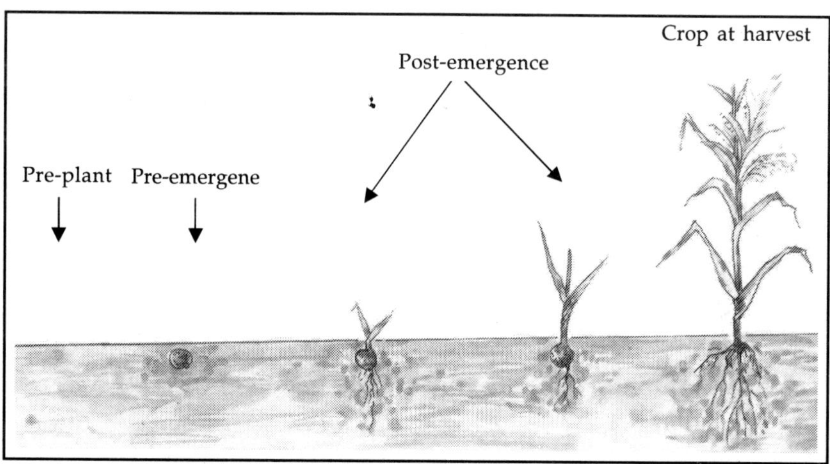

Fig. 10. Herbicide application at different stages of plant growth.

Factors affecting herbicide activity

1. Foliage applied herbicide

Foliar herbicides may also be selective or non-selective and are classified into two groups, those that kill on contact or those that are translocated through plant tissue into the root system and kill the whole plant. Examples of non-selective foliar applied herbicides are glyphosate, MSMA (both translocated) paraquat and diquat (contact herbicides). Selective translocated herbicides include 2, 4-D, dalapon, dicamba and picloram; contact selective herbicides include propanil and dinoseb. Contact herbicides usually act quickly and are useful in controlling annual weeds or perennial weed seedlings. They are less effective in established perennial weeds because regrowth can occur from roots or underground stems. Translocated herbicides will kill annual weeds but their major use is in controlling perennial herbaceous weeds and woody weeds. All of the contact herbicides and some of the translocated herbicides are only effective when applied to leaves because they are rapidly inactivated in the soil. Foliar uptake depends on a number of factors:

Type of herbicide

Molecular structure may effect penetration of a herbicide into the plant tissue (Sargent and Blackman 1969; Buta and Steffens 1971, Robertson *et al.*, 1971). Penetration of phenoxy acetic acids into bean leaf tissue increased with progressive chlorination of the parent molecule (Sargent and Blackman 1969). On the other hand, chlorination of benzoic acid depressed penetration. Difference between the rates of penetration of these particular derivatives arises from differences in their lipid solubility. Thus in general, modification of molecular structure which results in increased lipid solubility will enhance foliar penetration.

Age of plant when treated and nature of leaf surface

The stage of plant development i.e. ratio of young to mature leaves, leaf stem ratio, may markedly influence spray retention (Davies *et al.*, 1967). Plant age affects the uptake of herbicide, its translocation and activity in the plant. Young actively growing plants are more susceptible than older plants.

Absorption through the leaves is affected by the hairiness of the surface, angle of leaf and presence or not of a waxy leaf surface. Plants with smooth leaf surfaces are readily wetted by water based solutions whereas hairy/waxy surfaces are not. Mature fully expanded leaves generally absorb less than immature expanding leaves. Leaves damaged mechanically or by insect are more permeable than non-damaged leaves. Adequate moisture favours absorption, probably by maintaining the cuticle in a highly hydrated state (Overbeek, 1956).

Environmental conditions

The environmental factors affecting herbicide action are those of soil moisture, rainfall, wind, relative humidity, temperature and light. Rainfall influences the ability of leaves to retain water soluble herbicides which are reduced in activity if rain occurs within a few hours of application. It does not affect oil-based herbicides. In addition if plants are under water stress they do not readily absorb herbicides. Very low relative humidity will increase herbicide evaporation from leaf surfaces and also reduces leaf moisture content, thus do not favour foliar uptake. Wind increases the drying of herbicide on the leaf surface and reduces uptake.

Light affects herbicide uptake in different ways depending on weed species but at high intensity reduces herbicide efficiency through photodecomposition of the herbicide. Relatively low intensities (5000 to 15000 lx) are adequate for maximal response (Greene and Bukovac, 1972). Some herbicides such as diquat and paraquat obtain greater penetration of leaf surfaces in the dark i.e. spray late in the day, whereas 2, 4-D is absorbed more in strong light than in darkness.

Foliar absorption of herbicide is temperature dependent. The effect of temperature is generally one of increasing absorption at higher temperatures, although volatilization of herbicide may be increased. Sands and Bachelard (1973) found in the light, increased penetration of picloram into *Eucalyptus viminalis* leaves with an increase in temperature but only a slight increase or no effect in the dark.

Foliar absorption of herbicides is generally favoured by high relative humidity. High relative humidity increases the drying time of the spray droplets (Prasad *et al.*, 1967), favoured stomatal opening, enhance transport and may increase the permeability of the cuticular membrane. (Overbeek 1956). Greater quantities of 2, 4-D absorbed and transported in bean plants at 70 to 74 % relative humidity than at 34 to 48 % relative humidity. This increase in absorption is correlated with degree of stomatal opening. Wind increases the drying of herbicide on the leaf surface and reduces uptake.

The environment under which the plant develops may markedly affect the absorption of a chemical subsequently applied to the foliage. Leaves expanding in full sunlight produce a heavier cuticle than those developing in shade (Skoss, 1955). Low root temperature and moisture stress reduce the foliar absorption (Skoss, 1955).

Use of additives to aid herbicide absorption

Addition of a surfactant or wetting agent will help to breakdown spray droplets into a smaller form that can more readily make contact with the leaf surface. Oil based sprays are more effective on plants with waxy leaf surfaces. Most frequently cited effect of surfactant is the lowering of the surface tension of the spray solution. Generally as the concentration of the surfactant is increased, the surface tension is lowered at a point beyond which further addition of surfactant is without effect. This critical micelle concentration (CMC), lies between 0.01 to 0.5 % for most efficient surfactants (Osipowm 1964). A similar relationship has been observed between surfactant concentration and wetting of the plant surface (Becher and Becher 1969). In contrast herbicidal effectiveness is often maximal at concentration 10 times the CMC concentration or greater (Foy and Penner, 1965). Chemical interaction between surfactant and the herbicide may also occur in the spray solution, in most instances resulting in reduced efficacy (Smith and Foy 1967). Lowering the surface tension improve wetting and consequently the area of contact between the applied herbicide and the leaf surface is increased, however, the relationship between the surfactant and wetting is often quite specific. A surfactant producing a given surface tension may improve wetting, hence retention of a difficult to wet plant surface, while run off from an easy to wet surface may be excessive and retention time less than in absence of a surfactant. Similarly the drying time and characteristic of the spray droplet could be considerably modified by the presence of a surfactant. Surfactant may modify the plant surface by solubilizing the waxes or may interact with the cutin matrix, thus altering its charge characteristic and swelling properties and hence the membrane resistance to diffusion of a specific herbicide.

Time course

The absorption of foliar applied herbicides from spray droplets is initially rapid (Singh *et al.*, 1972; Sands and Bachelard 1973). With increasing time after treatment, penetration decreases at an increasing rate. The progressive reduction in penetration with time has been associated with the rate at which the droplet evaporates. Penetration apparently continues from the residue on the leaf surface, since there is a slight positive slope to the absorption curve. Continued penetration from the residues is more pronounced if the chemical is hygroscopic or if a surfactant or hemectant is added e.g. glycerin is added to the spray solution. Rewetting of the residue either experimentally or through the action of dew may markedly enhance penetration.

pH and concentration

pH plays a significant role in the penetration of weak organic acid type herbicides. The undissociated molecule is more lipids soluble and penetrates more readily than the anion. The effect of pH appears primarily on the penetratnt. Although there are dissociate groups at the cuticular surface, they do not pose an insurmountable obstacle to penetration. Patteen (1959) reported enhanced penetration of 2,4-D at pH levels above the pKa, particularly with ammonium and dihydrogen phosphate ions. The dissociate group within the cuticle even in the case of isolated cuticles appears to be little affected by the pH of the spray solution. This is undoubtedly because the spray solution does not penetrate sufficiently to influence these groups. pH may also indirectly influence penetration by modifying membrane potential or the metabolic activity of cell involved in the uptake transport processes.

At high concentration of the herbicides physiological changes may be induced in the uptake and transport process thus altering subsequent penetration.

2. Soil-applied herbicides

Factors influencing herbicide activity include application rate, application technique, plant maturity, and environmental conditions. In addition, soil characteristics can affect soil-active herbicides. For a soil-applied herbicide to be effective, the herbicide needs to be available for uptake by the germinating weed seedlings. The soil-applied herbicide must be absorbed into the germinating weed seedling to provide adequate control. Herbicides do not prevent weed-seed germination; rather, they are first absorbed by the root or shoot of the seedling and then exert their phytotoxic action. This generally happens before the seedling emerges from the soil. For a herbicide to be absorbed by the germinating seedlings, the herbicide must be in the soil solution or vapour phase. The most common methods for herbicides to become dissolved into the soil solution are by mechanical incorporation or precipitation.

Many early preplant applications in no-till systems attempt to increase the likelihood that sufficient precipitation will be received before planting to incorporate the herbicide. If, however, no precipitation is received between application and planting, mechanical incorporation, where feasible, will in most instances adequately move the herbicide into the soil solution. Many weed species, in particular small-seeded species, germinate from fairly shallow depths in the soil. The top 2.5 to 5 cm of soil is the primary zone of weed-seed germination and should thus be the target area for herbicide placement. Shallow incorporation can be achieved by mechanical methods or by precipitation.

Annual plants are usually more susceptible to herbicides when they are small than when they are mature. As they mature, plants develop thicker wax layers on leaf surfaces, reducing herbicide absorption. In addition, it is harder to achieve thorough spray coverage on large plants than on small plants. Established perennial weeds tend to be more susceptible to herbicides if applied during the early flowering stage of growth or to actively growing plants in the fall, probably because application at these times results in the greatest translocation of the herbicide to the roots. However, true seedlings are much easier to control than established perennial weeds.

Rainfall provides for a fairly uniform incorporation, but mechanical incorporation reduces the absolute dependence on receiving timely precipitation. How much precipitation is needed and how soon after application the precipitation should be received for optimal herbicide performance depends on many factors, but generally 1.25 to 2.5 cm of precipitation within 7 to 10 days after application is sufficient. Mechanically incorporated herbicides tend to provide more consistent weed control than surface-applied herbicides because the herbicide is in place, and adequate moisture usually is present in the soil to activate the chemical. However, incorporation too deep may dilute the herbicide so weed control is poor. Improperly adjusted equipment or incorporation when, soil is too wet, may result in streaking and poor weed control. To ensure good results, incorporation should be done with two perpendicular passes, 24 hours after application. The second incorporation pass usually should be done more shallowly so that untreated soil will not be moved into the herbicide zone. The label instructions must be followed concerning incorporation, depth and adjustment of the equipment based on soil characteristics, crop residue, and preplant tillage. If the field has more than 40 to 50 percent crop residue cover, it may be necessary to till it prior to herbicide application and incorporation. Regardless of when or how a herbicide is applied to the soil, the effectiveness of soil-applied herbicides is influenced by several factors.

Soil moisture

Soil moisture influences activity of soil-applied herbicides. Precipitation is essential to move surface-applied or pre-emergence herbicides into the soil and activate them. Soil moisture is important because it influences herbicide adsorption to soils. Therefore, the herbicide is unavailable for plant uptake. Adsorption occurs when herbicide molecules adhere to soil particles and organic matter. While adsorbed, herbicide molecules are unavailable for absorption by plants. Water molecules compete with herbicide molecules for adsorption sites on soil particles and organic matter. Therefore, herbicide adsorption is higher under dry soil conditions, and lower in moist soils. Consequently, weed control is generally satisfactory with moist soil conditions because more herbicide is available for plant uptake in the soil solution or gaseous phase. In addition, excess moisture from heavy rainfall can cause herbicides to leach or concentrate in sufficient quantities in the crop germination zone to damage the crop.

Temperature

Temperature affects the activity of soil-applied herbicides primarily because of its influence on the rate of seed germination, emergence, and growth. Seedlings tend to be more susceptible to soil-applied herbicides under cool conditions than under warm temperatures because plant emergence is delayed and metabolism is slowed. On the other hand, extremely high temperatures sometimes increase crop injury simply by placing the plant under multiple stresses.

Soil characteristics

Soil characteristics affecting herbicide activity are texture, organic matter, and pH. Herbicide adsorption is greater in fine-textured soils, high in organic matter than in coarse-textured soils low in organic matter. Thus, a lower proportion of herbicide is available in the fine-textured soils, so a higher herbicide application rate is required to provide the same level of weed control as in a coarse-textured soil. At the same time, the chance of crop injury is greater on coarse-textured soils, low in organic matter because a higher proportion of the applied herbicide is available for plant uptake. Soil-applied herbicide rates usually need to be adjusted according to soil texture and organic matter content.

Soil pH

Soil pH influences the availability and persistence of certain herbicides in the soil. Soil pH can alter the ionic nature of the herbicide molecule, which influences adsorption, solubility, and rate of herbicide breakdown. The triazine herbicides (atrazine, metribuzin, and simazine) and some of the sulfonylurea herbicides (chlorsulfuron, chlorimuron-ethyl, primisulfuron-methyl, prosulfuron, sulfometuron-methyl, and triasulfuron) are more active and more persistent in high pH soils (> 7.0) than in low pH soils.

Environmental conditions

Environmental conditions can have a two-fold effect on the performance of post-emergence herbicides. Higher humidity and favorable temperatures generally result in greater herbicide absorption and activity in plants. Environment also influences herbicide efficacy by affecting plant growth. Plants are generally most susceptible to post-emergence herbicides when actively growing. Extreme environmental conditions that slow plant growth and thicken leaf cuticles often increase plant tolerance to a herbicide. Crop injury from a herbicide, however, can increase during poor growing conditions because of slower metabolism and detoxification of the herbicide. Thus, if crop tolerance is based on the ability of the crop to rapidly metabolize the herbicide, the potential for crop injury may increase and weed control decrease if a herbicide is applied when plants are not growing actively. For this reason, most herbicide labels caution against application during extreme environmental conditions.

Herbicides remaining on the soil surface or those placed too deeply in the soil may not be intercepted by the emerging weed seedlings. Herbicides on the soil surface are subjected to several processes that reduce their availability. Volatility and photolysis are two common processes that can reduce the availability of herbicides that remain on the soil surface. Volatilization can be reduced through the incorporation of the herbicide into the soil by mechanical incorporation, irrigation, or precipitation. Proper incorporation ensures that volatile herbicides such as trifluralin can penetrate germinating weed seedlings as a gas. Care should be taken so that the herbicide does not move into the crop germination zone. For example, if herbicides are incorporated too deeply or wheat is seeded too shallow, the herbicide may come in contact with the developing wheat root system and inhibit its growth.

Soil-applied herbicides can also be lost through photodegradation, microbial degradation and chemical degradation. Photodegradation is the breakdown of herbicide by the action of sunlight. For example, herbicides such as trifluralin and eptam (EPTC) are photosensitive and can be lost from the soil if left on the surface.

Dry soil conditions may be conducive for planting but may also reduce the effectiveness of soil-applied herbicides. If applications are made prior to planting and no precipitation is received between applications and planting, a shallow mechanical incorporation may help to preserve much of the herbicide's effectiveness. Microbial degradation and chemical degradation processes are widely influenced by soil temperature and humidity.

The conditions that affect herbicide efficacy can also affect crop injury. For example, trifluralin can severely inhibit root growth of wheat if it is incorporated in the spring so deeply that the herbicide comes in contact with the developing root system. High herbicide application rates can also injure crops. Physical and environmental factors that enhance rapid crop emergence and reduce the time that a plant is exposed to the treated soil will reduce the potential for crop injury.

Herbicides vary in their ability to translocate within a plant. For example, trifluralin, is not mobile within the plant. Plant injury from this type of soil-applied herbicide would be confined to the site of uptake. Other herbicides are mobile and move within the plant. For example, atrazine is absorbed by plant roots and moves upward in the xylem and becomes concentrated in the leaves. Generally, plant injury symptoms associated with mobile herbicides will be most conspicuous at the location where the herbicides concentrate.

Application equipments

Numerous herbicides are available to assist in controlling weeds in different conditions. Herbicides are applied over the weeds, soil or water in such a way so as to be effective but without harm to crop and other desirable vegetation. Safe and effective use of herbicides requires proper selection, calibration and operation of the application equipment. A wide range of equipments is used to apply herbicides. Selection of the specific type of equipment depends primarily on the weed, crop, herbicide formulation and the advanced technology available. Other consideration include whether the herbicide will be applied as broadcast covering the entire area, in narrow bands, as individual spot treatment or a particular part of the plant. These are superimposed with physiographic, edaphic and climatic factors.

Most of the active ingredients of herbicide chemicals are formulated to be applied commonly as liquid sprays and in some cases as particles of solid materials as dusts. Granular formulations are applied to soil by mechanical spreaders similar to those used for broadcast seed or fertilizer. Herbicides application may also involve mechanical incorporation into the top 2.5 to 15 cm of soil, subsurface horizontal layering a few centimeter below the soil surface or injection into soil or aquatic system. The herbicides are classified on the basis of their use into selective and non-selective types and on the basis of mode of action into contact and translocated types. The selectivity of herbicides is not absolute and may depend on the amount of active ingredient of chemical to be applied, the way it is applied, the degree of wetting of the foliage, the tolerance of different plants to a specific chemical, difference in growth habit of crop and weed plants, the amount of rainfall or irrigation following the application. Because, selectivity of herbicides depends upon all these factors, therefore, its application methodology is important as the same chemical may either be selective or non-selective depending on the use made of it.

Herbicides are primarily applied as spray. Several types and makes of sprayer are available from small hand operated to large ground and aerial sprays but only a few of these have come in popular use. Spraying techniques are broadly classified as high volume, low volume and ultra-low volume spraying according to the total volume of spray solution to be applied per unit of ground area. The volume rates of spray solution for soil application or field crops under the different spraying techniques are given in Table 2.

Table 2. Spraying techniques based on spray volume (l/ha) (Matthews, 2000)

Spraying technique	Herbicides application (l/ha)	
	Soil/field crops	Bushes and trees
High volume spray	>600	> 1000
Medium high volume	200-600	500-600
Low volume spray	50-200	200-500
Very low volume spray	5-50	50-200
Ultra low volume spray	<5	< 50

Spaying is the most common method of application of a herbicide formulation. Sprays can be applied uniformly to the target, with the spray volume varying from 1 to 500 gal/acre. Lower volumes (1 to 5 gal/acre) are usually aircraft applications and very high volumes (>100 1500 litre/ha) may be required for thorough coverage in postemergence application to dense vegetation in right of way and some non cropped applications. Spray volumes for both soil and foliar application by ground rigs are usually in the range of 150-900 litre/ha. The most frequently used equipment to apply herbicide spray on farm is a low pressure sprayer. Such sprayers can deliver from 0 to 200 psi as contrasted with high pressure air blast sprayer that can develop pressure of 600 psi and are used to deliver insecticides to tall tress. Typical droplet size for targeted spray is presented in Table 3.

Table 3. The targets in crop protection

Target spraying	Optimum size of spay droplet (μ)
Flying insects	10-50
Insects on surfaces	30-150
Plant diseases	30-150
Weeds	100-300

Although sprays can be applied by hand-pump-type sprayers, most sprayer applications are made using tractor, jeep, truck, trailer or aircraft mounted equipment. This type of equipment has several essential components, including a tank, a pump, nozzles (generally on a boom), filters or strainers, pressure gauge, pressure regulator, shutoff valves and connecting hoses. A high volume spraying technique is used commonly in herbicide application also besides other pesticides. High volume spraying is effective because large volume of spray is applied to wet the target surfaces. If the wastage of costly chemicals is to be reduced then a narrower droplet spectrum is need to minimize losses caused by the excessive droplets (above 300 microns) and very small droplets. Choice of mean droplet size will depend on the particular target to be treated. Large droplets of about 250 microns is continued to be recommended for herbicide applications where downward movement of droplets are avoided.

Types of spraying appliances

Sprayers are designed according to the type of formulations and application technique used. The spraying appliances are available in a variety of sizes ranging from small to big unit keeping in view the application capacity and source of energy. These sprayers are either manually operated or motorized ones. They can be further classified into movable and portable types. Portable equipment can be carried by one person whereas moveable can be moved around on wheels or lifted by two or more persons.

Hydraulic energy sprayers

In these sprayers, hydraulic pressure is applied upon the liquid by the hand or foot operated or engine-powered pumps. As a result the liquid is forced through the nozzle in the form of spray droplets, most of which are normally varying in diameter. Sprayers of this type are high volume, high pressure and are suitable for complete coverage of both ground and field crops.

Manually operated sprayers

Stirrup pump sprayer

Stirrup pump sprayers are mostly of single barrel type and in some cases are double barrel types (Fig. 11). The liquid is kept in an open container. Two operators are required in its operation. One operator to pump the liquid from container and other to direct the spray from the spray nozzle. The lower end of the pump is immersed in the spray liquid of bucket or container. The outlet of the pump is near the top and is connected to a 6 metre long hose, lance and nozzle. Spraying is continuous because an air chamber is incorporated in the spray line. In single barrel pumps, the hollow plunger acts as a pressure chamber, whereas in double barrel, one bigger barrel acts as pressure chamber.

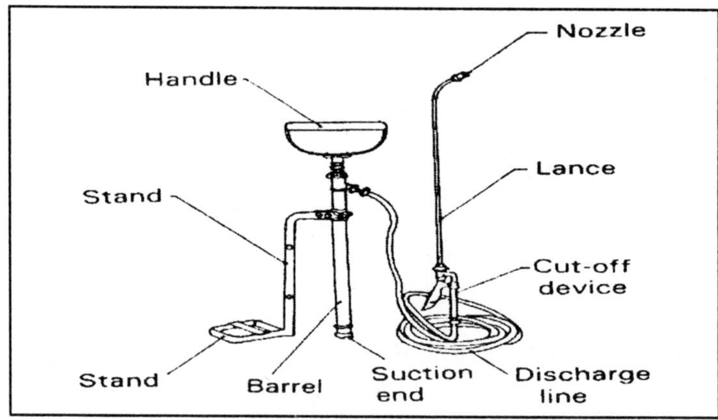

Fig. 11. Stirrup pump sprayer

Knapsack sprayer (lever operated)

Small field can be treated with knapsack sprayer (Fig. 12) easily for low crop, vegetable, nursery stocks, shrubs and trees up to 2.5 metre height. Pressure over the spray liquid is maintained by operating the hand lever during spraying at about 1-3 kg/cm^2. A piston or a diaphragm pump forces liquid from the tank into an air cylinder and then to discharge line and hydraulic nozzle. Air tripped in the cylinder is pressurized and forces the liquid via a cut-off valve to the nozzle. The spray tank is not pressurized and operator must continue pumping while spraying. Various types of hydraulic nozzles can be used. These sprayers are available in small tank of capacity 1 litre to big tank of even 18 litres. Usually tank capacities of 16 litres, 14 litres, 12 litres, 9, litres, 6 litres and 3.5 litres are common. Knapsack sprayers have tank shape either flat or bean shaped. In cases, where such sprayers have detachable containers, the frame of a sprayer is shaped so as to conveniently fit on the back of the operator. Tanks are generally made of galvanized iron, brass or stainless steel.

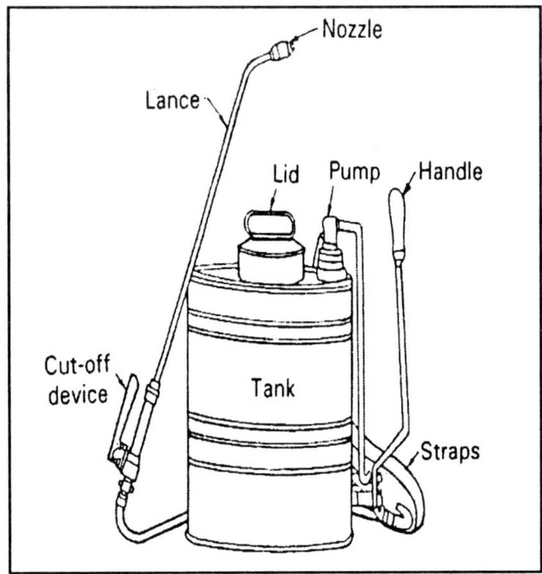

Fig. 12. Lever operated knapsack sprayer

Sprayer tanks should have a large filling hole to facilitate easy filling, inspection, and cleaning of the tank. A strainer is needed to filter out debris, which is easily introduced during loading. Tanks should also have drains to facilitate fast, complete drainage, without

all the spray liquid. A pressure regulating valve is needed to provide a more uniform pressure at the nozzle. The sprayer may be fitted with various types of hydraulic energy nozzle. A pressure gauge, safety valve and a pressure regulator is fitted to the tank so that spraying could be done at uniform pressure. The compression/pneumatic knapsack sprayers are useful tools to spray herbicides in odd situation like rice field, hilly terrain, etc. The field capacity of compression knapsack sprayer is usually 0.4 ha/day (Fig. 13). The tank capacity of compression knapsack sprayers is also varying in wide range from 1 litre to 16 litres. Small capacity sprayers of this type are used for small spot application, nursery and household kitchen gardens and are also known as hand sprayers.

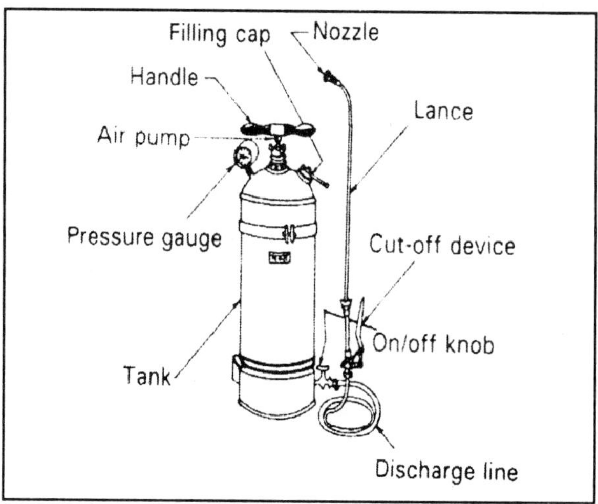

Fig. 13. Hand compression sprayer

Twin knapsack sprayer with 8-nozzle *boom*

Two knapsack sprayers with tank capacities of 16 litres each and boom with 8 nozzles can enhance field coverage to a substantial amount to about 0.4 ha/hour. The effective swath width covered by a 8 nozzle boom is about 4 metre. For effective spraying, walking speeds required is approximately 25 m/min (Fig. 14).

Foot sprayer

Foot sprayers are popular for application of herbicides on comparatively large holding. The sprayer has a provision for 1 or 2 long delivery hoses each fitted with either a lance or a nozzle boom

(Fig. 15). A foot pump sprayer uses an external spray tank through an intake pipe fitted with a stainer at its intake end. The foot sprayer consists of plunger assembly, a stand, a suction hose, 1 or 2 delivery hose and lances with nozzles. It is operated by upward and downward movements of pedal. This sprayer does not have an inbuilt tank therefore chemical solution has to be carried in buckets or containers. It develops a pressure up to 17 to 21 kg/cm^2 therefore can be used for spraying tall crops or fruit trees up to height of 4-5 metre. One foot sprayer with single nozzle boom can treat about 1 hactare of cropped area per day. Foot sprayers are suitable for wide-row crops.

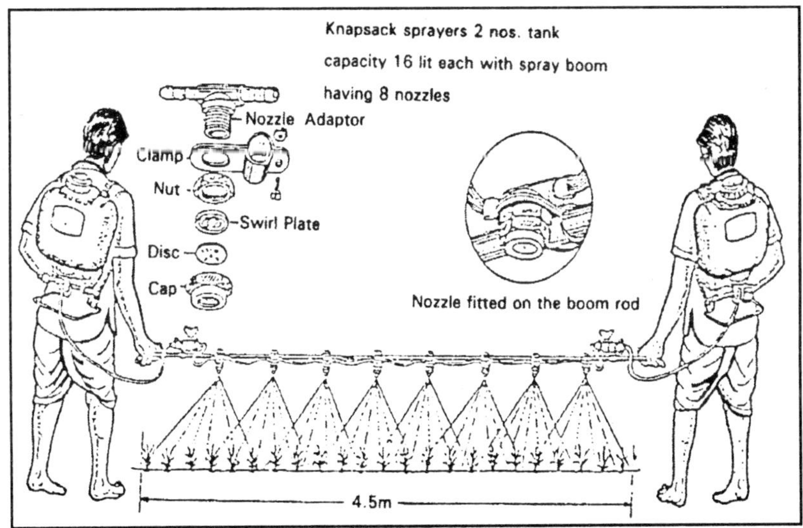

Fig. 14. Twin knapsack sprayer with 8-nozzle boom

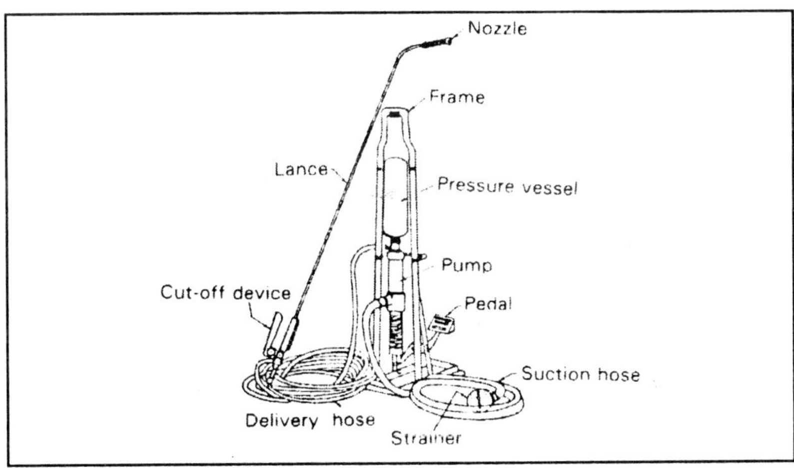

Fig. 15. Foot sprayer

Rocker sprayer

The rocker sprayer consists of a pump assembly, a platform, and operating lever, a pressure chamber, a suction hose, a delivery hose with lance and nozzle (Fig. 16). The rocking movement of the handle works the pumps and this results in building up of pressure in the pressure chamber. There is no built-in tank. Therefore a separate liquid container is required. A high pressure of 14 - 18 kg/cm^2 can be built-up in the pressure vessel which makes the sprayer suitable for spraying tall field crops and trees up to a height of 6 metre. The rocking sprayer are suitable for wide-row crops.

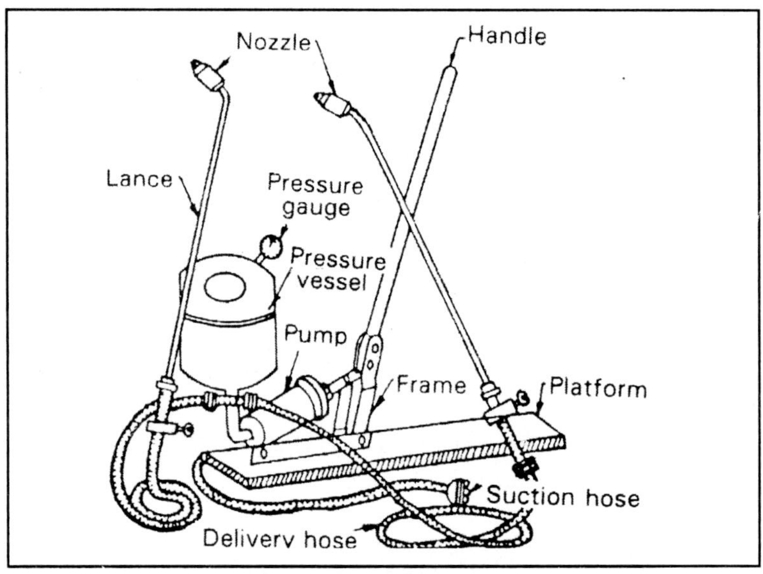

Fig. 16. Rocker sprayer

Engine powered sprayers (hydraulic energy)

Portable power sprayer

Portable power sprayer consist of an engine, a pump, frame with wheels or holding arrangement, a pressure chamber, a pressure regulator with relief valve, a pressure gauge, suction and delivery hose, a spray gun with nozzle, and overflow return pipe (Fig. 17). The delivery hose is 7-30 metre or more in length and is provided with a hand operated spray gun which has a control valve and mechanism for adjusting the spray pattern. These sprayers may be fitted with various types of boom with the nozzle frequently fixed. Portable power sprayers are suitable for field crops, orchards and shed trees.

Stretcher type portable sprayers are light and weigh only about 20 kg when empty. The sprayer is fixed on a frame having a pair of handles. There is no tank, the suction hose is designed for use with a separate chemical container. The wheelbarrow type power sprayer are designed to carry a built-in tank of capacity 50-100 litres. The four-stroke petrol engine provides power for operating the hydraulic sprayer. The small capacity power sprayer have discharge rate between 7 and 141/min at a pressure ranging up to 18 kg/cm^2. The spray fluid in the tank may be mechanically or hydraulically agitated.

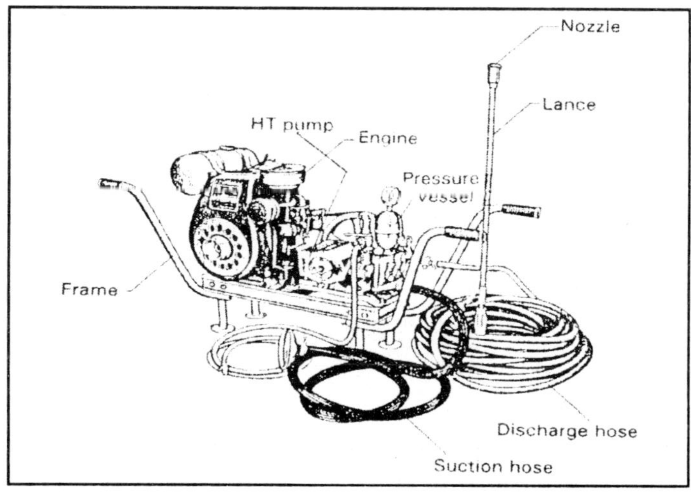

Fig. 17. Portable power sprayer

Tractor-mounted power sprayer

The tractor mounted power sprayer (Fig. 18) is attached to the tractor as a single unit and is completely carried on the 3 point linkage of tractor. Power for the spraying unit is provided by the tractor power take off.

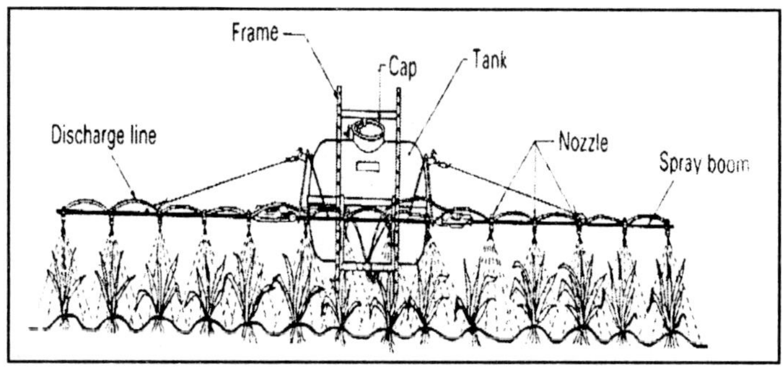

Fig. 18. Tractor mounted sprayer

The sprayer fluid containers have a capacity of 600 litres. The plunger pumps are used mostly in these sprayers for developing high pressure up to 42 kg/cm^2 and usually there are 3 plunger pumps to give higher discharge output. The spray fluid are filled in the mounted tanks with an auto-filling hose that can be suitably coupled with the sprayer. Generally tractor mounted hydraulic sprayers are used with spray boom having 14 to 20 nozzles. The use of tractor mounted hydraulic sprayers is limited to low-growing crops and to the early stages of tall crops. However, these crops must be grown in rows at proper distance, otherwise crops damage from the use of these sprayers would be more. Hydraulic sprayers incorporating air blowers are also available. The spray is produced from hydraulic nozzles in the normal manner but is carried to the target by the air blast.

Types of hydraulic nozzles and their use

Droplets produced by hydraulic nozzles are never uniform in size and spray contains both fine and coarse droplets. Table shows the types of hydraulic nozzles, their suitability and recommended use. Coarse and fine sprays are defined and measured in terms of their volume median diameter (VMD) and numerical mean diameter (NMD). Optimum droplet size for sprays on specific targets are usually defined only within a range of droplet diameters. Coarse droplets are largely influenced by gravity contrary to fine droplets which will be influenced by wind and have got tendency to drift.

Nozzles are comprised of a spray tip, a filter/strainer, and a nozzle body and cap.

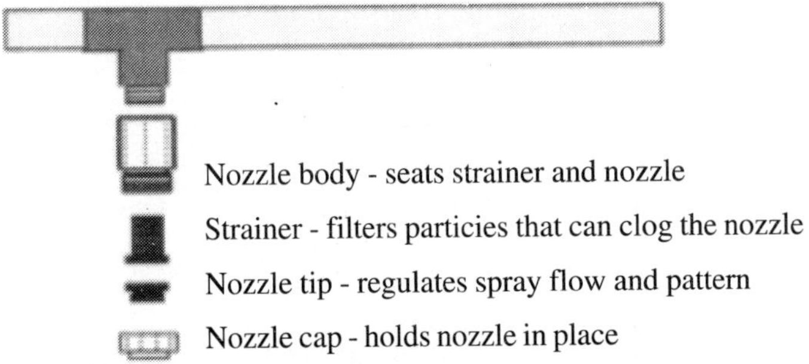

There are 4 basic types of nozzles (Fig. 19) which are most commonly used in pesticide applications. Different types of nozzles and their use is represented in Table 4.

Table 4. Types of nozzles and their use

Type	Suitability	Recommended use
Flat-fan nozzle	For ground and field crop application	Herbicides
Fan-nozzle	For ground and field crop applications	Herbicides
Solid-cone nozzle	For field crop and spot applications	Insecticides, fungicides and some herbicides
Hollow-cone nozzle	For field crops, (For row sown and broadcast) ensuring reliable coverage	Insecticides, fungicides
Triple-action nozzle	For spot, soil and field crop applications	Insecticides, fungicides and herbicides

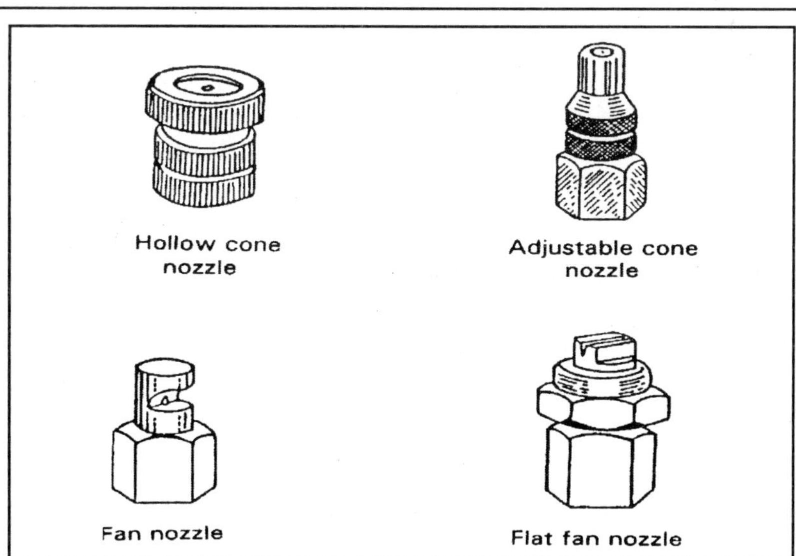

Fig. 19. Types of nozzles

Nozzles meter flow, atomize the liquid in a targeted range of droplet sizes, and disperse the droplets in a specific pattern for proper impact with plants or soil. Nozzles are available in brass, polymer, stainless steel, hardened stainless steel, and ceramic. Brass and polymer nozzles are most popular, primarily due to low cost. Stainless steel and hardened stainless steel nozzles last three to 15 times longer than brass.

Ceramic nozzles last about 100 times longer than brass. Herbicide nozzles provide uniform distribution of spray solution and develop a large droplet to minimize drift. Nozzles commonly used for broadcast herbicide applications include the regular flat fan, extended-range flat fan, drift-reduction flat fan, turbo flat fan, twin flat fan, and air-induction flat fan.

Operating pressure varies with nozzle type. Tips such as the extended-range flat fan and turbo flat fan nozzles are designed for operating pressures from 15 to 30 psi, which reduce drift by producing large droplets that are dispersed under low spray pressure. Drift-reduction flat fan and air-induction flat fan nozzles are operated at pressures of 30 psi or more. These tips are designed to produce larger spray droplets to minimize drift at their specified operating pressures. Operating nozzle tips at other than the specified pressures may result in a poor pattern and less than desirable pattern overlap.

Spray angle of nozzles also impacts drift potential. Tips that produce a broader pattern (110°) are preferred because they can be operated closer to the application target, which minimizes drift by shortening the distance spray solution travels. Flat fan nozzles should be uniformly spaced to provide 50 to 60 percent overlap. Table 5 suggests boom heights for flat fan nozzles with different angles. Spray particles that encounter one another while in flight to the ground are drift prone. To avoid this disruption, the orientation of each flat fan along the boom should be angled, or canted, at an angle of 5 degrees from the center line of the boom (Fig. 20) to produce the desired 50 to 60 percent overlap in spray on the ground without intersection of spray from individual nozzles (Fig. 20). For better efficiency suggested minimum spray height for flat fan nozzle is given in Table 5.

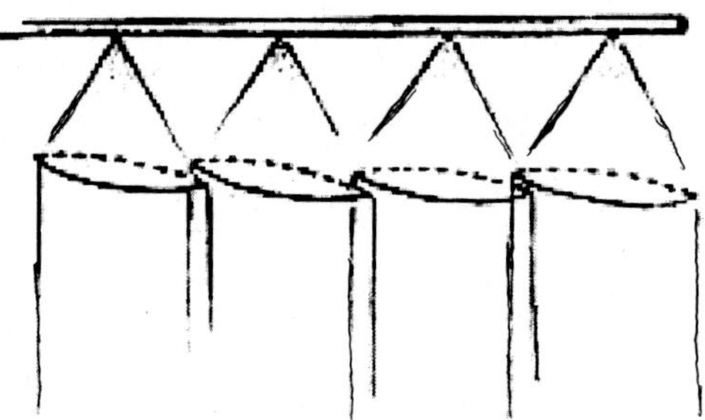

Fig. 20. Angling nozzles assures a similar deposition of spray material all along the spray band, without the particles shearing and drift potential seen when the edges of spray patterns collide in the air. Graphics by Will Hampton.

Table 5. Suggested minimum spray height for flat fan nozzle

Spray Nozzle	Height (Inches)	
	Spray tip angle	Nozzle height
	20″ spacing	30″ spacing
65	12.24	33-35
80	17-19	26-28
110	10-12	14-18

Factors affacting rate of spray

Spraying pressure

Spraying should be done at the same tank pressure as when calibrating. If, for example, a small compression tank is calibrated at 25-30 p.s.i. pressure, spraying should be done at that pressure. Spraying at a higher or lower pressure will result in more or less herbicide being applied as the sprayer passes over the ground.

When using a backpack type sprayer a pressure gauge or pressure regulator will enable you to maintain constant pressure. Of the two, the regulator is more convenient to use. It will automatically control the pressure. When using a gauge, which only indicates the pressure level, pressure must be controlled manually by the rate at which operator pump.

Nozzle type and size

Generally, even flat spray nozzles are most desirable for band or spot spraying. They provide the most even distribution of spray material across the spray pattern. Sprayers must be calibrated for each nozzle type and size. The larger the nozzle, the more spray is put out as the sprayer passes over the ground. Thus, to maintain a particular rate of application, when larger nozzles are used, smaller amounts of herbicide will be added to the water.

Nozzle types vary in droplet sizes produced at various spray pressures and litres per minute output (Table 6). Flat fan, flood and hollow cone nozzles produce similar-size droplets and a similar volume of small droplets when compared at equal spray pressure and equal delivery rate. The flood nozzle tends to produce slightly larger droplets than the flat fan, while the flat fan produces slightly larger droplets than the hollow cone.

Table 6. Influence of nozzle type and spray pressure on droplet size.

Nozzle type micron diameter	Delivery rates (gal/min)	Spray pressure (lb/sq in)	Spray angle degrees	Volume median diameter (microns)	Volume with less than 100 (%)
Flat fan	0.12	15	65	239	
	0.17	30	76	194	
	0.20	40	80	178	17.5
Flood fan	0.12	15	90	289	
	0.17	30	115	210	
	0.20	40	125	185	15.5
Hollow cone	0.12	15	-	228	
	0.17	30	-	185	
	0.20	40	70	170	15.5

Source: Delavan Manufacturing Company

Speed of travel

Spraying should be done at a uniform speed and at the speed at which the sprayer was calibrated. Slower or faster speeds will result in higher or lower rates of herbicide application. For this reason, hand sprayers, such as the small compression tank sprayers, should be calibrated for use by each individual who uses them. Each person has a different speed and pattern of spraying. One person may, for example, use approximately a litre to spray 30 sq. m square feet while another uses 3/4 of a litre and another 1/2 litre. Because the terrain affects speed of travel, calibration should be done on the actual area to be sprayed to achieve the most accurate application rate. When spraying with a backpack-type sprayer, a pressure regulator (A) or a pressure gauge (B) will enable operator to maintain constant pressure.

Ideal patterns based on treatment efficacy

- 25-30 or more droplets per sq. cm.
- most droplets 0.7 to 2.2 mm
- no droplets > 7 mm
- few droplets < 3 mm
- outside buffer zone, no droplets on non-target area cards

Table 7. Nozzle tip used in different spray application (Newton & Knight 1981)

Spray application	Nozzle type
Aerial Broadcast	Conventional type Hollow cone (Disc core)
Ground broadcast	Flat fan Flooding flat fan Straight (solid) stream Cluster or boomless
Band/ strip	Even flat fan
Bark treatment	
Thin line spraying	Straight (solid) stream
One-sided low volume	Even flat fan
Conventional volume	Solid (full) Cone
Spot	
Grid pattern or individual vegetation	Straight (solid) stream
Tree centered spots	Hollow cane
Herbaceous vegetation	Solid Cone
Cut stump	Flatfan Solid (full) Cone

Gaseous energy sprayers

Manually operated sprayers (hand atomizer): The hand atomizer consists of a small cylinder or chamber (0.5 litre or less) and a plunger pump to pressurize air at high velocity towards its outlet. In compression, liquid is drawn by drop of pressure inside the ventural and liquid is drawn which is broken into fine droplets as it comes out by force of air delivered out of the pump. The sprayer is suitable for experimental work or on individual plants.

Motorized knapsack mistblower-cum-duster: Knapsack mistblower-cum-duster are fitted with 35 or 45 or 70 cc petrol engine (Fig.21). The small engine is preferred because it is light but large engine is needed to provide sufficient power to spray the tall target. A direct drive connects the engine by anti-vibration mountings to a strong L-shaped knapsack frame. The most common atomizer fitted to these sprayers is air-shear nozzle/ atomizer, but various adaptations have been tried to improve spray droplet spectrum. An even flow of air over the liquid orifice assists droplet formation. The spray tank is mounted above the engine/blower unit and normally holds about 10-12 litres of spray liquid.

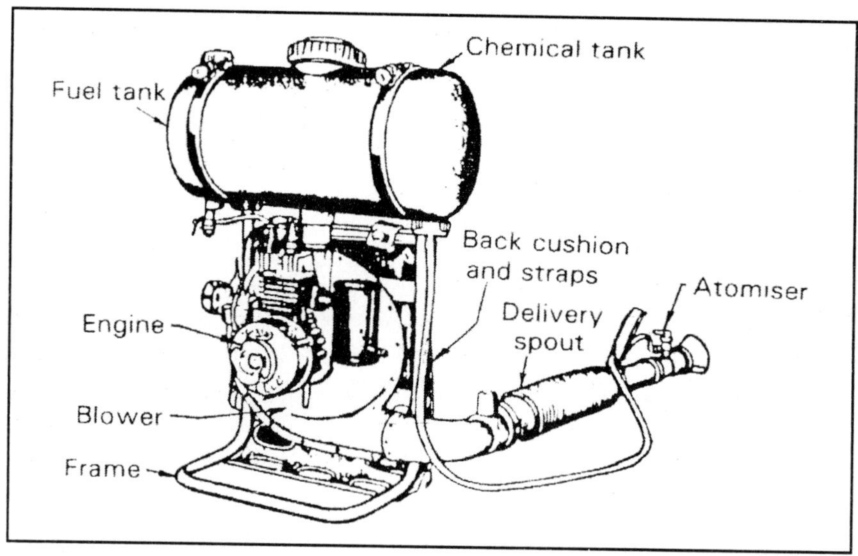

Fig. 21. Motorized knapsack mist-blower-cum-duster

The spray liquid is metered in atomizer by an adjustable barrel type valve, or a sleeve with 4 different orifices or a set of restrictors. The weight of a motorized knapsack mistblower-cum-duster may be as much as about 14 kg when empty, so they are much heavier to carry than other types of knapsack sprayers. The straps are, therefore, usually provided with pad over the shoulder and a back rest to improve operators's comfort.

Most of these motorized knapsack mistblower can be converted for dust application duster by removing the spray hose from the tank and inserting a wider tube of feed particles directly down from the hopper through a metering orifice into air stream. The outlet tube is rotated to stop the flow of materials. The dust output can be adjusted from 0 to 1.5 kg/min by rotating the pleated hose on the blower elbow which carries the discharge hole.

Centrifugal energy sprayers

These sprayers are designed to reproduce fine spray droplets of 50 to 100 microns size in the form of mists. A high speed rotating disc or cups are producing the fine mist of spray. A fan or blower of axial flow or centrifugal type supplies air at high velocity and throws the small droplets to the target. The sprayer may be hand-held type of battery operated and light in weight. A container is placed at the top of the rotor with the help of suitable frame causing the flow of liquid by gravity.

Hand held ultra low volume sprayer: The hand-held ULV sprayers have HDP plastic spray head with small DC motor which drives the rotating disc at 7000-9000 rpm. The sprayer consists of a liquid reservoir (a screw on bottle), a handle, a power supply unit, i.e. D.C. motor and the spray head (Fig. 22). In these sprayers, the pesticide is fed by gravity from the plastic bottle to the spinning disc which throws the concentrated pesticide through its toothed edge in the form of fine threads to the periphery by centrifugal force. Because of the high speed of the emerging liquid, the threads get disintegrated in the air to form fine spray droplets. To prove effective the droplets must land on the target. The wind is used to transport the spray droplets. The wind must be neither too gentle (less than 3 km/hour) nor too strong (more than 22 km/hour).

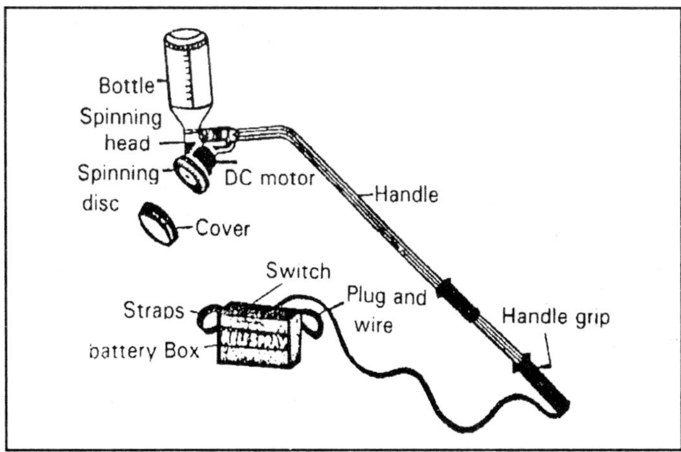

Fig. 22. Hand-held ULV sprayer (battery operated)

Herbicides' can be applied in low volume rates or in ultra low volume rates (1-5 litre/ha) with these units. The ULV application requires much skill with these sprayers. Height of the sprayhead must be much as not to cause unacceptable drift. Swath width of spray is effected by wind velocity, plant height and canopy. The area covered is influenced by the swath width and the walking speed of the operator. The controlled sized droplets of mean diameter of 250 microns (VMD) are effective for herbicide sprays. For herbicide application, hand-held ULV sprayer are designed slightly in a different working posture and are fitted with D.C. motor for producing the desired droplet size (Fig. 13). The CDA spinning disc sprayers are very handy for treating small holdings, rough topography land and wet paddy fields. The sprayer is also good for use in research experiments. The field capacity of hand-held ULV sprayers are around 0.15 to 0.2 ha/hour with swath width of 1-1.5 metre.

Twin/multiple spinning disc knapsack sprayer

Twin Spinning disc and multiple spinning disc knapsack sprayer have also been developed (Fig. 23, 24). In such sprayers the chemical tank of 10 litres is used with shoulder straps on the back of the operator. A 1.5 metre wide spray boom is used to hold 2 or 3 spinning disc atomizers so as to cover 3 m or 4.5 swath width. The height of boom and spacing between atomizers can be adjusted. The chemical solution flows from tank to spinning disc through a cut- off device and flow regulating' valves. The 2 or 3 number of spinning discs are operated by 6 or 12 V battery (rechargeable or dry cell type).

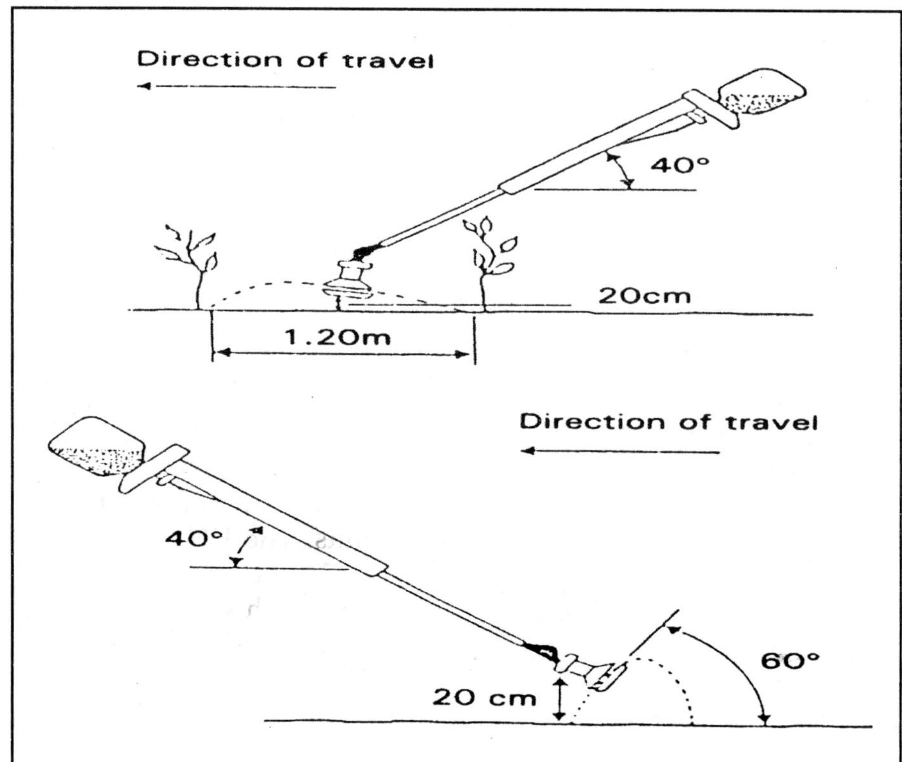

Fig 23. Hand held ULV sprayer for herbicides application (working posture)

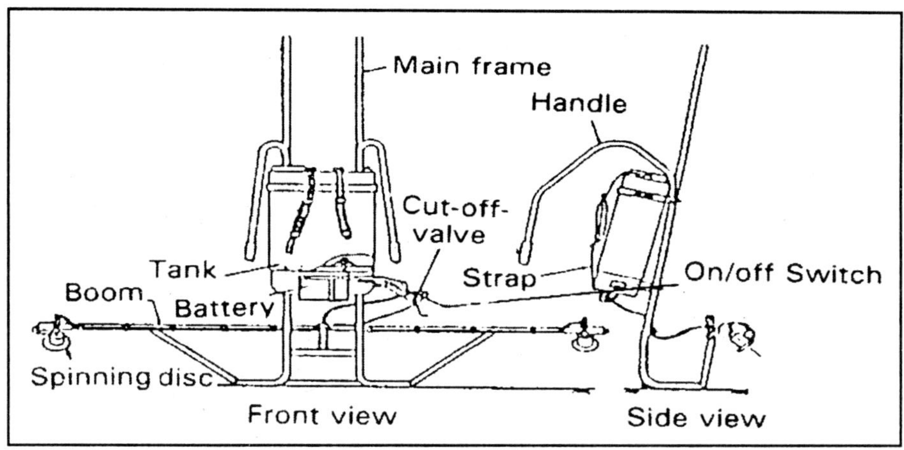

Fig. 24. Twin spinning disc ULV knapsack sprayer

Tractor mounted CDA crop sprayer

Controlled droplet application (CDA) techniques producing only the optimum sizes of spray droplet for the particular application. This is achieved by specifically designed spinning disc rotary atomisers which break up the spray liquid very evenly, resulting in a narrow range of spray droplet sizes.

The controlled droplet atomizer (CDA) (Fig. 25) uses a spinning disc rotary atomizer to produce optimum sized, evenly distributed spray droplet (Fig 26). This allow low pressure application which minimize chemical waste, environmental contamination and the opening down time necessary to refill tanks. Rotary atomisers are used to generate the droplets. Fluid meters forces cause the fluid to rise up the grooved walls of the housing where it is broken into droplets. Carefully designed fingers throw the droplets horizontally outward from the spinning atomiser. They then fall to the target under the force of gravity. Droplet size is determined by the speed of the rotation of the atomisers. Most common are atomisers driven with small electric motors. These are used on both large and small boom spraying applications.

For a given spray volume CDA sprayers will cover 8-10 times more area than pressure nozzles when applying herbicides and more than 20 times more area when applying insecticides and fungicides. This is arguably the greatest advantage of CDA.

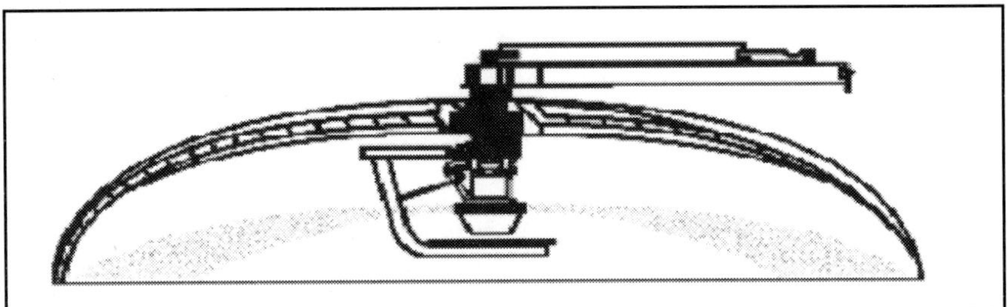

Fig. 25. CDA uses a spiininng disc rotary atomizer

Most spraying machines still break up spray liquid by squirting it under pressure through a hole - commonly referred to as an close-up of hydraulic nozzle detailing variation in droplet size hydraulic nozzle. The hydraulic nozzles, introduced over a hundred years ago break up the spray liquid very unevenly, resulting in a very wide ,range of spray droplet sizes. Although some of the spray droplets produced will be of the required sizes, most will not - with vast variations in the amount of chemical between the smallest and largest spray droplets (commonly over one million to one).

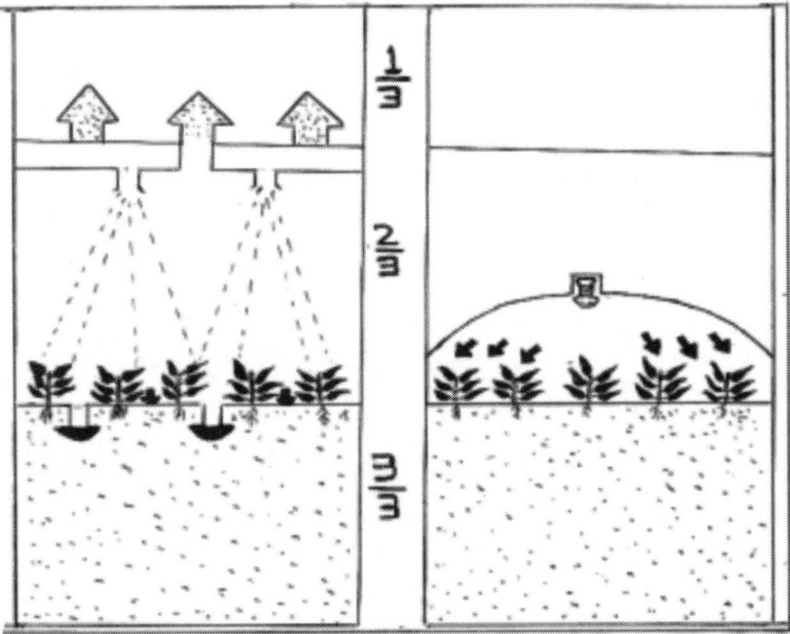

Fig. 26. Herbicide lost is high in conventional spraying as compared to controlled droplet application (CDA) a0 conventional spraying , b) CDA spraying (Source: www.bubco.com/turf.htm)

Very large and very small spray droplets are both wasteful, and will not generally even hit the target. The very large spray droplets end up on the ground and the very small spray droplets drift - either into neighbouring areas or into the atmosphere - thus wasting much of the spray and contaminating the environment. This wasteful inefficiency also means that hydraulic nozzle sprayers have to use much more spray liquid than is actually necessary to do the job. To fetch, carry and spray this extra liquid costs time, effort and money.

The waste and environmental contamination resulting from spraying pesticides through hydraulic nozzles is causing increasing world-wide concern. CDA offers the way forward in improving spraying efficiency, eliminating waste and safeguarding the environment, as shown by studies in many countries. The efficiency of CDA allows low and ultra-low volumes of spray liquid to be used. This logistical advantage can be of critical importance in crop protection, allowing effective spraying programmes to be undertaken quickly in non-mechanised agriculture and greater spraying productivity in mechanised agriculture. Being able to apply pesticides accurately, quickly and cost - effectively when needed gives the opportunity of fully implementing integrated pest management programmes, with reduced overall pesticide usage. Usually 250 μ size is considered best for herbicide application in CDA techniques, while the smaller droplets are preferred for fungicides and insecticides.

With the use of controlled droplet atomizers (CDA), development of a tractor mounted CDA crop sprayer was carried out at CIAE, Bhopal for spraying crops up to height of 1.0 metre. The CDA atomizers have 3 rotational speed, i.e. 2000, 3500 and 5000 rpm to produce droplets of250, 130 and 70 microns size (VMD) suitable for application for herbicides, insecticides and fungicides respectively. The field capacity of the tractor mounted sprayer was 0.76 ha/hour with effective swath width of 5.5 metre. The sprayer consists of a tank with air pump, pressure gauge, on-off valve and flow control valve. The liquid is fed to 4 separate feed lines from the distributor to the CDA atomizers. The 4 CDA atomizers were mounted on the boom which is of folding type. The sprayer being light in weight is mounted on the iron wheels and to the tree point linkage of tractor (Fig. 27).

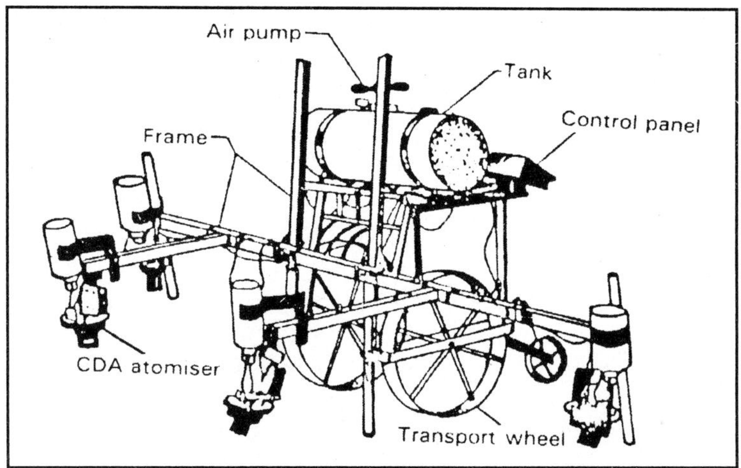

Fig. 27. Tractor mounted CDA crop sprayer (CIAE design)

Selection of spraying equipment

The purchase of spray application equipment is a long-tern investment, therefore, due care should be taken to select right type of spray application equipment are available in a variety of sizes rang a small to big sprayer keeping in view the capacity of the equipment and source of power. The following factors should be and kept in view while selecting spraying equipment.

(i) Type of sprayer
(ii) Nature of crops canopies where it is to be used
(iii) Size of equipment and its field capacities.
(iv) Technical skill required for its operation.
(v) Make and cost of the equipment.
(vi) Repair and replacement of parts, facility available, etc.

Calibration of sprayer

Calibration of sprayer is important before it is used in field spraying operations. For calibration, fill tank to indicate maximum level of liquid and pressurize the sprayer. Adjust the operating pressure to a level specified for the nozzle and measure spray width covered in meters at specified height of the boom/nozzle from ground. Walking speed/ speed of operation in metre/minutes and volume of spray liquid emitted by nozzle/nozzles in one minute are recorded. From this data, application rate in litres/ha can be calculated by formula (1).

$$\text{Application rate} = \frac{\text{Nozzle output (litres/min.)}}{\text{Walking speed (m/min.)} \times \text{swath width (m)}} \times 10{,}000 \quad (1)$$

Check the calibrated application rate in small measured plot/field area and measure the following.

(i) Fill the tank on horizontal ground level to the brim or a convenient mark of tank with clean water.

(ii) Measure the marked area of plot both length and width in meters.

(iii) Spray water on measured area of plot with the sprayer and find amount of water again required to top up to its previous mark.

Now actual application rate will be determined from the formulae at (2).

$$\text{Application rate} = \frac{\text{Volume (litres) to top-up}}{\text{Measured area (m}^2\text{)}} \times 10{,}000 \quad (2)$$

Now determine the dose at which the actual application rate achieved during field areal plot spraying correspond to that calibrated application rate calculated at (1). Adjust the chemical concentration tank filling according to the calibration. Adjustment in application rate calculated up to 10% can be made by increasing or decreasing speed of operation. Increase/decrease in application rate beyond 10% limit should be made by selection of suitable size of nozzle.

Calculation of herbicide requirement

The product label and the literature supplied with the herbicide generally provide details of herbicide name, active ingredient (a.i.), date of expiry, directions for use etc. It must be read before using the herbicide. It is particularly important to note the strength of the product (a.i.) as the same herbicide may be sold under different trade names with varying amounts of active ingredient. For example, isoproturon is available at 50 and 75% formulations. For this reason only, the recommendations are normally made on kg a.i. basis. Even in liquid formulations the herbicide present is mentioned in g/l.

The amount of commercial formulation of the herbicide required can be calculated by the following formula:

$$\text{Commercial product (kg/ha)} = \frac{\text{Dose in kg a.i. / ha}}{\text{\% a.i. in the product}} \times 100$$

Example 1

Atrazine is available as 50% WP. If the recommended rate of application is 1.0 kg ai/ha then the amount of commercial product required is :

$$50\%\ \text{WP product} = \frac{1.0 \times 100}{50} = 2.0\ \text{kg/ha}$$

Example 2

Paraquat is to be applied at 0.5 kg ai/ha. The herbicide is available as Gramoxone (commercial name of paraquat) which contains 25% paraquat

The quantity of Gramoxone required is :

$$\frac{0.50 \times 100}{25} = 2.0\ \text{l/ha}$$

Making stock solution

In order to apply herbicide uniformly in the entire required area, it is advisable to prepare stock solution of the herbicide. Suppose;

Area to be treated = 1 ha (2.5 acres)

Sprayer capacity = 15 litre

Sprayer calibration = 450 litres / ha

Then,

$$\frac{450}{15} = 30$$

that is 30 refills are required to spray one hectare area.

In which case, it is advisable to dissolve the required amount of herbicide as obtained in calculations 1, 2 or 3 in 30 measures (could be a glass tumbler, plastic mug or a container) of water which becomes the stock solution. Now add 1 measure of this stock solution to sprayer tank containing 15 l water, stir it and spray as suggested earlier. Alternatively one can dissolve the entire quantity of herbicide in 450 liters of water contained in a big container and use this solution directly for spraying.

Maintenance of sprayer

Sprayers require very little maintenance. However, a little care as mentioned below can considerably lengthen their life and enhance their functional efficiency.

(i) Wash inside of the pump and tank of the sprayer until only very clean water is delivered after use.

(ii) For off-season storage, the sprayer should preferably be kept in cool and dry place.

(iii) Motorized knapsack mistblowers need engine in perfectly tuned condition so that they can be operated without field problems. Proper calibration of the spraying unit is required.

(iv) The small DC motor of ULV spinning disc sprayers be serviced in accordance with the instruction of manufacturers.

Safety precautions

Operation	Precaution
(i) Opening container/diluting, mixing or transferring from one container to another, washing containers, washing of equipment	Overall and rubber apron, rubber boots, rubber gloves face shields (respirators)
(ii) Spraying ground or glass house: crops	As at (i) except overall and aprons
(iii) Spraying bushes and climbing: plants	As at (i) but wear rubber / waterproof coat and omit rubber apron
(iv) Spray applied to soil or injection	Overalls, rubber gloves, boots and rubber apron

Improving efficiency of herbicides

Application efficiency

Efficiency of spray which also includes economy of herbicides application depends upon the spray droplet size. Emphasis has been laid on atomization parameters such as droplet size, number of droplet per unit area and chemical concentration (a.i.) in the droplets. The spray droplets (MD) diameter is determined by the volume median diameter (VMD). It is measured in microns. The VMD is the number which divides a spray in two equal parts by volume, one half containing droplets smaller than the VMD diameter and the other containing droplets which are bigger. Herbicide spray pattern evaluation based on efficacy is depicted in Fig 28.

Distribution of herbicides

Weed control is most effective when all target plants or soil areas are reached by and retained an optimum dose of herbicide. To achieve this, the sprayed liquid must be distributed uniformly which depends on factors, viz., uniform discharge of spray across the treated width, uniform concentration of the active ingredient within the spray, correct overlaps of spray (in booms) and uniform speed of operation. The amount of herbicide intercepted and retained by the crop or weeds depends upon the number and size of spray droplets and their momentum and surface tension induced by the spraying system used.

Herbicidal dusts are less popular, because the dose of herbicide/ ha sometimes has to be higher than that applied as sprays to achieve comparable results. Although, the adherence of dusts may be enhanced by additives, but their distribution may be more variable than majority of sprays and involved more drift hazards. Only few herbicides are now available in dust formulations currently.

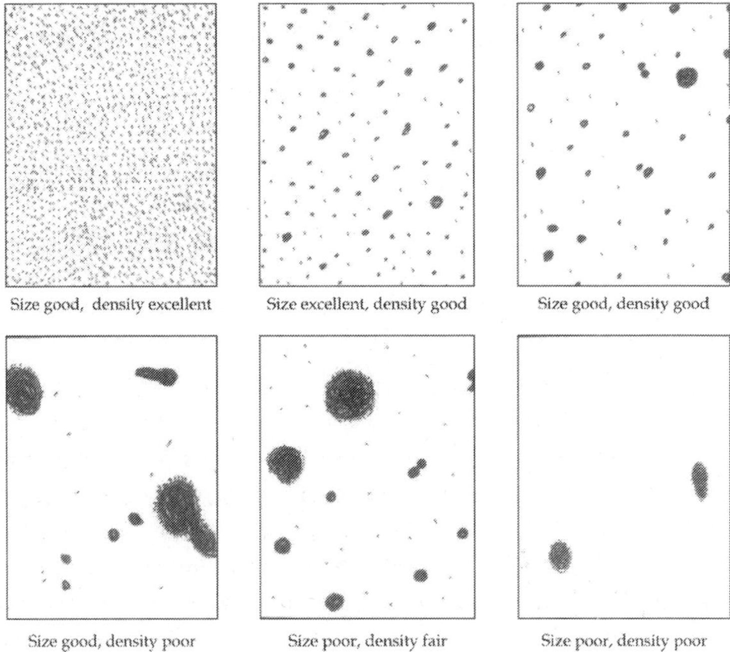

Fig. 28. Herbicide spray pattern (droplet) evaluation based on efficacy

Target, spray application area and spray volume

A pesticide needs to be applied to particular target area occupied by pests and weeds. If herbicides are to be used more judiciously, then best time and place to deposit the chemical can be deemed to achieve adequate control over the weeds. Generally spray application area differs from the land area except in the case where pre-sowing treatment is required in soil where land area equals the area to be sprayed. The area to be sprayed varies with the plant spacing, row width and growth of the crop. In an ideal situation where both optimum spray droplet diameter and most desirable droplet density are known, the minimum amount of spray solution could be calculated. More wastage of herbicides is undoubtedly due to very large droplets which have terminal velocity and fall rapidly.

Summary

Herbicide can be applied in a variety of different ways to the crop. These can be applied as single component product or as combination products, using the different properties and weed spectrum of the components of the mixtures. An improper method of application of a herbicide results in poor weed control and/or severe crop injury. Soil applied herbicides can be applied as surface application and are commonly applied to the surface of the soil where these may be either left undisturbed or incorporated into the soil. Subsurface layering is the application of a herbicide in a concentrated band about 7-10 cm below the soil surface. Broadcast application of herbicide is their application over the full surface area, without leaving and intentional gaps. Whereas, band application usually means treating a narrow strip directly over or on the crop row. Depending upon the nature of soil fumigant it can be applied either (1) by soil injection e.g., chloropicrin, (2) by releasing it under sealed plastic covers e.g., methyl bromide or (3) by direct soil surface application.

Foliage application of herbicides can be applied as blanket application to standing crops with disregard to the location of the crop plants. Directed spraying is done to save the crop plants from herbicide injury and improves weed kill. In protected spraying, non-selective herbicides can be employed to obtain selective weed control in distantly planted vegetables and ornamentals by covering the non-target plants before application of the herbicide with plastic or metallic covers. Whereas, spot treatment is usually done on small areas of serious weed infestation to kill and to prevent spread of weeds.

Brush weeds and unwanted trees are treated with herbicides by different methods, depending upon the situation. Foliage treatment is the most common method of treating brush. However, basal bark treatment is common to control big trees. But cut stump treatment comprises sawing of the tree above the ground followed by liberal application of the herbicide on the cut surface.

Most of the active ingredients of herbicide chemicals are formulated to be applied commonly as liquid sprays and in some cases as particles of solid materials as dusts. Granular formulations are applied to soil by mechanical spreaders similar to those used for broadcast seed or fertilizer. Herbicides are primarily applied as spray. Several types and makes of sprayer are available from small hand operated to large ground and aerial sprays but only a few of these have come in popular use. Although sprays can be applied by hand-pump-type sprayers, most sprayer applications are made using tractor, jeep, truck, trailer or aircraft mounted equipment. This type of equipment has several essential components, including a tank, a pump, nozzles (generally on a boom), filters or strainers, pressure gauge, pressure regulator, shutoff valves and connecting hoses. A high volume spraying technique is used commonly in herbicide application also besides other pesticides. Stirrup pump, knapsack, syringes, compression knapsack, twin knapsack, rocker and foot sprayer are operated manually. Power sprayers are also common for herbicides spraying. Flat-fan nozzles, fan nozzle, solid coe nozzle named hollow cone nozzles are 4 basic types of nozzles which are most commonly used in pesticide applications. Nozzles meter flow, atomize the liquid in a targeted range of droplet sizes, and disperse the droplets in a specific pattern for proper impact with plants or soil. Spray angle of nozzles also impacts drift potential. Tips that produce a broader pattern (110°) are preferred because they can be operated closer to the application target, which minimizes drift by shortening the distance spray solution travels. Controlled droplet application (CDA) techniques producing only the optimum sizes of spray droplet for the particular application. This is achieved by specifically designed spinning disc rotary atomisers which break up the spray liquid very evenly, resulting in a narrow range of spray droplet sizes.

❑❑❑

References

Abou-Waly, H., Abou-Setta, M. M., Nigg, H. N. and Mallory, L. L. (1991). Dose-response relationship of *Anabaena flosaquae* and *Selenastrum capricornutum* to atrazine and hexazinone using chlorophyll (a) content and ^{14}C uptake. Aquatic Toxicol. 20:195-204.

Adams, J. B. and Drew, M. E. (1965). Grain aphids in New Brundsdwick. IV. Effect of malathion and 2,4-D amine on aphid population and on yield of oats and barley. Can J. Zool. 47: 423-426.

Ahuja, K.N. and Yaduraju, N.T. (1992). Weed control in groundnut with herbicides. Abstract, Indian Soc. Weed Sci., Conf. HAU, Hisar p 66.

AICRP on Weed Control Annual Report reports, NRCWS, 1995-2006.

Akopyan, E. A. and Agarnonyan, A. G. (1968). Khimiya sel. Khoz. 6: 850-851.

Aldrich, R. J. (1984). Weed crop ecology. Principles in weed management. Breton Publishers, North Scituate, MA.

Ali, A. M. (1985). *Amaranthus polygonum, Flaveria australasia* and *Parthenium hysterophorus* control in brinjal. Abts. Ann. Conf. ISWS, Anand pp.23.

Allen, E. B. and Neil, W. (1993). Non-target effects of the herbicide tebuthiuron on mycorrhizal fungi in sagebrush semidesert. Mycorrhiza 3: 75-78.

Amarjeet, S., Punia, S., Yadav, A. and Malik, R. K. (2003). Effect of pH on degradation of chlorsulfuron in soil. Indian J. Weed Sci. 35: 97-99.

Ananthapadmanbhan, K. P., Goddard, E. D. and Chandler (1990). A study of solution interfacial and wetting properties of silicone surfactants. Colloids and Surfaces 44: 281-297.

Anderson, W. P. (1977). Weed Science: Principles. West Publishing, Co. St. Paul, MN.

Anonymous (1979). Methazole. In herbicide Handbook, Weed Science Soc. of America, Champaign, III pp 168-274.

Anonymous, (1986). Annual Report, All India Co-Coordinated Research Project weed Control, JNKVV, Jabalpur.

Antogini, J. and Day, B. E (1955). Geigy 444, a pre-emergence and post-emergence herbicide for cotton. Proc. 8th Southern Wed control Conf. pp 92-98.

Apt, W. J., Austenson, H. M. and Courtney, W. D. (1960). Use of herbicides to break the life cycle of the bentgrass nematodes, *Anguina agrosis.* Pl Dis Reptr 44: 524-526.

Araújo, A. S. F., Monteiro, R. T. R. and Abarkeli, R. B. (2003). Effect of glyphosate on the microbial activity of two Brazilian soils. Chemosphere 52: 799-804.

Armbruster, B.L., Molin W.T., and Budd, M.W. (1991). Effect of the herbicide dixhiophyr on cell division in wheat root tips. Pest Biochem. Physiol. 39: 110-126.

Arruda, J.S., Lopes, N.F. and Bacarin, M.A. (1999). Crescimento de plantas de soja em função de doses de sulfentrazone. Planta Daninha 17: 375-386.

Asare-Boamah, N. K. and Flewhcter, R.A. (1983). Physiological and cytological effects of BAS 9052 OH on corn (*Zea mays*) seedlings. Weed Sci. 31: 49-55.

Ashton, F. M. and Helfgott, S. S. (1966). Uptake and distribution of CIPC in seeds during inhibition. Proc. Calif. Weed Conf. 18: 8.

Ashton, F.M. (1958). Absorption and translocation of radioactive 2,4-D in sugarcane and bean plants. Weeds 6 (3): 257-262.

Aston, F. M. and Craft, A. S. (1981). Mode of action of herbicides Second Edition Wiley-Inter science publication, John Wiley & Son, Canada.

Aston, J. L. and Reynaert, J. (1974). The current and potential uses of the triazine herbicides in water. Proceedings of the European Weed Research Council Symposium on Aquatic Weeds, Vienna, pp.244-252, European Weed Research Society. Secretariat Wageniagen, The Netherlands.

Athens, J. F., Leonard, O. A. and Townley (1970). Chemical control of tree roots in sewer lines. J Water Poll. Control Federation 42: 1643-1655.

Audus, L. J. (1970). The action of herbicide and pesticide on the microflora. Meded. Fac. Landbouw Gent. 35: 465-492.

Ayansina, A. D. V. and Oso, B. A. (2006). Effect of two commonly used herbicides on soil microflora at two different concentrations. African J. Biotechnol. 5: 129-132.

Balajee, S. and Mahadevan, A. (1993). Biodegradation of 2, 4-dichlorophenoxyacetic acid in soil by *Azotobacter chroococcum*. Toxicol. Environ. Chem. 39: 169-172.

Balasubramanian, R., Veerabadran, V. and Kannathasan, M. (1999). Influence of continuous use of herbicides in rice based system on residue accumulation and on performance of succeeding pulse crop. Pestici. Res. J. 11: 200-203.

Balicka, N. and Sobieszczanski, J. (1969). The effect of soil microorganism in a field experiment. Acta Microbial. Pol. 1: 3-6.

Balyan, R. S. and Malik, R. K. (1992). Chemical control of wild oat (*Avena ludoviciana*) and other weeds in wheat (*Triticum aestivatum*). J. Tropical Agril. Veterinary Sci. 30: 169-175.

Bandhe, S. B. and Nalamwar, R. V. (1984). A critical study of crop-weed competition in hybrid sorghum. Weed Abstract 33: 174-187.

Barka, E. A, Belarbi, A., Hachet, C., Nowak, J. and Audan, J. (2000) Enhancement of *in vitro* growth and resistance to gray mould of *Vitis vinifera* co-cultured with plant growth-promoting rhizobacteria. FEMS Microbiology Letters 186: 91-95.

Barnwell, P. and Cobb, A. H. (1993). An investigation of aryloxy phenoxy propionate antagonism of auxin type herbicides action on proton efflux. Pesticide Biochem. Physiol. 47: 83-87.

Barrett, M. (1995). Metabolism of herbicides by cytochrome P_{450} in corn. Drug Metabolism and Drug Interactions 12: 299–315.

Bartha. R., Lanzilotta R. P. and Pramer, D. (1967). Stability and effect of some pesticides in soil. Appl. Microbiol. 15: 67-75.

Basavarajappa, D. N. and Nanjappa, H. V. (1994). Residual effect of herbicides applied to sunflower on succeeding crops. Indian J. Weed Sci. 26: 97-98.

Bavaresco, L. and Fogher, C. (1996). Lime induced chlorosis of grapevine as affected by rootstock and root infection with *Arbuscular mycorrhiza* and *Pseudomonas fluorescens*. Vitis 35: 119-123.

Bayer, D. E. and Foy, C. L. (1982). Action and fate of adjuvants in soils. In: Adjuvants for herbicides, WSSA, Champaign, IL. Pp. 84-92.

Becher, P. and Becher, D. (1969). In Pesticide formulation research: Physical and Colloidal Chemical abstracts, Adv. Chem. Ser., No 86 (Ed. Gould R. F.), pp 15-23, American Chemical Soc. Washington.

Beck, T. (1970). Der Mikrobielle Abbum von herbiziden und ihr einflurs auf die mikroflora des Bodems. Z. Bakteriol II. 124: 304-313.

Bell, A. R. and Nalewaja, J. D. (1968). Effect of duration of wild oat competition in flax. Weed Sci. 16: 509-512.

by the herbicides sethoxydim and haloxyfop, Biochem. and Biophy. Res. Comm. 148: 1039-1044.

Busse, M. D., Ratcliffe, A. W., Shestak, C. J. and Powers, R. F. (2001). Glyphosate toxicity and the effects of long-term vegetation control on soil microbial communities. Soil Biol. Biochem. 33: 1777-1789.

Buta J. G. and Steffens G. L. (1971). Plant growth inhibition and membrane penetration by a homologous series of dichloro benzoate esters. Physiologica Plantarum 24: 431-435.

Camper, N. D., Moherek, E. A. and Huffman, J. (1973). Changes in microbial populations in paraquat-treated soils. Weed Res. 13: 231-233.

Carter, M.C. and Naylor, A.W. (1961). Metabolism of 3-amino-l, 2,4-triazole-15 14C in plants. Bot. Gaz. 122 (2): 138-143.

Caseley, J. C. and Eno, C. F. (1966). Survival and reproduction of two species of earthworm and a rotifer following herbicide treatments. Proc. Soil Soc. Am. 30: 246-350.

Chakraborty, T. K. (1967). Study of the effect of pre and post-emergent application of herbicides on seed production, weed control and soil microflora in peas. Ph D Thesis, IARI, New Delhi.

Chakravarty, P and Chatarpaul, L. (1989). Non-target effect of herbicides: I. effect of glyphosate and hexazinone on soil microbial activity. Microbial population, and *in vitro* growth of ectomycorrhizal fungi, Pestic. Sci. 3: 233-241.

Challen, S. B. (1962). J. Pharm. Pharmac. 14: 707-714.

Chandra, P. (1964). Herbicidal effects on certain soil microbial activities in some brown Soils of Saskatchewan. Weed Res.4: 54-63.

Chang, F .Y. (1972). Comparative effects of three EPTC antidotes. Weed Sci. 21 (4): 292-294.

Chase, C. A., Bewick, T. A. and Shilling, D. G. (1998) Characterization of paraquat resistance in *Solanum americanum* Mill. II. Evidence for a chloroplast mechanism. Pesticide Biochem. Physiol. 60: 23-30.

Chauhan, U. K., Mahindra, A. and Suryanarayana, V. (1994). Influence on soil microflora in response to application of some pesticides. J. Ecotoxicol. Environ Monit. 4: 133-136.

Che, M., Loux, M.M., Traina, S.J., and Logan, T.I. (1992). Effect of pH on sorption and desorption of imazaquin and imazethapyr on clay and humic acid. J. Envion. Quality 21: 698-703.

Chiron, S., Abian J., Ferrer, M., Sanchez-Baeza, F., Messeguer A. and Barcelo, D. (1995). Comparative photodegradation rates of alachlor and bentazone in natural water and determination of breakdown products. Environ. Toxicol. Chem. 14:1287–1298.

Choudhary, G. G. and Webster, G. R B. (1985). Protocol of guidance for the investigations of photochemical fate of pesticide in water, air and soils. Res. Rev 96: 79.

Chow, P. N. P. and Taylor (1983). Wild oat herbicides studies, Physiological and biochemical bases for interaction of barban and growth regulator herbicides in wild oat. J. Agric. Food Chem. 31: 575-578.

Christiansen, K. (1970). Int Symp. Pesticide in the soil, Michigan Sate Uni 8-24.

Chulakov, Sh. A. and Zharasov Sh U. (1971). Vest Sel'-hoz Nauki Alma-Ata 10: 122-124.

Cliath M. M, Wencer W. F, Farmer W. F., Shoup T. D. and Grover R. (1980). Volatilization of S-etheyl N, N dipropylthiocarbamate from water and wet soil during and after flood irrigation of alfalfa. J. Agril. Food Chem. 28: 610-613.

Cole, D. J. (1994). Detoxification and activation of agrochemicals in plants. Pestic. Sci. 42: 209–222.

Cole, D. J. (1994). Detoxification and activation of agrochemicals in plants. Pestic. Sci. 42: 209-222.

Collier, J., Saunders, C., Jaworski, E. G. and Tiemeier, D. C. (1986). Messenger RNA encoding a glutathione-S-transferase responsible for herbicide tolerance in maize is induced in response to safener treatment.

Comai, L, Facciotti, D., Hiatt, W. R., Thompson, G., Rose, R. E. and Stalker, D. M. (1985). Confers tolerance to glyphosate. *Nature* (London) 317: 741-744.

Coret, J., Bambonnet, B., Brabet, F. and Chamel. A. (1993). Diffusion of three ethoxylated octylphenols across isolated plant cuticles. Pestic. Sci. 38: 201-209.

Cornish, P. S. (1992). Glyphosate residues in a sandy soil affect tomato transplants. Australian J. Exptl. Agri. 32: 395-399.

Cox, D.R. (1958). Planning of experiments. Willey and Sons, N. York, pp. 91-153.

Craft A. S. and Drever, H. (1960). Experiments with herbicides in soil Weeds 8(1): 12-18.

Craft, A. S. (1966). Bidirectional movement of labeled traces in soybean seedlings. Hilgardia 37: 625-638.

Crafts, A. S. (1945). A new herbicide, 2,4-dinitro, secondary butyl phenol. Science 101: 4147-418.

Croll, B. T. (1976). Herbicides and potable water supplies', *Technical Memorandum* No. 119, Water Research Centre, Medmenham.

Droog, F. (1997). Plant glutathione S-transferase, a tale of theta and tau. Journal of Plant Growth Regulation 16: 95–107.

Dubey, P.K., Srivastava A., Sand, N.K. and Gupta, K.C. (1998). Analysis of Metribuzin in soil, grain and straw of wheat by HPLC. Indian J. Weed Sci., 10:1, 101-104.

Duke, S. O. and Kenyon, W. H. (1988). Polyclic alkanoic acids. In Herbicides: Chemistry, degradation and mode of action'. (Kearney, P. C and Kaufman, D. D. Ed.) Dekker, New York, III pp 71-116.

Durand, G., Abad, J. L., Sanchez-Baeza F., Messeguerand, A. and Barcelo. D. (1994). Unequivocal identification of compounds formed 124:161–169.

Durner, J., Knorzez, O. C. and Boeger, P. (1994). Ketol acid reductoisoerase from barley (*Hordeum vulgare*): purification, properties and specific inhibition. Plant Physio. 103: 903-910.

Dusky, J. A., Davis, D. G. and Shimabukuro, R. H. (1980). Metabolism of diclofop-methyl (methyl-2-[4-2(2', 4'-dichlorophenoxy) phenoxyl] propionate) in cell suspensions of diploid wheat (*Triticum monococcum*). Physiol. Plant 49: 151-156.

Dybing, C. D. and Currier, H. B. (1961). Foliar penetration by chemicals. Pl. Physiol. Lancaster 36: 169-174.

Eastin, E. F. (1971). Fate of fluorodifen in resistant peanut seedlings. Weed Sci. 19:261-265.

Eberlein, C.V., Guttieri, M.J., Berger, P.H., Fellman, J. K., Mallory-Smith, C.A., Thill, D.C. Baerg, R.J. and Belkmap, W.R. (1999). Physiological consequences of mutation for ALS- inhibitor resistance. Weed Sci. 47 : 383-392.

Eberlin, C. V., Guittieri, M. J., Berger, P.H., Fleming, J. K., Mallory-Smith, C. A., Thil, D.C., Baerg, R. J. and Belkmap, W. R.(1999). Physiological consequences of mutation for ALS-inhibitor resistance. Weed Sci. 47: 383-392.

Edwards, C. A. (1970). Effect of herbicides on the soil micro-fauna. Proc. 10th Br. Weed Contr. Conf. 1052-1062.

Edwards, C. A. and P. J. Bohlen. (1996). Biology and Ecology of Earthworms. New York: Champman and Hall.

Edwards, R. (1996). Characterization of glutathione transferase and glutathione peroxidases in pea (*Pisum sativum*). Physiologia Plantarum 98: 594–604.

Eerd, L.L., Van, Hoagland, R.E., Zablotowicz, R.M. and Hall, J.C. (2003). Pesticide metabolism in plants and microorganisms. Weed Sci. 51: 472-495.

Eleftherohorinos, I. G. (1987). Phytotoxicity and persistence of chlorsulfuron as affected by activated charcoal. Weed Res. 27: 443-452.

Encheva, H. and Rankov, V. (1990) Pochvoznanie-i-Agrokhimiya 25: 66-73.

Epstein, E. (1973). Roots. Science of America 228: 48-58.

Erickson, L. E. and Lee, K. H. (1989). Degradation of atrazine and related S-triazines. Critical Reviews in Environmental Control 19: 1-14.

Farago, S., Brunold, C. and Kreuz, K. (1994). Herbicide safeners and glutathione metabolism. Physiologia Plantarum 91: 537–542.

Faulkner, J. (1982). Breeding herbicide-tolerant crop cultivars by conventional methods. In Herbicide Resistance in Plants (Eds LeBaron, H. M. and Gressel, J.), pp. 235-256, Wiley, New York.

Fawcett, C. H., Ingram, J. M. A. and Wain, R. L. (1954). The b-oxidation of a-phenoxyalkycarboxylic acids in the flex plant in relation to their plant growth regulating activity. Proc. Roy. Soc. (London) B 142: 60-72.

Fedtke, C. (1982): Biochemistry and physiology of herbicide action. Springer-Verlag, Berlin,Germany, 202 pp.

Fedtke, C. and Schmidt, R. R. (1983). Behaviour of metribuzin in tolerant and susceptible soybean varieties, in Pesticide Chemistry: Human Welfare and the Environment, Vol. 3, Mode of Action, Metabolism and Toxicology (Eds. Matsunaka, S., Hutson, D. H. and Murphy, S. D.), pp. 177-182, Pergamon Press, Oxford.

Finke, R. L. Warner R. L. and Muzik, T. J. (1977). Effect of herbicides on *in vivo* nitrate and nitrite reduction. Weed Sci. 25: 18-22.

Folmar L. C., Sanders, H. O. and Julin, A. M. (1979). Toxicity of the herbicide glyphosate and several of its formulations to fish and aquatic invertebrates. Arch. Environ. Contam. Toxicol. 8:269-278.

Folmar, L. C., Sanders, H. O. and Julin, A. M. (1979). Toxicity of the herbicide glyphosate and several of its formulations to fish and aquatic invertebrates. Arch. Environ. Contam. Toxicol. 8: 269-278.

Fox, C. J. S. (1964). The effect of five herbicides on the number of invertebrates animal in grassland soil. Canadian J. Pl. Sci. 44: 405-409.

Foy, C. L. and Penner, D. (1965). Effect of inhibitors and herbicides on tricarboxylic acid cycle substrate oxidation by isolated cucumber mitochondria. Weeds 13: 226-231.

Francois, D., Robinson, G., (1990). Indices of triazine toxicity in *Chlamydomonas geitlei* Ettl. Aquatic Toxicol. 16: 205-228.

Frear, D. S. and Still G. G. (1968). The metabolism of 3,4-dichloropropionanilide in plants. Partial purification and properties of an aryl acylamidase from rice. Phytochemistry 7: 913-920.

Frear, D. S. and Swanson, H. R. (1970). Biosynthesis of S-(4-ethylamino-6 isopropylamino-2-s-triazino) glutathione: partial purification and properties of a glutathione S-transferase from corn. Phytochemistry 9: 2123-2132.

Freitas, R. S., Berger, P. G., Tomaz, M. A., Ferreira, L. R., Pereira, C. J. and Cecon, P. R. (2005b). Efeitos do herbicida trifloxysulfuron-sodium no crescimento do algodoeiro. In: Congresso Brasileiro De Algodão, 5., Salvador, 2005. Anais. Salvador, CD-ROM.

Freitas, R. S., Carneiro, P. A. M., Cecon, P. R., Ferreira, L. R., Berger, P. G. and Sediyama Tolerância T. (2005a). do algodoeiro em três estádios de desenvolvimento submetido ao herbicida trifloxysulfuron-sodium. In: Congresso Brasileiro De Algodão, 5., Salvador,. Anais. Salvador, 2005a. CD-ROM.

Freitas, R., Tomaz, M. A., Ferreira, L. R., Berger, P. G., Pereira, C. J. and Cecon, P. R. V. (2006). Growth of cotton treated with trifloxysulfuron-sodium. Planta Daninha 24: 123-129, 2006.

Gaillard, C., Dufaud, A., Tommasini, R., Kreuz, K., Amrhein, N. and Martinoia, E. (1994). A herbicide antidote (safener) induces the activity of both the herbicide detoxifying enzyme and of a vacuolar transporter for the detoxified herbicide. FEBS Letters 352: 219–221.

Gaillardon, P., Gaudry, J. C. and R. Calvet. (1983). Effet des matières organiques ajoutées au sol sur l'adsorption des herbicides. Influence de la composition des matières organiques. Weed Res. 23:333–338.

Gaskin, R .E. and Holloway, P. (1992). Some physicochemical factors influencing foliar uptake enhancement of glyphosate-mono (isopropylaminonium) by polyoxyethylene surfactants. Pestic. Sci. 34: 195-206.

Gaskin, R. E. and Stevens, P. J. G. (1993). Antagonism of the foliar uptake of glyphosate into grasses by organosilicone surfactants. Pestic. Sci. 38: 185-200.

Gast, A., Knuesli, E. and Gysin, H. (1955). Uber Pffanzenwachs-tumsregulatoren chlorazin, eine phytotoxisch wirksame Substanz. Experimentia 11: 107-108.

Gautam, O. P., Arora, P. N., Mani, V. S., and Gautam, K. C. (1965). Herbicidal and stimulatory role of tenoron in carrots. Proc. Second Weed Control Seminar, Hisar pp 14-17.

Gautam, R. C. (1970). Weed control in Rabi season pearl millet through herbicides and physical system. Indian J. Weed Sci. 11: 20-25.

Gaynor, J. D., MacTavish, D. C. and Findlay, W. I. (1995). Atrazine and metolachlor loss in surface and subsurface runoff as affected by cultural practices. J. Environ. Qual. 24:246–256.

Getzin, L. W. (1981). Dissipation of chloropyrifos from dry soil surfaces. J Econ Entomol 74: 707.

Ghang, F. Y., Bandeen, J. D. and Stephenson, G. R. (1972a) A selective antidote for prevention of EPTG injury in com. Can. J. Pl. Sci. 52: 707-714.

Ghosh, D. C and Bera, P. S (1983). Weed control in rapeseed (*Brassica compestris* L.), Annual Conf., Abstract Indian Soc. Weed Sci., Conf., JNKVV, Jabalpur.

Gigliotti, C. and Allievi L. (2001). Differential effects of the herbicides bensulfuron and cinosulfuron on soil microorganisms. Journal of Environmental Science and Health, Part-B, 36: 775–782.

Gill, H. S. and Brar, L. S. (1974). Atrazine, simazine nitrofen, and alachlor for weed control in maize (*Zea mays* L.) and their residual effect on wheat. J Res., PAU, Ludhiana 10: 269-273.

Gill, H. S., Sandhu, K. S., Mehra, S. and Singh, T. (1983). Studies on chemical control of weed in Indian mustard (*Brasica juncea* Coss). Abstract Indian Soc. Weed Sci., Conf., HAU, Hisar

Gill, H. S., Singh, C. A. and Brar, L. S. (1976). Comparison efficacy of some herbicides for weed control in potato. Proc. Of ISWS/APAU Weed Sci. Conf., Hyderabad, p 217.

Gillberg, B. O. (1971). Arch Mikrobiol. 75: 203-208.

Gish T. J., Mekee L., Wienhold V.J., Shirmohamadi A (1995). Climate and tillage effect on herbicide volatilization pp 77-79 in clean water clean environment 21 century team agriculture working to protect water resources. Vol I pesticide proceeding Kansas city, USA

Glotfelty, D. E., Leech, M. M., Jersey, J. and Taylor, A.W. (1989). Volatilization and wind erosion of soil surface atrazine, simazine alachlor and oxaphene. J. Agric. Food Chem. 37: 546

Gogoi, A.K., Sondhia S. and Yaduraju N.T. (2006). Status paper on butachlor, for review of pesticide, plant protection quarantine and storage, Ministry of Agriculture, India, p. 16.

Gogoi, A.K., Sondhia S. and Yaduraju N.T. (2006). Status paper on pendimethalin, for review of pesticide plant protection quarantine and storage, Ministry of Agriculture, India, p. 11.

Gogoi, A.K., Sondhia S. and Yaduraju N.T. (2006). Status paper on atrazine, for review of pesticide. Plant protection quarantine and storage, Ministry of Agriculture, India, p. 16.

Gohre, K. and Miller, G. C. (1983). Singlet oxygen generation on soil surfaces. J. Agric Food Chem. 31: 1104

Gopal, M., Jha, K. J. and Shukla, L. (2005). Identifying bacteria for degradation atrazine, metribuzin and pendimethalin. Proceedings of the National Biennial Conference of Indian Soc. of Weed Sci., held at Ludhiana w.e.f. 6-8 April, p289-290.

Gorlach-Lira K., Stefaniak, O., Slizak, W. and Owedyk, I. (1997). The response of forest soil microflora to the herbicide formulations fusillade and Roundup. Microbiological Res. 152: 319-329.

Gowda, A. S. J., Devendra, R., Gowda, R. C. and Gowda, M. K. M. (1993). Mobility of pendimethalin under different soil moisture regimes: quantification by bioassay Integrated weed management for sustainable agriculture. Proceedings of an Indian Society of Weed Science International Symposium, Hisar, Haryana, 18-20 November Vol. II, 113-116.

Goyal, Madan, V. K. and Phogat, V. K (2003). Persistence of trifluralin in soils under intermittent and continuous ponding condition. Pesticide Res. J., 15: 181-184.

Green, J. M. (1992). Increasing efficiency with adjuvants and herbicide mixtures. Proceedings of the First International Weed Control Congress, Melbourne, AU. pp. 187-192.

Greene D. W. and Bukovac M. J. (1972). Pl. Penctration of napthalene acetic acid into pear (*Pyrus cummcinis* L.) leaves Cell Physiol. Tokyo 13: 321-330.

Greene, D. W. and Bukovac, M. J. (1974). Stomatal penetration: effect of surfactant role in foliar absorption of Am. J Bot. 61: 100-106.

Gressel, L., Ezra, G. and Jain, S. M. (1982). Genetic and chemical manipulation of crops to confer tolerance to chemicals. In chemical manipulation of crop Growth and development (Ed. McLaren, J. S.), pp. 79-91, Butterworth, London.

Griffiths, B. S., Bonkowski, M., Roy, J. and Ritz, K. (2001). Functional stability, substrate utilisation and biological indicators of soils following environmental impacts. Applied Soil Ecology 16: 49-61.

Groffman, P. M. and Bohlen, P. J. (1999). Soil and sediment biodiversity. Bioscience 49:139-148.

Grossbard, E. (1975). In Int Symp. The Interaction of herbicides, Micro-organism and plants, Worclaw 1973, Soil Sci. Annual 26: 117-130.

Grossbard, E. and Marsh, J. A. P. (1974). Effect of seven substituted herbicides on the soil microflora. Pestic. Sci. 609-623.

Grover, R., Shewchuk, R. S., Cassena, A. J., Smith, A. E. and Hunter, J. H. (1985). Fate of 2,4-D iso octyl ester after application to wheat field J. Enviorn Qual. 14: 203-210.

Grover, R., Wolt, J., Cessna, A. and Schiefer, H. (1997). Environmental fate of trifluralin. Reviews of Environ. Cont. and Toxico., 153: 1-16.

Gulick, S. H., Grimes, D. W., Munk, D. S. and Goldhamer, D. A. (1994) Cover-cropped-enhanced water infiltration of a slowly permeable fine sandy loam. Soil Science Society of America Journal 58: 1539-1546.

Gunkel, G. and Streit, B. (1980). Mechanisms ofbioaccumulation of a herbicide (atrazine, s-triazine) in a freshwater mollusc *(Ancylusjluviatilis* Mi1I.) and a fish *(Coregonusfera* Jurine). Water Res. 14: 1573-1584.

Gupta, O. P. (1984). Scientific Weed Management. Revisededn. Modern Weed Science, Today and Tomorrows Printers & Publishers, New Delhi.

Gupta, O. P. and Niranwal, N. K. (1976). Increasing herbicide selectivity in maize and cowpea by seed treatment with activated carbon and NA. PANS 22 (1): 86-89.

Gupta, O. P. and Singh, R. (1958). Herbicidal efficacy of EPTC, chloroxuron, amiben, diquat and MCPB on *Pisum sativum* (pea) on tarai soil, Pantnagar. Indian J. Horticulture 25: 202.

Gupta, S., Gajbhiye, V. T. and Gupta, S. (2000) Simple methods for dithiopyr residue analysis in soil, water and plant materials. Annals of P. Prot. Sci. 8: 79-83.

Gustafson, D. I. (1995). Development of novel active ingredients M. Vaghi and E. Funari, Eds. In: Pesticide risk in groundwater, Boca Raton, Fl CRC pp 153-161.

Gysin, H. and Knusli, E. (1960). The chemistry and herbicidal properties of triazine derivatives. In Advances in Pest Control Research, Metcalf, R. L. (Editor). Inter-science publishers, N. Y. pp 289-358.

Hahlbrock, K. Chappell, J. and Scheel, D. (1984). Genes involved in resistance in higher plants: possible candidates for gene transfer', in The Impact of Gene Transfer Techniques in Eukaryotic Cell Biology (Eds Schell, J. J. and Starlinger, P.), pp. 155-166, Springer Verlag, Berlin.

Hall, K. J., Holloway, P. J. and Stock, D. (1997). Factors affecting the efficacy of spray delivery onto foliage using oil-based adjuvants. Aspects Applied Biol. 48:113-120. .

Hamilton, R. H. (1964). Tolerance of several grass species to 2-chloro-s-triazine herbicides in relation to degradation and content of benzoxazinone derivatives. J. Agril. Food Chem. 12: 14-17.

Haney, R. L., Senseman, S. and Hons, F. M. (2002). Effect of Roundup Ultra on microbial activity and biomass from selected soils. J. Environ. Qual. 31: 730-735.

Harper, A., White, A. W. Jr., Bruce, R. R., Thomas, A. W. and Leonard R. A. (1976). Soil and microclimate effect on trifluralin volatilization. J. Environ. Qual. 5: 236

Hartman, W. A. and Martin, D. B. (1984). Effect of suspended bentonite clay on the acute toxicity of Roundup (glyphosate) to *Daphnia pulex* and *Lemna minor*. Bull. Environ. Contam. Toxicol. 33: 355-36l.

Haskell, D. A. and Rogers, B. J. (1960). Proc. N. Cent. Weeds Comntr. Conf. 17:39.

Hatton, P. J., Cole D. J. and Edwards, R. (1996). Influence of plants age on glutathione levels and glutathione transferases involved in herbicide detoxification in corn (*Zea mays* L.) and giant foxtail (*Setaria faberi* Herm). Pesticide Biochem. Physiol. 54: 199-209.

Hatton, P. J., Dixon, D., Cole, D. J. and Edwards, R. (1996). Glutathione transferase activities and herbicide selectivity in maize and associated weed species. Pestic. Sci. 46: 267-275.

Hatzios, K. K. (1983). Herbicide antidotes: Development, chemistry and mode of action. Adv. in Agron. 36: 265–316.

Hatzios, K. K. (1983). Herbicide antidotes: development, chemistry and mode of action. Adv. Agron. 36: 265-316.

Hatzios, K. K. (1997). Regulation of enzymatic systems detoxifying xenobiotics in plants. NATO ASI Series 3. High Technology - Vol. 37, Kluwer Academic Publishers, Boston, 385 pp.

Hauke, Pacewiczowa T. (1971). The effect of herbicides on the activity of soil micro-flora. Pam pulawski 46: 5-48.

Hawkes, T. R., Howard, J. L and Pontin, S. E. (1989). Herbicides that inhibits the biosynthesis of branched chain amino acids. In herbicides and plant metabolism, (Ed: Dodge A.D.) Cambridge University Press, pp 113-136.

Hayes, H. M. (1970). Adsorption of triazine herbicides on soil organic matter, including a short review on soil organic matter chemistry. p. 131–174. In Gunther, F.A. and Gunther, J.D. (Ed.) Residue reviews: Residues of pesticides and other foreign chemicals in foods and feeds. Springer-Verslag, Berlin.

Hayes, M. H. B. (1970). Adsorption of triazine herbicides on soil organic matter including a short review on soil organic chemistry. Residues Rev. 32: 131-168.

Hazen, J. L. (2000). Adjuvants - terminology, classification, and chemistry. Weed Technology 14: 773-784.

Heap, I. M. (2002). International survey of herbicide resistant weeds. Online Internet. Accessed on 15 February 2002. Available www.weedscience. com.

Heap, I. M. (2007). International Survey of Herbicide Resistant Weeds online. Weed Science Society of America. http://www.weedscience.org/In.asp.

Hebert, V. R. and Miller, G. C. (1990). Depth dependence of direct and indirect photolysis on soil surfaces. J. Agric. Food. Chem. 38: 913–918.

Helfgott, S. (1969). Ph.D. Dissertation, University of California, Davis 207 pp.

Helling, C. S., Kearney, P. C. and Alexander, M. (1971). Behaviour of pesticides in soil. Adv. Agron. 23:147-240.

Hernández, F., Fernandez, C., Díaz, J. C., Díaz, J. J. and Cortegaza, P. L. (2004). Herbicide efficacy and selectivity in sugarcane of herbicide Krismat DG 75 (trifloxysulfuron+ametryn).III Congreso 2004 Sociedad Cubana de Malezología, Memorias, Jardín Botánico Nacional, Ciudad Habana, 28, 29 y 30 de abril del

Hess, F. D. (1991). Wetting and penetration of plant surfaces. Sandoz Crop Protection, Walnut Creek, CA.

Hess, F. D. and Foy, C. L. (2000). Interaction of surfactants with plant cuticles. Weed Technol. 14: 807-813.

Hess, F.D. (1999). Surfactants and additives. Proc. of the California Weed Sci. Soc. 51: 156-172.

Hewitt, A. J. (1998). The effect of tank mix and adjuvants on spray drift. In: McMullan, P.M. (Ed.) Adjuvants for Agrochemicals: Challenges and Opportunities. Proceedings of the fifth International symposium on adjuvants for Agrochemicals, chemical producers distributors association, Memphis, TN. Pp. 451-462.

Hewitt, A. J., Miller, P. C. H., Dexter, R. W. and Bagley, W. E. (2001). The influence of tank mix adjuvants on the formation, characteristic and drift potential of agricultural spray pp 397-402 Proc 6th Int. Symp. On adjutants for agrochemicals ISAA 2001 ASAA Foundation the Netherlands.

Hill, B. D., Stobbe, E. H. and Jones, B. L. (1978). Hydrolysis of the herbicide benzoylprop-ethyl by wild oat exterase. Weed Res. 18:149-54.

Hiltibran, R. C. (1967). Effects of some herbicides on fertilized fish eggs and fry. Trans. Am. Fish. Soc. 96: 414-416.

Hiltibran, R. C., Underwood, D. L. and Fickle, J. S. (1972). Fate of diquat in the aquatic environment. WRC Research Report No. 52. University of Illinois, Urbana, IL.

Hilton, J. T., Monaco, D. E., Moreland, D. E. and Gentner, W. A. (1964). Mode of action of substituted uracil herbicides. Weeds 12: 129-131.

Hoda, S. A., Singh, S. P., Jain, B. P., Mandal, G. and Yadav, K. D (1988). Residual effect of different herbicidal treatment given on potato, succeeding green gram and sesamum after potato. Pestology 12: 17-19.

Hogland, R. E. and Graf, G. (1972). Enzymatic hydrolysis of herbicides in plants. Weeds Sci. 20: 303-305.

Hogland, R. E., Graf, G. and Handel, E. D. (1974). Hydrolysis of 3,4-dichloropropionanilide (propanil) by plant aryl acylamidases. Weed Res. 14: 371-374.

Holmes, J. D. and Hess, F. D. (1984). Growth inhibition and disruption of mitosis by DCPA in oat (*Avena sativa*) root. Weed Sci. 32:732-738.

Jones, R. O. (1985). Tolerance of the fry of common warm-water fishes to some chemicals employed in fish culture. Proc. South. Assoc. Game Fish Comm. 16: 436-445.

Jordan, L. S. and Cudney, D. W. (1987). Fate of pesticides in the environment. DANR University of California Publication 3320.

Jury, W. A., Grover, R., Specer, W. F. and Farmer, W. J. (1980). Modeling vapour loses of soil incorporated triallet, J. Soil Sci Soc. Am 3: 445

Jury, W.A. (1986). Chemical movement through soil. In S.C. Hern and S.M. Melancon (ed.) Vadose zone modelling of organic pollutants. Lewis Publ., Chelsea, MI.1. p. 135–158.

Kaushik, S. N. and Moolani, M. K. (1974). Persistence of atrazine in soil cropped with corn (*Zea mays* L.). Indian J. Weed Sci. 6: 53-60.

Kea, J., Neuderfer, Symula, G. J., Aylesworth, G. and Miccoli, J. (1987). The toxicity of Ortho-Diquat to the fry of several species of warm water fish. Technical Report No. 87-1 (BEP), Department of Environmental Conservation, Albany, NY.

Kearny, P. C. and Karns J. S. (1987). Microbial Metabolism. In: Fate of pesticides in the environment. J. W. Biggar and J. N. Seiber, eds. University of California, Publication 3320.

Kewat, M. L. and Pandey, J. (2002). Persistence of metribuzin in soil in soybean-wheat sequence. Ann. Agri. Res. 23: 651-653.

Khan, S. U. (1972). Adsorption of pesticide by humic substances. A review. Environ. Lett. 3:1–12.

Khan, S. U. (1974). Adsorption of bipyridium herbicides by humic acid. J Envion. Qual. 3: 202-206.

Khan, S. U. and Mazuurkewich, R. (1974). Adsorption of linuron on humic acid. Soil Sci. Soc. Am. Proc. 118: 339-344.

Khan, S. U. and Young, J. C. (1977). N-Nitrosamine formation in soil from the herbicide glyphosate. J. Agric. Food Chern. 25: 1430-1432.

Kirkwood, R. C. (1999). Recent developments in our understanding of the plant cuticle as a barrier to the foliar uptake of pesticides. Pestic. Sci. 55: 69-77.

Kiss L. (1967). Erdesz Kutat., 63: 249-258.

Knuteson, S. L., Whitwell, T. and Klaine, S. J. (2002). Influence of plant age and size on simazine toxicity and uptake. J. Environ. Qual. 31: 2090-2103.

Koblek, K. Focke, M. and Lichtenthaler, H. R. (1988). Fatty acid biosynthesis and acetyl CoA carboxylase as a target of diclofop, fenoxaprop and other aryoxyphenoxy-propionic acid herbicides. Z. Naturforsch 43c: 47-54.

Konstantinou, I. K., Zarkadis, A. K., and Albanis, T. A. (2001). Photodegradation of selected herbicides in various natural waters and soils under environmental conditions organic compounds in the environment. J. Environ. Qual. 30:121–130.

Koopman, H. and Daams, J. (1960). 2, 6-dichlorobenzonitrile: A new herbicide, Nature 186: 89-90.

Kotrikla, A., Lekkas, T. H. and Bletsa, G. (1997). Toxicity of the herbicide atrazine, two of its degradation products and the herbicide metolachlor on photosynthetic microorganisms. Fresenius Environ. Bull. 6: 502-507.

Kozak, J., Weber, J. B. and Sheets, T. J. (1983). Adsorption of prometryn and metolachlor by selected soil organic matter fractions. Soil Sci. 136:94–101.

Kreuz, K., Gaudin, J., Stingelin, J. and Ebert, E. (1991). Metabolism of the aryloxyphenoxypropanoate herbicide CGA 184927, in wheat, barley and maize: Differential effects of the safener CGA

Kreuz, K., Tommasini, R. and Martinoia, E. (1996). Old enzymes for a new job: Herbicide detoxification in plants. Plant Physiol. 111:349-353.

Kudsk (Eds.) Herbicide Bioassays. CRC Press, Boca Raton, FL. pp. 99-116.

Kulshrestha, G. (1982). Hydrolysis of isoproturon in aqueous medium and its persistence in soil and plants. Indian J. Weed Sci. 14: 96-102.

Kulshrestha, G. and Yaduraju, N. T. (1987). Persistence of pendimethalin in soil following pre-emergence application to wheat. Indian J. Agronomy 32: 271-274.

Kulshrestha, G., Singh, S. B. and Gautam, K. C. (1995). Residues of fluazifop-p-butyl following application to soybean. Bull. Environ. Cont. Toxicol. 55: 276-282.

Kulshrestha, G., Singh, S. B., Lal, S. P. and Yaduraju, N. T. (2000). Effect of long-term application of pendimethalin: enhanced degradation in soil. Pest Manage. Sci. 56: 202-206.

Kulshrestha, G., Yaduraju, N. T. and Mani, V. S. (1977). Relative toxicity of S-triazines to certain rabi crops. In Abstracts of papers Annual Conference ISWS, Prabhani.

Kumar, J., JaiPrakash, Kumar, J. and Prakash, J. (1993). Persistence of thiobencarb and butachlor in soil incubated at different temperatures. Integrated weed management for sustainable agriculture. Proceedings of an Indian Society of Weed Science International Symposium, Hisar, India, 18-20 November, Vol. II, 123-124.

Kumar, K., Kolar, J. S. and Bhandari, S. C. (1985). Effect of terbutryne, methabenzthiazuron and pendimethalin on soil microflora in Bengal gram

(*Cicer arietinum* L.) and lentil (*Lens culinaris* Medic.). Abstracts of papers, annual conference of Indian Society of Weed Sci. undated, 50.

Kumaran, K. and Devi, L. S. (1996). Adsorption desorption of atrazine in some selected soils of Karnataka. Pestic. Res. J. 8: 146-151.

Kumari, D., Kumari, B., Kathpal, T. S., Yadav, A. and Malik, R. K. (2004). Determination of 2, 4-D sodium salt residues in soil and wheat using HPLC. Annals of Agri. Bio Res. 9: 59-61.

Labeda, D. P, Kang-Chien, L. Casida, L. E. (1976) Colonization of soil by Arthrobacter and Pseudomonas under varying conditions of water and nutrient availability as studied by plate counts and transmission electron microscopy. Applied Environ. Microbiol. 31: 551-561.

Lal, S. S. (1990). Direct and residual effect of varying levels of herbicides for controlling the weeds in potato-black gram crop sequences. Indian J. Weed Sci. 22: 1-5.

Lambert, (1967). Functional relationship between sorption in soil and chemical structure. J. Agril. Fd. Chem. 15: 572-576.

Langkramer O. (1970). Zentbl. Bakt. ParasitKde. 125: 713-721.

Larcher, W. (2000). Ecofisiologia vegetal. São Carlos: RiMa, 531p.

Larson, R. A. and Weber, E. J. (1994). Reaction mechanisms in environ mental organic chemistry. CRC Press, Boca Raton, FL.

Lartiges, S. B. and Carrigues, P. P. (1995). Degradation kinetics of or ganophosphorus and organonitrogen pesticides in different waters under various environmental conditions. Environ. Sci. Technol. 29:1246–1254.

Latchanna, A. (1987). Persistence of phytotoxic residues of atrazine in soils of rainfed rabi sorghum. Indian J. Weed Sci. 19: 166-170.

Lavelle, P., (2001). SOM management in the tropics: Why feeding the soil macro -fauna? Nutr. Cycl. Agroecosys. 61:53-61.

Le Baron, H. M. (1991). Distribution and seriousness of herbicide resistant weed infestations world wide. In: Herbicide resistance in weeds and crops, (Casely, Z. C., Cussans G. W. and Atkin, R. K Eds.). Butterwarth-Heinemann, Oxford. pp. 27-55.

Leela D. (1987). Persistence of alachlor and butachlor in different soils in pot culture. Pesticides 21: 23-26.

Leela, D. (1982). Weed Control by herbicides in cabbage, bottle gourd and watermelon. Indian J. Weed Sci. 39: 252-256.

Leela, D. (1993). Weedicides for vegetables Indian-Horticulture 38: 13-15.

Lefebvre A.D., Maizonnier J.C., Gaudy D., Clair and Scalla R. (1987). Some effects of the herbicides EL-107 on cellular growth and metabolism. Weed Res. 27: 125-134.

Lehnen, L P., Sherma, N T. D., Becerril, J. M. and Duke, S. O. (1988). Tissue and cellular localization of acifluorfen induced porphyrins in cucumber cotyledons. Pesti. Biochem. Physio. 37: 239-248.

Leonard, O. A., Bayer, D. E., and Glenn, R. K. (1966). Translocation of herbicides and assimilates in red maple and white ash. Botany Gazette, 127: 193-201.

Leung, T. S., Naqvi M. S. and Leblanc. C. (1983). Toxicities of two herbicides (Basagran, diquat) and an algicide (Cutrine-plus) to mosquitofish *Gambusia affinis* Environ. Pollut.(Ser. A) 30: 153-160.

Li, G. C and Felbeck, G. T. (1972). A study on the mechanism of atrazine adsorption by humic acid from muck soils. Soil Sci. 113: 140-148.

Lichtenthaler, K. H., Koblek, K. and Ishii, K. (1987). Inhibition of sethoxydim of pigment accumulation and fatty acid biosynthesis in chloroplasts of Avena seedlings. Z Naturforsch, 42c: 1275-1279.

Linscott, D. L and Hagin, R. D. (1967). Protecting alfalfa seedlings from a triazine with activated charcoal. Weeds 15: 304-306.

López-Ovejero, R. F., Christoffoleti, P. J., Nicolai, M. and Barela, J. F. (2003). Manejo de plantas daninhas na cultura do milho. In: Fancelli, A.L; Dourado-Neto, D. (Ed.) Milho: estratégias de manejo para alta produtividade. Piracicaba: USP/ESALQ, p 47-79.

Loux, M. M., and Reese, K. D. (1992). Effect of soil pH on adsorption and persistence of imazaquin. Weed Sci. 40: 490-496.

Luckwill, L. C. and Lloyd-Jones, C. P. (1960). Metabolism of plant growth regulators. II. Decarboxylation of 2, 4-dichlorophenoxyacetic acid in the leaves of apple and strawberry. Annals. Applied Biol: 626-636.

Luisi, P. L. and Straub, B. (1984). Reverse Micelles -Technological and Biological Relevance, Plenum Press, New York.

Lund-Hoie, K. (1969). Uptake, translocation, and metabolism of simazine in Norway spruce (*Picea abies*). Weed Res. 9: 142-147.

Lydon, J. and Duke, O. S. (1988). Glyphosate induction of elevatin of hydrobenzoic acids in higher plants. J. Agril Food Chem. 36: 813-818.

MacIsaac, S. A., Paul, R. N. and Devine, M. D. (1991). A scanning electron microscope study of glyphosate deposits in relation to foliar uptake. Pestic. Sci. 31: 53-64.

Maèek J. I. and Milharèiè L. (1972) Prog. Rep Tex. Agric. Exp. Sin 11.

Mahajan, G and Bvar, L. S. (2001). Studies on herbicide resistance in *Phalaris minor* under Punjab conditions. Indian J. Weeds Sci. 33:1-4.

Malik, R. K. and Singh, S. (1995). Little seed canary grass (*Phalaris minor*) resistancetoisoproturon in India. Weed Tech. 9: 419-425.

Mani, V. S. (1977). Weed research in India, status, problem and strategies. Proc. of Weed Sci. Conf., Indian Soc. Weed Sci. pp 27-36.

Mani, V. S., Arora, P. N., Gsutam, K. C. and Singh, R. (1967). Weed control in potato through pre and post emergence application of herbicides. Unpublished data, IARI, New Delhi.

Manthey, F. A. and Nalewaja, J. D. (1992). Relative wax solubility and phytotoxicity of oil to green foxtail. pp. 464-470 in C. J. Foy (Ed.). Adjuvants for Agrochemicals. CRC Press, Boca Raton, FL.

Marilley, L. and Aragno, M. (1999). Phylogenetic diversity of bacterial communities differing in degree of proximity of *Lolium perenne* and *Trifolium repens* roots. App. Soil Ecol. 13: 127-136.

Martin, J. P. (1964). Influence of pesticide residues on soil microbial and chemical properties. Residue Rev. 4: 96-129.

Mass, G. and Kresel, G. (1988). Dirkte abt rift van herbiziden bei verwendung verscheidener Dusentypen und Zursatzstoffe Z. Pfelnzenkar 11: 241-247.

Mathews, G.A. (1979). Pesticide Application Metlwds. Longman, London, New York.

Matringe, M., Camadro, J. M., Labbe, P. and Scalla, R. (1989a). Proto-porphyrinogen oxidase as molecular target for diphenyl ether herbicides. Pestic. Biochem. Physio. 260: 231-235.

Matringe, M., Camadro, J. M., Labbe, P. and Scalla, R. (1989b). Proto-porphyrinogen oxidase inhibition by three peroxidizing herbicides: oxadiazon Ls 82:556 and M & B 39279. Febs letter 245: 35-38.

Matsuguchi, T. and Ishizawa, S. (1969). J. Sci. Soil Manure, Tokyo, 40: 20-25.

Matsunaka, S. (1969). Activation and inactivation of herbicides by higher plants. Residue Rev. 25: 45-58.

Matthews, G. A. and Bateman, R. P. (2004) Classification criteria for fog and mist application of pesticides. Aspects of Applied Biology 71: 55-60

Matysiak, R. and Nalewaja, J. D. (1999). Temperature and UV light affect sethoxydim phytotoxicity. Weed Technol. 13:94-99.

Mayasich, J. M., Karlander, E. P. and Terlizzi, Jr D. E. (1986). Growth responses of *Nannochloris oculata* Droop and *phaeodactylum tricornutum* Bohlin to the herbicide atrazine as influenced by light intensity and temperature. Aquatic Toxicol. 8: 175-184.

McCall, P. J. and Gavit, P. D. (1986),. Aqueous photolysis of triclopyr and its butoxyethyl ester and calculated environmental photodegradation rates. Environ. Toxicol Chem 5: 879.

McHenry, W. B. and Norris, R. F. (1972). Study guide for agricultural control advisers on weed control, University of California Publication 4050. Berkeley, CA.

McMullan, P. M. (1994). Effect of sodium bicarbonate on clethodim or quizalofop efficacy and the role of ultraviolet light. Weed Technol. 8: 572-575.

McMullan, P. M. (2000). Utility adjuvants. Weed Technol. 14: 792-797.

McWhorter, C.G. (1982). The use of adjuvants. In: Adjuvants for Herbicides, WSSA, Champaign, IL. Pgs. 10-25.

Mekwatanakarn and Sivasithamparam, K (1987). Effect of certain herbicides on soil microbial populations and their influence on saprophytic growth in soil and pathogenicity of take-all fungus. Biol. and Fertility of Soils 5: 175-180.

Mickovski, M. D. (1966). Goldisen Zb zemjod.-šum. Fak. Univ. Skopje 19: 5-25.

Miller, G. C. and Crossby, D. G. (1982). Pesticide photoproducts: generation and significance. J. Toxicol Clinical Toxicol. 19:707.

Miller, G. C. and Hebert, V. R. (1987). Environmental photodegradation of pesticides in Fate of pesticides in the environmental (Biggar, J.W. and Sieber, J. N. Eds.) Agricultural Experiment Station Pub.

Miller, G.C. and Zepp, R. G. (1979). Effect on suspended sediment on the photolysis of dissolved pollutants. Water Res. 13:453–459.

Monaco, T. J., Weller, S. C., and Ashton, F. M. (2002). Weed Science: principles and practices forth edition, John Wiley and Son, New York USA.

Moreland, D. E., Corbin, F. T., Fleischmann, T. J. and McFarland, J. E. (1995). Partial characterization of microsomes isolated from mung bean cotyledons. Pestic. Biochem. Physiol., 52: 98–108.

Morton, C. A., Harvey, R.G., Kells, J. J., Lueschen, W. E. and Fritz, V. A. (1991). Effect of DPX-V9360 and terbufos on field and sweet corn (*Zea mays*) under three environments. Weed Tech. 5: 130-136.

Moskallenko, A. A. and Karapetyan, N. V. (1996). Structural role of caratonoids in photosynthetic membrane Z. *Naturforch*. 51c: 763-858.

Mosleh, Y. Y., Paris-Palacios S., Couderchet, M. and Vernet, G. (2002). Fate and sub lethal effects of isoproturon on mature earthworm (*Lumbricus terrestris* L.). Meded Rijksuniv Gent Fak Landbouwkd Toegep Biol. Wet. 67:69-77.

Mukerjee, P. and Mysels, K.J. (1971). Critical micelle concentrations of aqueous surfactant systems, NSRDS-NBS, 36.

Mukhopadhyaya, S. K (1959). Control of weeds in sugarcane by chemical and cultural methods. Assoc. IARI (Agronomy), Thesis, Indian Agril. Res. Institue, New Delhi.

Mukhopadhyaya, S. K. and Haldar, H. K. (1979). Chemical weed control studies in hybrid maize. Viswa Bharti J Res. (Science) 4: 104-112.

Müller, G. (1970). Laboruntersuchungen zur Wirking von Herbiziden auf Carabiden. Arch PflSchutz 7: 351-364.

Murphy, K. J. (1982). The use of methylthio-triazines in freshwater systems; a review', Proceedings of the European Weed Research Society 6th International Symposium o Aquatic Weeds, Novi Sad, European Weed Research Society. Secretariat, Wageniagen,The Netherlands pp. 263-277.

Murphy, K. J., Hanbury, R. G. and Eaton, J. W. (1981). The ecological effects of 2methylthiotriazine herbicides used for aquatic weed control in navigable canals. I. Effects on aquatic flora and water chemistry. Arch. Hydrobiol. 91: 294-331.

Muthukaruppan, G., Suman Janardhanan,S. and Vijayalaks, G. (2005). Sublethal Toxicity of the Herbicide Butachlor on the Earthworm *Perionyx sansibaricus* and its Histological Changes, J. Soil Sediment 5: 82-86.

Nair, P. K. R., Hariwaltabh and Bhalla P. L. (1980). Study on the effect of certain herbicides on onions (*Allium sepa* L.). Abtr. Ann. Conf., ISWS, Bhbaneswar.

Nair, P. K. R., Tiwari, R. P. and Rahman, N. (1982). Influence of herbicide on weed control efficieny, growth and yield of tomato (*Lycopersicon esculentum* Mill). Abstract of paper, Annual Conf, ISWS Khandawa, India pp 34-35.

Nalayini, P., Sankaran, S. and Kempuchetty, N. (1992). Evaluation of herbicides in sunflower (*Helianthus annus*) under different moisture regimes. Abstract Indian Soc. Weed Sci., Conf., HAU, Hisar p 71.

Nalewaja, J. D. (1994). Esterified seed oil adjuvants. Proc. North Central Weed Sci. Soc. 49:149-156.

Nalewaja, J. D. (2002). Oils as and with herbicides. pp. 290-300 in Beattie, G.A.C. Watson, D. M. Stevens, M. L. Rae, D. J. and Spooner-Hart R. N. (Ed.). Spray Oils Beyond 2000. Univ. Western Sydney, Australia.

Nalewaja, J. D. and Matysiak, R. (1993). Optimizing adjuvants to overcome glyphosate antagonistic salts. Weed Technol. 7: 337-342.

Nalewaja, J. D. and Matysiak, R. (1993a). Optimizing adjuvant to overcome glyphosate antagonism salts. Weed Technol. 7: 154-342.

Nalewaja, J. D. and Matysiak, R. (1993b). Spray carrier salts affect herbicide toxicity to Kochia (Kochia scoparia). Weed Technol. 7: 154-158.

Nalewaja, J. D. and Matysiak, R. (1995). Ethoxylated linear alcohol surfactants affect glyphosate and fluazifop absorption and efficacy. pp. 291-296 in R. Gaskin E. (Ed.). 4th International Symposium on Adjuvants for Agrochemicals. Rotorua, NZ

Nalewaja, J. D. and Matysiak, R. (2000). Spray deposit from nicosulfuron with salts that affect efficacy. Weed Technol. 14:740-749.

Nalewaja, J. D. and Matysiak, R. (2001). Nicosulfuron response to adjuvants, salts, and spray volume. pp. 304-314 in H. deRuiter (Ed.) 6th International Symposium on Agrochemicals. Amsterdam, Netherlands.

Nalewaja, J. D., Matysiak, R. and Woznica, Z. (2001). Optimum surfactant HLB for nicosulfuron in salt deposits. pp. 131-140 in Viets A. K., Tann R. S., and Mueninghoff J. C. (Ed.). Pesticide Formulation and Application Systems. ASTM STP 1400, Conshohocken, PA.

Nalewaja, J. D., Praczyk, T. and Matysiak, R. (1995). Salts and surfactants Influence nicosulfuron activity. Weed technol. 9: 587-593.

Nankur, J. T. (1980). Weed problem in relation to season and location in potato in hills and plains of North Western India, Abstract of papers, ISWS/OUAT Wed Sci. Conf., Bhubaneshwar.

Neumann, S. and Jacob, F. (1968). von α-Aminoisobuttersäure durch die Blätter von. *Vicia faba* L. Naturwissenschaften 55: 89-90.

Newbold, C. (1975). Herbicides in aquatic systems. Bioi. Conserv. 7, 97-118.

Norris, L A. and Freed, V. H. (1966). The absorption and translocation characteristics of several phenoxylkyl acid herbicides in bigleaf maple. Weed Res. 6: 203-211.

Norris, L. A. and Freed, V. H. (1966a). The absorption and translocation characteristics of several phenoxyalkyl acid herbicides in bigleaf maple. Weed Res. 6(3): 203-211.

Norris, L. A. and Freed, V. H. (1966a). The metabolism of a series of chlorophenoxyalkyl acid herbicides in big leafmaple *(Acer macrophyllum* Pursh.) Weed Res. 6: 212-220.

Norris, L. A. and Freed, V. H. (1966b). The metabolism of a series of chlorophenoxyalkyl acid herbicides in bigleaf maple. Weed Res. 6: 212-220.

Norris, L. A. and Freed, V. H. (1966c). The absorption, translocation, and metabolism characteristics of 4-(2,4-dichlorophenoxy)butyric acid in bigleaf maple. Weed Res. 6(4): 283-291.

Norris, R.F. (1982). Action and fate of adjuvants in plants. In: Adjuvants for Herbicides, WSSA, Champaign, IL. Pgs. 68-83.

Ohkawa, H., Tsujii, H. and Ohkawa, Y. (1999). The use of cytochrome P_{450} genes to introduce herbicide tolerance in crops: a review. Pesti. Sci. 55: 867–874.

Oliveira Júnior, R. S. (2001). Seletividade De Herbicidas Para Culturas E Plantas Daninhas. In: Oliveira Júnior, R. S., Constantin, J. Plantas daninhas e seu manejo. Guaíba: Agropecuária, p.291-313.

O'Loughlin, E. M. (1975). Predicting the concentration of aquatic herbicide residues in waterways. Natl. Conf. Publ. Inst. Eng. Aust. 75 (3): 56-60.

O'Loughlin, E. M. and Bowmer, K. H. (1975). Dilution and decay of aquatic herbicides in flowing channels. J. Hydrol. (Aust.) 26: 217-235.

Osipown, L. I. (1964). Surface Chemistry, Rheinhold Publishing Corp., New York.

Overbeek, J van. (1956). Absorption and translocation of plant regulators. A Rev. Pl. Physiol. 7: 355-372.

Owen, M. D. K. and Zelaya, I. (2004). Herbicide-Resistant Crops and Weed Resistance to Herbicides. Pest Manage. Sci. 61 (3): 301-311.

Pahwa, S. K. and Bajaj, K. (1997). Persistence of trifluralin and pendimethalin in soils incubated at different temperature. Indian Journal of Weed Sci. 29: 187-182.

Palaniappan, S. P. and Balasubramanian, A. (1985). Influence of two pesticides on certain soil enzymes. Agricultural Res. J. Kerala. 23: 189-192.

Pallett, K. E., Little, J. P., Sheekey, M. and Veerasekaran, P. (1997). The mode of action of isoxaflutole I, Physiological effects, metabolism and selectivity. Pestic. Biochem. Physio. 62: 113-124.

Pallett, K.E., Little, J.P., Veerasekaran, P. and Sheekey, M. (1998). Inhibition of 4-hydroxyphenyl pyruvate dioxygenase: the mode of action of the herbicides RPA 2017772 isoxanflutole. Pestic. Sci. 50:83-84.

Pandey, C. S. and Kumar, A. (1984). Efficiency of weed control measurement in Indian mustard (*Brassica juncea* L.) planted at different dates. Indian J. Weed Sci. 26: 273-275.

Panwar, R. S., Rathi, S. S., Malik, R. K. and Malik, R. S. (1994). Effect of clopyralid alone and as tank mixture with isoproturon on weed control in wheat. Indian J. Weed Science 26: 1-5.

Parmar, N. B., Maraviya, G. V., Shah, P. G., Patel B. K., Ghelani, L. M. and Patel, A. M. (1998). Pendimethalin residues in tobacco plant. Tobacco Res. 24: 57-59.

Parr, J.F. (1982). Toxicology of adjuvants. In: Adjuvants for herbicides, WSSA, Champaign, IL. Pgs. 93-114.

Parr, J.F. and Norman, A. G. (1965). Considerations in the use of surfactants in plant systems: A review. Botanical Gazette 126: 86-96.

Patel S. S., Gungate, R. T., and Lad, B. L. (1985). Chemical weed control in mango bud wood nurseries. Pesticides 19: 32-35.

Patel, C. L. and Pacet, Z. C. (1989). Residual effect of herbicides on succeeding legumes, Indian J. Weed Sci. 21: 80-81.

Patel, R. B., Patel, B. K., Shah, P. G., Raj, M. F. and Patel, J. A. (1996). Pesticide Res. J. 8: 182-185.

Paterson, K. G. and Schnoor, J. L. (1992). Fate of alachlor and atrazine in a riparian zone field site. Water Environ. Res. 64: 274-283.

Patteen, B. S. (1959). Influence of ammonium phosphate buffers on foliar absorption of organic acids. Bot. Gaz. 120: 137-144.

Paul, E. A., Howard, A. Simonin, John Symula, and Robert, W. Bauer (1994). The Toxicity of diquat, endothall, and fluridone. The early life stages of fish, J. Freshwater Ecology, Vol. 9, Num. 3 - Sept 1994 pp.229-239

Payá-Pérez, A. B., Cortés M. N. Sala A. and Larsen B.. (1992). Organic matter fractions controlling the sorption of atrazine in sandy soils. Chemosphere 25:887–898.

Penner, D. (2000a). Activator adjuvants. Weed Technol. 14: 785-791.

Peterson, H.G., Boutin, C., Martin, P., Freemark, K. E., Ruecker, N. J. and Moody, M. J. (1994). Aquatic phyto-toxicity of pesticides, applied at expected environmental concentrations. Aquatic Toxicol. 28: 275-292.

Phillips, R. E., Egli, D. B. and Thompson, Jr (1972). Absorption of herbicides by soybean seeds and their influence on emergence and seedling growth. Weed Sci. 20: 506-510.

Pinamonti, F. Stefanini, M. Dalpiaz, A. (1996) Soil management effects on nutritional status and grapevine performance. Wein-Wiss. 51: 76-82.

Pitt, J. I. and Hocking, A. D. (1985) Fungi and food spoilage. CSIRO Division of Food Processing: North Ryde, NSW, Australia.

Plimmer, J. R. (1971). The photochemistry of halogenated herbicides. Residue Rec. 33:47-74.

Ponnuswany, P., Santhi, P., Gopalasamy, G. and Mohamed, Ali. (1997). Interaction of herbicides a with soil microbes cassava (*Manihot esculanta* Crantz). Indian J. Weed Sci. 29: 147-150.

Porpiglia, P. J., Rawls, E. K., Gillespie, G. R. and Peek, J. W. (1990). A method to evaluate the differential response of corn (*Zea mays*) to sulfonylureas. WSSA Abstracts 30: p.86.

Prakash, N. B., Devi, L. S. and Siddaramappa, R. (1999). Mobility of butachlor and pendimethalin in soil columns. Pesticide Res. J. 11: 143-149.

Prasad, H., Kumari, B. and Kathpal, T. S. (2006). Dissipation kinetics of trifluralin in water at different pH levels under laboratory conditions. Pesticide Res. J. 18: 208-210.

Prasad, R., Foy, C. L. and Craft, A. S. (1967). Weeds 15: 149-156.

Pruger, J. H., Hatfield, J. L. and Saucer, T. J. (1999). Field scale metolachlor volatilization flux estimates from broadcast and banded application method in central Iowa. J. Enviorn Qual. 28: 75-81.

Que Hee, S. S. and R. G. Sutherland. The phenoxyalkanoic herbicides. Vol 1, CRC series in pesticide chemistry. CRC Press, Boca Raton, Fla.

Radhamani, S., Sankaran, S. and Jayakumar, R. (1997). Degradation and persistence of herbicides in direct seeded puddled rice. Madras Agri. J. 84: 194-196.

Radosevich S.R., Steinbach K.E. and Arutzen C.J. (1979). Effect of photosystem II inhibitors on thylakoid membranes of two common groundsel biotypes. Weed Sci. 27: 216-217.

Rahman, A. and Ashford, R. (1970). Selective action of trifluralin for control of green foxtail in wheat. Weed Sci. 18(6): 754-759.

Rai, A. K., Chhonkar, P. K. and Agnihotri, N. P (2000). Persistence and degradation of pendimethalin and anilofos in flooded versus non-flooded soils. J. Indian Society Soil Sci. 48: 57-62

Raj, M. F., Patel, B. K. and Shah, P. G. (1999). Adsorption and desorption of pendimethalin, fluchloralin and oxadiazon on soils. Pesticide Res. J. 11: 162-167.

Rajana, K. B., Devi, L. S. and Jagannath, M. K. (1991). Movement of S-triazines in Karnataka soils. Current Research University of Agricultural Sci. Bangalore. 20: 254-255

Ramesh, A. and Maheshwari, S. T. (2003). Dissipation of sulfosulfuron in soil and wheat plant under predominant cropping conditions and in a simulated model ecosystem. J. Agril. Food Chem. 51: 3396-3400.

Ramesh, A. and Maheshwari, S. T. (2004). Dissipation of alachlor in cotton plant soil and water and its bioaccumulation in fish. Chemosphere 54: 547-652.

Rana, M. C. and Angiras, N. N. (1993.). Studies on persistence of imazethapyr applied in soybean and its residual effect on wheat + pea cropping system Integrated weed management for sustainable agriculture. Proceedings of an Indian Society of Weed Science International Symposium, Hisar, India, 18-20 November, Vol. II, 106-108.

Randhawa S. K. and Sandhu K. S. (1992). Persistence of isoproturon and methabenzthiazuron applied for weed control in onion (*Allium cepa*). Indian J. Agronomy 37: 620-622.

Randhawa, K. S. and Bhalla , P. L. (1976). The effect of some herbicides on the weed flora and soil microflora in the onion crop (*Allium cepa* L.) and persistence of their residues in soil. Indian J Weed Sci. 3: 38-43.

Randhawa, K. S., Sandhu, K. S. and Daljit, K. S. (1979). Studies on the effect of pre and post emergent herbicides in combination with cultural practices on the weed flora in okra (*Abelmoschus esculentus* (L) Moench). Abstract of paper, ISWS/MAU Weed Sci. Conf., Parbhani.

Randhawa, S. K. and Sandhu, K. S. (1989). Persistence in soil of isoproturon applied for weed control in wheat. Indian J. Weed Sci. 21: 80-88.

Randhawa, S. K. and Sandhu, K. S. (1997). Persistence of metoxuron in soil. J. Res., Punjab Agricultural University 34:136-137.

Randhawa, S. K., Walia, U. S., Walia, S. S., Kler, D. S. and Goel, M. (2000). Persistence of herbicides in soil, crop produce and underground water -a review. Environ. Ecol. 2000 18: 61-77.

Rani, S., Kumari, B. and Tripathi, T. S. (2006). Effect of pH on the dissipation of flufenacet (FOE-5043) in water. Pesticides Res. J. 18: 201-204.

Rao K.N., and Gupta K.M., (1980). From cesthetic to part (*Eichornnia crassipes* (mart) som S.). Meeting Indian Soil Weed Sci. Bhubaneswar, India. No. 206.

Rapoport, E. H. and Cangioli, C. (1963). Herbicides and soil micro flora. Pédobiologia 2: 235-238.

Rathod, B. M., Patel, J. C. and Malvia, D. B. (1983). Efficacy of herbicides in pearl millet. Indian J. Weed Sci. 15: 214-216.

Raut, A. K. and Kulshrestha, G. (1991). Persistence of fluazifop-p-butyl in an inceptisol. Pesticides Res. J. 3: 67-72.

Rban, O. J. and Cook, N. J. (1986). Hazard evaluation division, standard evaluation procedure, and ecological risk assessment. EPA 540/9-85-001. Final Report, U. S. Environmental Protection Agency, Washington, DC.

Reddy, D. V. V., Koteswara Rao, A. S., Sridhar and Soundara, Raja M. S. (1983). Chemical weed control in rainfed and irrigated groundnut for achieving effective control of nut sedge (*Cyperus rotundus* L.). Abstract Indian Soc. Weed Sci., Conf., HAU, Hisar p 50.

Reddy, K. N., Rao, B. N., Sultan, M. A., Reddy, D. J. and Babu, T. R. (1998). Residues of butachlor in paddy. J. of Res. ANGRAU. 26: 48-49.

Reid, C. P. P. and Hurtt, W. (1969). Translocation and distribution of picloram in bean plants associated with nastic movement. Plant Physiol. 44: 1393-1396.

Retzinger, Jr., E. J. and C. Mallory-Smith. (1997). Classification of herbicides by site of action for weed resistance management strategies. Weed Technol. 11:384-393.

Reuppel, M. L., Brightwell, B. B., Schaefer, J. and Marvel, 1. T. (1977). Metabolism and degradation of glyphosate in soil and water. J. Agric. Food Chem. 25: 517-528.

Rilley and Eagle (1990). Herbicides in soil and water, In: Weed Control Handbook principles (Eds. Hance, R.J. and Holly, K.), Blackwell Scientific Publications, UK.

Ró¿ycki, H., Dahm, H., Strzelczyk, E, Li C. Y. (1999). Diazotrophic bacteria in root-free soil and in the root zone of pine (*Pinus sylvestris* L.) and oak (*Quercus robur* L.). Applied soil Ecology 12: 239-250.

Roberts, H. A. (1982). Weed Control Handbook Principles. 7th Edn. British Crop Protection Council, Blackwell Scientific Publications, Oxford, London.

Robertson, M.M., Parham, P.H. and Bukovac, M.J. (1971). J. Agric. Fd. Chem. 19.

Robinson, D. W. (1965). The use of adsorbents and simazine on newly planted strawberries. Weed Res. 5: 43-51.

Rogers R. L. (1971). J. of Agri. Chem. 19: 32-36.

Roggenbuck, F. S., Penner, D., Burow, R. F. and Thomas, B. (1993). Study of the enhancement of herbicide activity and rainfastness by an organosilicone adjuvant utilizing radio-labelled herbicide and adjuvant. Pesticide Sci. 37: 121-125.

Rose, M. A and Lembi, C. A. (2000). Applied weed science second edition Pentice Hall Upper Saddler River, New Jersey.

Rosen J. D. (1967). Bull. Environ. Contam. 2: 349-354.

Rosen M. J. (1978). Surfactants and Interfacial Phenomena, Wiley, New York

Ross, M. A and Lembi, C.A. (1985). Applied Weed Science. Burgess Publishing Company, Minneapolis, MN.

Ross, M. A. and Lembi, C.A. (1999).Applied Weed Science, Publisher Prentice-Hall, Inc, United States of America.

Sabba, R. P. and Vaughn, K. C. (1999). Herbicides that inhibit cellulose biosynthesis. Weed Sci. 47: 757-763.

Sacher, R. M. (1978). Safety of Round-up in the aquatic environment. Proceedings of the European Weed Research Society 5th International Symposium on Aquatic Weeds, Amsterdam, pp. 315-322, European Weed Research Society. Secretariat, Wageniagen, The Netherlands.

Saikia, T. P. and Pandey, J. (1999). Dissipation of fluchloralin in chickpea (*Cicer arietinum* L.). Journal of the Agricultural Science Society of North East India. 12: 236-238.

Sandermann, H. (1985). Herbicide resistance through gene transfer? Biochemical and toxicological aspects', in The Impact of Gene Transfer Techniques in Eukaryotic Cell Biology (eds J.S. Schell and P. Starlinger), pp. 167-179, Springer-Verlag, Berlin.

Sandermann, H. (1992). Plant metabolism of xenobiotics. TIBS 17:82-84.

Sandhu K.S. and Randhawa K.S. (1980). Chemical weed control in onion in Punjab India. Tropical Pest Manage. 26: 41-44.

Sandhu, K. S., Randhawa S. K., Bhatia and R. K. (1994). Atrazine persistence studies in berseem for fodder. Journal of Research, Punjab Agricultural University. 1: 399-401.

Sandhu, K.S and Gill, G. S. (1973). Studies on critical period of weed competition in maize. Indian J Weed Sci. 5: 1-5.

Sandhu, K.S., Singh Tarlok , Gill, G. S., and Thind I.J.S. (1987). Bioefficacy of some herbicides for weed control in sorghum (*Sorghum bicolor* (L.) Morecnch.) grown for fodder and grain. Res., PAU, Ludhiana, 24: 376-381.

Sandhu, P. S., Randhawa, S. K. and Walia, U. S (1994). Persistence of atrazine in soil applied to sugarcane. Journal of Research, Punjab Agricultural University. 31: 6-8.

Sands R. and Bachelard E. P. (1973). New Phytol. 72: 656-663.

Sannino, F. and Gianfreda, L. (2001). Pesticide influence on soil enzymatic activities. Chemosphere 45: 417-425.

Sanyal, D. and Kulshrestha, G. (2003). Degradation of metolachlor in soil inoculated with a mixed fungal culture. Biology and Fertility of Soils. 38: 253-256.

Sanyal, D., Yaduraju, N. T. and Kulshrestha, G. (2003). Metolachlor persistence in laboratory and field soils under Indian tropical conditions. Journal of Environmental Science and Health. Part-B, Pestic., Food Cont. Agril. Wastes. 35: 571-583.

Saraswat, V. N. (1988). Consolidated Report of AICRP on Weed Control: Phase I 1978-79 to 1983-84. Indian Council of Agril. Res., Central Rice Res. Institute, Cuttack.

Sargent, J. A. and Blackman, G. E. (1969). Studies on foliar penetration: iv. mechanisms controlling the rate of penetration. J. Exp Bot. 20: 542-555.

Sarkar, R. (2003). Effect of herbicides on earthworm. M. Sc. Dissertation.

Sarmah, A. K., Kookana, and Alston (1999). Degradation of chlorsulfuron and triasulfuron in alkaline soils under laboratory conditions. Weed Res. 39: 83–94.

Schafer, D. E. and Chilcote, D. O. (1970). Translocation and degradation of bromoxynil in a resistant and a susceptible species. Weed Sci. 18(6): 729-732.

Schuler, M. A. (1996). Plant cytochrome P450 mono-oxygenases. Critical Reviews in Plant Sci. 15: 235–284.

Schulz, A., On, O., Beyer, P. and Kleinig, H. (1993). SC-0051, a 2-benzoyl-cyclohexane 1,3 dione bleaching herbicide is a potent inhibitor of enzyme p-hydroxyphenylpyruvate dioxygenase. Febs letters 318: 162-166.

Scott, H. D. D. and Phillips, R. E. (1971). Diffusion of herbicides to seeds. Weed Sci. 19: 128-132.

Selvamani, S. and Sankaran, S. (1989). A comparative study of different herbcides for weed control in irrigated groundnut. Indain J. Weed Sci. 21: 37-40.

Selvamani, S. and Sankaran S. (1991). Residual effect of herbicides in irrigated groundnut (*Arachis hypogaea*). Indian J. Weed Sci. 21: 7-9.

Senapati, B. K., et al. (1999). In-soil earthworm technolo- C. A. Harvey, C. Gascon, H. L. Vasconcelos and A. N. Izac, 198-226. Washington: Island Press.

Senesi, N. (1992). Binding mechanisms of pesticides to soil humic substances. Sci. Total Environ. 123/124:63–76.

Senesi, N., Loffredo, E., D'Orazio, G., Brunette, V., Miano, T. M. and La Cava, P. (2001). Adsorption of pesticides by humic acids from organic amendments and soils. In C.E. Clapp et al. (ed.) Humic substances and chemical contaminants. SSSA, Madison, WI. p. 129–176.

Senesi, N., P. La, Cava, and Miano, T. M. (1997). Organic chemicals in the environment: Adsorption of imazethapyr to amended and nonamended soils and humic acids. J. Environ. Qual. 26: 1264–1270.

Sethi, R. K. and Chopra, S. L. (1975). Desorption, degradation and leaching of alachlor in some soils. Journal of the Indian Soc. Soil Sci. 23: 184-194.

Sharma, A.B., Asrar, S. M. and Bhalla, B. L. (1980). Evaluation of effective, selective and economical herbicides for garlic) *Allium sativum* L. Abstr. Ann. Conf. ISWS, Bhubaneswar.

Sharma, N. and Angiras, N. N. (1997). Studies on persistence of herbicides applied in maize by bioassay technique. World Weeds 4: 37-39.

Sharma, R. K., Brar, H. S., Khera, A. S. and Dhillon, B. S. (1988). Weed control in transplanted maize in winter. Indian J. Weed Sci. 10: 103.

Sharton, A. et al., (1993). Extended terbutryne persistence and herbicidal activity affected by superphosphate fertilizer. Phytoparasitica 21: 111-122.

Sheets, T. J. (1961). Uptake and distribution of simazine by oat and cotton seedlings. Weeds 9: 1-13.

Shimabukuro, Ma, Shimbukuro, R. H., Nord, W. S. and Heauf. (1978). Physiological effects of methyl 1, 2 [4-(2,4 dichlorophenoxy) phenyl] propionate on oat, wild oat and wheat. Biochem. Physiol. 8: 199-207.

Shimabukuro, R H. (1985). Detoxification of herbicides pp 215-240 in S. O. Duke (Ed.), Weed Physiology, Vol 2, Herbicide Physiology. CRC press, Inc. Boca Raton, FL.

Shimabukuro, R. H. (1967). Atrazine metabolism and herbicidal selectivity. Plant Physiol. 42: 1269-1276.

Shinoda, K. (1967). Solvent Properties of Surfactant Solutions, M. Dekker, New York.

Shinoda, K. and Friberg, S. (1986). Emulsion and Solubilization, Wiley, New York.

Shrivastava, C and Palta, R. K. (2002). Effect of atrazine and oxyflorfen against earthworm (*Eisenia fetida*). Annals Pl. Prot. Sci. 10: 145-146.

Sikka, H. C. and Davis, D. E. (1969). Effect of prometryne on $14CO_2$ fixation in cotton and soybean. Weed Sci. 17: 122-123.

Simonin, H. A. and Skea, J. C. (1977). Toxicity of diquat and cutrine to fingerling brown trout. NY Fish and Game J. 24: 37-45.

Sindhu, P. S., Bhandari, S. C., Mehra, S. P. and Gill, H. S. (1985). Effect of fluchloralin, oxadiazon and pendimethalin on soil microflora and nodulation in groundnut. Journal of Res. Punjab Agricultural University. 22: 33-38.

Singh G., Bhan, V. M. and Tripathi, S. S. (1984). Weed control in French bean. Indian J. Weed Sci. 16: 124-126.

Singh S. B. and Kulshrestha G. (1995). Degradation of fluchloralin in soil under predominating anaerobic conditions J. Environ. Sci. Health. Part-B, Pesticides, Food Cont. Agril. Wastes, 30: (3) 307-319.

Singh S. B. and Vaishya (1994). Studies on residual effect of herbicides applied in transplanted rice on succeeding wheat corp. Indian J. Weed Sci. 26: 129-131.

Singh S. P. and Sharma H. C. (1984). Effect of 2, 4-D and hand weeding weed population, growth and yield of wheat. J. Res., HAU, Hisar, 17: 341-345.

Singh, B. and Kulshrestha, G. (2003). Soil persistence of triasulfuron herbicide as affected by biotic and abiotic factors. Environmental Science and Health, Part B: Pestic. Food Cont. Agril. Wastes 2006: 635-645.

Singh, D. P., Singh, J. N., Singh, K. N. and Singh, G. (1985). Efficacy of certain herbicides on weed controlling maize and their residual toxicity by bioassay. Indian J. Weed Sci. 17: 14-21.

Singh, J. N., Baasler, E and Santlemann, P. W. (1972). Pestic. Biochem. Physiol. 2: 143-152.

Singh, K. P. and Malik, Y. S. (1993). Chemical weed control in carrot seed crop. Indian J. Weed Science 15: 12-16.

Singh, M., Orsenigo J.R. and Shah D.O. (1984). Surface tension and contact angle of herbicide salutions affected by surfactant JAOCS 61: 275-287.

Singh, R. and Mani, V. S. (1981). Effects of atrazine and alachlor in maize-pulse rotation. Proc. 8 th Asian Weed Science Conference, Bangalore, p 127-130.

Singh, R. K., Singh, S.S. and Vaishya, R. D. (1996). Studies on herbicide residues in post harvested soils of lentil under field conditions. Indian J. Weed Sci. 28: 198-199.

Singh, S. B., Maheswari, B. K., Prasad, J. and Singh, R. N. (1974). Studies on weed competition in sugarbeet (*Beta vulgaris* L.). Indian J. Weed Sci 6: 115-120.

Singh, S. B., Yaduraju, N. and Kulshrestha, G. (1997). Residues of metolachlor herbicide in soil and potato tubers under Indian tropical conditions. Bulletin Environ.Contamin. Toxicol. 59: 216-221.

Singh, S. B., Yaduraju, N. T. and Kulshrestha, G. (1999). Terminal residues of fluazifop-p-butyl in soybean. Annals of Plant Prot. Sci. 7: 87-90.

Singh, S. S., Shri Parkash and Vaisya, R. D. (1996). Residual effect of thiobencarb applied on rice on succeeding wheat crop. Indian J. Weed Sci. 28: 90-81.

Sinha, S. N., Agnihotri, N. P. and Gajbhiye, V. T (1996). Field evaluation of pendimethalin for weed control in onion and persistence in plant and soil. Annals of Pl. Prot. Sci. 4: 71-75.

Sivannarayana, G., Bhanumurthy, V. B. and Chandra Shehara Reddy, S. (1990). Effect of different methods and time of application of fluchloralin on weed control and yield of groundnut. Abstract, Biennial Conf., Indian Soc. for Weed Sci. JNKVV, Jabalpur.

Skea, J., G. Neuderfer, Symula, J. Aylesworth, G. and Miccoli. J. (1987). The toxicity of ortho-diquat to the fry of several species of warm water fish. Technical Report No. 87-1 (BEP), Department of Environmental Conservation, Albany, NY.

Slade, P. (1965). Photochemical degradation of paraquat. Nature, London, 207: 515-516.

Slade, P. (1967). The fate of paraquat applied to plants. Weed Res. 6: 158-167.

Slife, F. W., Key,J. L., Yamaguchi, S. and Grafts, A. S. (1962). Penetration, translocation and metabolism of 2,4-D and 2,4,5-T in wild and cultivated cucumber plants. Weeds 10: 29-35.

Smit E., Leeflang P., Gommans S., van den Broek J., van Mil S. and Wernars K (2001). Diversity and seasonal fluctuations of the dominant members of the bacterial soil community in a wheat field as determined by cultivation and molecular methods. Applied Environ. Microbiol. 67: 2284-2291.

Smith A. E. (1969). Factors affecting the loss of triallate from soils. Weed Research 9: 306-313.

Smith, A. E., Kerr, L. A. and Caldwell, B. (1997). Volatilization of ethalfluralin, trifluralin and triallate from a field following surface treatment with granular formulation. J. Agri Food Chem 43: 1473-1478.

Smith, L. W. and Foy, C. L. (1967). Interaction of several paraquat surfactant mixtures. Weeds 15:67.

Sobotka, A. (1970). Zembl. Bakt. Parasitenk Infektionskn Hyg. Abt 2 Natural, 125: 723-730.

Sollins, P., Homann, P. and Caldwell, B. A. (1996). Stabilization and destabilization of soil organic matter: mechanisms and controls. Geoderma 74: 65-105.

Sondhia, S. (2005). Annual Report, National Res. Centre for Weed Sci.23-24.

Sondhia, S. (2006a). Determination of terminal residues of haloxyfop-p-ethyl in onion. Indian J. Pl. Prot. 34: 258-259.

Sondhia, S. (2008). Determination of terminal residues of imazethapyr in Soybean grain straw and soil. Pest. Res. J. 20(1): 128-129.

Sondhia, S. (2008). Imazosulfuron residues in rice crop and soil using diode array detector. Pesticide Res. J. 19: 251-253.

Sondhia, S. (2008a). Evaluation of potential risk of herbicides bioaccumulation in fishes. (Editors: Sengupta L. and Dalwani R.) Preceding of TAAL The 12th World Lake Conference, p. 149-151.

Sondhia, S. (2008b). Herbicide residues in soil, water and food chain: an Indian perspectives in Biennial Conference of Indian Society of Weed Science held at Patna w.e.f. 27-28 February 2008, pp. 31-36.

Sondhia, S. (2008c). Leaching behaviour of metsulfuron-methyl in two texturally different soil. Environ. Monitoring and Assess.

Sondhia, S. and Singhai B. (2007). Phytotoxicity and persistence of sulfosulfuron under wheat cropping system. In Sustainable Agriculture for food, Bio energy and livelihood Security held at JNKVV, Jabalpur.

Sondhia, S. and Dixit, A. (2008). Persistence of flumioxazin residues in soybean crop and soil. Indian J. of Agri. Sci. 78: 68-70.

Sondhia, S. and Dubey, R.P. (2006b). Detection of butachlor and pendimethalin residues in onion. Abstract in of 93rd Science Congress in Agriculture and Forestry held at Hyderabad, India.

Sondhia, S. and Singhai, B. (2008). Persistence of sulfosulfuron under wheat cropping system. Bull. Environ. Contamination and toxicol. 40(5): 423-427.

Sondhia, S. and Singhai, B. (2008). Persistence of sulfosulfuron under wheat cropping system. International Conference on Agrochemicals Protecting Crop, Health & Natural Environment, New Delhi, January.

Sondhia, S. (2001). Determination and Assessment of atrazine residues in potato (*Solanum tuberosum* L.) soil. Geobios 28: 140-142.

Sondhia, S. (2002). Ultra-violet spectroscopic analysis of triazine herbicides in soil samples. Proceedings of the Eastern Analytical Symposium, held at, New Jersey, USA, w.e.f. 18-21 November 2002, P 25.

Sondhia, S. (2002a). Final report on, "Studies of herbicides residues in Maize-potato cropping system, NRCWS.

Sondhia, S. (2002b). Final report on, Studies of herbicides residues in soybean-wheat cropping system, NRCWS.

Sondhia, S. (2004). Annual report on, "herbicides residues in soil, water and food chain, NRCWS.

Sondhia, S. (2005). Phytotoxicity and persistence of metribuzin residues in black soil. Toxicol. Environ. Chem. 87: 387-389.

Sondhia, S. (2006). Ist Annual Report of the project, "behaviours of sulfosulfuron in sub soil under the influence of wheat cropping system-identification and quantification of potential transformation products responsible for the phytotoxicity and their bioaccumulation in fish 2006-2007 p 30.

Sondhia, S. (2006). Terminal residues of imazethapyr in soybean grains, straw and soil. Weed Newsletter, 6: 3-4.

Sondhia, S. (2006b). Annual Report, herbicide residues in soil water and food chain. NRCWS.

Sondhia, S. (2006e) Herbicide residues in soil, water and food chain, NRCWS, Annual Report.

Sondhia, S. (2007). Determination of Imazosulfuron persistence in rice crop and soil. Environ. Monit. Assess. 137: 205-211.

Sondhia, S. (2007). Evaluation of leaching potential of pendimethalin in clay loam soil. Pesticide Res. J. 19: 119-121.

Sondhia, S. (2007a). Fluazifop-butyl residues in soybean crop and soil. Pestic. Res. J. 19: 248-250.

Sondhia, S. (2007b). IInd Annual Report of the project, "behaviours of sulfosulfuron in sub soil under the influence of wheat cropping system-identification and quantification of potential transformation products responsible for the phytotoxicity and their bioaccumulation in fish 2006-2007 p 30.

Sondhia, S. (2007b). Phytotoxic effect of sulfosulfuron on succeeding pea and lentil. Weed Newsletter 8: 3.

Sondhia, S. (2008). Persistence of metsulfuron-methyl in wheat crop and soil. Environ. Monit. Assess. 147: 463-469.

Sondhia, S. (2008d). Persistence of oxyfluorfen in soil and detection of its residues in rice crop. Toxicol. Environmental Chem. (in press).

Sondhia, S. and Dixit, A. (2006c). Oxyfluorfen residues in onion. Weed Newsletter No. 6: 1-2.

Sondhia, S. and Dixit, A. (2007). Determination of terminal residues of oxyfluorfen in onion. Annals Pl. Prot. Sci. 15: 232-234.

Sondhia, S. and Dubey, R. P. (2006d). Determination of terminal residues of butachlor and pendimethalin in onion. Pesticide Res. J. 18: 85-86.

Sondhia, S. and Mishra, J. S. (2005). Determination of terminal residue of clodinafop propargyl in soil, wheat grains and straw. Indian J. Weed Sci. 37: 296-297.

Sondhia, S. and N. T. Yaduraju (2005). Evaluation of leaching of atrazine and metribuzin in tropical soil. Indian J. of Weed Sci. 37: 298-300.

Sondhia, S. and Saraswat, V. N. (2000a). Dissipation and assessment of atrazine residues. Proceeding of the National seminar on bio-diversity conservation, management and utilization for sustainable development, India pp 58.

Sondhia, S. and Saraswat, V. N. (2000b). Spectrophotometric determination of atrazine residues in maize grains and soil. Proceeding of the International Conf. on natural resource management held at New Delhi, India w.e.f. 14-18 February, pp 226-227.

Sondhia, S. and Singh, V. P. (2005). Persistence of isoproturon in soil and detection of its residues in wheat grain and straw. Geobios 33: 209-211.

Sondhia, S. and Singhai, B. (2006). Dissipation of metsulfuron-methyl in soil and its residues in wheat grains and straw by high performance liquid chromatography method, In Second International symposium on green/ sustainable chemistry, 10-13 January, New Delhi, abstract 145.

Sondhia, S. and Varsheney, Jay G. (2008). Evaluation of potential risk of metsulfuron-methyl persistence in rice crop and soil international Conference on Agrochemicals Protecting Crop, Health & Natural Environment, New Delhi, January 8-11, 2008.

Sondhia, S. and Yaduraju, N. T. (2004). Determination of metribuzin leaching in sandy clay soil, abstract in International conference on sustainable crop production in stress management and genetic options, Feb. 9-12, Jabalpur.

Sondhia, S, Singh V. P. and Yaduraju, N. T. (2006). Persistence of butachlor in sandy clay loam soil and detection of its residues in rice grains and straw. Annals Plant Prot. Sci. 14: 206-209.

Sondhia, S., Singh, V. P. and Yaduraju N. T. (2005). Dissipation of butachlor in sandy clay loam soil and detection of its residues in rice grains and straw. Proceedings of the National Biennial Conference of Indian Soc. of Weed Sci., held at, Ludhiana, w.e.f. 6-8 April p298-299.

Taiwo, L. B. and Oso, B. A. (1997). The Influence of some pesticides on soil microbial flora in relation to changes in nutrient level, rock phosphate solubilization and P-release under laboratory conditions. Agric.Ecosystem & Environ. 65: 59-68.

Taiz, L. and Leiger, E. (1991). Plant physiology benzamin/lumnings publishing co; Inc. Redwoodcity CA.

Takahashi, N., Mikami, N., Matsuda, T. and Miyamoto, J. (1985). Photo degradation of the pyrethroid insecticide cypermethrin in water and on soil surface. J. Pestic. Sci. 10:629–642.

Taylor, A. W. and Glotfelty, D. E. (1988). Evaporation from soils and crops. Chapter 4 in Environmental chemistry of herbicides, Vol I. (Grover R., Ed.) CRC Press, Boca Raton, Fla. p. 89.

Taylor, A. W. and Glotfetlty, D. E., Galss B. L., Freeman H. P. and Edwards W. M. (1976). Volatilization of deildrin and heptachlor from a maize field. J. Agric. food Chem. 24: 625.

Templeman, W. A. and Sexton, W. A. (1945). Effect of some aryl carbonic esters and relatd compounds upon and other plant species. Nature 156: 630.

Thakur, C. and Negi, N. S. (1954). Chemical control of Nilkalmi A menace to sugarcane crop. Proc. of 2nd Biennial Conf. (Jullundur) pp 549-554.

Thelen, K. D., Jackson, E. P. and Penner, D. (1995). The basis for the hard-water antagonism of glyphosate activity. Weed Sci. 43: 541-548.

Thomson, H. E., Swanson, C.P. Normaon, A. G. (1946). New growth regulating compounds. In summary of growth inhibitory activities of some organic compounds. Bot. Gaz. 107: 476-507.

Tiwari, J. P. and Bisen, C. R. (1985). Competitive impact of Phalaris weed ecosystem on sink co-efficient of some Rabi crops. Indian J. Weed Sci., 17: 21-27.

Tiwari, J. P., Kerchania, S. P. and Trivedi, K. K. (1988). Herbicidal selectivity to control weeds in mustard. Annual Conf., Indian Soc. Weed Sci., Conf., AAU, Jorhat.

Tooby, T. E. (1978). A scheme for the evaluation of hazards to non-target aquatic organisms from the use of chemicals. Proceedings of the European Weed Research Society 5th International Symposium on Aquatic Weeds, Amsterdam, pp.287-294, European Weed Research Society. Secretariat, Wageniagen, The Netherlands.

Tooby, T. E. (1985). Fate and biological consequences of glyphosate in the aquatic environment. In the herbicide glyphosate (Eds. Grossbard, E. and Atkinson, D.), pp. 206-217, Butterworths, London.

Torstennson, L. (1985). Behaviour of glyphosate in soils and its degradation. In the herbicide glyphosate (Eds Grossbard, E. and Atkinson, D.), pp. 137-150, Butterworths, London.

Tosh, G. C. and Jena, B. C. (1983). Herbicides evaluation for inter-cropped wheat with mustard and linseed. Indian J. Weed Sci. 15: 55-57.

Tosh, G. C., Patro, G. K. and Misra, A. (1977). Comparative efficiency of herbicides for weed control in groundnut. Indian J. Weed Sci. 15: 62-64.

Tripathi, B., Kumar, C. P. C., Rengarajan, T., Chandurkar, P. S. and Vijayalakshmi (1993). Pendimethalin, metolachlor, residues in soil and their effect on soil microflora, weed spectrum and green peas (*Pisum sativum* L.) yield integrated weed management for sustainable agriculture. Proceedings of an Indian Society of Weed Science International Symposium, Hisar, India, 18-20 November, Vol. II, 98-100.

Tripathi, J. P., Kurchania, S. P., Paradkar, N. R. and Bhalla, C. S. (1996). Direct and residual effect of herbicides in a soybean (*Glycine max*)-wheat *Triticum aestivatum*) cropping system, Indian J. Weed Sci. 28: 323-334.

Tuitert, G., Szczech, M. and Bollen, C. J. (1998). Suppression of *Rhizoctonia solani* in potting mixtures amended with compost made from organic Household Waste. Phytopathology 88: 764-772.

Tulabaev, B. D. (1972). Khimiya sel Khoz 10: 776-780.

Turner, B. C., Glotfetlty, D. E., Taylor, A. W. and Wastson, D. R. (1978). Volatilization of microencapsulated and conventionally applied chloropham in the field. J. Agronomy. 70: 933.

Tyler, M. J. (1997a). Herbicides kill frogs. Newsletter of the declining amphibians population task force 21.

Tyler, M. J. (1997b). Environmentally friendly: A false sense of security? Species. Newsletter of the Species Survival Commission, IUCN, The World Conservation Union. 29:20-21.

U. S. Environmental Protection Agency R.E.D. Facts, April (1996).

Umrani, N. K. and Bio, P. G. (1982). Studies on weed control in bajra under dryland conditions. J. Maharastra Agril. University 7: 145-147.

Urban, O. J. and Cook, N. J. (1986). Hazard evaluation division, standard evaluation procedure, ecological risk assessment. EPA 540/9-85-001. Final Report, U.S. Environmental Protection Agency, Washington, DC.

Vairavan, K., Sankaran, S. and Jeyaraman. (1992). Performance of herbicide under line source sprinkler irrigation systems I groundnut. Abstract, Indian Soc. For Weed Sci., Conf. HAU, Hisar p 63.

Valent. U. S. A. Corp. (1989). Diquat herbicide - H/A: for aquatic plant management. Product Report. Walnut Creek, CA.

Valverde, B and Greesel, J. (2006). Dealing with the evolution and spread of *Sorghum helepense* Glyphosate resistance in Argentina. A consultancy Report to SENASA (Confidential).

Valverde, B. E., Riches, C. R. and Casely, Z. C. (2000). Prevention and management of herbicide resistant weeds in rice: Experiences from Central America with *Echinochloa colona.*

Van Bruggen, A. H. C. and Semenov, A. M. (1999). A new approach to the search for indicators of root disease suppression. Australasian Plant Pathology. 28: 4-10.

Van Valkenburg, J. W. (1982). Terminology, classification, and chemistry. In: Adjuvants for Herbicides, WSSA, Champaign. p 1-9.

Vargas, R. N. (2001). Herbicide Resistance. Cooperative Exten. Bulletin. UC Davis. USA.

Vargas, R. N. and Wright, S, D. (2004). Herbicide Resistance Management. In Cotton Field Check. Coop. Extn. Bulletin UC Davis. USA.

Varshney J.G., (2005). Annual Report. Indian Institute of Pulses Research.

Vasmani, L. K. and Vijayakumar (1978). Herbicidal activity of diuron for weed control in cotton and its residual toxicity to oats and wheat. Abstract, Annual Conference of Indian Society of Weed Science, TNAU, Coimbatore p 18.

Vaughan, P. P. and Blough. N. V. (1998). Photochemical formation of hydroxy radical by constituents of natural waters. Environ. Sci. 32: 2947–2953.

Vaughn, K. C., Hoffman, J. C., Hahn, M. G. and Staehelin, L. A. (1996). The herbicides dichlobenil disrupts cell plate formation: immunogold characterization. Protoplasma 194:117-132.

Verma, R. D. and Bhardwaj R. B. L. (1958). A better way of talking weeds in sugarcane. Indian Farming III 3: 29-31.

Vindard, G. (1952). C. R. Head. Séance. Acad. Agric. Fr. 38: 417-418.

Voets, J .P., Meerschman, P. and Verstraete, W. (1974). Soil microbiological and biochemical effects of long-term atrazine applications. Soil Biol. Biochem. 6: 149-152.

Voos, G. and Groffman, P. M. (1997). Relationships between microbial biomass and dissipation of 2, 4-D and dicamba in soil. Biol. Fertil. Soils 24:106-110.

Vostral, H. J., Buchholtz, K. P. and Kust, C. A. (1970). Effect of root temperature on absorption and translocation. WSSA Abstract. pp. 39-40.

Wade, P. M. (1981). The long term effects of aquatic herbicides on macrophyte flora of freshwater habitats. Proceedings of the Conference on Aquatic Weeds and their Control, Oxford, National Vegetable Research Station, Wellesbourne; Association of Applied Biologists, London pp.234-240,.

Wain, R. L. (1955). A new approach to selective weed control. Ann. Appl. Biol. 42: 151-157.

Wain, R. L. (1957). Selective weed control with MCPB. Agriculture 63: 757-759.

Waldrep, T. W. and J. F. Freeman. (1964). EPTC injury to corn as affected by depth of incorporation in the soil. Weeds 12: 315-317.

Waldrep, T.W. and Taylor, H.M. (1976). 1-methyl-3-phenul-5-[(3-trifluromethyl) phenyl]-4- (1H)-pyridinone a new herbicide J. Agric. Food Chem. 24: 1250-1251.

Walia, U. S., Singh, K., Singh R., Kler, D. S., Singh K. and Singh R (2000). Environ. and Ecol. 18: 39-42.

Walker, A. (1987). Herbicide persistence in soil. Review Weed Sci. 3: 1-18.

Walker, K. A., Ridley, M., Lewis, T. and Harwood, J. L. (1989). Action of aryloxyphenoxy carboxylic acids on lipid metabolism. Rev Weed Sci. 4: 71-84.

Walter, B. (1970). Z. PflKrankh. PflPath. PFlshutz Sonderh. 5: 29-131.

Wan, H.B., Wong, M. K. and Mok, C. Y. (1994). Comparative study on the quantum yields of direct photolysis of organophosphorus pesticides in aqueous solution. J. Agric. Food Chem. 42: 2625–2630.

Wanamarta, G., Kells, J. J. and D. Penner. (1993). Overcoming antagonistic effects of sodium bentazon on sethoxydim absorption. Weed Tech. 7: 322-325.

Waschkies, C., Schropp, A. and Marschner, H. (1994). Relations between grapevine replant disease and root colonisation of grapevine (Vitis sp.) by fluorescent pseudomonads and endomycorrhizal fungi. Plant and Soil 162: 219-227.

Wauchope, R. D. (1978). The pesticide content of surface water draining from agricultural fields, a review. J. Environ. Qual. 7: 459–472.

Weaver, Mark, A., Krutz, L. Jason., Zablotowicz, Robert M., Reddy and Krishna N. (2007). Effects of glyphosate on soil microbial communities and its mineralization in a Mississippi soil Pest Management Science, Volume 63: 388-393.

Weber, J. B (1972). Interaction of organic pesticides with particulate matter. Adv In Chem. Ser. No. 111: 55-150.

Weber, J. B and Scott, D. C. (1966). Availability of a cationic herbicide adsorbed on clay minerals to cucumber seedlings. Weed Sci. 152: 1400-1402.

Weber, J. B. (1980). Use of activated carbon to clean up herbicides in greenhouse. North Carilina Floer Growers Bull. 24: 9-10.

Wed, S. B. and Wever, J. B. (1972). Pesticide organic matter interaction. Paper No 3840 Agri. Expt. Sta, NCS, Releigh, USA, 39-65.

Werck-Reichhart, D., Hehn, A. and Didierjean, L. (2000). Cytochromes P_{450} for engineering herbicide tolerance. Trends In Plant Sci. 5: 116–123.

Western, N. M., D. Coupland, V. Breeze, and Bieswal, M. (1998). Evaluation of different vegetable oil as possible replacements for mineral oil adjuvants. pp. 352-358 in P. M. McMullan, (ed.). 5th International Symposium on Adjuvants for Agrochemicals. Memphis, TN.

Whipps, J. M. (2001). Microbial interactions and biocontrol in the rhizosphere. J. Experi. Bo. 52:487-511.

White, L. A. Jr., Harper, R. A., Ball, L. and Turnbull, J. W. (1977). Trifluralin volatilization losses from a soybean field. J. Environ. Qual 6: 105.

Wichman, J. and Byrnes, W. (1975). Uptake, distribution and degradation of simazine by black walnut and yellow poplar seedlings. Weed Science 23: 448-453.

Wiegand, R. C., Shah, D. M., Mozer, T. J., Harding, E. I. and Diaz-Wilkinson, R. E. (1981). Metolachlor influence on growth and terpenoid synthesis. Pestic. Biochem. and Physiol. 16: 63-71.

Wildlife and Fisheries Commission, Bulletin No. 13. Baton Rouge, LA.

Williams, E. H., Mather, E. L. and Carter, S. M. (1984). Toxicity of herbicides endothall and diquat to benthic crustacea. Bull. Environ. Contam. Toxicol. 33: 418-422.

Wilson, D. C. and Bond, C. E. (1969). The effect of the herbicide diquat and dichlobenil (casoron) on pond invertebrates, part I: aquatic toxicity. Trans. Am. Fish. Soc. 98: 438-443.

Wingfield, G. I., and Johnson, J. M. (1981). Deoxygenation of water following use of the herbicide terbutryne simulated in a batch culture system. Bull. Environ. Contam. Toxicol. 26: 65-72

Witkowski, D. A. and Halling, B. P. (1989). Inhibition of plant protophyrinogen oxidase by the herbicide acifluorfen-methyl. Plant Physiol., 90: 1239-1242.

Witt, W.W. (2001). Adjuvants. University of Kentucky College of Agriculture, Agripedia.

Woestemayer, V. W. and Zick, W. H. (1960). Response of Bermuda grass to herbicides applied as pre-paving treatments. Western Weed Control conf. Res. Prog. Rept. Denver, Colo p-66.

Wojtaszek, T., Cherry, J. H. and Warren, G. F. (1966). The effect of 4, 6-dinitro-O-sec butylphenol on phosphate accumulation and incorporation in tomato leaf discs. Plant Physiol. 41: 34-38.

Wolfe, N. L., Mingelgrin, U. and Miller, G. C. (1990). Abiotic transformation in water, sediments and soil in pesticides in the Soil Environment: Processes, Impacts and Modeling (HH Cheng Ed.,) Soil Science Society of Americ, Madison, Wisconsin p 103.

World Health Organization. (1984). 2, 4-Dichlorophenoxyacetic acid (2,4-D), Environmental Health Criteria 29. United Nations Environment Programme, Geneva. 151 pgs.

Wrischer, M., Ljubesic, and Salopek. (1998). The role of caratonoids in the structural and functional stability of thylakoids in plastic of dark grown spruce seedlings. J. Plant Physiol. 153: 46-52.

WSSA. (1982). Adjuvants for Herbicides. Weed Science Society of America, Champaign. 144 pgs.

Yadav, A., Faroda, A. S., Malik, R. K., Kumar, V. and Dharam, Bir (1995). Persistence of pendimethalin applied in wheat under varying irrigation levels. Crop Research Hisar. 9: 399-402.

Yadav, D. B., Singh, S., Malik, R. K., Yadav, A. and Kumar, R. (1995). Residual effects of metsulfuron-methyl on onion (*Allium cepa* L.) Integrated weed management for sustainable agriculture. Proceedings of an Indian Society of Weed Science International Symposium, Hisar, India, 18-20 November 1993, Vol. II, 117-119.

Yaduraju, N. T., Kulshrestha, G., Sharma, R. P. and Ahuja, K. N. (1993). Isoproturon for weed control in potato (*Solanum tuberosum*) and its residue in soil and tubers. Indian J. Agril. Sci. 63: 731-733.

Young, B. G., Young, J. M., Matthews, J. L., Owen, M. D. K., Zelaya, I. A., Hartzler, R. G., Wax, L. M., Rorem, K. W. and Bollero, G. A. (2003). Soybean development and yield as affected by three postemergence herbicides. Agronomy J. 95: 1152-1156.

Yuan, J. S., Tranel, P. J. and Stewart Jr., C.N. (2007). Non-target-site herbicide resistance: a family business. Trends in Plant Sci. 12: 6-13.

Zafiriou, O.C., Joussot-Dubien, J., Zepp, R. G. and Zika, R. G. (1984). Photochemistry of natural waters. Environ. Sci. Technol. 18: 358–371

Zepp, R. G., Baughman, G. L. and Schlotzhauer, P. F. (1981). Comparison of photochemical behaviours of various humic substances in water, I Sunlight induced reactions of aquatic pollutants photosensitized by humic substances, Chemosphere 10: 109.

Zepp, R. G., Schlotzhauer, P.F. and. Sink, R.M (1985). Photosensitized transformations involving electronic energy transfer in natural waters: Role of humic substances. Environ. Sci. Technol. 19:74–81.

Zharasov S. H. U. (1969). Vest sel'-khoz, Nauki, Alma-Ata 12: 25-29.

Zimdhal, R. L. (1980). Crop weed competition-A review, published in the United Sates by the International Plant Protection Centre, Oregon State University, Oregon, USA.

Zimdhal, R. L. (1993). Fundamental of Weed Science, Academic Press, San Diego, CA.

Zimmerman, P.W. and Hitchcoch, A.E. (1942). Substituted phenoxy and benzoic acid growth substances and the relation of structure to physiological activity. Contr. Boyce. Thompson Inst. 12: 321-344.

Zweig, G. and Ashton, F. M. (1962). The effect of 2-chloro-4-ethylamino-6-isoprpylamino-s-triazine (atrazine) on distribution of $^{14}CO_2$ compound following $^{14}CO_2$ fixation in excised kidney bean leaves. J. Experimental Bot. 13: 5-11.

For Further Readings

Anderson, W. P. (1996). Weed Science: Principles and applications. 3rd Edition. West Publishers Co., St Paul, MN.

Applied weed science second edition (Ross, M. A. and Lembi, C. A.) Prentice Hall, New Jersey (1999).

Carroll, B. (2001). Selecting the right adjuvant. Helena Chemical Company's Guide to Adjuvants,http://www.helenachemical.com/proprietary/products/adjuvants/introduction.htm#

Dekker, M. (1967). Nonionic Surfactants (Schick, M.J. Ed.), New York.

Dekker, M. (1970). Cationic Surfactants (Jungermann, E. Ed.), New York.

Dekker, M. (1973). Anionic Surfactants (Linfield, W.M. Ed.), New York (1973).

Dekker, M. (1985). Ion-Pair Chromatography (Hearn, M.T. Ed.), New York.

DuPont (2000). Guidelines to qualify adjuvant for use with DuPont row crop and cereal herbicides. DuPont Agricultural Bulletin H-87285.

DuPont (2001). Approved adjuvant list for use with DuPont row crop and cereal herbicides. DuPont Agril. Bull. H-91222.

Extension Agronomy. http://www.weeds/iastate.edu/mgmt/qtr98-2/cropoils.htm

Fletcher, W. W. (1960). Herbicide and the soil, (Woodford, E.K. and Saga, G.R. Eds.). pp 20-62. Blackwell Scientific Publications, Oxford.

Green, J. M. (2000). Adjuvant outlook for pesticides. Pesticide Outlook. October: 196-199

Green, J. M. (2001). Herbicide adjuvants. In: UC Davis WRIC Weed Science School, September 26- 28 Woodland, CA.

Green, J. M. and Hazen, J. L. (1998). Understanding and using adjuvants properties to enhance pesticide activity. In: McMullan, P.M. (Ed.) Adjuvants for Agrochemicals: Challenges and Opportunities. Proceedings of the Fifth International Symposium on Adjuvants for Agrochemicals, Chemical Producers Distributors Association, Memphis, TN. Pp. 25-36.

Gupta, O. P. (1998). Modern weed management publisher, Agro Botanica.

Hall, J. C., Hoagland, R. E. and R. M. Zablotowicz, Editors. (2001). Pesticide biotransformation in plants and microorganisms - similarities and divergences. ACS series 777. American Chemical Society, Washington DC. 432 pp

Hartzler B. (2001). Role of AMS with glyphosate products. Iowa State University Extension Agronomy. http://www.weeds.iastate.edu/mgmt/2001/ams.htm

Herbicides (1994). WSSA, Champaign, IL. Pp. 84-92.

Herbicides: chemistry, degradation and mode of action. (Editors: Kearney and Kaufman, D. D.), Marcel Dekker, Inc., New York (1988).

Herbicides: Physiology, biochemistry, ecology (Editors; Audas, L.J.), Academic Press, Inc. (1976).

http://www.ca.uky.edu/agripedia/pls404/adjuvant.htm

Interaction between herbicides and the soil, (Editors Hance, R. L.) Academic Press, London (1980).

Kudsk P. and Streibig J. C. (1993). Formulation and adjuvants. In: Strigib, J.C.

Liss, A. R. (1984). Membranes, detergents and receptor solubilization (Venter, J.C. and Harrison, L.C. Ed.), New York.

Micellization, solubilization and micro-emulsions (K.L. Mittal Ed.), (1977). Plenum Press, New York.

Miller, P. and Westra, P. (1996). Herbicide surfactants and adjuvants, no. 0.559. Colorado State University Cooperative Extension, Production Crop Series.

Miller, P. and Westra, P. (1998). How surfactants work, no. 0.564. Colorado State University Cooperative Extension, Crop Fact Sheet. http://www.ext.colostate.edu/pubs/crops/00564.html

Mode of action of herbicides (1981). Second Edition, (Aston, F. D. and Craft, A. S.), Wiley-Interscience publication, New York.

Modern trends of colloid science in chemistry and biology (1985). (Eicke, H.E. Ed.), Birkhduser, Basel.

Physiology of herbicide action (1993). (Devine, M. D., Duke, S. O. and Fedtke, C). PTR prentice Hall, Englewood Cliffs, New Jersey.

Pringnitz, B. (1988) Adjuvants, 5th edition. Southern Illinois University, Carbondale. Clearing up confusion on adjuvants and additives. Iowa State University.

Shimabukuro R. H. (1967). Atrazine Metabolism and Herbicidal Selectivity *Plant Physiology* 42: 1269-1276.

Solution chemistry of surfactants (Mittal, K.L. Ed.), Plenum Press, New York (1979).

Acid equivalent (A.E.) : The amount of active ingredient expressed in terms of the parent acid.

Acid rain : Rainwater that contains sulfur dioxide and other pollutants from industrial plants. There has been considerable damage done to the forests of the US and Canada.

Acidic soil : Soil with a pH value less than 7.0 in highly acid soils some nutrients and elements become toxic, while others are unavailable resulting in poor plant growth.

Acidification : The process by which soils become increasingly acidic due to leaching of minerals such as Ca, Mg, K and nitrate nitrogen.

Acidifier : A material that can be added to spray mixtures to lower the pH.

Activator : A material that increases the biological efficacy of agrochemicals.

Active transport : An energy-expending mechanism by which a cell moves a chemical across a cell membrane from a point of lower concentration to a point of higher concentration, against the diffusion gradient.

Acute : Single or short-term exposure; used to describe brief exposures and effects which appear promptly after exposure.

Additive effect : Combined effect of two or more chemicals equal to the sum of their individual effects.

Adjuvant : Substance added to a formulated pesticide product to act as a wetting or spreading agent, sticker, penetrant, or emulsifier in order to enhance the physical characteristics of the product.

Adsorption coefficient (Koc) : A measure of a material's tendency to adsorb to soil particles. High Koc values indicate a tendency for the material to be adsorbed by soil particles rather than remain dissolved in the soil solution. Strongly adsorbed molecules will not leach or move unless the soil particle to which they are adsorbed moves (as in erosion). Koc values of less than 500 indicate little or no adsorption and a potential for leaching.

$$\text{Koc} = \frac{\text{conc. adsorbed/conc. dissolved}}{\text{\% organic carbon in the soil}}$$

Adventitious presence : Unintended presence of genetic.

Aeration : The loosening of soil by digging or other mechanical means to allow air to pass freely, usually done on lawns.

Aerobic : A process requiring oxygen or free air; conditions in which oxygen are present.

Aesthetic damage : The amount of damage to the aesthetics of a planting which will justify the cost of applied control measures.

Agricultural (or plant) biotechnology : A tool to make seeds with special qualities that can allow farmers to grow plants that are more nutritious, more resistant to pests and more productive.

Agrochemical : Generic term for the various chemical products used in agriculture.

Alkaline soil : Opposite to acidic.

Allelopathy : The release of chemicals by certain plants that will prevent or encourage/stimulate the growth of other plants nearby. Walnut trees are very well known to do this.

Alien weed : Weed, not native to a country.

Amelioration of soil : To correct or improve the condition of the soil through some practice such as liming or cultivation etc.

Anaerobic : Describes organisms living or occurring when oxygen is absent. Usually term used when talking about compost heaps.

Anionic surfactant : A surface-active agent in which the active portion of the molecule containing the lipophilic segment forms exclusively a negative ion (anion) when placed in aqueous solution.

Annual : A plant that completes its lifecycle within one year and then dies.

Antifoaming agent : A material used to inhibit or prevent the formation of foam.

Application rate : The amount of pesticide product applied to a given area.

Aqueous : Watery; pertaining to water.

Arable land : Land which is capable of the production of cultivated crops.

Attractant : A material that attracts specific pests.

B

Band application : An application in which a material such as fertilizer or herbicide is applied in strips, usually to the bed or seed row.

Basic alkaline: having a pH of greater than 7.

Basin : A portion of a rice field bounded by levees.

Beneficial insects : Those insects which are useful or helpful to people (e.g., predators and parasites of pests).

Biennials : Plants that live for two growing seasons.

Bioassay : Test which determines the effect of a chemical on a population of living organisms.

Bio-concentration : The accumulation of a chemical in tissues of an organism (such as fish) to levels that are greater than the level in the medium (such as water) in which the organism resides; movement of a substance such as a pesticide from the surrounding environment (abiotic) into living organisms, especially *via* passive absorption.

Bioconcentration factor : A measure of the tendency for a chemical to accumulate. The ratio of the concentration of a substance in a living organism (mg/kg) to the concentration of that substance in the surrounding environment (mg/L for aquatic systems).

$$\text{BCF} = \frac{\text{conc. in organism} \quad (\text{mg/kg})}{\text{conc. in environment} \quad (\text{mg/l})}$$

Biodegradable : Able to be broken down by micro-organisms in the environment.

Biodiversity : The wide diversity and interrelatedness of earth organisms based on genetic and environmental factors, the variety of flora and fauna in the natural environment.

Biological control : The purposeful manipulation of natural enemies to reduce pest populations.

Biomagnification : Process by which substances such as pesticides or heavy metals move up the food chain, becoming more concentrated with each succeeding step up the chain.

Broad spectrum herbicides : Herbicides that kill a wide range of plant types.

Broadcast : A method by which seeds or fertilizer are scattering randomly to cover an area.

Broadcast application : The uniform application of a pesticide to an entire field or area.

Broadleaf : A dicotyledon or dicot plant usually characterized by the following: two seed leaves (cotyledons), leaves with net like veins and root systems with tap roots.

Bt Cotton : Genetically engineered to control tobacco budworms, bollworms, and pink bollworms.

Bt Crops : Genetically engineered to carry the gene from the soil bacterium bacillus thuringiensis. The bacteria produce a protein that is toxic when ingested by certain lepidopteran insects. Crops containing the bt gene are able to produce this toxin, thereby providing protection throughout the entire plant.

Buffer or buffering agent : A compound or mixture that, when contained in solution, causes the solution to resist change in pH. Each buffer has a characteristic limited range of pH over which it is effective.

Buffer zone (or 'separation distance') : A strip of land separating different types of crop, wide enough to minimize the level of cross-pollination between them.

Buffering capacity (cereal or grass) : The capacity of a soil to resist changes in pH. A monocotyledon or monocot plant usually characterized by the following: one seed leaf (cotyledon), leaves with parallel veins and diffuse (fibrous) root systems.

Bulb : An underground storage organ composed principally of enlarged and fleshy leaf bases.

C

Calibration : Measurement of the delivery rate of application equipment.

Canker : A localized, dead, often sunken or cracked area on a stem, twig, limb, or trunk.

Canopy penetrating agent : An adjuvant that increases the penetration of the spray material into the crop canopy.

Cartagena protocol on biosafety : It seeks to protect biological diversity from the potential.

Cationic surfactant : A surface-active agent in which the active portion of the molecule containing the lipophilic segment forms exclusively a positive ion (cation) when placed in aqueous solution.

Certified applicator : A person certified by the WDATCP to use pesticides. A private or a commercial not-for-hire or hire applicator must be certified to use or direct the use of restricted-use pesticides or metam sodium.

Chemical name : The scientific name of an active ingredient, which complies with, accepted guidelines established by chemists.

Chlorophyll : The green photosynthetic substance in plants which allows them to capture solar energy.

Chlorosis : The yellowing of green plant tissue.

Codex Alimentarius : This is the "food code," consisting of standards, codes of practice, guidelines, and recommendations for producing and processing food. It is administered by the Codex Alimentarius Commission.

Colorant : A material used to alter the color of the tank mix.

Common name : A commonly used name of a particular species; unlike scientific names, there may be a number of common names for the same species.

Compatibility agent : A surface-active agent that allows simultaneous application of liquid fertilizer and agrichemical, or two or more agrichemical formulations, as a uniform tank mix, or improves the homogeneity of the mixture and the uniformity of the application.

Concentration : The amount of active ingredient or pesticide equivalent in a quantity of diluent, expressed as lb/gal, ml/liter, etc.

Concentration factors : Concentration factors indicate whether or not a compound is accumulated in the tissue of an organism. It is calculated by dividing the concentration of the compound in the tissue by the concentration ingested in the diet or taken up from the surrounding medium.

Contact herbicide : A herbicide that kills primarily by contact with plant tissue rather than as a result of translocation; only the portions of the plant which actually come in contact with the herbicide are affected.

Cotyledon : The first set of leaves to grow after a seed has germinated.

Cover crop : A crop that is planted to add humus to the soil or to control weeds (i.e. winter rye). Usually done between normal planting seasons. Or A quick growing crop used specifically as mulch to protect the soil.

Crop oil (Emulsifiable) : An emulsifiable, petroleum oil-based product containing up to 5% w/w surfactant and the remainder of a phytobland oil.

Crop oil (Nonemulsifiable) : Phytobland oil.

Crop oil concentrate : An emulsifiable, petroleum oil-based product containing 15% to 20% w/w surfactant and a minimum of 80% w/w phytobland oil.

Crop protection products (pesticides) : Crop protection products are both naturally occurring and man made (synthetic) chemicals, that control the diseases, insects and weeds which harm or destroy our food crops. Insecticides, fungicides and herbicides are all crop protection products.

Cross resistance : In pest management, resistance of a pest population to a pesticide to which it has not been exposed that accompanies the development of resistance to a pesticide to which it has been exposed.

Cultivars : A named agricultural variety of a plant as distinguished from a natural variety of a species. The word is used more in the profession than in the public.

D

Decomposition : Alternation of the toxicant molecule structure in soil.

Degradation : Alteration of herbicide molecule structure in plant or soil, leading to its in activation.

Defoaming agent : A material that eliminates or suppresses foam in the spray tank.

Degree-day : One degree of temperature above an insect's developmental threshold, lasting for one day. It is the most common method of expressing heat units.

Dessicant : A compound that promotes dehydration of plant tissue.

Deposition aid : A material that improves the ability of agrichemical sprays to deposit on target surfaces.

Diagnosis : The positive determination of the nature and cause of plant damage.

Dicot : A plant with two cotyledons or seed leaves. Dicots are also called broadleaves and usually have leaf veins in a netlike pattern and a tap root.

Disease : A condition of a continuous nature in which any part of an organism is abnormal as the result of an infectious or noninfectious agent.

Dormancy : Temporary suppression of seed germination or growth of an organ which may be of advantage in surviving ultimately unfavourable conditions.

Dormant oil : A horticultural spray oil applied during the dormant phase of the targeted plant.

Dose (a measure of exposure) : Dose is often expressed in milligrams per kilogram (mg/kg) or parts per million (ppm).

Dose response : A quantitative relationship between the dose of a chemical and the degree/severity of an effect caused by the chemical.

Drift control agent : A material used in liquid spray mixtures to reduce spray drift.

Dry matter : The portion of plants or vegetation which is not water.

E

Ecosystem : The organisms in a plant population and the biotic and abiotic factors which impact each other in a system.

Effective dose : The ED_{50} is the effective dose for 50 percent of tested subjects.

EL : Lowest effect level in a series of dose levels tested, it is the lowest level at which an effect is observed in the species tested.

Emergence : The event in seedling establishment when a shoot becomes visible by pushing through the soil surface.

Encapsulated : A pesticide formulation in which the active ingredient is encased in extremely small capsules made of inert synthetic polymers. The pesticide is released gradually over a period of time.

Entomology : The science that deals with the study of insects.

Environmental fate : The destiny of a chemical after release to the environment; involves considerations such as transport through air, soil and water, bioconcentration, degradation, etc.

EPA : United States Environmental Protection Agency.

Epicotyl : The part of an embryo or seedling above the attachment point of the cotyledon(s).

Erosion : The movement of soil and associated materials, principally by water and wind.

Esterification : The process of reacting fatty acids with an alcohol. Although methanol is the most commonly used alcohol (methylated), ethanol (ethylated), n-butanol and isopropyl alcohol could be used.

Evaporation reduction agent : A material that reduces the evaporation rate of a spray mix during, after, or both during and after, application.

Evapotranspiration : The loss of soil moisture due to evaporation from the soil surface and transpiration by plants.

Exposure : Contact with a chemical. Some common routes of exposure are dermal (skin), oral (by mouth) and inhalation (breathing).

Extender : A material that increases the effective life of an agrichemical after application.

F

FAO : Food and Agriculture Organization of the United Nations.

FAO code of conduct : The Food and Agriculture Organization's code of conduct on the distribution and use of pesticides. The objectives of the code are to set forth responsibilities and establish voluntary standards of conduct for all public and private entities associated with the distribution and use of pesticides, particularly where there is not yet adequate legislation to regulate pesticides.

FDA monitoring : The collection and analysis for pesticide residues carried out by the food and drug administration. The primary components of the monitoring are the total diet studies, which entail examining ready-to-eat foods, and the compliance programs, special assignments, and surveys on both domestic and imported foods, which include both surveillance and compliance examinations of fresh fruits and vegetables, grains, milk and dairy products, fish, and a variety of processed products and by-products.

Foaming agent : A material that increases the volume or stability of the foam formed in a spray mixture.

Foliar herbicide : Herbicide which is applied to the foliage of plants and is taken up through the leaves.

Food chain : A sequence of species in which each species serves as a food source for the next species. Food chains usually begin with species which consume detritus or plant material (herbivores) and proceed to larger and larger carnivores. Ex: grasshopper eaten by snake eaten by owl.

Formulated pesticide product : A mixture of one or more pesticidal active ingredients with inert ingredients such as carriers, diluents, etc.; a packaged, ready-for-sale pesticide product. Types of formulations include liquids, dusts, emulsifiable concentrates, ultra-low volume and granular products.

Formulation : The pesticide product as purchased, usually consisting of a mixture of active and inert ingredients.

Free radical : High reactive molecule containing an unpaired electron, which would be likely to damage any biological systems it contacts (these are odd-electron molecules).

Functional food : Food enriched or fortified with components or ingredients offering a specific health benefit other than a strict nutritional effect.

Fungus : Small, often microscopic plants that live either as a saprophyte or parasite. It contains no chlorophyll, has no vascular system, and is not differentiated into roots, stems or leaves.

G

Gall : A pronounced swelling or outgrowth on a plant.

Genetic engineering : Altering the genetic material of cells or organisms in order to make them capable of making new substances or performing new functions.

Genetic modification : A technique whereby individual genes can be copied and transferred to another living organism to alter its genetic make up and thus incorporate or delete specific characteristics into or from the organism.

Genomics : The study of genes and their function.

Germination : The initiation of growth in seeds.

Granular formulation : A dry, ready-to-use pesticide product which consists of an active ingredient mixed with or impregnated into a carrier such as coarse particles of clay.

Grasses : Plants that usually have parallel veins and a fibrous root system. Grasses have one seed leaf (cotyledon).

Green manure crop : Crop grown for the purpose of improving soil fertility and structure and which is returned to the soil.

GRGL : Groundwater Residue Guidance Level.

Groundwater : Water located in saturated zones below the soil surface. Many wells and springs are fed by groundwater.

H

Half-life : The time required for half of the residue to lose its analytical identity whether through dissipation, decomposition, metabolic alteration or other factors. The half-life concept may be applied to residues in crops, soil, water, animals or specific tissues.

Harmful organisms : Comprehensive term to include vertebrate or invertebrate pests, pathogen and weeds.

Harvest aid chemical : Material applied to a plant before harvest to facilitate harvesting by reducing plant foliage.

Harvest intervals : The period between the last application of a pesticide to a crop and the harvest, usually defined by law.

Hazard : The probability that a substance will cause harm under conditions of exposure.

Hazard prediction/identification : The process of recognising a potential risk, involving both toxicity assessment and exposure assessment. Toxicity assessment determines the nature and extent of adverse health effects that the chemical in question can exert related to dose. Exposure assessment characterizes exposed populations and where possible estimates exposure levels or doses incurred. The major objective of this process is the integration of toxicity dose responses and exposure data to provide dose-response relationships.

Health advisory (HA or LHA) : A non-enforceable guideline set by EPA and used to evaluate the health significance of a contaminant in drinking water when no maximum contaminant level has been set by EPA for that contaminant. HA's are set at a concentration of contamination that can be consumed daily in drinking water over the course of a lifetime with no adverse health effects. HA's may or may not include consideration of cancer risks posed by the contaminant. Also known as "lifetime health advisory."

Herb : Any vascular plant which is not woody.

Herbaceous plant : A plant that does not develop woody tissue.

Herbicide : Any substance or mixture of substances intended for preventing, destroying or controlling any weed, including any algae or aquatic weed.

Herbicide : A pesticide used to kill or inhibit plant growth.

Herbicide chronicity : The concept where long-term exposure to low concentrations of a herbicide is as, or more phytotoxic than short exposure to a high concentration of herbicide.

Herbicide drift : The drift of a herbicide off-target.

Herbicide tolerant : A plant that is tolerant of (specific) herbicides. Herbicide-tolerant crops were developed to survive certain herbicides that previously would have destroyed the crop along with the targeted weeds, and allow farmers to use them as post-emergent herbicides, providing an effective weed control. The most common herbicide-tolerant crops (cotton, corn, soybeans, and canola) are roundup ready (RR) crops resistant to glyphosate, a herbicide effective on many species of grasses, broadleaf weeds, and sedges. Other genetically modified herbicide-tolerant crops include liberty link (ll) corn resistant to glufosinate-ammonium, and BXN cotton resistant to bromoxynil.

Herbicide, hormone-type : Herbicide in which the active ingredient has growth-regulating properties.

Herbicide, residual : Herbicide showing persistent effect when applied to soil.

Herbicide, selective : Herbicide possessing differential toxicity to certain weeds and crop plants.

Herbicide, total : Herbicide affecting all plants.

Human equivalent dose : A dose which, when administered to humans, produces an effect equal to that produced by a dose in animals.

Humectant : A material which increases the equilibrium water content and increases the drying time of an aqueous spray deposit.

Humic : Moist.

Humid climate : A climate with sufficient precipitation to support forest vegetation.

Humidity : The condition of the atmosphere in respect of its water vapour content. The word 'humidity' used alone generally signifies relative humidity, but various other measures are employed such as humidity mixing ratio, vapour concentration, vapour pressure, etc.

Humus : The well-decomposed, more or less stable part of the organic matter of the soil.

Husking or dehusking : The process of removing the husk from the hulling or shelling grain during milling.

Hybrid : Plant or animal resulting from a cross between parents which are genetically different.

Hydrology : The study of the properties, distribution, behavior and effects of water on the earth's surface, in the soil and underlying rocks and in the atmosphere.

Hydrolysis : Chemical process of breakdown or decomposition involving a splitting of the molecule and addition of water molecule. e.g. during pesticide breakdown.

Hydrolytic : Causing hydrolysis or the reaction between a compound and the hydrogen and hydroxyl ions.

Hydrophilic : Describes the molecule or group of atoms having affinity towards water, because of hydrogen bonding, a strong affinity for water i.e. water attracting.

Hydrophilic/lipophilic balance : The ratio of the water/oil affinities of an emulsifier.

Hydrophobic : Opposite of hydrophilic. Water repelling.

Hydrophyte : Hydrophyte.

Hydroponics : The growing of plants in aqueous chemical solutions.

Hypocotyl : The stem of the embryo or young seedling below the cotyledons.

I

Immobilisation of nitrogen : The conversion of N to an organic form in microbial or plant tissue which is unavailable for uptake or use by plants.

In vitro studies : Studies of chemical effects conducted in tissues, cells or subcellular extracts from an organism (i.e., not in the living organism).

In vivo studies : Studies of chemical effects conducted in intact living organisms.

Incorporation : Mixing a chemical into the surface soil by mechanical means, rainfall or irrigation.

Inert ingredient : Ingredients in a formulated pesticide product which will not prevent, destroy, repel or mitigate any pest and which is intentionally included in the product. It includes carriers and materials which dilute the active ingredient.

Infiltration : The rate at which water is taken into the soil, usually measured in mm/hr.

Injury : A sudden, one-time adverse change to a plant.

Insect : A small, invertebrate animal with three body regions and six jointed legs; usually winged.

Integrated pest management (IPM) : A system where two or more pest control methods are used.

Integrated weed management (IWM) : Using a range of weed management tactics in conjunction with each other to reduce weed populations in sustainable system for whole farm management of weeds.

K

Kd : Soil-water adsorption coefficient, calculated by using measurements of pesticide distribution between soil and water.

Koc : Adsorption coefficient.

Kow : Octanol-water partition coefficient.

L

Label : The directions for using a herbicide approved as a result of the registration process.

Landscape : Any turf, ornamental, or mulched area, including an area being prepared for such a purpose that is in or around residential premises, public or commercial facilities, parks, workplaces, care facilities, recreational areas, or public lands.

Lateral movement : Movement of a herbicide through soil, generally in a horizontal plane, from the original site of application.

Lateritic soil : Soils of various textures containing (10-50%) ironstone gravel in the surface layers or within the profile.

Lawn care : Fertilization and herbicide treatments to a lawn.

LC : Lethal concentration.

LC_{50} : The concentration of toxicant necessary to kill 50 percent of the organisms being tested. It is usually expressed in parts per million (ppm).

$LC_{50/96}$ *HR* : Lethal concentration in a 50/96 hour test.

LD : Lethal dose.

LD_{50} : The amount of a chemical that is lethal to one-half (50%) of the experimental animals exposed to it. LD_{50} are usually expressed as the weight of the chemical per unit of body weight (mg/kg). It may be fed (oral LD_{50}), applied to the skin (dermal LD_{50}), or administered in the form of vapors (inhalation LD_{50}).

Leaching : The downward removal of nutrients and chemicals in solution through the soil.

Licensing : The issuance of a license by the WDATCP to individual commercial pesticide applicators and to commercial application businesses so that they can legally use pesticides in *wisconsin*.

Life cycle : Complete succession of developmental stages in the life of an organism or various stages in the life of a plant e.g. emergence, flowering, seed set.

Liming : Application of lime (calcium oxide CaO) or similar neutralising material to reduce soil acidity.

Lithophyte : Plants that grow on rocks or other areas that do not need soil. They will receive their nourishment from the air. i.e. orchids and lichens.

Loam : Good quality soil. Adequate supplies of clay, sand, and fiber must be present. Crumbly to the touch. Ideal for most gardening.

LOEL : Lowest Observed Effect Level; the lowest dose in an experiment which produced an observable effect.

Lowest Effect Level : It is the lowest level at which an effect is observed in the species tested.

Lowest Observed Adverse Effect Level (LOAEL) : The lowest dose in an experiment which produced an observable adverse effect.

M

Manure : Any animal droppings with a high content of nitrogen, these should be composted and aged before use. Take special note not to use cat or dog droppings. All to often these contain disease organisms.

MATC : Maximum Acceptable Tolerance Concentration.

Material Safety Data Sheets (MSDS) : Data sheets that contain information on toxicity, first aid, personal protection controls, storage and handling precautions spill and leak cleanup and disposal practices, transportation, physical data and reactivity data. MSDS'S are available from manufacturers.

MCL *(Maximum contaminant level)* : The highest amount of a contaminant allowed by EPA in water supplied by a municipal water system; also referred to as "drinking water standard."

Metabolism : The process of chemical change by which energy is provided in living cells.

Metabolite : Any product of metabolism, especially a transformed chemical.

mg/Kg : Milligrams per kilogram.

mg/Kg/day : Milligrams per kilogram per day.

mg/L : Milligrams per liter.

Micrograms/Liter : Usual unit of measurement for contaminants of water; μg/L; equivalent to parts per billion (ppb).

Micro-nutrient : Nutrients or elements required in minute.

Microorganisms : Animals and plants that are too small to be seen clearly with the naked eye but are the soil enablers helping to improve the dirt.

Microbial decomposition : Decomposition of a substance into more elementary compounds by the action of microorganisms such as bacteria.

MLD : Minimum Lethal Dose; the smallest of several doses which kills one of a group of test animals.

Mode of action : How a herbicide controls weeds.

Modified vegetable oil : An oil extracted from seeds that has been chemically modified (for e.g., methylated).

Modified vegetable oil concentrate : An emulsifiable, chemically modified vegetable oil product containing 5% to 20% w/w surfactant and the remainder chemically modified vegetable oil.

Monitoring : Regular inspection of crops to identify insect and disease activity in order to determine whether control measures are necessary. Monitoring is an integral part of integrated pest management.

Monocot : A plant having a single cotyledon or seed leaf. Monocots include grasses and sedges.

MPI : Maximum Permissible Intake. A limit established by EPA, usually expressed as mg/day for a 60 kg person, which is used to determine the level of pesticide residues permitted on crops for human consumption.

MTD: Maximum Tolerated Dose; the highest dose of a chemical that does not alter the life span or severely affect the health of an animal.

Mulch : A thin layer of organic material covering the soil surface. Shredded hardwood mulch is processed mulch made mainly of the "bark" byproducts of trees, usually double ground for finer consistency and usually treated with a "rinse" process to provide even color. Wood chip mulch is chipped tree byproducts usually including bark as well as actual branch wood chipped into usable sized mulch. It contains more extraneous materials and is usually used for more rustic situations, in parks, as pathway surfaces, etc.

Multiple tolerance (to herbicides) : Of a plant, being able to tolerate a number of different herbicides.

N

Naphtha-based oil : A petroleum oil containing a majority of the naphtha fraction.

Natural control : The reduction of pest populations by natural enemies without human assistance.

Necrosis : Death of plant cells or entire portions of a plant, usually resulting in darkening of the affected areas.

NEL : The No-Effect-Level of a pollutant is the concentration at or below which there will be no defined effect, either deleterious or beneficial, on a member of a population exposed to the pollutant in question.

NOAEL : No-Observed-Adverse-Effect Level; the highest dose in an experiment which did not produce an observable adverse effect.

Nodulation : The process by which legume roots are colonised by rhizobium bacteria and nodules are formed.

NOEL : The dosage or exposure level at which no toxicologically significant adverse effect(s) can be detected. The NOEL has been used interchangeably with the NEL (No Effect Level). However, there is a distinction which is based on the interpretation of the occurrence of an effect. An NEL denotes that at a particular dose, there was absolutely no effect. In reality, an effect may have occurred but went undetected for a variety of reasons. Thus, the more accepted terminology is the NOEL, which indicates that while an effect was not observed under a particular set of test conditions, it does not preclude the possibility that some effect may have occurred.

Nonionic surfactant : A surface-active agent having no ionizable polar end groups but comprised of hydrophilic and lipophilic segments.

Non-selective herbicide : An herbicide which is non-specific and controls most vegetation to which it is applied. Glyphosate is an example of a nonselective herbicide.

Non-target species : Any plant, animal, or pest which is not intended to be treated with a pesticide application.

Non-target vegetation : Vegetation which is not expected or not planned to be affected by an herbicide application.

No-tillage : A system of establishing crops and pastures without prior cultivation. Specialized planters are used and weeds are controlled using herbicides.

Noxious weed : A plant defined by law as being especially undesirable, troublesome, and difficult to control.

Nuisance weeds : Weeds, such as purple loosestrife and multiflora rose, which no one may sell, distribute, or cultivate in *wisconsin.*

Nutrient : Element which is taken up and used by plants for growth and reproduction.

O

Octanol-water partition coefficient (Kow) : Measurement of how a chemical is distributed at equilibrium between octanol and water. It is an important parameter and is used often in the assessment of environmental fate and transport for organic chemicals. Additionally, *Kow* is a key variable used in the estimation of other properties.

Organic farming : Farming without the use of chemical fertilizers and pesticides.

Organic Matter (OM) : Soil particles created by the decomposition of plant tissues. Soil OM has a high adsorptive capacity (higher than clay) and offers resistance to leaching of soil water and dissolved molecules.

P

Paraffinic oil : A petroleum oil (derived from paraffin crude oil) whose paraffinic-carbon type content is typically greater than 60%.

Penetrant : A material that enhances the ability of an agrochemical to enter a substrate or penetrate a surface.

Percolation : The downward movement of water through saturated soil.

Perennial : A plant which continues to grow from year to year or plants that live for more than two years.

Permeability of a soil : The ease with which water enters and passes though the soil.

Persistent organic pollutants : Class of chemicals that persist in the environment, are capable of long-range transport, bioaccumulate in human and animal tissue, and have significant impacts on human health and the environment.

Pest : Any organism that competes with people for food, or space; presents a threat to the health of people or domestic animals; or, with respect to ornamental plants, detracts from our visual environment by damaging such plants (either during production or end use).

Pesticide : A chemical used to directly control pest populations or to prevent or reduce pest damage (e.g., insecticides, herbicides, fungicides).

Petroleum oil : Oil derived from petroleum: contains a mixture of hydrocarbons that are broadly classified as paraffins, naphthenes, aromatics, or other unsaturates, or combinations thereof.

pH : A measure of the acidity or alkalinity of a solution. pH values below 7.0 indicate acidity; above 7.0 indicate alkalinity pH 7.0 being neutral.

Phytobland oil : A highly refined paraffinic material with minimum unsulfonated residue of 92% v/v.

Phytotoxic : Injurious or lethal to plants.

Plant science industry : Invents, develops, manufactures and sells products and services designed to improve.

Plant-back period : Recommended period of time, after the use of herbicides that will avoid damage to succeeding crops.

Podsolic soil : Soils with texture profiles which change from light surfaced to bleached, heavier clay layers at depth. Usually poorly drained and subject to water logging in the wet season.

POPS : Persistent organic pollutants.

Post-emergent : The group of weed control herbicides that are applied directly to the unwanted plant's foliage, killing the plant systematically. (e.g. Round-up).

ppb : Parts per billion; a measure of concentration. May represent the concentration of a residue in soil, water or whole animals.

ppm : Parts per million; a measure of concentration. May represent the concentration of a residue in soil, water or whole animals. For example, one ppm is equivalent to one minute in two years.

Pre-emergent : The group of weed control herbicides which are applied to turf areas or planting bed areas as a preventative measure. Weed seeds are typically prevented from germinating due to the chemical barrier created by the pre-emergent chemical. (e.g. crabgrass preventer, preen, and snapshot.)

Pre-plant application : Applied before planting a crop, either as a foliar application to control existing vegetation or as a soil application.

Preplant incorporated (PPI) : Applied and blended into the soil before seeding.

Pruning : Usually involves a more thorough, structural approach. Thinning of major stems, removal of crossing or rubbing limbs, reducing size substantially, rejuvenating or renewing growth by major thinning or cutting nearly to ground level.

Prussic acid : Hydrocyanic acid (HCN) formed in sorghum plants as a result of stress or rapid growth. Causes poisoning in livestock.

R

Rainfastness : Refers to the period other material must be on the plant before a rainfall or irrigation to retain its effectiveness.

Rate : The amount of active ingredient or acid equivalent applied per unit area or other treatment unit.

Red Earth : Class of soils which are the massive structured red clay loam and sandy loam soils of the top end.

Reduced tillage : Synonymous with minimum tillage.

Rejuvenation of pastures : Renovation of 'run-down' pastures by undertaking weed control, reseeding, fertilizing or other practice to improve productivity.

Residual herbicide : Herbicide which is soil-acting and controls germinating weeds over an extended period.

Residue : That quantity of pesticide, its degradation products, and/or its metabolites remaining on or in the soil, plant parts, animal tissues, whole organisms, and surfaces.

Restricted use pesticide (RUP) : A designation given to a pesticide by the EPA which restricts purchase and use of that pesticide to certified applicators. The designation is given to pesticides which pose a potential hazard to applicators, the general public or to the environment or may be highly toxic.

Reverse xylem mobility : Movement of herbicide or other solvent from shoot to root through xylem in plant.

RFD : Reference dose.

Rhizobia : Rhizobia a genus of symbiotic bacteria which fixes nitrogen in the leguminous plants.

Rhizome : Underground stems formed by many plants as a means of vegetative spread.

Risk : The potential for realization of unwanted negative consequences or events.

Risk assessment : A qualitative or quantitative evaluation of the environmental and/or health risk resulting from exposure to a chemical or physical agent (pollutant); combines exposure assessment results with toxicity assessment results to estimate risk.

Rotation : A management system where crops and pastures are grown in alternate years, to control weeds, insect pests and diseases and maintain soil structure and fertility.

RRFD : Risk reference dose.

Runoff : Rainfall in excess of the infiltration rate which is lost as surface flow.

Safener : Substance capable of antagonising specific herbicide phytotoxicity to plant.

Sand : A soil component consisting of coarse particles (0.05 to 2.0 mm diameter). Sand particles provide relatively little surface area for adsorption of molecules. Soil water and dissolved materials move/leach easily through sandy soils. Soil texture and many other soil characteristics are determined by the relative amounts of sand, silt, clay and loam in a soil.

Scientific name : The latin name of genus and species of an organism. Scientific names are used to avoid confusion that can result from the use of common names, which may vary from one area to another.

Sediment : Suspended soil particles which are carried away in runoff water and are deposited further down the slope.

Selective herbicide : A herbicide which controls a specific group of plants but is inactive against others, i.e. grass herbicides.

Soil microbes : Beneficial and harmful species of bacteria, fungi and other microscopic organisms which are a part of the soil's ecology.

Soil mobility : Movement of a compound through soil from the treated area by leaching, volatilization, adsorption and desorption, or dispersal by water. Leaching is of particular concern because of the potential for contamination of groundwater.

Soil moisture : The amount of water in the soil (wet weight minus dry weight).

Soil organic matter : Partially decayed and partially re-synthesized plant and animal residues in the soil.

Soil structure : The physical arrangement of soil particles into units or aggregates.

Solubility : The concentration of a substance that dissolves in a given solvent. High solubility: readily dissolves. Low solubility: does not dissolve very well.

Solvent solubility : Concentration of a material that dissolves in a given solvent.

Sorption : A surface phenomenon which may be either absorption or adsorption, or a combination of the two; often used when the specific mechanism is not known.

Species shift : The selection and increase of naturally tolerant weed species.

Spot spray : Application of herbicide by handheld spray gun directly to individual weeds.

Spot treatment : Application of a pesticide to small, discrete areas.

Spreader : A material which increases the area that a droplet of a given volume of spray mixture will cover on a target.

Spreader/Sticker : A material that has the properties of both a spreader and a sticker.

Sticker : A material that assists the spray deposit to adhere or stick to the target and may be measured in terms of resistance to time, wind, water, mechanical action or chemical action.

Strip application : Uniform application of herbicide to a limited portion of the orchard floor; i.e., a narrow strip along either side of the tree row.

Stubble : Plant residue left over from a crop.

Subchronic : Intermediate between acute and chronic toxicities; sub chronic toxicity studies involve repeated daily exposures of animals to a chemical for part (not exceeding 10%) of a lifespan. In rodents, this period extends up to 90 days of exposure.

Summer annuals : Plants that germinate from seed in spring, flower and produce seed during the summer, and die in the summer or fall.

Suppression : Off target movement of a pesticide.

Surface agent : A material that when added to a liquid agent medium, modifies the properties of the medium at a surface or interface. This is a general term that includes soluble detergents in liquid medium, dispersing agents, emulsifying agents, foaming agents, penetrating agents and wetting agents.

Surfactant : Chemical additive which improves the dispersion, spreading and wetting action of herbicides or other chemicals.

Susceptibility : Capacity to be adversely affected by pesticide treatment or the sensitivity to, or degree to which, a plant is injured by a herbicide treatment.

Suspension : A formulation in which very fine, solid herbicidal particles are dispersed, but not dissolved, in a liquid.

Sustainable agriculture : Whereby the farmer is able to increase or maintain crop yields long term, whilst conserving natural resources and protecting the environment.

Synergism : The effect of two substances in combination which has a multiplier rather than an additive effect.

Synergist : For herbicides – a non-herbicidal compound used to increase the efficacy of a herbicide by a physiological mechanism.

T

Tap Root : A single, relatively large central root which gives rise to smaller, lateral branches; usually found in dicots.

TD : Toxic dose; the dose of a chemical that produces signs of toxicity.

TDLO : The lowest dose which produces signs of toxicity.

Teratogenic : The ability to produce birth defects. A teratogenic agent has the ability to induce or increase the incidence of congenital malformations, i.e., permanent structural or functional deviations arising during embryogenesis.

Teratogenicity : The ability of a substance to produce irreversible birth defects or anatomical or functional disorders as a result of an effect on the developing embryo or fetus.

Theoretical maximum residue contribution (TMRC) : The amount of a substance that would be present in a 1.5 kg "average" daily diet if all commodities with established tolerances bore residues at the tolerance level. The percentages of various commodities in the "average" diet are those used by EPA and are based on the "1965-66 household food consumption survey" conducted by USDA.

Threshold : The lowest dose of a chemical at which a specified measurable effect is observed and below which it is not observed.

Tindall clay loam : Class of sandy clay loam and clay loam red earth soils with structured by horizon. Agriculturally important and found in the katherine-daly basin.

TLM TL_{50} : Median Tolerance Limit; for e.g. the concentration of chemical in water necessary to kill 50 percent of the test aquatic organisms during a specific exposure period. The TLM is usually expressed as parts per million parts of water for 24, 48, 72, or 96 hours of exposure.

TLV : Threshold Limit Value. The highest allowable air concentration of a chemical in which workers may work for many years (8 hours a day, 40 hours per week) without negative health effects.

Expressed as milligrams (mg) per cubic meter of air (mg/m^3). The level to which persons may be exposed for an 8-hour workday without adverse effects.

Tolerance : Ability to continue normal growth or function when exposed to a potentially harmful agent.

Tolerance : (1) A legal limit, established by EPA, for the maximum amount of a pesticide residue which may be present in or on a food. Pesticide tolerances on raw agricultural commodities are listed in part 180, CFR 40; tolerances on processed foods are listed in part 193, CFR 21; and tolerances on processed animal feeds are listed in part 561, CFR 21. Temporary tolerances, which cover residues resulting from an experimental use, generally expire after one year. (2) Capacity to withstand pesticide treatment without adverse effect on normal growth and function.

Toxic : Harmful; poisonous.

Toxicity : (1) The capacity or property of a substance to cause adverse effects. (2) The specific quantity of a substance which may be expected, under specific conditions, to do damage to a specific living organism.

Trace elements : Also called micro-nutrients. Elements required by plants in minute quantities but essential for healthy plant growth.

Trade name : The specific, registered name given by a manufacturer to a pesticide product.

Transition period : The period between the dry season and the onset of the wet season when sporadic storms cause spoilage of dry feed.

Translocation : The process whereby a chemical is absorbed into the plant, via the leaves or roots, and is then moved to other parts of the plant.

Trim/Shape : (also see pruning) refers to a cosmetic type of pruning where most plants being trimmed have an established form and merely need light tipping or shearing of recent growth in order to maintain their established form (hedges, etc.).

Tropics : Geographical regions between the tropics of capricorn and cancer.

True leaf : Any leaf produced after the cotyledons.

U

Ultra-low volume (YLV) : A liquid pesticide product which is applied undiluted at a rate of one-half gallon per acre or less.

V

Vapour pressure : A relative measure of the volatility of a chemical in its pure state. The pressure exerted by a gas that is in equilibrium with its solid or liquid form.

Vegetable oil : Oil extracted from seeds; typically those of corn, cotton, peanut, rapeseed, sunflower, canola or soybean.

Vegetable oil concentrate : An emulsifiable vegetable oil product containing 5% to 20% w/w surfactant and a minimum of 80% w/w vegetable oil.

Volatile : Capable of vaporizing or evaporating readily.

VSD : Virtually Safe Dose.

W

Water holding capacity : The amount of water a soil can hold and make available for plant use. Sands and clays have low and high water holding capacities, respectively.

Water solubility : The maximum concentration of a chemical that dissolves in pure water at a specific temperature and pH.

Water soluble granules : A formulation consisting of granules to be applied as a true solution of the active ingredient after dissolving in water, but which may contain insoluble inert ingredients.

Water soluble powder : A powder formulation which forms a true solution of the active ingredient when dissolved in water but which may contain insoluble inert ingredients.

Waterlogged : Saturated with water. Conditions detrimental to plant growth. Common problem in some soils during the wet season.

Waterway : A structure designed to collect runoff water from diversion and contour banks and carry it safely to a discharge area or natural water course.

Weed : An unwanted plant.

Weed allelopathy : The damage to crops by weed toxic exudates.

Weed assessment scales : Descriptive scales relating weed and crop densities.

Weed control : The process of limiting weed infestations so that crops can be grown profitably or other operations can be conducted efficiently.

Weed density : The total amount of weeds per unit area as determined by counting or weighing.

Weed killer : A chemical that will kill or inhibit the growth of weeds, a herbicide.

Weeds, perennial : Herbaceous plants that live for more than two years. Those living only for two years are called biennials.

Wettable powder : Finely particulate inert material (e.g. kaolin) containing pesticide that forms a stable (semi-permanent) suspension when added to water (may require tank agitation to maintain suspension).

Wetter, wetting agent : Material that reduces contact angle of liquid on given surface. Used in pesticide formulations (see adjuvant and surfactant). Added to a spray to reduce the surface tension of the applied droplets.

Wetting agent : Wetting agents can be considered synonymous with spreading agents in function.

WHO : World Health Organization.

Wild type : The phenotype characteristic of individuals of a species that predominate under natural conditions.

X

Xylem : Plant tissue that conducts water and nutrients from the roots up through the plant.

Y-leaf : The most recently matured leaf.

Appendix I

Common name, chemical name and toxicity level of commonly used herbicides (WSSA, Herbicide Handbook).

Common name	Chemical name	Rat oral LD_{50} (mg/kg)	Toxicity rating
1	2	3	4
Acifluorfen	5-[2-chloro-4-(trifluoromethyl) phenoxy]-2-nitrobenzoic acid	4790	III
Alachlor	2-chloro-2′, 6′-diethyl-N-(methoxymethyl) acetanilide	930	III
Ametryn	*N*-ethyl-*N*-(1-methylethyl)-6-methylthio) -1,3,5-triazine-2, 4-diamine	1950	III
Amitrole	1*H*-1, 2,4-triazol-3-amine	>5000	IV
Asulam	Methyl [(4aminophenyl) sulfanyl] carbamate	>5000	IV
Atrazine	6-chloro-*N*-ethyl-*N*-(1-methylethyl)-1,3,5-triazine-2,4-diamine	1886	III
Benefin	*N*-butyl-N-ethyl-2, 6-dinitro-4- (trifluoro methyl) benzenamine	>5000	IV

Contd...

Contd...

1	2	3	4
Bensulfuron-methyl	methyl 2 (((((4,6-dimethoxy pyrimidin-2-yl) amino) carbonyl) amino) sulfonyl) methyl) benzoate	> 5000	IV
Bensulide	S-(0, 0-diisopropyl phosphorodithisate) ester of N-(2-mercaptoethyl benzene sulfonamide)	704	III
Bentazon	3-isopropyl-1H-2, 1, 3-benzothiadiazin-(4) 3H-one 2,2-dioxide	2063	III
Bromacil	5-bromo-6-methyl-3-(1 methylpropyl)-2, 4 (1*H*,3*H*) pyrimidinedione	5175	IV
Bromoxynil	3,5-dibromo-4-hydroxybenzonitrile	779	III
Butylate	*S*-ethyl bis (2-methylpropyl) carbamothioate	5366	IV
Buturon	(3-(4-chlorophenyl)-1-methyl-1- (1 methylprop-2-ynyl) urea	3000	III
Butachlor	N-butoxymethyl-2-chloro-2′6′-dimethylacetanilide	3300	III
Carfentrazone	2-dichloro-5- [4-(difluoromethyl)-4,5-dihydro-3-methyl-5-oxo1*H* -1, 2,4-triazole-1-yl)-4-fluorobenzenepropanoic acid, ethyl ester	5143	IV
Chlorimuron	Ethyl-2- [[[[(4-chloro-6-methoxy-2-pyrimidinyl) amino] carbonyl] amino] sulfonyl] benzoate	>5000	IV
Chlorobromuron	3-(4-Bromo-3-chlorophenyl) -1-methoxy-1-methylurea	2150	III

Contd...

Contd...

1	2	3	4
Chlorfenprop-methyl	Methyl 2-chloro-3- (4-chlorophenyl) propionate	1190	III
Chlorsulfuron	2-chloro-N-[(4-methoxy-6-methyl-1,3,5-triazin-2-yl) amino carbonyl]benzene sulfonamide	> 5000	IV
Chloroxuron	3-[4-(4-chlorophenoxy) phenyl]-1,1-dimethylurea	3000	III
Chlorotoluron	3-(3-chloro-p-tolyl)-1,1-dimethylurea	10000	IV
Clodinofop-propargyl	*R*)-2-[4-(5-chloro-3-fluoro-2 -pyridyloxy) phenoxy] propionic acid	>2276	III
Cloransulam	3-chloro-2- [[(5-ethoxy-7-fluoro [1,2,4] triazolo [1, 5-*c*] pyrimidin-2yl) sulfonyl] amino] benzoic acid, methyl ester	>5000	IV
Clethodim	*(E, E)-*(±)-2-[1-[[(3-chloro-2-propenyl) oxy] imino] propyl]-5-[2- (ethylthio) propyl]-3-hydroxy-2-cyclohexen-1-one	3610	III
Clomozone	2-[(2-chlorophenyl) methyl]-4, 4-dimethyl-3-isoxazolidinone	2343	III
Clopyralid	3,6-dichloro-2-pyridinecarboxylic acid	4300	III
Chloramben	3-amino-2, 5) dichlorobenzoic acid	5620	IV
DCPA	Dimethyl 2, 3, 5, 6-tetrachloro-1, 4-benzene di carboxylate	>10000	IV

Contd...

Contd...

1	2	3	4
Dicamba	3,6-dichloro-0-anisic acid	2629	III
Dalapon sodium	Sodium 2, 2-dichloropropionate	6600-8100	IV
Dichlobenil	2,6-dichlorobenzonitrile	>5000	IV
Dichlofop	(±)-2- [4- (2, 4-dichlorophenoxy) phenoxy] propanoic acid	2020-2176	III
Diclosulam	*N*-(2,6)-dichlorophenyl)- 5-ethoxy-7-fluro (1,2,4) triazolo-(1,5c)-pyrimidine- 2-sulfonamide	5000	III
Diquat	6-7-dihydrodipyrido [1, 2-a 2′, 1′-c] pyrazinediium ion	230	II
Dithiapyr	*S-S*-dimethyl 2- (difluoromethyl)-4-(2- methylpropyl)- 6(trifluoromethyl)-3,5- pyridinedicarbothioate	>5000	IV
Diuron	3-(3,4-dichlorophenyl)-1, 1-dimethylurea	3400	III
Dinoseb	2-sec-butyl-4, 6-dinitrophenol	40-50	I
DSMA	Disodium methanearsonate	1936	III
Endothall	7-oxabicyclo [2.2.1] heptane-2, 3-dicarboxylic acid	38-51	I-II
Ethoxysulfuron	1-(4,6-dimethoxypyrimidin- 2-yl)-3-(2-ethoxyphenoxys- ulfonyl) urea	3270	III
Ethalfluralin	*N*-ethyl-*N*- (2-methyl-2 - propenyl)-2,6-dinitro- 4 - (trifluoromethyl) benzeamine	>10000	IV

Contd...

Contd...

1	2	3	4
EPTC	S-ethyl dipropylthiocar bamate	1652	III
Fenoxaprop-ethyl	(±)-2-[4-[(6-chloro-2-benzoxazolyl) oxy] phenoxy] propanoic acid	3310	III
Flumetsulam	*N*- (2,6-difluorophenyl)-5-methyl [1,2,4] triazolo [1,5-*a*] pyrimidine-2-sulfonamide	>5000	IV
Flufenacet	N-(4-fluorophenyl)-N-(1-methylethyl)-2-[[5-(trifluoromethyl) -1, 3, 4 -thiadiazol-2-yl]oxy] acetamide	371-1365	III
Flumiclorac pentyl ester	[Pentyl [2-chloro-4-fluoro-5-(1,3,4,5,6,7-hexahydro-1, 3-dioxo-2H-isoindol-2-yl) phenoxy] acetate]	3, 200	III
Flumioxazin	2-[7-fluoro-3, 4-dihydro-3-oxo-4- (2-propynyl)-2*H*-1, 4-benzoxazin-6-yl]-4,5,6,7 -tetrahydro-1*H*-isoindole-1, 3(2*H*)-dione	>5000	IV
Fluometuron	*N*, *N*-dimethyl-*N* -[3-trifl uoromethyl) phenyl] urea	6416	IV
Fluazifop-P	(*R*)-2-[4-[[5-(trifluoromethyl) -2pyridinyl] oxy] phenoxy] propanoic acid	>5000	IV
Fosamine	Ammonium salt of fosamine [ethyl hydrogen (amino carbonyl) phosphoate]	>5,000	IV
Fluchloralin	N- (2-chloroethyl) 2, 6-dinitro - N-propyl-4 -(trifluoromethyl) aniline	5580	IV
Glufosinate	2-amino-4- (hydroxymethyl phosphinyl) butanoic acid	2170	III

Contd...

Contd...

1	2	3	4
Glyphosate	N- (phosphonomethyl- glycine	>5000	IV
Halosulfuron	Methyl-6[((4,6-dimethoxy -2-pyrimidinyl)-amino) carbonylaminosulfonyl]-4- chloro-1-methyl-1H-pyrazole -4-carboxylate	8866	IV
Haloxyfop	*RS*)-2-[4-(3-chloro-5-trifluo- romethyl-2-pyridyloxy) phenoxy]propionic acid	2870	III
Hexazinone	3-cyclohexyl-6 - (dimethyla- mino)-1-methyl-1, 3,5- triazine-2, 4(1*H*, 3H)-dione	1690	III
Imazapic	(±)-2-[4,5-dihydro-4-methyl- 4- (1-methylethyl)-5-oxo- 1*H*-imidazol-2-yl]-5-methyl- 3-pyridinecarboxylic acid	>5000	IV
Imazapyr	(±)-2-[4,5-dihydro-4-methyl- 4-(1-methylethyl)-5- oxo-1*H*-imidazol-2-yl]-3- pyridinecarboxylic acid	>5000	IV
Imazaquin	2-[4,5-dihydro-4-methyl-4- (1-methylethyl)-5-oxo -1*H*-imidazol-2-yl]-3-quinol- inecarboxylic acid	>5000	IV
Imazethapyr	(±)-2-[4,5-dihydro-4-methyl- 4- (methylethyl)-5-oxo- 1H-imidazol-2-yl]-5-ethyl-3- pyridinecarbolic acid	>5000	IV
Imazosulfuron	1-(2-chloroimidazo[1,2-*a*] pyridin-3-ylsulfonyl)-3- (4,6-dimethoxypyrimidin- 2-yl)urea	>5000	IV
Isoxaben	*N*-[3-(1-ethyl-1-methyl propyl)-5-isoxazolyl]-2, 6- dimethoxybenzamide	>10000	IV

Contd...

Contd...

1	2	3	4
Isoxaflutole	5-cyclopropyl-4-(2-methyl-sulfonyl-4-trifluoromethyl-benzoyl) isoxazole	>5, 000	IV
Lactofen	(±)-2-ethoxy-1-methyl-2-oxoethyl 5-[2-chloro-4-(trifluoromethyl) phenoxy] -2-nitrobenzoate	2533	III
Linuron	3-(3, 4-dichlorophenyl)-1-methoxyl-1-methylurea	1254	III
Metolachlor	2-chloro-N- (2-ethyl-6-methylphenyl)-N- (2-methoxy-1-methylethyl) acetamide	2877	III
Metribuzin	4-amino-6- (1,1-dimethylethyl)-3-(methylthio)-1, 2, 4-triazine-5 (4*H*) one	1090	III
Metsulfuron	Methyl 2-[[[[(4-methoxy-6-methyl-1, 3,5-triazin-2-yl) amino] carbonyl] amino [sulfonyl] benzoate	>5000	IV
MSMA	Monosodium methanearsonate	2833	III
Metobromuron	3-(4-bromophenyl)-1-methoxy-1-methyurea	3000	III
Mecoprop	±-2-(4-chloro-o-tolyloxy) propionic acid	700-1500	III
Monuron	3-(4-chlorophenyl)-1, 1-dimethylurea	3600	III
Mapropamide	*N, N*-diethyl-2-(1-naphthalenyloxy) propanamide	>5000	IV
Mesosulfuron	2-[(4,6-dimethoxypyrimidin -2-ylcarbamoyl) sulfamoyl] -á-(methanesulfonamido)-p-toluic acid (parent substance)	> 5000	IV

Contd...

Contd...

1	2	3	4
Nicosulfuron	2-[[[[(4,6-dimethoxy-2-pyrimidinyl) amino] carbonyl] amino] sulfonyl]-*N, N*-dimethyl-3-pyridinecarboxamide	>5000	IV
Norflurazon	4-chloro-5- (methylamino) -2-(3-trifluoromethyl) phenyl)-3(2*H*)-pyridazinone	1140	III
Naptalum	N-1-napthylphthalamic acid	8200	IV
Nitrofen	2,4-dichloro 4-nitrophenyl ether	2630	III
Oryzalin	4-dipropylamino)-3, 5-dinitrobenzenesulfonamide	>500	III
Oxadiazon	[2-tert-buty 1-4- (2, 4-dichloro-5-isopropoxypheny 1)-1,3,4-oxadiazolin-5-one]	>5,000	IV
Oxyfluorfen	2-chloro-1- (3-ethoxy-4-nitrophenoxy)-4-trifluoromethyl) benzene	>2000	IV
Paraquat	1,1'-dimethyl-4, 4'-bipyridinium ion	40-150	I-II
Pebulate	*S*-propyl butylethylcarbamothioate	1400	III
Pendimethalin	*N*- (1-ethylpropyl)-3, 4-dimethyl-2, 6-dinitrobenzenamine	4075	III
Pretilachlor	2-chloro-2',6'-diethyl-*N*-(2-propoxyethyl) acetanilide	6099	IV
Prometon	6-methoxy-*N, N* -bis (1-methylethyl)-1,3, 5-triazine-2, 4-diamine	4345	III
Prometryn	*N, N* -bis (1-methylethyl)-6-(methylthio)-1, 3-5-triazine-2, 4-diamine	4550	III

Contd...

Contd...

1	2	3	4
Propazine	2-chloro-4, 6-bis (isopropylamino) 1,3, 5-triazine	5000	III
Propyzamide	3,5-dichloro-N- (1, 1-dimethyl-2-propyl) benzamide	5620-8350	IV
Proximpham	2-prpanone O- [(phenylamino) carbonyl] oxime	1540	III
Pronamide	3,5-dichloro (N-1, 1-dimethyl-2-propyny) benzamide	>5000	IV
Propanil	*N*-3, 4-dichloropropionanilide	3269	III
Pyridate	*O*- (6-chloro-3-phenyl- 4-pyridazinyl)-S-octyl- carbohothioate	4690	III
Pyrithiobac	Sodium 2-chloro-6- (4, 6-dimethoxypyridmidin- zylthio benzoate)	4,000	III
Quizalofop-P	(*R*)-2-[4-(6-chloro-2- quinoxalinyl) oxy) phenoxy] propanoic acid	6600	IV
Rimsulfuron	*N*-[[(4,6-dimethoxy-2- pyrimidinyl)amino] carbonyl]-3-(ethylsulfonyl)- 2-pyridinesulfonamide	>5000	IV
Sethoxydim	2-[1-(ethoxyimino) butyl]- 5-[2-(ethylthio) propyl]-3- hydroxy-2-cyclohexen-1-one	>5000	IV
Simazine	6-chloro-*N*, *N'*-diethyl-1, 3,5-triazine-2, 4-diamine	>5000	IV
Sulfometuron	Methyl 2-[[[[(4,6-dimethyl- 2-pyrimidinyl) amino] carbonyl] amino] sulfonyl] benzoate	>5000	IV

Contd...

Contd...

1	2	3	4
Sulfosulfuron	1-(4,6-dimethoxypyrimidin -2-yl)-3-(2- ethylsulfony-limidazo [1,2-*a*]pyridin-3-ylsulfonyl)urea	> 5000	IV
Tebuthiuron	N-[5-(1,1-dimethylethyl)-1,3,4-thiadiazol-2-yl]-N, N -diamethylurea	644	III
Terbumeton	2-tert-buytlamino-4-ethylamino-6-methoxy-1, 3,5-triazine	483	II
Terbacil	5-chloro-3- (1,1-dimethylethyl)-6-methyl-2, 4-(*1H, 3H*)-pyrimidinedione	1255	III
Thiobencarb	*S*-[(4-chlorophenyl) methyl] diethylcarbamothioate	1033	III
Triclopyr	[(3,5,6-trichloro-2-pyridinyl) oxy] acetic acid	1581	III
Trifluralin	2,6-dinitro-*N, N*-dipropyl-4- (trifluoromethyl) benzenamine	>5000	IV
Terbutryne	2-tet-2-butyl-4-ethylamino-6-methylthio-1, 3,5-triazine	2980	III
Triasulfuron	1-[2-(2-chloroethoxy) phenylsulfonyl] -3 - (4-methoxy-6-methyl-1, 3, 5-triazin -2-yl)urea	>5000	IV
Thiazafluron	1,3-dimethyl-1- (5-trifluoromethyl-1, 3, 4-thiadiazol-2-yl) urea	278	II
2,4-D	(2,4-dichlorophenoxy) acetic acid	>1000	III
2,4-DB	2,4-dichlorophenoxy butyric acid	1960	III
Vernolate	S-propyl dipropylcarbamate	1780	III

❑❑❑

Appendix II

Maximum residue limits (MRLs) of herbicides in Food Commodities (EPA).

S. No.	Herbicides	Commodity	MRL (mg/kg)
1.	**Ametryn**	Cotton seed	0.05
		Edible offal (mammalian)	0.05
		Meat [mammalian)	0.05
		Milks	0.05
		Pineapple	0.05
		Pome fruits	0.1
		Sugar cane	0.05
2.	**Amitrole**	Avocado	0.01
		Banana	0.01
		Cereal grains	0.01
		Citrus fruits	0.01
		Edible offal (mammalian)	0.01
		Grapes	0.01
		Meat [mammalian]	0.01
		Milks	0.01
		Papaya	0.01
		Passion fruit	0.01
		Pecan	0.01
		Pineapple	0.01
		Pome fruits	0.01
		Potato	0.05
		Stone fruits	0.02
		Sugar cane	0.01
3.	**Asulam**	Apple	0.1
		Edible offal (mammalian)	0.1
		Hops, dry	0.1

Contd...

Contd...

S. No.	Herbicides	Commodity	MRL (mg/kg)
		Meat [mammalian]	0.1
		Milks	0.1
		Poppy seed	0.1
		Potato	0.4
		Sugar cane	0.1
4.	**Atrazine**	Edible offal (mammalian)	0.1
		Lupin (dry)	0.02
		Maize	0.1
		Meat [mammalian] T	0.01
		Milks T	0.01
		Potato	0.01
		Rape seed	0.02
		Sorghum	0.1
		Sugar cane	0.1
		Sweet corn (corn-on-the-cob)	0.1
5.	**Chlorsulfuron**	Cereal grains	0.05
		Edible offal (mammalian)	0.05
		Meat [mammalian]	0.05
		Milks	0.05
6.	**Clodinofop acid**	Edible offal (mammalian)	0.1
		Eggs	0.1
		Meat [mammalian]	0.1
		Milks	0.1
		Poultry, Edible offal of	0.1
		Poultry meat	0.1
		Wheat	0.1
7.	**Clodinafop-propargyl**	Edible offal (mammalian)	0.05
		Eggs	0.05
		Meat [mammalian]	0.05
		Milks	0.05
		Poultry, Edible offal of	0.05
		Poultry meat	0.05
		Wheat	0.05
8.	**2,4-D**	Cereal grains	0.2
		Citrus fruits	5
		Edible offal (mammalian)	2

Contd...

Contd...

S. No.	Herbicides	Commodity	MRL (mg/kg)
		Eggs	0.05
		Legume vegetables	0.05
		Lupin (dry)	0.05
		Meat [mammalian]	0.2
		Milks	0.05
		Oilseed	0.05
		Pear	0.1
		Poultry Edible offal of	0.05
		Poultry [in the fat]	0.05
		Pulses	0.05
		Sugarcane	5
9.	2, 4 -DB	Cereal grains	0.02
		Edible offal (mammalian)	0.2
		Eggs	0.05
		Meat [mammalian]	0.2
		Milks	0.05
		Poultry, Edible offal of	0.05
		Poultry meat	0.05
10.	Diazinon	Cereal grains	0.1
		Citrus fruits	0.7
		Edible offal (mammalian)	0.7
		Eggs	0.05
		Fruits [except citrus fruits; grapes; olives; peach]	0.5
		Grapes T	2
		Kiwifruit	0.5
		Meat [mammalian] [in the fat]	0.7
		Milks [in the fat]	0.5
		Olive oil, crude	2
		Parsley	0.7
		Peach	0.7
		Poultry, Edible offal of	0.05
		Poultry meat	0.05
		Shallot	0.5
		Spring onion	0.5
		Sugar cane	0.7
		Sweet corn (corn-on-the-cob)	0.7
		Tree nuts	0.1

Contd...

Contd...

S. No.	Herbicides	Commodity	MRL (mg/kg)
		Vegetable oils, crude [except olive oil, crude]	0.1
		Vegetables	0.7
11.	Dicamba	Cereal grains	0.05
		Edible offal (mammalian)	0.05
		Eggs	0.05
		Meat [mammalian]	0.05
		Milks	0.1
		Poultry, Edible offal of	0.05
		Poultry meat	0.05
		Sugar cane	0.1
		Sugar cane molasses	2
12.	Diquat	Barley	5
		Beans, except broad bean and Soya bean	1
		Broad bean (green pods and immature seeds)	1
		Cotton seed	1
		Cotton seed oil, crude	0.1
		Edible offal (mammalian)	0.05
		Eggs	0.01
		Fruits	0.05
		Lentils (dry)	0.5
		Linseed	0.01
		Lupin (dry)	0.5
		Maize	0.1
		Meat [mammalian]	0.05
		Milks	0.01
		Oats	5
		Onion, Bulb	0.1
		Peas	0.1
		Poppy seed	5
		Potato	0.2
		Poultry, Edible offal of	0.05
		Poultry meat	0.05
		Rape seed	2
		Rape seed oil, crude	0.1

Contd...

Contd...

S. No.	Herbicides	Commodity	MRL (mg/kg)
		Rice	5
		Rice, polished	1
		Rye	2
		Sesame seed	5
		Sesame seed oil, crude	0.1
		Sorghum	2
		Soya bean (dry)	1
		Sugar beet	0.1
		Sugar cane	0.05
		Sunflower seed	1
		Sunflower seed oil, crude	2
		Tree nuts	0.05
		Wheat	2
		Triticale	2
		Vegetables [except beans; broad bean; lupin (dry); onion, bulb; peas; potato; soya bean (dry); sugar beet]	0.05
13.	**Diuron**	·Asparagus	2
		Cattle, Edible offal of	3
		Cattle meat	0.1
		Cattle milk	0.1
		Cereal grains	0.1
		Cotton seed oil, crude	0.5
		Field pea (dry)	0.05
		Fruits	0.5
		Sugarcane	0.2
		Pineapple	05
		Oilseed	0.5
14.	**Fluazifop-butyl**	Assorted tropical and sub-tropical fruits — inedible peel [except avocado and banana]	0.05
		Avocado *	0.02
		Banana	0.02
		Berries and other small fruits	0.2
		Brassica (cole or cabbage) vegetables, Head cabbages, flowerhead brassicas	1

Contd...

Contd...

S. No.	Herbicides	Commodity	MRL (mg/kg)
		Carrot	0.1
		Celery	0.02
		Citrus fruit	0.02
		Coffee beans	1
		Edible offal (mammalian)	0.05
		Eggs	0.05
		Endive	0.05
		Fruiting vegetables, Cucurbits	0.1
		Garlic	0.05
		Ginger, root	0.05
		Herbs	1
		Hops, dry	0.05
		Leek	0.5
		Legume vegetables	0.1
		Lettuce, Head	0.05
		Lettuce, Leaf	0.05
		Lupin (dry)	0.1
		Meat [mammalian]	0.05
		Milks	0.1
		Oilseed	0.5
		Olives	0.05
		Onion, Bulb	0.05
		Parsnip	0.1
		Peppers, Sweet [capsicums]	0.02
		Poultry, Edible offal of	0.05
		Pome fruits	0.01
		Potato	0.05
		Pulses	0.5
		Rhubarb	0.02
		Shallot	0.05
		Spring Onion	0.05
		Stone fruits	0.05
		Sugar cane	0.1
		Tomato	0.1
15.	**Fluroxypyr**	Cereal grains	0.2
		Edible offal (mammalian)	2
		Eggs	0.01
		Meat [mammalian]	0.1
		Milks	0.1

Contd...

Contd...

S. No.	Herbicides	Commodity	MRL (mg/kg)
		Poultry, Edible offal of	0.05
		Poultry meat	0.05
		Sugar cane [in the juice]	0.2
		Sweet corn (corn-on-the-cob)	0.2
16.	Fenoxaprop-ethyl	Barley	0.01
		Chick-pea (dry)	0.01
		Edible offal (mammalian)	0.2
		Eggs	0.02
		Meat [mammalian]	0.05
		Milks	0.02
		Poultry, Edible offal of	0.1
		Poultry meat	0.01
		Rice	0.02
		Rye	0.01
		Triticale	0.01
		Wheat	0.01
17.	Glufosinate and Glufosinate ammonium	Assorted tropical and sub-tropical fruits - inedible peel	0.2
		Berries and other small	0.1
		Citrus fruits	0.1
		Coffee beans	0.05
		Edible offal (mammalian)	5
		Eggs	0.05
		Meat [mammalian]	0.1
		Milks	0.05
		Olives	0.1
		Pome fruits	0.1
		Poultry, Edible offal of	0.1
		Poultry meat	0.05
		Rape seed	0.05
		Stone fruits	0.05
		Tomato	0.05
		Tree nuts	0.1
18.	Glyphosate	Adzuki bean (dry)	10
		Avocado	0.05
		Banana	0.2
		Barley	10
		Berries and other small fruits	

Contd...

Contd...

S. No.	Herbicides	Commodity	MRL (mg/kg)
		Bulb vegetables [alliums]	0.01
		Cereal grains (except wheat and barley)	0.1
		Citrus fruits	0.5
		Coffee beans	0.2
		Cotton seed	10
		Cotton seed oil, crude	0.1
		Cowpea (dry)	10
		Custard apple	0.05
		Edible offal (mammalian)	0.05
		Eggs	0.05
		Fig	0.1
		Fruiting vegetables, Cucurbits	0.1
		Fruiting vegetables, other than Cucurbits	0.1
		Guava	0.05
		Hops, dry	0.1
		Kiwifruit	0.05
		Leafy vegetables	0.1
		Legume vegetables	0.1
		Litchi	0.2
		Mango	0.05
		Meat [mammalian]	0.1
		Milks	0.1
		Monstero	0.05
		Mung beans (dry)	10
		Oilseed [except cotton seed and rape seed]	0.1
		Olives	0.1
		Papaya	0.1
		Passion fruit	0.05
		Peanut	3
		Persimmon, American	0.1
		Persimmon, Japanese	0.05
		Pome fruits	0.05
		Poultry, Edible offal of	1
		Poultry meat	0.1
		Pulses [except soybeans (dry), cowpeas (dry), adzuki beans (dry), mungbeans (dry)]	5

Contd...

Contd...

S. No.	Herbicides	Commodity	MRL (mg/kg)
		Rape seed	2
		Rollinia	0.05
		Root and tuber vegetables	0.1
		Soya bean (dry)	10
		Stalk and stem vegetables	0.01
		Stone fruits	0.2
		Sugar cane	0.3
		Sugar cane molasses	5
		Tree nuts	0.2
		Wheat	5
		Wheat bran, unprocessed	20
19.	**Haloxyfop**	Assorted tropical and sub-tropical fruits — inedible peel	0.05
		Berries and other small fruits	0.05
		Citrus	0.05
		Cotton seed	0.1
		Cotton seed oil, crude	0.2
		Edible offal (mammalian)	0.5
		Eggs	0.01
		Garlic	0.05
		Linola seed	0.1
		Linseed	0.1
		Meat mammalian [in the fat]	0.02
		Milks	0.02
		Onion, Bulb	0.05
		Peanut	0.05
		Persimmon, Japanese	0.05
		Pome fruits	0.05
		Poultry, Edible offal of	0.05
		Poultry meat [in the fat]	0.01
		Pulses	0.1
		Rape seed	0.1
		Stone fruits	0.05
		Sugar cane	0.03
		Sunflower seed	0.05
		Tree nuts	0.05
20.	**Imazapyr**	Edible offal (mammalian)	0.05
		Maize	0.05
		Meat (mammalian)[in the fat]	0.05

Contd...

Contd...

S. No.	Herbicides	Commodity	MRL (mg/kg)
		Milks	0.01
		Rape seed	0.05
		Wheat	0.05
21.	Imazethapyr	Edible offal (mammalian)	0.1
		Eggs	0.1
		Legume vegetables	0.1
		Maize	0.05
		Meat mammalian	0.1
		Milks	0.1
		Peanut	0.1
		Poultry, Edible offal	0.1
		Poultry meat	0.1
		Pulses	0.1
22.	Linuron	Cereal grains	0.05
		Coriander, seed	0.2
		Edible offal (mammalian)	1
		Eggs	0.05
		Herbs	0.05
		Kaffir lime leaves	0.05
		Leek	0.2
		Lemon balm	0.05
		Lemon grass	0.05
		Lemon verbena	0.05
		Meat [mammalian]	0.05
		Milks	0.05
		Poultry meat	0.05
		Poultry, Edible offal of	0.05
		Turmeric, root	0.05
		Vegetables (except leek)	0.05
23.	MCPA	Cereal grains	0.02
		Edible offal (mammalian)	0.05
		Eggs	0.05
		Meat [mammalian]	0.05
		Milks	0.05
		Poultry, Edible offal of	0.05
		Poultry meat	0.05
24.	MCPB	Cereal grains	0.02
		Edible offal (mammalian)	0.05

Contd...

Contd...

S. No.	Herbicides	Commodity	MRL (mg/kg)
		Eggs	0.05
		Legume vegetables	0.02
		Meat [mammalian]	0.05
		Milks	0.05
		Poultry, Edible offal of	0.05
		Poultry meat	0.05
25.	Metribuzin	Asparagus	0.2
		Cereal grains	0.05
		Edible offal (mammalian)	0.05
		Eggs	0.05
		Meat [mammalian]	0.05
		Milks	0.05
		Peas, shelled	0.05
		Potato	0.05
		Poultry, Edible offal of	0.05
		Poultry meat	0.05
		Pulses [except Soya bean (dry)]	0.01
		Soya bean (dry)	0.05
		Tomato	0.1
26.	Metsulfuron-methyl	Edible offal (mammalian)	0.01
		Eggs	0.01
		Meat (mammalian)	0.01
		Milks	0.01
		Poultry, Edible offal of	0.01
		Poultry meat	0.01
		Wheat	0.02
27.	Metolachlor	Cotton fodder	0.1
		Maize forage	0.02
		Maize fodder	0.1
		Primary feed commodities	5
		Sorghum forage (green)	0.2
		Sorghum straw and fodder, dry	0.2
28.	Oxyfluorfen	Assorted tropical and sub-tropical fruits (inedible peel)	0.01
		Brassica (cole or cabbage) vegetables,	0.05
		Head cabbages, Flowerhead brassicas	0.05

Contd...

Contd...

S. No.	Herbicides	Commodity	MRL (mg/kg)
		Bulb vegetables	0.05
		Cereal grains	0.05
		Coffee beans	0.05
		Cotton seed	0.05
		Edible offal (mammalian)	0.01
		Eggs	0.05
		Grapes	0.05
		Meat (mammalian)[in the fat]	0.01
		Milks *	0.01
		Pome fruits	0.05
		Poultry meat [in the fat]	0.2
		Poultry, Edible offal of	0.01
		Stone fruits	0.05
		Tree nuts	0.05
29.	**Paraquat**	Cereal grains [except maize, rice]	0.05
		Cotton seed	0.2
		Cotton seed oil, edible	0.05
		Edible offal (mammalian)	0.5
		Eggs	0.01
		Fruits [except olives]	0.05
		Hops, dry	0.2
		Maize	0.1
		Meat [mammalian]	0.05
		Milks	0.01
		Olives	1
		Peanut	0.01
		Peanut, whole	0.01
		Potato	0.2
		Poultry, Edible offal	0.05
		Poultry meat	0.05
		Pulses	1
		Rice	10
		Rice, polished	0.5
		Sugar cane	0.05
		TN 0085 Tree nuts	0.05
		Vegetables [except potato; pulses]	0.05

Contd...

Contd...

S. No.	Herbicides	Commodity	MRL (mg/kg)
30.	Pendimethalin	Assorted tropical and sub-tropical fruits inedible peel	0.05
		Barley	0.05
		Berries and other small fruits	0.05
		Brassica (cole or cabbage) vegetables, Head cabbages, Flowerhead brassicas	0.05
		Bulb vegetables [alliums]	0.05
		Citrus fruits	0.05
		Edible offal (mammalian)	0.01
		Eggs	0.01
		Leafy vegetables	0.05
		Legume vegetables	0.05
		Maize Meat (mammalian)	0.05
		Milks	0.01
		Oilseed	0.01
		Olives	0.05
		Pome fruits	0.05
		Poultry, Edible offal of	0.05
		Poultry meat	0.01
		Pulses	0.01
		Root and tuber vegetables	0.01
		Rice	0.05
		Stone fruits	0.05
		Sugar cane	0.05
		Sweet corn (corn-on-the-cob)	0.05
		Tree nuts	0.05
		Tomato	0.05
		Wheat	0.05
31.	Propanil	Cattle, Edible offal of	0.1
		Cattle meat	0.1
		Eggs	0.1
		Milks	0.01
		Poultry, Edible offal of	3
		Poultry meat	0.1
		Rice	2
		Sheep, Edible offal of	0.1
		Sheep meat	0.1
32.	Prometryn	Cattle milk	0.05
		Cereal grains	0.1

Contd...

Contd...

S. No.	Herbicides	Commodity	MRL (mg/kg)
		Cotton seed	0.1
		Edible offal (mammalian)	0.05
		Meat [mammalian]	0.05
		Peanut	0.1
		Sunflower seed	0.1
		Vegetables	0.1
33.	**Quizalofop-ethyl**	Beetroot	0.02
		Cabbages, Head	0.01
		Carrot	0.02
		Cauliflower	0.05
		Common bean (pods and/or immature seeds)	0.02
		Cucumber	0.02
		Edible offal (mammalian)	0.2
		Eggs *	0.02
		Grapes	0.02
		Meat [mammalian]	0.02
		Melons, except Watermelon	0.02
		Milks	0.1
		Onion, Bulb	0.02
		Peanut	0.02
		Pineapple	0.05
		Potato *	0.01
		Poultry, Edible offal of	0.05
		Poultry meat *	0.02
		Pulses 0	0.2
		Pumpkins	0.02
		Radish	0.02
		Rape seed	0.02
		Sunflower seed	0.05
		Tomato	0.02
34.	**Sethoxydim**	Asparagus	1
		Beans, except broad bean and soybean	0.1
		Bergamot	0.1
		Broad bean (green pods and immature seeds)	0.1
		Brassica (cole or cabbage) vegetables, Head cabbages, Flowerhead brassicas	0.2

Contd...

Contd...

S. No.	Herbicides	Commodity	MRL (mg/kg)
		Celery	0.1
		Chervil	0.1
		Coriander (leaves, stems, roots)	0.1
		Coriander seed	0.1
		Cotton seed	0.2
		Dill seed	0.1
		Edible offal (mammalian)	0.05
		Eggs	0.05
		Endive	0.05
		Fennel, Bulb	0.01
		Fennel seed	0.1
		Fruiting vegetables, Cucurbits	0.1
		Garlic	0.3
		Herbs [except thyme]	0.1
		Kaffir lime leaves	0.1
		Leeks	0.3
		Lemon grass	0.1
		Lemon verbena [fresh weight]	0.1
		Lettuce, Head	0.1
		Lettuce, Leaf	0.1
		Lupin (dry)	0.2
		Meat [mammalian]	0.05
		Milks	0.05
		Mizuna	0.1
		Onion, Bulb	0.3
		Peanut	2
		Peanut oil, crude	2
		Peas	0.1
		Poppy seed	0.2
		Poultry, Edible offal of	0.05
		Poultry meat	0.05
		Pulses [except lupin (dry)]	0.1
		Rape seed	0.5
		Root and tuber vegetables	1
		Rose and dianthus (edible flowers)	0.1
		Rucola (rocket)	0.1
		Salad burnett	0.1
		Spinach	0.1
		Strawberry	0.1

Contd...

Contd...

S. No.	Herbicides	Commodity	MRL (mg/kg)
		Sunflower seed	0.1
		Thyme	0.5
		Tomato	0.1
		Turmeric, root	1
		Wheat	0.1
35.	Simazine	Asparagus	0.0.1
		Broad bean	0.01
		Broad bean (green pods and immature seeds)	0.01
		Chick-pea (dry)	0.05
		Chick-pea (green pods)	0.05
		Edible offal (mammalian)	0.01
		Eggs	0.01
		Fruits	0.1
		Leek	0.01
		Lupin (dry)	0.05
		Meat [mammalian]	0.01
		Milks	0.01
		Poultry, Edible offal of	0.01
		Poultry meat	0.02
		Rape seed	0.1
		Tree nuts	
36.	Sulfosulfuron	Edible offal (mammalian	0.005
		Eggs	0.005
		Meat (mammalian)	0.005
		Milks	0.005
		Poultry, Edible offal of	0.005
		Poultry meat	0.005
		Triticale *	0.01
		Wheat	0.01
		Milks	0.01
37.	Thiodicarb	Brassica leafy vegetables	1
		Cotton seed	0.1
		Cotton seed oil, crude	0.1
		Edible offal (mammalian)	0.05
		Eggs	0.02
		Maize	0.1
		Meat [mammalian]	0.05

Contd...

Contd...

S. No.	Herbicides	Commodity	MRL (mg/kg)
		Milks	0.05
		Peppers, sweet [Capsicum] T	5
		Potato	0.1
		Pulses	0.1
		Sorghum T	0.5
		Sweet corn (corn on-the-cob)	0.1
		Tomato	2
38.	Triallate	Cereal grains	0.05
		Edible offal (mammalian) (except kidney)	0.1
		Kidney of cattle, goats, pigs and sheep	0.2
		Legume vegetables	0.05
		Mammalian fats	0.2
		Meat [mammalian]	0.1
		Milks *	0.1
		Oilseed	0.05
		Poultry, Edible offal of	0.2
		Poultry fats	0.2
		Poultry meat	0.1
		Pulses	0.05
39.	Triasulfuron	Cereal grains	0.02
		Edible offal (mammalian)	0.05
		Eggs	0.05
		MM 0 Meat [mammalian]	0.05
		Milks	0.01
40.	Tribenuron-methyl	Barley	0.01
		Maize	0.05
		Oats	0.01
		Sorghum	0.01
		Wheat	0.01
		Chick-pea (dry)	0.01
		Mung bean (dry)	0.01
		Soya bean (dry)	0.01
		Rape seed *	0.01
		Cottonseed	0.05
		Sunflower seed	0.01
		Meat [mammalian]	0.01

Contd...

Contd...

S. No.	Herbicides	Commodity	MRL (mg/kg)
		Edible offal (mammalian)	0.01
		Milks	0.01
41.	**Triclopyr**	Cattle, Edible offal of	5
		Cattle meat [in the fat]	0.2
		Eggs	0.05
		Goat, Edible offal of	5
		Goat meat [in the fat]	0.2
		Milks [in the fat]	0.1
		Poultry, Edible offal of	0.05
		Poultry meat [in the fat]	0.05
		Sheep, Edible offal of	5
		Sheep meat [in the fat]	0.2
		Sorghum	0.1
42.	**Trifluralin**	Adzuki bean (dry) bergamot	0.05
		Broad bean (dry) [faba bean]	0.05
		Carrot	0.05
		Cereal grains	0.5
		Chick-pea (dry)	0.05
		Coriander (leaves, stems, roots)	0.05
		Coriander seed T	0.05
		Cowpea (dry)	0.05
		Dill seed T	0.05
		Edible offal (mammalian)	0.05
		Eggs	0.05
		Fennel, bulb	0.05
		Fennel seed T	0.05
		Fruits	0.05
		Herbs	0.05
		Hyacinth bean (dry)	0.05
		Kaffir lime leaves	0.05
		Lemon grass	0.05
		Lemon verbena (fresh weight)	0.05
		Lupin (dry)	0.05
		Meat [mammalian]	0.05
		Milks	0.05
		Mung	0.05
		bean (dry)	0.05
		Oilseed	0.05

Contd...

Contd...

S. No.	Herbicides	Commodity	MRL (mg/kg)
		Parsnip	0.5
		Poultry, Edible offal of	0.05
		Poultry meat	0.05
		Sugarcane	0.001
		Shrimp or prawn	0.5
		Turmeric root	0.05
		Vegetables (except carrot)	
43.	Terbutryne	Straw, fodder (dry) and hay of cereal grains and other grass-like plants	30
		Pastures (mixed grasses/ leguminous)	75
		Field pea forage and fodder	30

Maximum residue limits (MRLs) of herbicides in drinking water (WHO).

S. No.	Herbicides	MRL (μg/L)
1.	Alachlor	20
2.	Atrazine	2
3.	2,4-D	30
4.	Bentazon	300
5.	Dalapon	200
6.	Dinoseb	7
7.	Diquat	10
9.	Glyphosate	700
10.	Metolachlor	10
11.	Pendimethalin	20
12	Propanil	20
13.	Simazine	2
14.	Trifluralin	20

□□□

Appendix III

Herbicide recommendation for different crops

Crop	Herbicide name	Trade name	Dose (g ai/ha)	Product (g or ml/ha based on formulation)	Application time
Rice	2,4-DEE	Knock weed fernoxone needmar	750-1000	2250-3000 (36EC) 2-2.5 kg (4 g)	20-25 DAT
	Anilofos	Arozin, Aniloguard	400	1200	3-5 DAT
	Butachlor	Machite	1000-1500	2000-3000 2-2.5 kg (5% g)	6-7 DAS
	Butachlor + propanil	Butanil	1120 + 1120	4000 (Ready mix)	10-12 DAS
	Chlorimuron + Metsulfuron ethyl	Almix	4	20	15-20 DAS
	Cyhalofop-butyl	Clincher	60-75	600-750	15-20 DAS
	Fenoxaprop p-ethyl	Whip super	80-120	800-1200	25-30 DAS
	Oxadiargyl	Raft, Topstar	90	1500	0-3 DAS
	Oxadiazon	Ronstar	500-700	2000-3000	Pre - emergence or early post - emergence
	Oxyfluorfen	Goal	150-250	600-1000	0-6 DAS
	Pendimethalin	Stomp	1000-1500	3000-4500	6-7 DAS
	Pretilachlor	Refit	750	1500	3-7 DAS
	Pretilachlor + safener	Sofit	750	1500	3-7 DAS
	Pyrazosulfuron	Saathi	150	3000	20-25 DAS

Contd...

Contd...

Crop	Herbicide name	Trade name	Dose (g ai/ha)	Product (g or ml/ha based on formulation)	Application time
	Thiobencarb	Saturn	1000-1500	2000-3000 2025 Kg (5% g)	6-7 DAS
Wheat	2,4-DEE	Knockweed Weedmar Fernoxone	400-600	Varies with formulation	30-35 DAS
	Clodinofop	Topic	60	400	4-6 WAS
	Dichlofop-methyl	Iloxan	1000	3000	30-35 DAS
	Fenoxaprop	Pumasuper	100-120	1000-1200	4-6 WAS
	Isoproturon	Isoguard Arelon	750-1000	1000-1250 (75 WP) 1500-2000 (50 WP)	25-30 DAS
	Metribuzin	Sencor	175-210	250-300	30-35 DAS
	Metsulfuron-methyl	Algrip	4-6	20-30	25-30 DAS
	Pendimethalin	Stomp	1000	3000	0-3 DAS
	Sulfosulfuron	Leader	25	33	25-30 DAS
Maize	2,4-DEE	Knockweed Weedmar Fernoxone	750	Varies with formulation	Post emergence
	Alachlor	Lasso	2000-2500	4000-5000	0-3 DAS
	Atrazine	Atrataf	750-1000	1500-2000	0-3 DAS
	Metolachlor	Dual	1000-1500	2000-3000	0-3 DAS
	Pendimethalin	Stomp	1000-1500	3000-4500	0-3 DAS
Sorghum	2,4-DEE	Knockweed Weedmar Fernoxone	750	Varies with formulation	5-6 WAS
	Atrazine	Atrataf	500-750	1000-1500	0-3 DAS
	Pendimethalin	Stomp	750-1000	2500-3000	0-3 DAS
Pearlmillet	2,4-DEE	Knockweed Weedmar Fernoxone	750	Varies with formulation	5-6 WAS
	Atrazine	Atrataf	250-500	500-1000	0-3 DAS
	Pendimethalin	Stomp	750-1000	2500-3000	0-3 DAS
Greengram /blackgram /pgeaipea	Alachlor	Lasso	2000-2500	4000-5000	0-3 DAS
	Fluchloralin	Basalin	750-1000	1500-2000	Pre-planting
	Metolachlor	Dual	1000-1500	2000-3000	0-3 DAS
	Oxadiazon	Ronstar	250	1000	0-3 DAS
	Oxyfluorfen	Goal	100-125	400-500	0-3 DAS
	Pendimethalin	Stomp	750-1000	2500-3000	0-3 DAS
	Quizalofop ethyl	Targa - super	40-50	800-1000	15-20 DAS

Contd...

Contd...

Crop	Herbicide name	Trade name	Dose (g ai/ha)	Product (g or ml/ha based on formulation)	Application time
Chickpea/lentil /field peas	Fluchloralin	Basalin	750-1000	1500-2000	Pre-planting
	Metolachlor	Dual	1000-1500	2000-3000	0-3 DAS
	Metribuzin (in peas)	Sencor	250	350	0-3 DAS
	Oxyfluorfen (in peas)	Goal	100-125	400-500	0-3 DAS
	Pendimethalin	Stomp	750-1000	2500-3000	0-3 DAS
	Quizalofop ethyl	Targa - Super	40-50	800-1000	15-20 DAS
Groundnut	Alachlor	Lasso	1500-2000	3000-4000	0-3 DAS
	Butachlor	Machete	1000-1250	2000-2500	Pre-emergence
	Fluchloralin	Basalin	750-1000	1500-2000	Pre-planting
	Imazethapyr	Pursuit	100	1000	0-3 DAS or early post em.
	Metolachlor	Dual	1000-1500	2000-3000	0-3 DAS
	Oxadiazon	Ronstar	750	3000	0-3 DAS
	Oxyfluorfen	Goal	250	1000	0-3 DAS
	Pendimethalin	Stomp	750-1000	2500-3000	Pre plant or pre em.
	Quizalofop ethyl	Targa - Super	40-50	800-1000	15-20 DAS
	Trifluralin	Treflan	1000	2000	0-3 DAS
Soybean	Alachlor	Lasso	1500-2000	3000-4000	0-3 DAS
	Chlorimuron	Klorben	8-12	40-60	15-20 DAS
	Fenoxaprop	Whips-uper	80-100	800-1000	20-25 DAS
	Fluchloralin	Basalin	750-1000	1500-2000	Pre-planting
	Imazethapyr	Pursuit	100	1000	Pre em. or early post em.
	Metolachlor	Dual	1500-2000	3000-4000	0-3 DAS
	Metribuzin	Sencor	350-525	500-750	0-3 DAS
	Oxadiazon	Ronstar	500	2000	Pre em. or early post em.
	Pendimethalin	Stomp	750-1000	2500-3000	Pre plant or pre em.
	Quizalofop ethyl	Targa - super	40-50	800-1000	15-20 DAS
	Trifluralin	Treflan	1000	2000	Pre planting

Contd...

Contd...

Crop	Herbicide name	Trade name	Dose (g ai/ha)	Product (g or ml/ha based on formulation)	Application time
Sunflower	Alachlor	Lasso	1000-1500	2000-3000	0-3 DAS
	Butachlor	Machete	1000-1500	2000-3000	0-3 DAS
	Fluchloralin	Basalin	1000	2000	Pre-planting
	Metolachlor	Dual	750-1000	1500-3000	0-3 DAS
	Oxadiazon	Ronstar	500-1000	2000-4000	0-3 DAS
	Oxyfluorfen	Goàl	250	1000	0-3 DAS
	Pendimethalin	Stomp	750-1000	2500-3000	0-3 DAS
	Trifluralin	Treflan	1000	2000	Pre-planting
Rapeseed/ mustard	Fluchloralin	Basalin	1000	2000	Pre-planting
	Isoproturon	Isoguard Arelon	750-1000	1000-1250 (75 WP) 1500-2000 (50 WP)	Pre em. or early post em.
	Oxadiazon	Ronstar	500	2000	Pre em. or early post em.
	Pendimethalin	Stomp	750-1000	2500-3000	Pre - em.
	Pendimethalin	Stomp	750-1000	2500-3000	15-20 DAS
Linseed/ sesame/ niger	Fluchloralin	Basalin	1000	2000	Pre-planting
	Isoproturon	Isoguard	1000-1500	1250-3000 (75 WP) 2000-3000 (50 WP)	Pre em. or early post em.
	Oxadiazon	Ronstar	500	2000	Pre em. or early post em.
	Pendimethalin	Stomp	750-1000	2500-3000	0-3 DAS
Jute	Fluchloralin	Basalin	1000	2000	Pre-planting
	Pendimethalin	Stomp	750-1000	2500-3000	0-3 DAS
Cotton	Butachlor	Machete	1000-1250	2000-2500	Pre-emergence
	Diuron	Klass	500-750	600-900	Pre-emergence
	Fluchloralin	Basalin	1000	2000	Pre-planting
	Glufosinate	Liberty	450	1000	Post em.
	Glyphosate	Roundup, Glycel	1000	2500	Post em.
	Oxadiazon	Ronstar	500-750	1000-1500	0-3 DAS
	Paraquat	Gamoxone	500	2000	Post em.
	Pendimethalin	Stomp	1000-1500	3000-5000	Pre plant or pre em. Or lay by application
	Trifluralin	Treflan	500-1000	1000-2000	0-3 DAS

Contd...

Contd...

Crop	Herbicide name	Trade name	Dose (g ai/ha)	Product (g or ml/ha based on formulation)	Application time
Sugarcane	2,4-DEE	Knock-weed Weedmar Fenoxone	1000	3000	Post emergence
	Atrazine	Atrataf	2000	4000	Pre em.
	Diuron	Klass	1500	2100	Pre-emergence
	Glyphosate	Roundup, Glycel	1000	2500	Post em.
	Metribuzin	Sencor	1000-1500	1500-2250	Pre em. or early post em.
	Paraquat	Gamoxone	500 or 0.30 % spray	2000	Pre em. or post em
Potato	Butachlor	Machete	750-1000	1500-2000	Pre-emergence
	Fluchloralin	Basalin	1000-1500	2000-3000	Pre-planting
	Isoproturon	Isoguard Arelon	750-1000	1000-1250 (75 WP) 1500-2000 (50 WP)	Pre em. or early post em.
	Metolachlor	Dual	750-1000	1500-3000	Pre em.
	Metribuzin	Sencor	525	750	Pre em. or early post em.
	Pendimethalin	Stomp	1000	3000	pre em.
Cabbage/ cauliflower	Fluchloralin	Basalin	1000	2000	Pre or post transplanting
	Pendimethalin	Stomp	1000	3000	Pre plant or pre em.
Tomato/ Brinjal/ Chilli/ Green pepper	Alachlor	Lasso	2000	4000	Pre-transplanting
	Fluchloralin	Basalin	1000	2000	Pre-transplanting
	Metribuzin	Sencor	525	750	Pre-transplanting or early post em.
	Oxadiazon (in chili)	Ronstar	500	2000	Pre em. or early post em.
	Oxadiazon	Ronstar	1000	3000	Pre em. or early post em.

Contd...

Contd...

Crop	Herbicide name	Trade name	Dose (g ai/ha)	Product (g or ml/ha based on formulation)	Application time
Onion/garlic	Butachlor	Machete	1000-1500	2000-3000	Post-transplanting
	Fluchloralin	Basalin	1000	2000	Pre or early post transplanting
	Haloxyfop-butyl	-	100	-	2 WAT
	Oxadiargyl	Raft, Torstar	90	1500	0-3 DAT
	Oxadiazon	Ronstar	500-750	2000-3000	Pre or post transplanting
	Oxyfluorfen	Goal	250	1000	Pre or post transplanting
	Pendimethalin	Stomp	1000	3000	Pre or post transplanting
Beetroot/carrot/radish	Fluchloralin	Basalin	1000	2000	Pre-planting or Pre or post transplanting
	Pendimethalin	Stomp	1000	3000	0-3 DAS or Pre or post transplanting
Cumin/Coriander/Fennel/ (Saunf) Fenugreek (methi)	Fluchloralin	Basalin	1000	2000	Pre-planting
	Pendimethalin	Stomp	1000-1500	3000-5000	Pre. em.

DAS- Days after sowing; DAT-days after transplanting; pre em. - Pre emergence; WAS-Weeks after sowing

□□□

Appendix IV

Physico-chemical properties of approved herbicides in India

Herbicide name	Chemical name	Molecular formula	Molecular weight	Vapour pressure (mbar)	Melting point (°C)	Boiling point (°C)	Solubility
Alachlor	2-chloro-N-(2,6-diethylphenyl)-N (methoxymethyl) acetamide	$C_{14}H_{20}ClNO_2$	269.8	3 x 10^{-5} at 25 °C	40	100 at 0.03 mbar	148 mg/l in water; soluble in common organic solvents
Anilofos	S-4-chloro-N-isopropylcarbaniloylmethyl O,O-dimethyl phosphorodithioate	$C_{13}H_{19}ClNO_3PS_2$	367.8	2.2 at 60 °C	52.5	-	Soluble in water 13.6 mg/I. Soluble in acetone, ethanol, ethyl acetate , dichloromethane, benzene, hexane
Atrazine	2-chloro-4- ethylamino-6-isopropylamino-1,3,5-triazine	$C_8H_{14}ClN_5$	215.7	4 x 17^{-7} at 20°C	173-175	-	In water 70 mg/I. Soluble in diethyl ether , methanol, ethyl acetate
Butachlor	N-butoxymethyl-2-chloro-2′, 6′diethylacetanilide.	$C_{17}H_{26}ClNO_2$	311.9	0.6 mPa at 25°C	<5	-	Water 20 mg/l. Soluble in most of the organic solvents such as acetone, ethanol, ethyl acetate and hexane
Chlorimuron-ethyl	ethyl 2-(4-chloro-6-methoxypyriraidin-2-ylcarbamoyisulfamoyl) benzoate	$C_{15}H_{15}ClN_4O_6S$	414.8	2 at 25°C	185 -187	-	In water 9 (pH 5), 1200 (pH 7) (both in mg/l, 25 °C). Low solubility in organic solvents.
Clodinafop	Prop 2-ynyl(R)-2-[4-(5-chloro-3-fluoropyridin-2-yloxy)phenoxy]propionate	$C_{17}H_{13}ClFNO_4$	269.8	3.19 x 10^{-3}	48.2-57.1	-	In water 4 mg/l, easily soluble in organic solvents
Clomazone	2-[(2-chlorophenyl) methyl]-4, 4-dimethyl-3-isoxazolidinone	$C_{12}H_{14}Cl NO_2$	239.70	1.44x10^{-4} (25°C).	25	275.4 at 740 mm Hg	Solubility in water 1100 mg/I, soluble in acetone, methanol, benzene

Contd...

Contd...

Herbicide name	Chemical name	Molecular formula	Molecular weight	Vapour pressure (mbar)	Melting point (°C)	Boiling point (°C)	Solubility
Cyhalofop-butyl	2-[4-(4-cyano-2-fluorophenoxy)phenoxy]propanoic acid, butylester	$C_{16}H_{12}FNO_4$	357.38	$5.3x10^{-8}$	45.5 - 49.5	-	0.44 mg/l in water, highly soluble in n-heptance, xylene, acetone and methanol
Dalapon	2,2-dichloropropionic acid	$C_3H_4Cl_2O_2$	143.0	$<10^{-5}$ at 20°C	166.5	185-190	50g/100 g in water .Soluble in acetone , ethanol, ethyl acetate, dichloromethane,carbon tetrachloride and benzene
Dicamba	3,6-dichloro-2-methoxybenzoic acid	$C_8H_6CI_2O_3$	221.04	5 x 10^{-3} at 100°C	114-116	-	0.79g/g in water, readily soluble in ethanol, acetone, and other ketone
2, 4-D	(2, 4-dichlorophenoxy) acetic acid	$C_8H_6CI_2O_3$	221	< 10^{-7} at 20°C	140.5	-	In water 900 mg/L. Insoluble in benzene, hexane and xylene
Diclofop-methyl	(RS)-2-[4-2,4-dichlorophenoxy) phenoxylpropionic acid	$C_{15}H_{12}CI_2O_4$	341.2	3.4x 10^{-7} at 20 °C	39-41	173-175	In water 50 mg/l. Soluble in acetone, ethanol, ethyl acetate , dichloromethane, carbon tetrachloride, benzene, hexane
Diuron	3-(3,4-dichlorophenoxy)-I, I-dimethylurea	$C_9H_{10}CI_2N_2O$	233.1	4.1 x 10^{-6} at 50°C.	158-159	-	water 42 mg/l. Soluble in acetone and benzene
Ethoxysulfuron	1-(4-6dimehtoxypyrimidin-2yl)-3-(2 ethoxyphenoxysulfonyl) urea	$C_{15}H_{18}N_4O_7S$	398.43	1.2 x 10^{-4}	144-147	-	In water 26.4 mg/L (pH 5); 1353 mg/L (pH 7); 9628 mg/L (pH 9)

Contd...

Contd...

Herbicide name	Chemical name	Molecular formula	Molecular weight	Vapour pressure (mbar)	Melting point (°C)	Boiling point (°C)	Solubility
Fenoxaprop	2-[4-[(6-chloro-2 benzoxazolyl)oxy]phenoxy] propanoic acid	$C_{16}H_{12}ClNO_5$	333.73	1.87 x 10^{-4}	89 -91	-	Less soluble in water (0.9 mg/l), soluble in acetone, methanol, cyclohexane, *n*-octanol, ethyl acetate, toluene
Fluchloralin	N-(2-chloroethyl)-2,6-dinitro-N-propyl-4-(trifluoromethyl)aniline	$C_{12}H_{13}ClF_3N_3O_4$	355.7	4 x10^{-5} at 20°C	42-43	-	Soluble in water<1 mg/l. Soluble in acetone , ethanol , ethyl acetate and cyclohexane
Glufosinate	2-[hydroxyl (methyl)phosphinoyl]DL-homoalanine	$C_5H_{12}NO_4P$	181.1	<0.1 at 20°C	251	99.5	Soluble in water 1370 g/l. Soluble in most of the organic solvents (acetone ethanol, ethyl acetate , hexane)
Glyphosate	N-(phosphonomethyl)glycine	$C_3H_8NO_5P$	169.08	Practically zero	200	-	Soluble in water at 1%, springly soluble in acetone, ethanol and xylene. The alkali metal and amine salts are readily soluble in water.
Haloxyfop-methyl	2-[4-[3-chloro-5-(trifluromethyl)-2-pyridinyl]oxy]phenoxy]propanoic acid	$C_{16}H_{13}ClF_3NO_4$	375.73	1.3 x 10^{-7} at 25°C	55-57	-	0.5 mg/1 in water, 43.3 mg/l(acid), 9.3 mg/1 (haloxyfop-methyl), soluble in xylene, acetone, toluene
Imazethapyr	2-[4.5-dihydro-4-methyl-4-(1-methylethyl)-5-oxo-IH-imidazol-2-yl]-5-ethyl-3-pyridinecarboxylic acid	$C_{15}H_{19}N_3O_3$	289.33	<1 x 10^{-7} at 60°C	169-173	Decomposes at 180	In water 1.4g/l. Soluble in acetone, dimethyl sulfoxide, heptane, methanol, and dichloromethane

Contd...

Contd...

Herbicide name	Chemical name	Molecular formula	Molecular weight	Vapour pressure (mbar)	Melting point (°C)	Boiling point (°C)	Solubility
Isoproturon	3(4-isopropylphenyl)-1, 1-dimethylurea	$C_{12}H_{18}N_2O$	206.3	3.15×10^{-4} at 77°C	151-153	-	Soluble in water at 70 mg/I. Soluble in acetone, methanol, dichloromethane, benzene, hexane and xylene
Linuron	3-(3,4-dichlorophenyl)-I-methoxy-I-methylurea	$C_9H_{10}C_{12}N_2O_2$	249.1	2×10^{-5} at 24°C	93-94	-	Soluble in water at 75 mg/I. Soluble in acetone, ethanol, benzene, heptane and xylene
Methabenzthiazuron	1-(1, 3-benzothiazol-2yl)-1, 3-dimethylurea	$C_{10}H_{11}N_3OS$	221.3	$< 10^{-6}$ at 20°C	119-120	-	in water 5.9 mg/100 ml. Soluble in ethanol, diethyl ether and toluene
Metolachlor	2-chloro-6′-ethyl-N-(2-methoxy-1-rnethylethyl) acet-o-toluidide	$C_{15}H_{22}CINO_2$	283.8	1.7×10^{-5} at 20°C	-	10 at 0.001 mbar	In water, 530 mg/I. Soluble inorganic solvents with the exception of petroleum ether
Metoxuron	3-(3-chloro-4-methoxyphyenyl)-1,1-dimethylurea	$C_{10}H_{13}CIN_2O_2$	228.7	4.2×10^{-8} at 20°C	126-127	-	Soluble in water at 678 mg/1
Metribuzin	4-amino-6-tert-butyl-4, 5-dihydro-3-methylthio-1,2,4-triazin-5-one	$C_8H_{14}N_4OS$	214.3	Less than 10^{-6} at 20°C	126-127	-	Soluble in water at 1.05 g/I (20°C). Soluble in acetone, ethanol, dichloromethane, benzene and hexane

Contd...

Contd...

Herbicide name	Chemical name	Molecular formula	Molecular weight	Vapour pressure (mbar)	Melting point (°C)	Boiling point (°C)	Solubility
Metsulfuron-methyl	Methyl-2-[[[[(4-methoxy-6-methyl-1,3,5-triazin-2-yl)-amino]carbonyl]amino] sufonyl] benzoate	$C_{14}H_{15}N_5O_6S$	381.36	3.3×10^{-7}	158	-	In water 0.55 (pH 5), 2.79 (pH 7), 213 g/l (pH 9), soluble in xylene, ethanol, methanol, acetone and dichloromethane
Oxadiazon	5-tert-butyl-3-(2,4dichloro-isopropoxyphenyl) -1,3,4-oxydiazol-2-one	$C_{15}H_{18}Cl_2N_2O_3$	345.2	Less than 10^{-6} at 20°C	90	-	In water, 0.7 mg/I. Soluble in organic solvents
Oxadiargyl	5-tert-butyl-3-(2, 4-dichloro-5-(prop-2ynyloxy)phenyl)-1, 3, 4-oxydiazol-2-(3H)-one	$C_{15}H_{14}Cl_2N_2O_3$	341.90	0.0025	131	-	Soluble in organic solvents, in water 0.37 mg /l
Oxyfluorfen	2-chloro-1,(3-ethoxy-4-nitrophenoxy)-4-(trifluoromethyl)benzene	$C_{15}H_{11}ClF_3N_4$	361.7	0.267 at 25°C)	85-90 (Tech 65-84)	358.2	Water 0.116 mg/I (25°C). Readily soluble in most organic solvents
Paraquat, Paraquat dichloride	1,1-dimethyl-4,4′-bipuriinium dichloride	$C_{12}H_{14}Cl_2N_2$	257.2	Less than 10^{-7} at 20°	300	-	Readily soluble in water. Slightly soluble in alcohols. Barely soluble in most other organic solvents
Pendimethalin	N--(I-ethylpropyl)-2,6-dinitro-3,4-xylidine	$C_{13}H_{19}N_3O_4$	281.3	4×10^{-5} at 25	54-58	Decomposes on distilation	In water 0.3 mg/I. Readily soluble in most organic solvents such as alcohol, acetone, benzene, toluene, chloroform and methyl-ethyl chloride
Pretilachlor	2-chloro-2′, 6′-diethyl-N-(2-propoxyethyl) acetanilide	$C_{17}H_{26}ClNO$	298.85	0.133 at 26°C.	-	135/0.01 (1 mmHg)	Soluble in methanol, dichloromethane, benzene and hexane. Solubility in water 50 mg/I

Contd...

Contd...

Herbicide name	Chemical name	Molecular formula	Molecular weight	Vapour pressure (mbar)	Melting point (°C)	Boiling point (°C)	Solubility
Propanil	3′,4′dichloropropionanilide	$C_9H_9Cl_2NO$	218.08	1.2 x 10- at 60	85-90 (tech 65-84)	-	In water 0.02 g/100 ml. Soluble in benzene, acetone, and ethanol. It is also rapidly soluble in, isopropanol, methyl ethyl ketone
Pyrazosulfuron	5-[(4,6dimethoxypyrimidin -2-ylcarbamoyl)sulfamoyl] -1-methylpyrazole-4-carboxylic acid	$C_{12}H_{14}N_6O_7S$	414.4	0.0147 at 20	181-182	-	In water 14.5 mg/l, soluble in organic solvents
Quizalofop-p-butyl	(R)-2-[4-[6-chloro-2-quinoxalinyl)phenoxy]propanic acid	$C_{19}H_{17}ClN_2O_4$	372.80	0. 000011 at 20 (6)	91.7-92.1	220 at 0.2 mm Hg	0.4 mg/l in water, soluble in acetone, ethanol, hexane, and xylene
Sulfosulforon	1-(2-ethylsulfonylimidazo[1,2-a]pyridine-3-(4,6-dimthoxypyrimidine-2-yl) ureas	$C_{16}H_{18}N_6O_7S_2$	470.48	8.81 x 10^{-8} at 25	201.1-201.7	-	17.6 (pH 5), 1,627 (pH 7), 482 (pH 9) (mg/l, 20°C) Soluble in organic solvents such as in acetone , methanol, ethyl acetate 1.01, dichloromethane, xylene and heptane
TCA-Sodium	Sodium trichloroacetate	$C_2Cl_3NaO_2$	185.4	<10^{-6} at 70	350(decomposes	-	Soluble in water at 1.2 kg/100 ml, soluble in acetone, methanol, and benzene
Thiobencarb	S-4-chlorobenzyl diethylthiocarbamate	$C_{12}H_{16}ClNO_5$	257.8	2.2 at 23	33	126-129 /0.008 mm Hg	Soluble in most of the organic solvents (acetone, ethanol. N-hexane and xylene). Solubility in water 30 mg/I.

Contd...

Contd...

Herbicide name	Chemical name	Molecular formula	Molecular weight	Vapour pressure (mbar)	Melting point (°C)	Boiling point (°C)	Solubility
Tri-allate	S-2,3,3-trichloroallyl-di-isopropyl(thiocarbamate)	$C_{10}H_{16}CI_3NOS$	304.7	1.6×10^{-4} at 25	29-30 (tech.)	117	In water 4 mg/l, readily soluble in organic solvents
Trifluralin	2--6-dinitro N,N-dipropyl-4 (trifluoromethyl) benzenamine	$C_{13}H_{16}F_3N_3S_4$	335.3	2.6×10^{-4} at 29.5	48.5	139-140 at 4.2 mm Hg	In water less than 1-mg/I. Readily soluble in common organic solvents (Acetone and, xylene)

□□□

Appendix V

Common name, trade name and weed control efficiency of approved herbicides in India in various crops

Herbicide name	Trade name	Stability	Compatibility	Formulation (a.i. %)	Weed control	Uses	Source
Alachlor	Lasso	Hydrolyzed by strong acid alkalis	Compatible with many other herbicides and liquid fertilizers	10 GR, 50 EC	*Echinochloa spp., Euphorbia hirta, Phyllanthus niruri, Trianthema* spp., *Digera arvensis, Portulaca oleracea,* and *Eleusine* spp	Pre-emergence to control annual grasses and many broad leaved weeds in cotton, brassicas, maize, oilseed rape, peanuts, radish, soybeans, beans and sugarcane	Monsanto
Anilofos	Arozin, Aniloguard	Stable at pH at 5-9 at 22 °C but decomposes at 150 °C. Non-sensitive to sunlight	Compatible with most pesticides	30 EC; 18 GR	*Echinochloa crus-galli, Echinochloa colona, Ischaemum rugosum, Leptochloa chinesis, Cyperus difformis, Fimbristylis* spp, *Eleusine indica, Monochoria vaginalis* etc.	Pre-emergence in transplanted rice and direct seeded rice in wet condition.	Bayer Garada, Dhanuka
Atrazine	Atrataf, . Solaro, Gesaprim	Stable in nature, weakly acidic and weakly media.	Miscible with other powdered herbicides and fertilizers at the normal application rates.	50 WP	*Trianthema* spp, *Digera arvensis, Eleusine* spp., *Echinochloa* spp., *Polygonum* spp., *Amaranthus* spp., *Digitaria* spp., *Chenopodium album* and *Spergula arvensis*	Used as pre- and post-emergence for grasses and board leaved weeds in maize, sorghum, vine yard, fruit and orchards, control grasses in young forest trees	Bayer, Dhanuka, Rallis

Contd...

Contd...

Herbicide name	Trade name	Stability	Compatibility	Formulation (a.i. %)	Weed control	Uses	Source
Butachlor	N-butoxymethyl-2-chloro-2′, 6′diethylacetanilide	-	Miscible with other herbicides and fertilizers at the normal application rates.	50% EC, 5% GR., 50% EW,	*Echiochloa* spp., *Fimbristylis spp.*, *Eleusine indica*, *Monochoria vaginalis* and *Cyperus iria*	Pre-emergence for the control of annual grasses and certain broad-leaved weeds in rice, both seeded and transplanted	Monsanto, Dhanuka, Rallis, Coromondal, Bayer
Chlorimuron-ethyl	Kloben	-	Compatible with linuron and metribuzin	25 WP	Cocklebur (*Xanthium* spp.), pigweed (*Amaranthus* spp.) and annual morning glory (*Convolvulus arvensis*), *Chenopodium album*, Carpetweed (*Mollugo verticillata*), *Polygonum* spp., *Commelina* spp. *Phyllanthus niruri*, *Euphorbia* spp. etc	Post-emergence for the control of important broad-leaved weeds in soybeans. Active at low use rates and has shown good crop tolerance to soybeans, peanut, wheat and rice at 8-12 g/ha Post-emergence for the control of important broad-leaved weeds in soybeans, peanut, wheat and rice at 8-12 g/ha	Dupont
Clodinafop	Topik	-	-	15 WP	*Agrostis capillaries*, *Alopecurus myosuroides*, *Avena* spp, *Arrbentherum*	Post-emergence to control annual grasses specially wild oat and *Phalaris minor*	Syngenta

Contd...

Contd...

Herbicide name	Trade name	Stability	Compatibility	Formulation (a.i. %)	Weed control	Uses	Source
					elatius, Cynosurus echinatus, Digitaria sanguinalis, Echinochloa crusgalli, Lolium spp, *Panicum milleaceum, Phalaris* spp., *Poa trivialis, Setaria glauca, Setaria viridis, Setaria verticillata, Elymus repens*	in wheat, mustard, pulses at 40-60 g/ha.	
Clomazone	Command	Relatively stable to UV light, decomposes at>200°C	Compatible with other herbicides	50 EC	*Agrostis tenuis, Alopecurus myosuroides, Abutilon theophrasti, Avena* spp., *Bromus mollis, Brachiaria platyphylla, Cenchrus echinatus, Chenopodium album, Eleusine indica, Chloris barbata, Dactyloctenium aegyptium, Digitaria sanguinalis, Echinochloa colona,*	Applied as pre-emergence or pre-plant incorporation to control many annual and broad-leaved weeds in soybeans, peppers, pumpkins, potato, sugarcane, tobacco and pea	Rallis

Contd...

Contd...

Herbicide name	Trade name	Stability	Compatibility	Formulation (a.i. %)	Weed control	Uses	Source
					Echinochloa crus-galli, Lolium multiflorum, Lolium rigidum, Panicum virgatum, Rottboellia exaltata, Setaria viridis.		
Cyhalofop-butyl	Clincher	Stable to metals and metal ions after stored at 50°C (2 weeks	-	10 % EC	*Elymus repens, Avena spp., Echinochloa* spp., *Digitaria sanguinalis, Setaria glauca, Panicum* spp., *Sorghum halepens* and others. All annual grasses and most perennial grasses are suspected to be sensitive	Used as pre and post-emergence herbicide in rice	Denocil
Dalapon	Devipon, Dowpon	Decomposes by hydrolysis at 25°C-50°C. Hence aqueous solution should not be kept for a long time. Contact with alkali causes dehydrochlorination above 120°C	Compatible with many other herbicides. Not recommended in combination with oils	85 W; SP; GR; EC; WP	*Cynodon dactylon, Digitaria* spp., *Eleusine indica, Paspalum* spp. *Pennisetum* spp., *Setaria* spp., *Sorghum halepens*	Used to control annual and perennial grasses on non-cropland, vineyards, forestry, bananas, sugarcane, tea, rubber, cotton and ornamental shrubs. Also controls semi-aquatic grass weeds in water courses	BASF

Contd...

Contd...

Herbicide name	Trade name	Stability	Compatibility	Formulation (a.i. %)	Weed control	Uses	Source
Dicamba	Banvel	The acid is very stable to hydrolytic and oxidative influences, but is decomposed in sunlight	Compatible with many other herbicides and liquid fertilizers	Aqueous solution as dimethyl amine salt (480 g/l).	*Amaranthus* spp., *Chenopodium album*, *Chenopodium murale*, *Cirsium arvense*. *Convolvulus arvensis*, *Lathyrus aphaca*, *Medicago denticulata*, *Polygonum convolvulus*	Post-emergence herbicides for controlling broad leaved weeds in wheat and barley at 500-750 g/ha	Huanuo Che, Co.
2, 4-D (salts and esters)	Weedmar, Veerkill	Stable up to 50°C for at least 2 years	Compatible with many other herbicides. Reduces the activity of fops and dims	38 EC, 34 EE, 80WP, 72 WSC, 58% SL 22.5% SL	*Amaranthus* spp., *Anagallis arvensis*, *Atriplex* spp., *Capsella bursa pastoris*, *Centaurea cyanus*, *Chenopodium album*, *Chenopodium murale*, *Galeopsis tetrahit*, *Galinsoga parviflora*, *Galium aparine*, *Ipomoea carnea*, *Kochia* spp., *Lantana camara*, *Lithospernum arvensis*, *Matricaria* spp., *Medicago*	Widely used for weed control mainly broad leaved weeds in cereals and other crops at 0.28-2.3 kg/ha	Rhone-Poulenc Herbicide India Ltd.

Contd..

Contd...

Herbicide name	Trade name	Stability	Compatibility	Formulation (a.i. %)	Weed control	Uses	Source
					denticulate, Mercurialis annua, Myosotis arvensis, Papaver rhoeas, Polygonum convolvulus, Polygonum persicari, Portulaca oleracea, Ranunsulus arvensis, Raphanus raphanistrum, Scleranthus annuus, Senecio valgaris, Sinapis arvensis, Solanum nigrum, Spergula arvensis, Stellaria media, Thlaspi arvensis, Veronica spp., *Viola tricolor*		
Diclofop-methyl	IIIoxan	Stable to light and in water	Incompatible with most other herbicides	50, 75WP	*Avena* spp., *Agrostis* spp., *Brachiaria platyphylla, Echinocloa curs-galli, Eleusine agricana, Eleusine indica, Eragrostis*	Used as post-emergence control of annual grasses in wheat, barley, rye and broad leaved crops such as soybean, legumes oil seeds, sunflower,	Bayer

Contd...

Contd...

Herbicide name	Trade name	Stability	Compatibility	Formulation (a.i. %)	Weed control	Uses	Source
					spectabilis, Leptochloa filiformis, Leptochloa multiflorum, Lolium perenne, Setaria feberii, Setaria verticillata, Setaria viridis, Poa trivialis, Lolium multiflorsm, Lolium rigidum, Panicum capillare, Panicum dichtomiflorum, Phalaris spp	brassica, carrots beetroot, lettuce, spinach, potatoes, cucumber, peas, bean and tomatoes *etc.* at 750 g/ha	
Diuron	Agromex, Diuron, Karmex, Klass, True	Very stable in neutral media at normal temperature but hydrolysis occurs at higher temperatures by acids and alkalies	Compatible with most other herbicides	80 WP	*Amaranthus* spp., *Chenopodium album, Convolvulus arvensis, Digitaria bificiculata, Portulaca oleuracca, Setaria glauca, Xanthium strumarium, Anagallis arvensis, Commelina benghalensis, Setaria* spp., *Coronopus didymus*	Control all weeds mosses on non-crop areas selective control of germination grasses and broad leaved weeds in many crops including tree fruits, bush-fruits, citrus fruits, vines, olives pineapples, sugarcane, cotton, legumes, cereals, maize and sorghum etc. at 500-750 g/ha	Aventis

Contd...

Contd...

Herbicide name	Trade name	Stability	Compatibility	Formulation (a.i. %)	Weed control	Uses	Source
Ethoxysulfuron	Sunrise	-	-	10 EC	*Amaranthus* spp., *Chenopodium album*, *Chenopodium murale*, *Cirsium arvensis*, *Convolvulus arvensis*, *Lathyrus aphaca*, *Medicago denticulate*, *Polygonum convolvulus*, *Cyperus* spp., *Commelina* spp	Used as a pre and post-emergence to control broad leaved weeds and sedges in rice, cereals and sugarcane at 15-30 g/	Bayer
Fenoxaprop	Whip super and Puma super	Slowly degraded by Ultra Violet light; decomposed by acids and alkalis	-	10 and 9 EC	*Alopecurus aequalis*, *Alopecurus myosuroides*, *Alopecurus pratensis*, *Apera spicaventi*, *Avena fatua*, *Avena ludoviciana*, *Avena sterilis*, *Avena strigosa*, *Brachiaria ciliatissima*, *Echinochloa colona*, *E. crus-galli*. *Phalaris minor*, *Phalaris paradoxa*	Used as a post-emergence to control annual grassy and perennial weeds in rice, wheat, soybean, cotton and sugarcane	Bayer

Contd...

Contd...

Herbicide name	Trade name	Stability	Compatibility	Formulation (a.i. %)	Weed control	Uses	Source
Fluchloralin	Basalin	Decomposed by Ultra Violet light though it is stable for at least 2 years when stored in original unopened containers at normal room temperatures	-	45 EC	*Echinochloa* spp., *Amaranthus* spp., *Eclipta alba*, *Cyperus iria*, *Trianthema* spp., *Portulaca* spp., *Panicum* spp., *Eleusine* spp., *Chenopodium album*, *Poa annaua*, *Chloris barbata*	Used as pre-emergence or pre-plant spray. Controls annual grasses and borad-leaved weeds in cotton, rice (transplanted), soybean, groundnut, lime, beans, okra, jute, potatoes and several vegetable crops. But sorghum, spinach and oats are sensitive to the application of fluchloralin	BASF
Glufosinate	Basta, Liberty	Highly stable	-	15 SL	All weeds	Non selective contact post-emergence herbicide effective against mono-and dicotyledons weeds in non-crop situations with some systemic activity at dose- 0.35 - 1.7 kg/ha	Bayer

Contd...

Contd...

Herbicide name	Trade name	Stability	Compatibility	Formulation (a.i. %)	Weed control	Uses	Source
					emergence application or pre-plant incorporation in to soil. *Echinochloa* spp., *Eleusine indica, Digitaria* spp., *Dactyloctenium aegytium, Panicum* spp., *Cyperus* spp., *Amaranthus virids.*		
Metoxuron	Dual	Very stable when stored under normal conditions. Strong acids and alkalis hydrolyze metoxuron. Solution is decomposed by UV light.	Compatible with many other herbicides	80WP; SC; WDG	*Phalaris minor, Avena* spp., *Poa annua, Melitotus alba, Anagallis arvensis, Chenopodium album*	Used in pre and post emergence control of some grasses and most annual broad-leaved weeds in wheat, barley, rye and carrots. Also useful for haulm destruction in stored potatoes and for pre-harvest defoliation in tomatoes	Novertis
Metribuzin	Tata metri, Lexone, Sencor	Stable to Ultra Violet light. At 20°C, stable to dilute acids and bases	Compatible with most other herbicides and liquid fertilizers	70 WP; liquid suspension; water-dispersible granules	*Amaranthus* spp., *Anagallis arvensis, Atriplex patula, Alopecurus myosuroides,*	Widely used for pre and post emergence control of many grasses and broad leaved weeds in	Bayer

Contd...

Contd...

Herbicide name	Trade name	Stability	Compatibility	Formulation (a.i. %)	Weed control	Uses	Source
					Brachiaria spp., *Commelina communis*, *Digitaria ischaemum*, *Digitaria sanguinalis*, *Echinochloa* spp., *Lolium temulentum*, *Panicum dichotomiflorum*, *Poa annua*, *Bidens* spp., *Capcella bursa-pastoris*, *Chenopodium album*, *Coronopus didymus*, *Datura stramonium*, *Eclipta alba*, *Fumaria officinalis*, *Galinsoga parviflora*, *Gnaphalium* spp., *Lamium purpureum*, *Linaria* spp., *Malva* spp., *Mollugo verticillata*, *Parthenium hysterophorus*, *Papaver rhoeas*, *Polygonum* spp.,	soybean, potato, tomato, sugarcane, maize and cereals	

Contd...

Contd...

Herbicide name	Trade name	Stability	Compatibility	Formulation (a.i. %)	Weed control	Uses	Source
					Sida spinosa, Sinapis arvensis, Sonchus arvensis, Spergula arvensis, Stellaria media, Thlaspi arvense, Tribulus terrestris		
Metsulfuron-methyl	Algrip	-	-	20 WP	*Amaranthus* spp., *Chenopodium album, Chenopodium murale, Cirsium arvensis, Convolvulus arvensis, Lathyrus aphaca, medicago denticulate, Polygonum convolvulus*	Used as post emergence to control broad leaved weeds in wheat and barley	Dupont
Oxadiazon	Ronstar	active ingredient and the commercial preparations are stable under normal storage conditions	-	25 EC	Many annual broad-leaved and grasses weeds including *Oxalis* spp., *pigweed,* crabgrass (*Digitaria sanguinalis*), carpetweed and green foxtail (*Setaria virids*	Used as pre-emergence to control grasses and post emergence control of broad-leaved weeds. Widely used for broad leaved weeds in carnations, gladiolus, ornamenthal trees	Rhone-Poulenc

Contd...

Contd...

Herbicide name	Trade name	Stability	Compatibility	Formulation (a.i. %)	Weed control	Uses	Source
						and shrubs, cotton, rice, soybean, sunflower at 500-750 g/ha	
Oxadiargyl	Topstar	Stable in 50°C	-	80 WP, 6 EC	*Eragrostis major, Eleusine indica, Dactyloctenium aegyptium* and *Asphodolus tenuifolius Chenopodium album, Chenopodium murale, Melilotus indica, Boerhavia diffusa* and *Digera arvensis*	Used as pre-emergence to control grasses and post emergence control of broad-leaved weeds. Widely used for broad leaved weeds in carnations, gladiolus, ornamenthal trees and shrubs, cotton, rice, soybean, sunflower at 500-750 g/ha	Bayer
Oxyfluorfen	Goal	Stable upto 50°C, Decomposes under Ultra Violet light	Compatible with most other pesticides	23.5 EC; 0.35 GR	*Chinochloa* spp., *Cyperus iria, Cyperus difformis, Eleusine indica, Eclipta alba, Fimbristyles* spp., *Digitaria* spp., *Imperata cylindrical, Paspalum conjugatum, Chenopodium*	Used as post or pre-emergence application to control annual broad-leaved weeds and grasses in tropical and subtropical crops. Used in crops including fruits, trees, vines, nuts, cereals, maize,	Indofil, Bayer

Contd...

Contd...

Herbicide name	Trade name	Stability	Compatibility	Formulation (a.i. %)	Weed control	Uses	Source
					pyxidaria, Ludwigia spp., *Monochoria vaginalis, Rotala indica, Scirpus juncoides, Scirpus mucronatus.*		
Propanil	Stam F- 34	Hydrolyzed in acidic and alkaline media	Incompatible with carbamate and organophosphate insecticides and liquid fertilizers	EC; ULV.	*Echinochloa* spp., *Cyperus difformis, Cyperus iria, Chenopodium album, Melitotus* spp., *Brachiaria* spp., *Fimbristyles miliacea, Lathyrus aphaca, Argemone maxicana*	Used as post-emergence spraying. Used in rice and vegetables to control broad-leaved and grassy weeds. Also effective when used in mixture with MCPA in wheat	Meghmani Organics Limited
Pyrazosulfuron	Sathi, Act, Agreen, Sparkstar	Stable	-	10 WP	*Echinochloa curs-galli, Cyperus* spp., *Monochoria* spp., *Scripus* spp	Used as post-emergence herbicide in rice	UPL
Quizalofop-p-butyl	Targa super	Moderately stable to Ultra Violet light; Decomposes at 320°C	-	5 EC	*Avena ludoviciana, Agropyron repens, Aloperues myosuroides, Cynodon dactylon, Digitaria sanguinalis, Dinebra Arabica,*	Used as post-emergence spraying. Used in rice and vegetables to control broad-leaved and grassy weeds. Also effective when used in mixture with	Dhanuka

Contd...

Contd...

Herbicide name	Trade name	Stability	Compatibility	Formulation (a.i. %)	Weed control	Uses	Source
					Echiniochloa crus-galli, Eleusine indica, Hemarthria compressa, Lolium multiforusm, Phalaris minor, Setaria faberi, Setaria viridis, Sorghum halepense	MCPA in wheat	
Sulfosulforon	Leader	Moderately stable to Ultra Violet light; Decomposes at 320°	-	75 WG	*Agropyron repens, Apera spica-venti, Amaranthus* spp., *Avena ludoviciana, Bromus* spp., *Capsella bursa-pastoris, Chenopodium album, Galium aprine, Phalaris minor, Poa trivialis, Polygonum convolvulus, Sinapsis arvensis, Stellaria media, Elytrigia repens*	Used as a post-emergence treatment to control annual grasses and broad-leaved weeds in wheat	Monsanto. Rallis
TCA-Sodium	NaTa, Tecane	Stable under dry conditions but decomposes in aqueous, strongly	-	SG; WG; SP	*Avena ludoviciana, Agropyron repens, Aloperues myosuroides,*	Used as a post-emergence herbicide for many annual and perennial grasses	

Contd...

Contd...

Herbicide name	Trade name	Stability	Compatibility	Formulation (a.i. %)	Weed control	Uses	Source
		alkaline media			*Cynodon dactylon, Digitaria sanguinalis, Dinebra Arabica, Echiniochloa crus-galli, Eleusine indica, Hemarthria compressa, Lolium multiforusm, Phalaris minor, Setaria faberi, Setaria viridis, Sorghum halepense*		
Thiobencarb	Saturn, Benthiocarb	Tsatble in water at pH 5-9 for 30 days (21°C). Stable in light	Compatible with most other pesticides	50 EC, GR	*Echiniochloa* spp., *Cyperus* spp., *Ischaemum rugosum, Leptochloa chinensis, Fimbristylis* spp., *Eleusine indica, Monochoria vaginalis*	Used as a post-emergence spraying. Used in rice and vegetables to control broad-leaved and grassy weeds. Also effective when used in mixture with MCPA in wheat	Pesticide India Ltd.
Tri-allate	Avadex BW, Far-Go	Stable to light; decomposition temperature > 200°C. Non-flammanble	-	50 EC	*Agrostis* spp., *Avena* spp., *Brachiaria platyphylla, Echinocloa crus-galli, Eleusine agricana, Eleusine*	It is used to control wild oats in barley, lentils, peas and spring and winter wheat	Monsanto

Contd...

Contd...

Herbicide name	Trade name	Stability	Compatibility	Formulation (a.i. %)	Weed control	Uses	Source
					indica, Eragrostis spectabilis, Leptochloa filiformis, Leptochloa multiflorum, Lolium multiflorum Lolium rigidum, Panicum capillare, Panicum dichotomiflorum, Phalaris spp., *Rottobellia exaltata*		
Trifluralin	Treflon, Triflurex	Very stable. Decomposed by Ultra Violet irradiation	Compatible with most of the pesticides and fertilizers (liquid, dry	48 EC	*Trianthema* spp., *Acalypha indica, Euphorbia* spp., *Commelina* spp., *Eleusine* spp., *Eragrostis tenella, phyllanthus niruri, Amaranthus* spp., *Echinochloa* spp., *Cyperus deformis, Cyperus iria; Euphorbia* spp., *Acalypha indica, Cyperus difformis, Cyperus iria, Phyllanthus niruri*	Widely used for pre-emergence control of many annual grasses and broad-leaved weeds in brasicas, beans, peas, carrots, cabbage, cucumber, turnips capsicums, tomatoes, onions garlic, vines, strawberries, citrus, groundnuts, soybean, sunflower, safflower, ornamentals, cotton, sugarcane and in forestry	De-nocil/Gharda, UPL

□□□

About the Author

Dr. Shobha Sondhia joined Agricultural Research Services (ARS) in 1997 at CIFE, Mumbai as scientist and worked on phyto-remediation techniques and heavy metal pollution. She transferred to Directorate of Weed Science Research, Jabalpur in 1999 and consistently elevated to higher position and currently working as a Senior Scientist.

She obtained MSc degree from the Prestigious Rani Durgavati University, Jabalpur in Chemistry with specialization in Organic Chemistry in 1994. She did Ph.D. from Rani Durgavati University, Jabalpur, on Natural Product Chemistry. Dr. Sondhia is recipient of many National and International Awards, fellowship and recognition. She is life/annual member of a dozen scientific and professional societies.

She has published over 50 research papers in high impact factor refereed international and national Journals. She has organized one national training, 'Advance instrumental training for the analysis of pollutants in the food commodity and water'. She has presented over 60 research papers in international and national conference/seminar/scientific meetings in the capacity of invited /oral speaker. She has completed several in-house and outside funded projects on herbicide chemistry/residues/natural products. She has recognized as a resource person in the field of herbicide chemistry/residues/natural product chemistry by many Universities and Research Institutes. She has guided several M.Sc. and M. Phill students.

Dr. Sondhia is on the editorial board of Indian Journal of Weed Science. She has worked in the capacity of referee/reviewer of international and national journals in the field of MS/NMR spectroscopy, herbicides/pesticide chemistry, residues, and allelopathy for the Ecotoxicology, Journal of Environmental Quality, Geobios, Weed Science, Journal of Pest Management, International Journal of Pesticide Science, Allelopathy Journal, Pesticide Research Journal, The Indian Journal of Agricultural Sciences, Indian Journal of Weed Science *etc.*

Dr. Sondhia has been involved in screening of herbicides/new molecules for residues and phytotoxicity. She has extensively worked on herbicide residues method & development, influence of biotic and physicochemical factors on herbicide efficacy and selectivity, effect of herbicides on water quality, herbicides residue management techniques, risk of ground water contamination by herbicides, allelopathy, isolation and identification of biomolecules/natural compounds from plant species. She has also conducting work on ecotoxicological research studying the effects of herbicides on non-target organisms such as microbes, beneficial insects and fishes.